Introduction to Theoretical and Mathematical Fluid Dynamics

Introduction to Theoretical and Mathematical Fluid Dynamics

Bhimsen K. Shivamoggi
University of Central Florida
Orlando, Florida, United States

Third Edition

This edition first published 2023

Registered Office
John Wiley & Sons, Inc., 111 River Street, Hoboken, NJ 07030, USA

Editorial Office
111 River Street, Hoboken, NJ 07030, USA

For details of our global editorial offices, customer services, and more information about Wiley products visit us at www.wiley.com.

Wiley also publishes its books in a variety of electronic formats and by print-on-demand. Some content that appears in standard print versions of this book may not be available in other formats.

Library of Congress Cataloging-in-Publication Data
Names: Shivamoggi, Bhimsen K., author.
Title: Introduction to theoretical and mathematical fluid dynamics /
Bhimsen Krishnarao Shivamoggi, University of Central Florida, Orlando, United States.
Description: Hoboken, NJ : John Wiley & Sons, 2023 | Includes bibliographical references and index.
Identifiers: LCCN 2021041747 (print) | LCCN 2021041748 (ebook) | ISBN 9781119101505 (hardback) | ISBN 9781119101512 (pdf) | ISBN 9781119101529 (epub) | ISBN 9781119765158 (obook)
Subjects: LCSH: Fluid dynamics. | Fluid dynamics–Mathematical models.
Classification: LCC QA911 .S462 2021 (print) | LCC QA911 (ebook) | DDC 532/.05–dc23
LC record available at https://lccn.loc.gov/2021041747
LC ebook record available at https://lccn.loc.gov/2021041748

Cover image: © Bocskai Istvan/Shutterstock
Cover design by Wiley

Set in 9.5/12.5pt STIXTwoText by Integra Software Services Pvt. Ltd, Pondicherry, India
SKY10036042_091322

In memory of my beloved mother, Pramila.

Contents

Preface to the Third Edition

Though fluid dynamics is primarily concerned with the study of causes and effects of the motion of fluids, it provides a convenient general framework to model systems with complicated interactions among their constituents. This is accomplished via a formalism that uses only a few macroscopic quantities like pressure while ignoring the practically intractable details of these particle interactions. A case in point was the use of a liquid drop model of an atomic nucleus in describing the nuclear fission of heavy elements. In fact, George Uhlenbeck (one of the two famous original proponents of electron spin) pointed out that "every physicist should have some familiarity with the field (fluid dynamics)."

This book is concerned with a discussion of the dynamical behavior of a fluid and seeks to provide the readers with a sound and systematic account of the most important and representative types of fluid flow phenomena. Particular attention has been paid, as with the Second Edition, to emphasize the most generally useful fundamental ideas and formulations of fluid dynamics.

This book is addressed primarily to graduate students and researchers in applied mathematics and theoretical physics. Nonetheless, graduate students and researchers in engineering will find the ideas and formulations in this book to be useful - as confirmed by book reviews on the previous two editions.

The effort to keep the length of the introduction to mainstream fluid dynamics reasonable makes it difficult to provide a thorough treatment of every topic. For instance, engineering details like the skin-friction calculations in boundary layer theory are not dealt with in this book. In the same vein, non-newtonian fluids (for which the coefficient of viscosity is dependent on the rate of deformation of the fluid) have also been excised from the discussion.

As with the Second Edition, flows of an incompressible fluid have been given especially large coverage in the text because of their central place in the subject. In the references provided at the end, I have tried to mention papers connected with the original developments and formulations given in this book. This list is not to be construed as being exhaustive in covering these details. Furthermore, a few exercises have been provided at the end of each chapter. In addition to being essential to understand the ideas and formulations discussed in this book, the solutions to these exercises are intended to encourage further exploration of those topics.

The Third Edition, constitutes an extensive revision and rewrite of the text, and entails the following main features:

- major reorganization of several topics to facilitate a smoother reading experience.
- refinement and detailed reworking of mathematical developments
- addition of several new topics:
  - Hamiltonian formulation of two-dimensional incompressible flows
  - Variational principle for surface waves in water
  - Tidal bores
  - The dam break problem
  - The shock tube
  - Burgers vortex;
- addition of several new exercise problems.

Since the discussion of incompressible, compressible and viscous flows constitute the teaching material for three graduate courses, I dropped the discussion of hydrodynamic stability and turbulence from the Third Edition. Special topics like hydrodynamic stability and turbulence are typically not included in the mainstream fluid dynamics graduate course package.

For the readership, an elementary background in fluid dynamics is helpful, but familiarity with the theory and the analytical methods (perturbation methods, in particular) of solutions to differential equations is an essential prerequisite.

In closing, I wish to place on record with gratitude, the immense ingenuity of several applied mathematicians, theoretical physicists, and engineers who developed the theoretical formulations of fluid flows. This great activity is not complete and several major theoretical developments are still to come. However, the most important message from the theoretical accomplishments so far is that analytical formulations are still feasible. Therefore, one may not resort too quickly to exploring on the computer as it may not necessarily provide a proper organization of knowledge and a clear insight into the dynamical behavior of fluids.

Orlando, 2023 *Bhimsen K. Shivamoggi*

Acknowledgments

I would first like to thank Professor Xin Li, our Chairman and former Graduate Coordinator, for his support in sustaining our Theoretical and Mathematical Fluid Dynamics graduate course sequence. His encouragement and effort to provide resources were crucial in bringing this third edition to fruition. I wish to thank Dr. Nelson Ying, Sr. for a grant that partially supported this book writing project. I wish to express my gratitude to Professor Katepalli Sreenivasan who has been a constant source of encouragement in my academic endeavors. I wish to thank my students who took this course sequence and my colleagues who rendered valuable assistance in this effort. I am thankful to Dr. Leos Pohl for his help with proofreading the book manuscript. My thanks are due to Gayathri Krishnan for her immense help with editing the book manuscript. The responsibility for any remaining defects is however mine alone. As with the previous editions, in preparing the third edition material, I have drawn considerably from ideas and formulations developed by many eminent authors. I have acknowledged the various sources in the Bibliography and References given at the end of this book and wish to apologize to any sources I may have inadvertently missed. I wish to express my gratitude to several publishers for granting permission to reproduce in this book the following figures from the original sources,

- 8.9-8.11, 9.2-9.4, 10.14, 10.18, 16.7,19.5 Cambridge University Press
- 10.11, 16.8 Elsevier Publishing Company
- 20.2, 21.4 Oxford University Press
- 6.10, 11.21, 11.22 Pearson
- 10.17 The Royal Society of London
- 14.6-14.10, 21.3 Springer Publishing Company
- 11.7, 11.8, 11.10, 11.11, 11.16-11.19 Wiley Publishing Company

My sincere thanks are due to Linda Perez-Rodriguez for the tremendous patience and dedication with which she typed this book and incorporated my numerous changes and to Joe Fauvel for his excellent job in doing the figures. I am very thankful to John Wiley and Sons, and Kimberly Monroe-Hill, the Managing Editor, for their tremendous cooperation in this project. I have tried to do justice to the great intellectual strides made by eminent pioneers in fluids during the nineteenth and twentieth centuries. My immense thanks are due to my wife Jayashree for her understanding and cooperation during this endeavor.

Finally, the writing of this book (or for that matter, even my higher education and creative pursuits) would not have been possible without the inspiration of my beloved late mother, Pramila, who could not attend college but always craved learning and knowledge. This book is dedicated to her memory.

Part I

Basic Concepts and Equations of Fluid Dynamics

1

Introduction to the Fluid Model

While dealing with a fluid, in reality, one deals with a system that has many particles which interact with one another. The main utility of fluid dynamics is the ability to develop a formalism which deals solely with a few macroscopic quantities like pressure, while ignoring the details of the particle interactions. Therefore, the techniques of fluid dynamics have often been found useful in modeling systems with complicated interactions (which are either not known or very difficult to describe) between the constituents. Thus, the first successful model of the nuclear fission of heavy elements was the *liquid drop model* of the nucleus, which treats the nucleus as a fluid. This replaces the many body problem of calculating the interactions of all the protons and neutrons with the much simpler problem of calculating the pressures and surface tension in this fluid.[1] Of course, this treatment gives only a very rough approximation to reality, but it is nonetheless a very useful way of approaching the problem.

The primary purpose of fluid dynamics is to study the causes and effects of the motion of fluids. Fluid dynamics seeks to construct a mathematical theory of fluid motion based on the smallest number of dynamical principles, which are adequate to correlate the different types of fluid flow as far as their macroscopic features are concerned. In many circumstances, the incompressible, inviscid fluid model is sufficiently representative of real fluid properties to provide a satisfactory account of a great variety of fluid motions. It turns out that such a model makes accurate predictions for the airflow around *streamlined* bodies moving at low speeds. While dealing with streamlined bodies (which minimize flow-separation) in flows of fluids of small viscosities, one may divide the flow field into two parts. The first part, where the viscous effects are appreciable consists of a thin boundary layer adjacent to the body and a small wake

1 The fluid model presupposes that there is a wide separation of scales between the macroscopic variables considered in the model and the microscopic motions that are ignored.

Introduction to Theoretical and Mathematical Fluid Dynamics, Third Edition.
Bhimsen K. Shivamoggi.

behind the body. The second part is the rest of the flow field that behaves essentially like an inviscid fluid. Such a division greatly facilitates the mathematical analysis in that the inviscid flow field can first be determined independent of the boundary layer near the body. The pressure field obtained from the inviscid-flow calculation is then used to calculate the flow in the boundary layer.

An attractive feature of fluid dynamics is that it provides ample room for the subject to be expounded as a branch of applied mathematics and theoretical physics.

1.1 The Fluid State

A fluid is a material that offers resistance to attempts to produce relative motions of its different elements, but deforms continually upon the application of surface forces. A fluid does not have a preferred shape, and different elements of a homogeneous fluid may be rearranged freely without affecting the macroscopic properties of the fluid. Fluids, unlike solids, cannot support tension or negative pressure. Thus, the occurrence of negative pressures in a mathematical solution of a fluid flow is an indication that this solution does not correspond to a physically possible situation. However, a thin layer of fluid can support a large normal load while offering very little resistance to tangential motion – a property which finds practical use in lubricated bearings.

A fluid of course, is discrete on the microscopic level, and the fluid properties fluctuate violently, when viewed at this level. However, while considering problems in which the dimensions of interest are very large compared to molecular distances, one ignores the molecular structure and endows the fluid with a continuous distribution of matter. The fluid properties can then be taken to vary smoothly in space and time. The characteristics of a fluid, caused by molecular effects, such as viscosity, are incorporated into the equation of fluid flows as experimentally obtained empirical parameters.

Fluids can exist in either of two stable phases – liquids, and gases. In case of gases, under ordinary conditions, the molecules are so far apart that each molecule moves independently of its neighbors except when making an occasional "collision." In liquids, on the other hand, a molecule is continually within the strong cohesive force fields of several neighbors at all times. Gases can be compressed much more readily than liquids. Consequently, for a gas, any flow involving appreciable variations in pressure will be accompanied by much larger changes in density. However, in cases where the fluid flows are accompanied by only slight variations in pressure, gases and liquids behave similarly.

In formulations of fluid flows, it is useful to think that a fluid particle is small enough on a macroscopic level that it may be taken to have uniform macroscopic properties. However, it is large enough to contain sufficient number of molecules to diminish the molecular fluctuations. This allows one to associate with it a macroscopic property which is a statistical average of the corresponding molecular property over a large number of molecules. This is the *continuum hypothesis.*

As an illustration of the limiting process by which the local continuum properties are defined, consider the mass density ρ. Imagine a small volume $\delta \mathcal{V}$ surrounding a point P, let δm be the total mass of material instantaneously in $\delta \mathcal{V}$. The ratio $\delta m/\delta \mathcal{V}$, as $\delta \mathcal{V}$ reduces to $\delta \mathcal{V}^*$, where $\delta \mathcal{V}^*$ is the volume of fluid particle, is taken to give the mass density ρ at P.

1.2 Description of the Flow-Field

The continuum model affords a *field* description, in that the average properties in the volume $\delta \mathcal{V}^*$ surrounding the point P are assigned in the limit to the point P itself. If q represents a typical continuum property, then one has a fictitious continuum characterized by an aggregate of such local values of q, i.e., $q = q(\mathbf{x}, t)$. This enables one to consider what happens at every fixed point in space as a function of time – the so-called *Eulerian* description. In an alternative approach, called the *Lagrangian* description, the dynamical quantities, as in particle mechanics, refer more fundamentally to identifiable pieces of matter, and one looks for the dynamical history of a selected fluid element.

Imagine a fluid moving in a region Ω (see Figure 1.1). Each particle of fluid follows a certain trajectory. Thus, for each point $\mathbf{x_0}$ in Ω, there exists a path line $\sigma(t)$ given by

$$\sigma(t) : \mathbf{x} = \boldsymbol{\phi}(t, \mathbf{x}_0) \text{ with } \boldsymbol{\phi}(0, \mathbf{x}_0) = \mathbf{x}_0 \text{ and } \boldsymbol{\phi}(t+\tau, \mathbf{x}_0) = \boldsymbol{\phi}(t, \boldsymbol{\phi}(\tau, \mathbf{x}_0)), \tag{1}$$

where the flow mapping $\boldsymbol{\phi} : \Omega \Rightarrow \Omega$, depends continuously on the parameter t. $\mathbf{x}(t, \mathbf{x}_0)$ and its inverse $\mathbf{x}_0(t, \mathbf{x})$ are both continuous, so this mapping is *one-to-one* and *onto.*[2]

The velocity of the flow is given by

$$\mathbf{v}(t, \mathbf{x}) = \frac{d\mathbf{x}}{dt} = \frac{d}{dt}\boldsymbol{\phi}(t, \mathbf{x}_0). \tag{2}$$

2 A mapping $T : \mathcal{V} \Rightarrow W$ is called *one-to-one* if $T(x) = T(y)$ implies $x = y, \forall x, y \in \mathcal{V}$. If $T : \mathcal{V} \Rightarrow W$ is a function with range W, i.e., $T(\mathcal{V}) = W$, then T is called *onto*. In other words, given $y \in W$, there is a $x \in \mathcal{V}$ such that $T(x) = y$; i.e., the images of $\mathcal{V}$ under T fill out all of W.

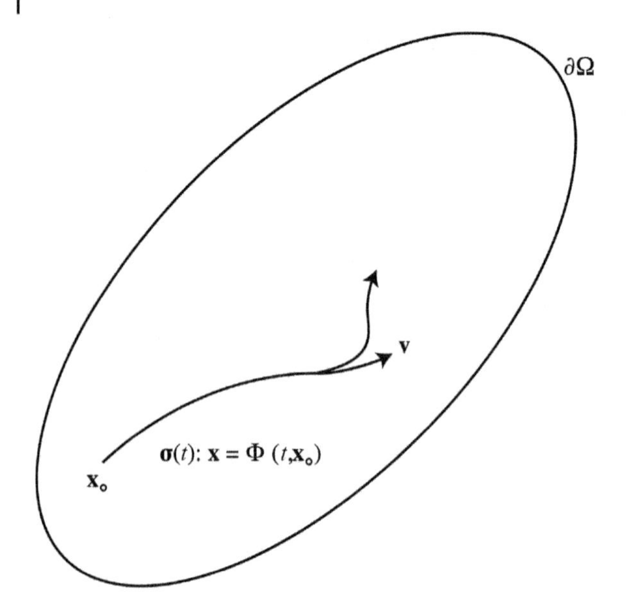

Figure 1.1 Motion of a fluid particle.

We then have the following results.

Theorem 1.1 (*Existence*) Assume that the flow velocity $\mathbf{v}(t, \mathbf{x})$ is a C^1 function of $\mathbf{x}$ and t. Then, for each pair $(t_0, \mathbf{x_0})$, there exists a unique integral curve – the *path line* $\boldsymbol{\sigma}(t)$, defined on some small interval in t about t_0, such that $\boldsymbol{\sigma}(t_0) = \mathbf{x_0}$.

Theorem 1.2 (*Boundedness*) Consider a region Ω with a smooth boundary $\partial\Omega$. If the flow velocity $\mathbf{v}$ is parallel to $\partial\Omega$, then the integral curves of $\mathbf{v}$, i.e., the path lines starting in Ω remain in Ω.

Streamlines $\mathbf{s}(t)$ are obtained by holding t fixed, say $t = t_0$, and solving the differential equation

$$\mathbf{s}'(t) = \mathbf{v}(t_0, \mathbf{s}(t)). \tag{3}$$

Streamlines coincide with pathlines if the flow is steady, i.e., if $\mathbf{v}(t, \mathbf{x}) = \mathbf{v}(\mathbf{x})$.

The transformation from the Eulerian to the Lagrangian description is given by

$$\mathbf{v}(t, \mathbf{x}) = \mathbf{v}(t, \boldsymbol{\phi}(t, \mathbf{x}_0)) = \mathbf{V}(t, \mathbf{x_0}) \tag{4}$$

Since the Lagrangian description makes the formalism cumbersome, we shall instead use the Eulerian description. However, the Lagrangian concepts of *material volumes, material surfaces,* and *material lines* which consist of the same fluid particles and move with them are still useful in developing the Eulerian description.

1.3 Volume Forces and Surface Forces

One may think of two distinct kinds of forces acting on a fluid continuum. *Long-range* forces such as gravity penetrate into the interior of the fluid, and act on all elements of the fluid. If such a force $\mathbf{F}(\mathbf{x}, t)$ varies smoothly in space, then it acts equally on all the matter within a fluid particle of density ρ and volume $\delta\mathcal{V}$. The total force acting on the particle is proportional to its mass and is equal to $\mathbf{F}(\mathbf{x}, t)\,\rho\,\delta\mathcal{V}$. In this sense, long-range forces are called *body* forces.

The *short-range* forces (which have a molecular origin) between two fluid elements, on the other hand, are effective only if they interact through direct mechanical contact. Since the short-range forces on an element are determined by its surface area, one considers a plane surface element of area δA in the fluid. The local short-range force is then specified as the total force exerted on the fluid on one side of δA by the fluid on the other side and is equal to $\boldsymbol{\Sigma}(\hat{\mathbf{n}}, \mathbf{x}, t)\,\delta A$ (Cauchy, 1827). The direction of this force is not known a priori for a viscous fluid (unlike in the case of an inviscid fluid). Here, $\hat{\mathbf{n}}$ is the unit normal to the surface element δA, and it points away from the fluid on which $\boldsymbol{\Sigma}$ acts. The total force exerted across δA on the fluid on the side into which $\hat{\mathbf{n}}$ points, by Newton's third law of motion, is $\boldsymbol{\Sigma}(-\hat{\mathbf{n}}, \mathbf{x}, t)\,\delta A = -\boldsymbol{\Sigma}(\hat{\mathbf{n}}, \mathbf{x}, t)$, so that $\boldsymbol{\Sigma}$ is an odd function of $\hat{\mathbf{n}}$. In this sense, short-range forces are called *surface* forces.

In order to determine the dependence of $\boldsymbol{\Sigma}$ on $\hat{\mathbf{n}}$, consider all the forces acting instantaneously on the fluid within an element of volume $\delta\mathcal{V}$ shaped like a tetrahedron (Figure 1.2). The three orthogonal faces have areas $\delta A_1, \delta A_2, \delta A_3$ and unit outward normals $-\hat{\mathbf{i}}, -\hat{\mathbf{j}}, -\hat{\mathbf{k}}$, and the fourth inclined face has area δA and unit outward normal $\hat{\mathbf{n}}$. Surface forces will act on the fluid in the tetrahedron, across each of the four faces, and their resultant is

$$\boldsymbol{\Sigma}(\hat{\mathbf{n}})\,\delta A + \boldsymbol{\Sigma}(-\hat{\mathbf{i}})\,\delta A_1 + \boldsymbol{\Sigma}(-\hat{\mathbf{j}})\,\delta A_2 + \boldsymbol{\Sigma}(-\hat{\mathbf{k}})\,\delta A_3$$

or

$$\left\{\boldsymbol{\Sigma}(\hat{\mathbf{n}}) - \left[\hat{\mathbf{i}}\cdot\hat{\mathbf{n}}\boldsymbol{\Sigma}(\hat{\mathbf{i}}) + \hat{\mathbf{j}}\cdot\hat{\mathbf{n}}\boldsymbol{\Sigma}(\hat{\mathbf{j}}) + \hat{\mathbf{k}}\cdot\hat{\mathbf{n}}\boldsymbol{\Sigma}(\hat{\mathbf{k}})\right]\right\}\delta A.$$

Now, since the body force and inertia force acting on the fluid within the tetrahedron are proportional to the volume $\delta\mathcal{V}$, they become negligible compared to the surface forces if the linear dimensions of the tetrahedron are made to approach zero without changing its shape. Then, applying *Newton's second law*

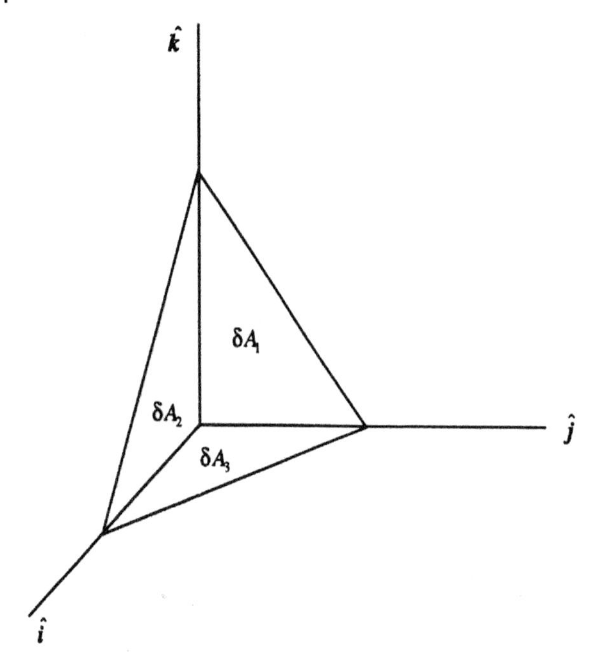

Figure 1.2 Tetrahedron-shaped fluid element.

of motion to this fluid element gives

$$\boldsymbol{\Sigma}(\hat{\mathbf{n}}) = \hat{\mathbf{i}} \cdot \hat{\mathbf{n}} \boldsymbol{\Sigma}(\hat{\mathbf{i}}) + \hat{\mathbf{j}} \cdot \hat{\mathbf{n}} \boldsymbol{\Sigma}(\hat{\mathbf{j}}) + \hat{\mathbf{k}} \cdot \hat{\mathbf{n}} \boldsymbol{\Sigma}(\hat{\mathbf{k}}) \equiv \boldsymbol{\tau} \cdot \hat{\mathbf{n}} \tag{5}$$

where

$$\boldsymbol{\tau} \equiv \hat{\mathbf{i}} \boldsymbol{\Sigma}(\hat{\mathbf{i}}) + \hat{\mathbf{j}} \boldsymbol{\Sigma}(\hat{\mathbf{j}}) + \hat{\mathbf{k}} \boldsymbol{\Sigma}(\hat{\mathbf{k}}) \tag{6}$$

which is a second-order tensor called the *stress tensor* and prescribes the state of stress at a point in the fluid. Thus, the resultant of stress (force per unit area) across an arbitrarily oriented plane surface element with a unit normal $\hat{\mathbf{n}}$ is related to the resultant of stress across any three orthogonal plane surface elements at the same position in the fluid as if it were a vector with orthogonal components $\boldsymbol{\Sigma}(\hat{\mathbf{i}}), \boldsymbol{\Sigma}(\hat{\mathbf{j}}), \boldsymbol{\Sigma}(\hat{\mathbf{k}})$. Note that $\hat{\mathbf{n}}$, $\boldsymbol{\Sigma}$, and $\boldsymbol{\tau}$ do not depend at all on the choice of the reference axes.

This leads to the following theorem.

Theorem 1.3 (*Cauchy*) There exists a matrix function $\boldsymbol{\tau}$, such that

$$\tau_{ij} \equiv \Sigma_i(\hat{\mathbf{n}}) \hat{n}_j$$

where τ_{ij} is the i-component of the stress exerted across a plane surface element normal to the j-direction. Thus, the state of stress at a point is characterized by *normal* stresses (τ_{ii}) and *shear* stresses $\left(\tau_{ij}, i \neq j\right)$ acting on three mutually perpendicular planes passing through the point, i.e., by nine cartesian components.

One may use an argument similar to the one above to demonstrate that the nine components of the stress tensor are not all independent. Now, let us consider the moments of the various forces acting on the fluid within a volume element $\delta \mathcal{V}$ of arbitrary shape. The ith component of the total moment, about a point O within this volume, exerted by the surface forces at the boundary of the volume is

$$\iint_{\delta \mathcal{A}} \varepsilon_{ijk} x_j \tau_{kl} \hat{\mathrm{n}}_l dA$$

where ε_{ijk} is the *Levi-Civita tensor* and $\mathbf{x}$ describes the position of the surface element $\hat{\mathbf{n}}\, dA$ relative to O. This can be rewritten using the *divergence theorem* as

$$\iiint_{\delta \mathcal{V}} \varepsilon_{ijk} \frac{\partial \left(x_j \tau_{kl}\right)}{\partial x_l} d\mathbf{x} = \iiint_{\delta \mathcal{V}} \varepsilon_{ijk} \left(\tau_{kj} + x_j \frac{\partial \tau_{kl}}{\partial x_l}\right) d\mathbf{x}.$$

Let us now reduce the volume $\delta \mathcal{V}$ to zero without changing its shape. We first equate the rate of change of angular momentum of the fluid instantaneously in $\delta \mathcal{V}$ to the total moment of the body and surface forces acting on $\delta \mathcal{V}$. Taking $\delta \mathcal{V}$ to be small, the terms of $O(\delta \mathcal{V})$ are dropped and we obtain

$$\varepsilon_{ijk}\, \tau_{kj} = 0$$

or

$$\varepsilon_{ijk}\, \tau_{kj} + \varepsilon_{ikj}\, \tau_{jk} = 0$$

or

$$\varepsilon_{ijk}\, \tau_{kj} - \varepsilon_{ijk}\, \tau_{jk} = 0$$

from which,

$$\tau_{jk} = \tau_{kj} \tag{7}$$

so that the stress tensor is *symmetrical* and has only six independent components.

In a fluid at rest, the shearing stresses which are set up by a shearing motion with parallel layers of fluid sliding relative to each other all vanish. Furthermore, the normal stresses are, then all equal because a fluid is unable to withstand any tendency by applied forces to deform it without change in volume. Thus, in a fluid at rest, the stress tensor becomes isotropic,

$$\tau_{ij} = -p\delta_{ij} \tag{8}$$

where p is called the hydrostatic pressure. Thus, in a fluid at rest, the stress exerted across a plane surface element with unit normal $\hat{\mathbf{n}}$ is $-p\hat{\mathbf{n}}$ which is a normal force of the same magnitude in all directions of $\hat{\mathbf{n}}$ at a given point.

1.4 Relative Motion Near a Point

The stress tensor at a point in a fluid depends on the local deformation of the fluid caused by the motion. Therefore, as a prelude to dynamical considerations, it is necessary to analyze the relative motion in the neighborhood of any point.

Consider the deformation of the fluid. Let $\mathbf{v} = \langle u, v, w \rangle$ denote the instantaneous velocity at the point $P(x, y, z)$. Then, the velocity at a neighboring point $(x + \delta x, y + \delta y, z + \delta z)$ is given by $\mathbf{v} + \delta\mathbf{v}$, where

$$\left.\begin{aligned} u + \delta u &= u + \frac{\partial u}{\partial x}\delta x + \frac{\partial u}{\partial y}\delta y + \frac{\partial u}{\partial z}\delta z + \cdots \\ v + \delta v &= v + \frac{\partial v}{\partial x}\delta x + \frac{\partial v}{\partial y}\delta y + \frac{\partial v}{\partial z}\delta z + \cdots \\ w + \delta w &= w + \frac{\partial w}{\partial x}\delta x + \frac{\partial w}{\partial y}\delta y + \frac{\partial w}{\partial z}\delta z + \cdots \end{aligned}\right\} \tag{9}$$

The geometrical character of the relative velocity $\delta\mathbf{v}$ regarded as a linear function of $\delta\mathbf{x}$ can be recognized by decomposing the second-order velocity gradient tensor $\nabla\mathbf{v}$ into

$$\left.\begin{aligned} \delta u &= \varepsilon_{xx}\delta x + \varepsilon_{xy}\delta y + \varepsilon_{xz}\delta z + \eta\delta z - \zeta\delta y \\ \delta v &= \varepsilon_{xy}\delta x + \varepsilon_{yy}\delta y + \varepsilon_{yz}\delta z + \zeta\delta x - \xi\delta z \\ \delta w &= \varepsilon_{xz}\delta x + \varepsilon_{yz}\delta y + \varepsilon_{zz}\delta z + \xi\delta y - \eta\delta x \end{aligned}\right\} \tag{10}$$

where

$$\left.\begin{aligned} &\varepsilon_{xx} \equiv \frac{\partial u}{\partial x},\ \varepsilon_{yy} \equiv \frac{\partial v}{\partial y},\ \varepsilon_{zz} = \frac{\partial w}{\partial z} \\ &\varepsilon_{xy} \equiv \frac{1}{2}\left(\frac{\partial v}{\partial x} + \frac{\partial u}{\partial y}\right),\ \varepsilon_{yz} \equiv \frac{1}{2}\left(\frac{\partial w}{\partial y} + \frac{\partial v}{\partial z}\right),\ \varepsilon_{zx} \equiv \frac{1}{2}\left(\frac{\partial u}{\partial z} + \frac{\partial w}{\partial x}\right) \\ &\xi \equiv \frac{1}{2}\left(\frac{\partial w}{\partial y} - \frac{\partial v}{\partial z}\right),\ \eta \equiv \frac{1}{2}\left(\frac{\partial u}{\partial z} - \frac{\partial w}{\partial x}\right),\ \zeta \equiv \frac{1}{2}\left(\frac{\partial v}{\partial x} - \frac{\partial u}{\partial y}\right) \end{aligned}\right\}.$$

If one imagines a rectangular fluid particle having the sides $\delta x, \delta y, \delta z$ initially, then the total motion of the particle may be decomposed into

- a pure *translation* with velocity components (u, v, w);
- a mean *rigid-body rotation* (ξ, η, ζ); each of these components defines the mean angular velocity of two mutually perpendicular fluid line-elements;
- a *dilatation* corresponding to the three strain rates $\varepsilon_{xx}, \varepsilon_{yy}, \varepsilon_{zz}$;
- a *shear deformation* corresponding to the three strain rates $\varepsilon_{xy}, \varepsilon_{yz}, \varepsilon_{zx}$.

In order to see these identifications, consider deformation of a line element of fluid *PQ* (see Figure 1.3) in a two-dimensional flow. Equation (10) now becomes

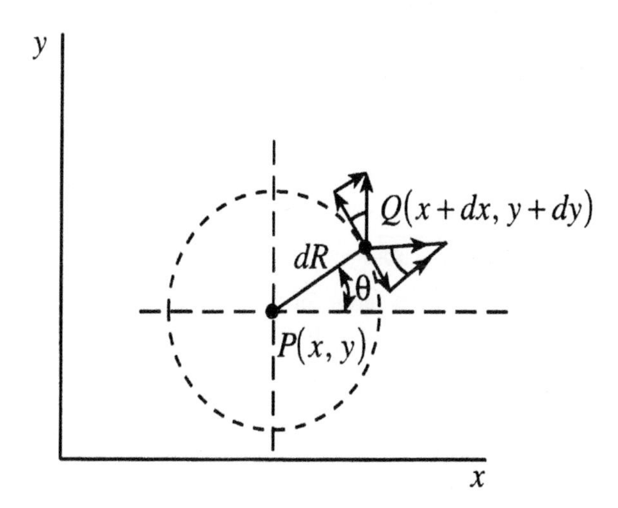

Figure 1.3 Deformation near a point in a fluid.

$$\left.\begin{aligned}\delta u &= -\frac{1}{2}\left(\frac{\partial v}{\partial x} - \frac{\partial u}{\partial y}\right)dy + \left[\frac{\partial u}{\partial x}dx + \frac{1}{2}\left(\frac{\partial v}{\partial x} + \frac{\partial u}{\partial y}\right)dy\right] \\ \delta v &= \frac{1}{2}\left(\frac{\partial v}{\partial x} - \frac{\partial u}{\partial y}\right)dx + \left[\frac{1}{2}\left(\frac{\partial v}{\partial x} + \frac{\partial u}{\partial y}\right)dx + \frac{\partial v}{\partial y}dy\right]\end{aligned}\right\}.$$

The velocity of Q relative to P is

$$\mathbf{v}_{QP} = \left(\frac{\partial u}{\partial x}dx + \frac{\partial u}{\partial y}dy\right)\hat{\mathbf{i}}_x + \left(\frac{\partial v}{\partial x}dx + \frac{\partial v}{\partial y}dy\right)\hat{\mathbf{i}}_y.$$

The contribution of the x-component of $\mathbf{v}_{QP}$ to the rotational speed of PQ about P is given by

$$-\left(\frac{\partial u}{\partial x}dx + \frac{\partial u}{\partial y}dy\right)\frac{\sin\theta}{dR}$$

or

$$-\left(\frac{\partial u}{\partial x}\sin\theta \cdot \cos\theta + \frac{\partial u}{\partial y}\sin^2\theta\right).$$

Similarly, the contribution of the y-component of $\mathbf{v}_{QP}$ to the rotational speed of PQ about P is given by

$$\left(\frac{\partial v}{\partial x}\cos^2\theta + \frac{\partial v}{\partial y}\sin\theta \cdot \cos\theta\right).$$

Overall, the rotational speed of PQ about P is

$$\frac{\partial v}{\partial x}\cos^2\theta - \frac{\partial u}{\partial y}\sin^2\theta + \left(\frac{\partial v}{\partial y} - \frac{\partial u}{\partial x}\right)\sin\theta \cdot \cos\theta.$$

The average rotational speed of PQ about P is then

$$\begin{aligned}\omega \equiv \frac{1}{2\pi}\int_0^{2\pi}\left[\frac{\partial v}{\partial x}\cos^2\theta - \frac{\partial u}{\partial y}\sin^2\theta + \left(\frac{\partial v}{\partial y} - \frac{\partial u}{\partial x}\right)\sin\theta \cdot \cos\theta\right]d\theta \\ = \frac{1}{2}\left(\frac{\partial v}{\partial x} - \frac{\partial u}{\partial y}\right)\end{aligned}$$

which characterizes the rotation in the neighborhood of P. Thus, ω represents the instantaneous average angular speed of line segment PQ within the neighborhood of P and describes a rigid-body rotation of the fluid element at P.

The vorticity Ω, given by

$$\Omega \equiv \frac{\partial v}{\partial x} - \frac{\partial u}{\partial y}$$

is therefore twice the angular velocity ω.

Thus, the velocity field variation in the neighborhood of a point relative to the point is made up of two parts - the rigid-body rotation and the straining rate of the element at the point.

The translation and the rigid-body rotation do not produce any change in the shape of the fluid particle and instead only bodily displace it. On the other hand, the strain rates ε's produce a deformation of the fluid particle, and since they obey the transformation laws of tensor, they form a *strain rate tensor* ε_{ij}, which is a *symmetric tensor* by construction.

$$\varepsilon_{ij} \equiv \frac{1}{2}\left(\frac{\partial v_i}{\partial x_j} + \frac{\partial v_j}{\partial x_i}\right). \tag{11a}$$

Indeed, one may decompose the velocity gradient tensor $\partial v_i/\partial x_j$ as follows:

$$\partial v_i/\partial x_j = \varepsilon_{ij} + \Omega_{ij}$$

where Ω_{ij} is the *antisymmetric tensor*,

$$\Omega_{ij} \equiv \frac{1}{2}\left(\frac{\partial v_j}{\partial x_i} - \frac{\partial v_i}{\partial x_j}\right). \tag{11b}$$

Note therefore that $\Omega \equiv \Omega_{12}$.

1.5 Stress–Strain Relations

In a fluid at rest, as we saw previously, there exist only normal stresses, that are equal in all directions at a point, and the stress tensor has the *isotropic* form

$$\tau_{ij} = -p\delta_{ij}. \tag{12}$$

For a fluid in motion, one may therefore write

$$\tau_{ij} = -p\delta_{ij} + d_{ij},\ d_{ii} = 0. \tag{13}$$

where the nonisotropic part d_{ij} represents the tangential stresses that arise only in a moving fluid. Note that,

$$p = -\frac{1}{3}\tau_{ii} \tag{14}$$

and p represents the average of the three normal stresses along any orthogonal set of axes, and reduces to the hydrostatic pressure when the fluid is at rest. We shall call it the pressure in a moving fluid. Since fluid in relative motion is not in a state of thermodynamic equilibrium, p here is not a *state variable* in thermodynamics.

In order to relate d_{ij} to the local strain rate, let us assume that the fluid is isotropic (so d_{ij} cannot depend on pure rotation), and has no memory effects. Furthermore, we assume that the strain rates are not too large, and write a linear relation,

$$d_{ij} = 2\mu\varepsilon_{ij} + \mu'\delta_{ij}\varepsilon_{kk} \tag{15}$$

where μ and μ' are scalar coefficients. Next, since $d_{ii} = 0$, equation (15) gives

$$2\mu + 3\mu' = 0$$

or

$$\mu' = -\frac{2}{3}\mu \tag{16}$$

so (15) becomes

$$d_{ij} = 2\mu\left(\varepsilon_{ij} - \frac{1}{3}\delta_{ij}\varepsilon_{kk}\right). \tag{17}$$

μ is called the shear viscosity coefficient. The effect of viscosity is to redistribute and equalize the momenta of different fluid particles.

Experiments involving a variety of fluid flows have shown that this linear relation between the strain rate and the nonisotropic part of the stress may hold over a wide range of values of the strain rates. The fluids, for which this is a valid model, are called *Newtonian* fluids. In this book, we shall exclusively deal with such fluids.

2

Equations of Fluid Flows

The laws governing the motion of fluids are

- the law of conservation of mass;
- Newton's laws of motion;
- the First Law of thermodynamics.

These laws typically refer to a system, i.e., a collection of matter of fixed identity.

In order to derive the equations governing the motion of fluids embodying these laws, one may consider a material volume W consisting of the same fluid particles, and hence, moving with the fluid (the *Lagrangian* description). The space coordinates defining W will then be functions of time since they depend on the changing locations of the fluid particles as time progresses. A typical macroscopic property is then represented by a *material integral* (one that always refers to the same fluid particles),

$$\hat{Q} = \iiint\limits_{W} q(t, \mathbf{x}) \rho \, d\mathcal{V} \tag{1}$$

where a property q has been associated with a fluid particle of fixed identity and mass $\rho \, d\mathcal{V}$.

Alternatively, one may consider a fixed region of space called the *control volume* through which the fluid flows so that different fluid particles occupy this region at different times (the *Eulerian* description). A typical macroscopic property Q is represented here by an integral over the control volume $\mathcal{V}$ fixed in space,

$$Q = \iiint\limits_{\mathcal{V}} q(t, \mathbf{x}) \, d\mathbf{x} \tag{2}$$

Introduction to Theoretical and Mathematical Fluid Dynamics, Third Edition.
Bhimsen K. Shivamoggi.

where $q(t,\mathbf{x})$ is a scalar field (for example, density or a velocity component) associated with the fluid.

Thus, the rate of change of $\hat{Q}$, namely, $D\hat{Q}/Dt$ associated with the material system instantaneously occupying the control volume receives contributions from the changes of q with time in W and the flux of q out of W.

2.1 The Transport Theorem

Lemma Let $J(t,\mathbf{x})$ be the Jacobian determinant of the flow $\phi(t,\mathbf{x})$, i.e.,

$$J(t,\mathbf{x}) \equiv det\left|\frac{\partial \phi_i}{\partial x_j}\right| > 0.$$

Let $\phi(t,\mathbf{x})$ be invertible as a function of $\mathbf{x}$, for $t \in I$, some interval in $\mathbb{R}$.

Then,

$$\frac{\partial}{\partial t}J(t,\mathbf{x}) = J(t,\mathbf{x})\,\nabla\cdot\mathbf{v}(t,\mathbf{x}),\, t \in I. \tag{3}$$

Proof: Observing that the determinant of a matrix is multilinear in the columns (or rows), we have

$$\frac{\partial}{\partial t}J(t,\mathbf{x}) = \sum_{all\ columns}\begin{vmatrix} \frac{\partial}{\partial t}\left(\frac{\partial \phi_1}{\partial x_1}\right) & \frac{\partial \phi_2}{\partial x_1} & \frac{\partial \phi_3}{\partial x_1} \\ \frac{\partial}{\partial t}\left(\frac{\partial \phi_1}{\partial x_2}\right) & \frac{\partial \phi_2}{\partial x_2} & \frac{\partial \phi_3}{\partial x_2} \\ \frac{\partial}{\partial t}\left(\frac{\partial \phi_1}{\partial x_3}\right) & \frac{\partial \phi_2}{\partial x_3} & \frac{\partial \phi_3}{\partial x_3} \end{vmatrix}.$$

Note that,

$$\frac{\partial}{\partial t}\left(\frac{\partial \phi_1}{\partial x_i}\right) = \frac{\partial}{\partial x_i}\left(\frac{\partial \phi_1}{\partial t}\right)$$

$$= \frac{\partial}{\partial x_i}v_1(t,\phi(t,\mathbf{x}))$$

$$= \frac{\partial v_1}{\partial x_j}\frac{\partial \phi_j}{\partial x_i},\ \text{summation over } j.$$

Using this, we obtain

$$\frac{\partial}{\partial t}J(t,\mathbf{x}) = \sum_{all\ columns} \begin{vmatrix} \frac{\partial v_1}{\partial x_j}\frac{\partial \phi_j}{\partial x_1} & \frac{\partial \phi_2}{\partial x_1} & \frac{\partial \phi_3}{\partial x_1} \\ \frac{\partial v_1}{\partial x_j}\frac{\partial \phi_j}{\partial x_2} & \frac{\partial \phi_2}{\partial x_2} & \frac{\partial \phi_3}{\partial x_2} \\ \frac{\partial v_1}{\partial x_j}\frac{\partial \phi_j}{\partial x_3} & \frac{\partial \phi_2}{\partial x_3} & \frac{\partial \phi_3}{\partial x_3} \end{vmatrix}$$

which vanishes unless $j = k$ (for the kth column).

Thus we have,

$$\frac{\partial}{\partial t}J(t,\mathbf{x}) = \sum_{all\ columns} \frac{\partial v_1}{\partial x_1}J(t,\mathbf{x})$$

$$= (\nabla \cdot \mathbf{v})\,J(t,\mathbf{x}).$$

□

Theorem 2.1 (*Transport Theorem*) Let $\mathbf{v}(t,\mathbf{x})$ be a C^2 vector field on a region Ω, parallel to the boundary $\partial\Omega$, with flow $\phi(t,\mathbf{x}_0)$, and let $q(t,\mathbf{x})$ be a scalar field associated with the fluid on Ω. If $\phi(t,\mathbf{x}_0)$ is invertible as a function of $\mathbf{x}_0$, for $t \in I$, then,

$$\frac{d}{dt}\iiint_{\phi(t,W)} q(t,\mathbf{x})\,d\mathbf{x} = \iiint_{\phi(t,W)} \left[\frac{\partial q}{\partial t} + \nabla\cdot(q\mathbf{v})\right] d\mathbf{x},\ t \in I. \tag{4}$$

Here, W is a subregion of Ω with ∂W as its boundary.[1]

---

1 Equation (4) may be rewritten, on using the divergence theorem,

$$\frac{d}{dt}\iiint_{\phi(t,W)} q(t,\mathbf{x})\,d\mathbf{x} = \frac{\partial}{\partial t}\iiint_{\phi(t,W)} q(t,\mathbf{x})\,d\mathbf{x} + \oint\!\!\oint_{\phi(t,\partial W)} q\mathbf{v}\cdot\hat{n}\,dS$$

$\hat{\mathbf{n}}$ being the outward normal to dS. Thus, the rate of change of the total q over a material volume $\phi(t,W)$ equals the rate of change for the total q over a fixed volume instantaneously coinciding with $\phi(t,W)$ plus the flux of q out of bounding surface $\phi(t,\partial W)$.

Proof: We have, on using equation (3), and noting that W does not change with t (so one may differentiate under the integral sign),

$$\frac{d}{dt}\iiint\limits_{\phi(t,W)} q(t,\mathbf{x})\,d\mathbf{x} = \frac{d}{dt}\iiint\limits_{W} q(t,\phi(t,\mathbf{x}_o))\,J(t,\mathbf{x}_o)\,d\mathbf{x}_o$$

$$= \iiint\limits_{W} \left(\frac{\partial q}{\partial t} + \nabla q \cdot \mathbf{v} + q\nabla \cdot \mathbf{v}\right) J(t,\mathbf{x}_o)\,d\mathbf{x}_o$$

$$= \iiint\limits_{\phi(t,W)} \left[\frac{\partial q}{\partial t} + \nabla \cdot (q\mathbf{v})\right] d\mathbf{x}.$$

□

2.2 The Material Derivative

The time rate of change of q as one follows a given fluid particle is called the *material derivative* Dq/Dt, given by

$$\frac{Dq}{Dt} = \frac{d}{dt} q(t,\phi(t,\mathbf{x}_o)) = \frac{\partial q}{\partial t} + \nabla q \cdot \mathbf{v}. \tag{5}$$

Thus, even in a steady flow, a fluid particle can change its properties by simply moving to a place where these properties have different values.

2.3 The Law of Conservation of Mass

Since the total mass of fluid in a material volume W is conserved throughout its motion $\phi(t,W)$, we have

$$\frac{d}{dt}\iiint\limits_{\phi(t,W)} \rho\,d\mathbf{x} = 0. \tag{6}$$

Using equation (4), equation (6) leads to

$$\iiint\limits_{\phi(t,W)} \left[\frac{\partial \rho}{\partial t} + \nabla \cdot (\rho\mathbf{v})\right] d\mathbf{x} = 0. \tag{7}$$

Since equation (7) must apply to an infinitesimal material volume as well, one obtains,

$$\frac{\partial \rho}{\partial t} + \nabla \cdot (\rho \mathbf{v}) = 0 \tag{8}$$

or

$$\frac{D\rho}{Dt} + \rho \nabla \cdot \mathbf{v} = 0. \tag{9}$$

2.4 Equation of Motion

While applying Newton's law of motion to a finite, extended mass of fluid, one typically equates the external resultant force to the rate of change of resultant momentum of the fluid occupying a material volume. Applying this procedure to a mass of fluid in a material volume W as it moves with the flow, we obtain

$$\frac{d}{dt} \iiint\limits_{\phi(t,W)} \mathbf{v}\rho d\mathbf{x} = \iiint\limits_{\phi(t,W)} \mathbf{F}\rho d\mathbf{x} + \oiint\limits_{\phi(t,\partial W)} \boldsymbol{\tau} \cdot \hat{\mathbf{n}} dS \tag{10}$$

where $\mathbf{F}$ is the body force per unit mass acting on the fluid.

Using equations (4) and (10) becomes

$$\iiint\limits_{\phi(t,W)} \left[\frac{\partial}{\partial t}(\rho \mathbf{v}) + \nabla \cdot (\rho \mathbf{v}\mathbf{v})\right] d\mathbf{x} = \iiint\limits_{\phi(t,W)} \mathbf{F}\rho d\mathbf{x} + \oiint\limits_{\phi(t,\partial W)} \boldsymbol{\tau} \cdot \hat{\mathbf{n}} dS. \tag{11}$$

Using the *Green's theorem*, equation (11) may be rewritten as

$$\iiint\limits_{\phi(t,W)} \left[\frac{\partial}{\partial t}(\rho \mathbf{v}) + \nabla \cdot (\rho \mathbf{v}\mathbf{v}) - \rho \mathbf{F} - \nabla \cdot \boldsymbol{\tau}\right] d\mathbf{x} = 0 \tag{12}$$

from which, one has

$$\frac{\partial}{\partial t}(\rho \mathbf{v}) + \nabla \cdot (\rho \mathbf{v}\mathbf{v}) = \rho \mathbf{F} + \nabla \cdot \boldsymbol{\tau}. \tag{13a}$$

Upon using equation (9), equation (13a) becomes

$$\frac{D\mathbf{v}}{Dt} = \mathbf{F} + \frac{1}{\rho} \nabla \cdot \boldsymbol{\tau}. \tag{13b}$$

2.5 The Energy Equation

Consider the energy balance of the fluid in the material volume W as it moves with the flow. Work is done on this mass of fluid by both body and surface forces, and heat may also be transferred across the surface S. According to the *First Law of thermodynamics*, the work done and the heat transferred lead to an increase in the kinetic energy and the *internal energy* of the fluid.[2]

A fluid flow necessarily occurs under thermodynamic non-equilibrium conditions. Therefore, a revision of the definition of some of the thermodynamic quantities is necessary before we proceed further to set up an equation embodying the conservation of energy of the fluid. Let us assume that a fluid particle is passing through a succession of states in which the departure from equilibrium is small. We may then define an internal energy E for this fluid particle at any instant as the value corresponding to a hypothetical equilibrium state. Such an equilibrium state can be attained by suddenly isolating the fluid particle from its surroundings, and making it reach equilibrium *adiabatically* without any work being done on it. We may next define density ρ in the standard way as the ratio of mass to instantaneous volume of the fluid particle. A Knowledge of the two properties of state E and ρ, at any instant, then enables one to determine other quantities as in equilibrium thermodynamics.

Though this procedure may look smooth, there appears to be a small difficulty. The thermodynamic pressure p_{th} calculated as per the above procedure may not be equal to the mechanical pressure p. Indeed, one may write

$$p - p_{th} = -\kappa \varepsilon_{kk} \tag{14}$$

where κ is a scalar coefficient called the *bulk viscosity* coefficient. The lag in the adjustment of the mechanical pressure to the continually changing values of ρ and E in a fluid flow is indicated by (14). Physically, this lag is consequent to the delay in the *equipartition of energy* between the various modes (translational, rotational, and vibrational) of a molecule.

However, in most cases the dilatation rates are very much smaller than the shear rates. So, the bulk viscosity will not play an important role. We shall, therefore, neglect it and take p and p_{th} as being identical everywhere, in the following discussion.

The First Law of thermodynamics states that the rate of change of the total energy of a material volume is equal to the rate at which work is being done on the material plus the rate at which heat is transferred into it. The equation expressing this is

2 The internal energy is associated with the thermal motion of the molecules. The additivity of kinetic and internal energies in the total energy of a fluid particle is valid if the fluid is only slightly dissipative.

$$\frac{d}{dt}\iiint\limits_{\phi(t,W)}\left(e+\frac{1}{2}\mathbf{v}^2+U\right)\rho d\mathbf{x}=\oiint\limits_{\phi(t,\partial W)}\mathbf{q}\cdot\hat{\mathbf{n}}dS+\frac{DW}{Dt} \tag{15}$$

where e is the internal energy of the fluid per unit mass, U is the potential energy of the fluid in a *conservative body-force* field $\mathbf{F}=-\nabla U$, $\mathbf{q}$ is the heat flux vector directed *into* the system, and W is the work done on the system.

Using Transport Theorem (4), and writing

$$\frac{DW}{Dt}\equiv\oiint\limits_{\phi(t,\,\partial W)}(\boldsymbol{\tau}\cdot\hat{\mathbf{n}})\cdot\mathbf{v}dS \tag{16}$$

equation (15) becomes

$$\iiint\limits_{\phi(t,W)}\left[\frac{\partial}{\partial t}\left\{\rho\left(e+\frac{1}{2}\mathbf{v}^2+U\right)\right\}\right. \\ \left.+\nabla\cdot\left\{\rho\mathbf{v}\left(e+\frac{1}{2}\mathbf{v}^2+U\right)-\boldsymbol{\tau}\cdot\mathbf{v}\right\}+\nabla\cdot\mathbf{q}\right]d\mathbf{x}=0 \tag{17}$$

from which, one has

$$\frac{\partial}{\partial t}\left\{\rho\left(e+\frac{1}{2}\mathbf{v}^2+U\right)\right\}+\nabla\cdot\left\{\rho\mathbf{v}\left(e+\frac{1}{2}\mathbf{v}^2+U\right)-\boldsymbol{\tau}\cdot\mathbf{v}\right\}=\nabla\cdot\mathbf{q}. \tag{18}$$

Using equations (8) and (18) becomes

$$\rho\left[\frac{\partial}{\partial t}\left(e+\frac{1}{2}\mathbf{v}^2+U\right)+(\mathbf{v}\cdot\nabla)\left(e+\frac{1}{2}\mathbf{v}^2+U\right)\right]-\nabla\cdot(\boldsymbol{\tau}\cdot\mathbf{v})=\nabla\cdot\mathbf{q}. \tag{19}$$

Using equations (13) and (14) in Chapter 1, equation (19) becomes

$$\rho\left[\frac{De}{Dt}+p\frac{D}{Dt}\left(\frac{1}{\rho}\right)\right]=\nabla\cdot\mathbf{q}+d_{ij}\frac{\partial v_j}{\partial x_i}. \tag{20}$$

Noting from equations (12) and (18) in Chapter 1 that

$$d_{ij}\frac{\partial v_j}{\partial x_i}=\varepsilon_{ij}d_{ij}=2\mu\left(\varepsilon_{ij}\varepsilon_{ij}-\frac{1}{3}\varepsilon_{kk}^2\right) \\ =2\mu\left(\varepsilon_{ij}-\frac{1}{3}\varepsilon_{kk}\delta_{ij}\right)^2 \tag{21}$$

equation (20) becomes

$$\rho\left[\frac{De}{Dt}+p\frac{D}{Dt}\left(\frac{1}{\rho}\right)\right]=\nabla\cdot\mathbf{q}+2\mu\left(\varepsilon_{ij}-\frac{1}{3}\varepsilon_{kk}\delta_{ij}\right)^2. \tag{22}$$

Note that the last term on the right-hand side in equation (22) is always positive and represents the *viscous heating*.

The molecular transport of heat is proportional to the local gradient of temperature T, which is a measure of the kinetic energy of the random molecular motion of the fluid. If the heat flux occurs as a consequence of molecular transport between contiguous portions of fluid across the material surface ∂W, then we have,

$$\mathbf{q} = K\nabla T \tag{23}$$

K being the *thermal conductivity* of the fluid. If K is constant, equation (23) leads to

$$\nabla \cdot \mathbf{q} = K\nabla^2 T. \tag{24}$$

Note that an *equation of state* of the form $p = p(\rho)$ is finally needed to close equations (8), (13a), (22) and (24).

2.6 The Equation of Vorticity

Upon taking the curl of equation (13a), one obtains for the evolution of *potential vorticity* $\mathbf{\Omega}/\rho, (\mathbf{\Omega} \equiv \nabla \times \mathbf{v})$, in an inviscid fluid,

$$\frac{D}{Dt}\left(\frac{\mathbf{\Omega}}{\rho}\right) = \left(\frac{\mathbf{\Omega}}{\rho} \cdot \nabla\right)\mathbf{v} - \frac{1}{\rho^3}\nabla p \times \nabla\rho + \frac{1}{\rho}\nabla \times \mathbf{F}. \tag{25}$$

The second term on the right-hand side represents the *baroclinic* generation of potential vorticity which is due to the misalignment of the density and pressure gradients. This mechanism is the cause of generation of vorticity by shock waves (see Chapter 16).

For a *barotropic* fluid for which $p = p(\rho)$, equation (25) becomes, in the absence of the body force $\mathbf{F}$,

$$\frac{D}{Dt}\left(\frac{\mathbf{\Omega}}{\rho}\right) = \left(\frac{\mathbf{\Omega}}{\rho} \cdot \nabla\right)\mathbf{v} \tag{26}$$

which implies that the potential vorticity lines are *frozen* in such a fluid (see Chapter 8).

2.7 The Incompressible Fluid

A fluid is said to be *incompressible* when the density of an element of the fluid is not affected by changes in pressure, and $D\rho/Dt = 0$.[3] Equations (8), (13a), (21) and (25), then become[4] the Navier-Stokes equations,

$$\nabla \cdot \mathbf{v} = 0 \tag{27}$$

$$\rho\left[\frac{\partial \mathbf{v}}{\partial t} + (\mathbf{v}\cdot\nabla)\mathbf{v}\right] = \rho\mathbf{F} - \nabla p + \mu\nabla^2\mathbf{v} \tag{28}$$

$$e = const. \tag{29}$$

$$\frac{\partial \boldsymbol{\Omega}}{\partial t} + (\mathbf{v}\cdot\nabla)\boldsymbol{\Omega} = (\boldsymbol{\Omega}\cdot\nabla)\mathbf{v} + \nabla\times\mathbf{F} + \nu\nabla^2\boldsymbol{\Omega} \tag{30}$$

where, ν is the *kinematic viscosity*,

$$\nu \equiv \frac{\mu}{\rho}.$$

It may be noted that, under the incompressibility constraint, the pressure field, instead of being determined by thermodynamics, plays no dynamical role. It simply adjusts instantaneously so as to keep the velocity field *solenoidal* and satisfy the given boundary conditions.

Now, in a two-dimensional flow, the fluid particles move in planar paths. All such planes are parallel and streamline patterns are identical in each plane. For

---

3 This condition merely implies that the mass density of each fluid element is conserved. Equivalently, the iso-density surfaces $\rho = const$ are frozen in the fluid and the volume $\mathcal{V}(\rho)$ inside a closed iso-density surface $\rho = const$ is conserved, as the surface moves with the flow.

4 In the Lagrangian formulation, the incompressibility condition (27) becomes

$$J(t,\mathbf{x}) = det\left|\frac{\partial \phi_i}{\partial x_j}\right| = 1$$

while equation of motion equation (28) for an *inviscid* fluid becomes

$$\frac{\partial^2}{\partial t^2}\phi(t,\mathbf{x}) = -\nabla p(t,\phi(t,\mathbf{x})).$$

a two-dimensional flow (in the xy-plane), letting

$$\mathbf{\Omega} = \omega \hat{\mathbf{i}}_z \tag{31}$$

and assuming that the body force **F** is conservative, equation (30), leads to

$$\frac{\partial \omega}{\partial t} + u\frac{\partial \omega}{\partial x} + v\frac{\partial \omega}{\partial y} = \nu \left(\frac{\partial^2 \omega}{\partial x^2} + \frac{\partial^2 \omega}{\partial y^2} \right). \tag{32}$$

where,

$$\left. \begin{aligned} \mathbf{v} &= (u, v, 0), \\ \omega &\equiv \frac{\partial v}{\partial x} - \frac{\partial u}{\partial y} \end{aligned} \right\}. \tag{33}$$

Introducing the *stream function* Ψ,[5] such that

$$u = \frac{\partial \Psi}{\partial y}, \; v = -\frac{\partial \Psi}{\partial x} \tag{34}$$

equation (33) gives,

$$\omega = -\nabla^2 \Psi \tag{35}$$

and equation (32) leads to

$$\frac{\partial \nabla^2 \Psi}{\partial t} + \frac{\partial \Psi}{\partial y}\frac{\partial \nabla^2 \Psi}{\partial x} - \frac{\partial \Psi}{\partial x}\frac{\partial \nabla^2 \Psi}{\partial y} = \nu \nabla^4 \Psi. \tag{36}$$

2.8 Boundary Conditions

The boundary conditions required in solving equations (8), (13a) and (22) are that at any solid boundary, there can neither be a penetration of the fluid into the boundary, nor a gap between the fluid and the boundary. This means that the fluid velocity component normal to the boundary must be zero at the boundary. When the boundary itself is in motion, the fluid velocity component normal to the boundary must equal the velocity of the boundary normal to itself. No restrictions are placed on the tangential component of fluid velocity at a boundary when inviscid fluids are considered. However, in a viscous fluid, even the tangential component of the fluid velocity at a boundary is the same as that of the boundary. This way, there is no slip of the fluid at the boundary.

5 The existence of the stream function Ψ incorporates equation (27) and hence automatically satisfies the mass conservation condition.

At a boundary given by $F(x, y, z, t) = 0$ moving with the fluid, one has the *kinematic condition*,

$$\frac{DF}{Dt} = 0 \tag{37}$$

so that such a boundary is always in contact with the same fluid particles.

When the fluid is set into motion by the body moving through it, a necessary condition is that the fluid at infinity remains at rest. If this were not so, it would imply that the finite forces exerted by the body had imparted infinite kinetic energy to the fluid in finite time, in violation of the principles of work and energy.

2.9 A Program for Analysis of the Governing Equations

The set of equations governing the fluid flows given in the previous section is too complicated to render a direct mathematical approach feasible. Furthermore, progress is possible only by isolating the various physical features represented by these equations, and then analyzing specific flow fields embodying these features separately. When assembled and interpreted appropriately, these special cases provide an insight into the totality of the phenomena described by the governing equations.

Thus, as a first step, one considers an inviscid, incompressible fluid as one that is endowed with inertia and no other physical properties (Part 2). Though apparently a very restricted model, it nonetheless covers a wide variety of flows of a real fluid. Next, one may allow for variations in density brought about by flow-velocity variations and obtain a compressible fluid (Part 3). Furthermore, including the nonzero resistance offered by a fluid to shearing deformations imposed on it results in a viscous fluid (Part 4).

3

Hamiltonian Formulation of Fluid-Flow Problems

Hamiltonian formulations have traditionally played an important role in both the classical and quantum mechanics of particles and fields. They not only offer a new perspective on familiar results, but also provide powerful tools like the conservation laws and stability theorems. Despite this, Hamiltonian formulations were not introduced into fluid-flow problems until recently. This is because of the fact that the Eulerian variables in fluid-flow problems are *non-canonical*, (Salmon, 1988) thanks to the particle-relabeling symmetry property of fluid flows. These Eulerian variables constitute a *reduced* set and are not related to the corresponding Lagrangian variables by a *canonical transformation*. The Lagrangian (particle-following) variables of a fluid flow are *canonical*, i.e., their time evolution is given by Hamilton-like equations. The Lagrangian equations of motion possess an additional particle-relabelling symmetry, in that the Hamiltonian is invariant under translations of fluid particles along lines of constant vorticity. Such symmetries are *invisible* in the Eulerian description. Consequently, some conserved Eulerian quantities are no longer related to *explicit* symmetries of the Hamiltonian and cannot be obtained from *Noether's Theorem*, but instead become *Casimir invariants* (see Section 3.1).

Introduction to Theoretical and Mathematical Fluid Dynamics, Third Edition.
Bhimsen K. Shivamoggi.

3.1 Hamiltonian Dynamics of Continuous Systems

For continuous systems (unlike discrete systems[1]), the Hamiltonian system in question is *infinite-dimensional* because the dynamical variables are now fields and the functions of state $F(u)$ become *functionals of state* $\mathscr{F}(u)$, and the partial

---

1 Consider a dynamical system,

$$\dot{\mathbf{x}} = \mathbf{v}(\mathbf{x}, t).$$

Let the system be Hamiltonian, i.e.,

$$\dot{\mathbf{x}} = \mathbf{J} \cdot \frac{\partial H}{\partial \mathbf{x}}$$

where $\mathbf{J}(\mathbf{x})$ is an antisymmetric matrix satisfying the *gauge condition*,

$$\frac{\partial}{\partial \mathbf{x}} \cdot \left(\mathbf{J}(\mathbf{x}) \cdot \frac{\partial H}{\partial \mathbf{x}} \right) = 0.$$

This leads to

$$\frac{\partial}{\partial \mathbf{x}} \cdot \dot{\mathbf{x}} = 0$$

which ensures that the dynamics generates a volume-preserving flow in the *phase space*. Furthermore, if $H = H(\mathbf{x})$, then

$$\dot{H} = \frac{dH}{d\mathbf{x}} \cdot \dot{\mathbf{x}} = \frac{dH}{d\mathbf{x}} \cdot \mathbf{J} \cdot \frac{dH}{d\mathbf{x}} = 0$$

so H is a constant of the motion. Introducing the *Poisson bracket* for this system, as per,

$$\{f, g\} \equiv \frac{\partial f}{\partial \mathbf{x}} \cdot \mathbf{J} \cdot \frac{\partial g}{\partial \mathbf{x}}$$

the equation of motion becomes

$$\dot{\mathbf{x}} = \{\mathbf{x}, H\}.$$

The time evolution of an arbitrary physical quantity $Q(\mathbf{x}, t)$ obeys,

$$\dot{Q} = \frac{\partial Q}{\partial t} + [Q, H].$$

In the Euclidean space R^{2n} with coordinates $(\mathbf{p}, \mathbf{q})$,

$$[f, g] = \sum_{i=1}^{n} \left(\frac{\partial f}{\partial \mathbf{q}_i} \frac{\partial g}{\partial \mathbf{p}_i} - \frac{\partial f}{\partial \mathbf{p}_i} \frac{\partial g}{\partial \mathbf{q}_i} \right) = \nabla f \cdot \mathbf{J} \cdot \nabla g$$

derivatives $\partial F/\partial u$ become the *functional* (or *variational*) *derivatives* $\delta \mathcal{F}/\delta u$ [2] defined by

$$\begin{aligned}\delta \mathcal{F} &\equiv \mathcal{F}(u+\delta u) - \mathcal{F}(u) \\ &= \left(\frac{\delta \mathcal{F}}{\delta u}, \delta u\right) + O(\delta u)^2\end{aligned} \tag{1}$$

for admissible but otherwise arbitrary variations δu, where $(\cdot, \cdot)$ is the relevant inner product for the function space $\{u\}$. If

$$\mathcal{F}(u) = \int_{x_1}^{x_2} F(x, u, u_x)\, dx \tag{2}$$

then

$$\delta \mathcal{F} = \int_{x_1}^{x_2} \left[\frac{\partial F}{\partial u} - \frac{d}{dx}\left(\frac{\partial F}{\partial u_x}\right)\right] \delta u \, dx. \tag{3}$$

So,

$$\frac{\delta \mathcal{F}}{\delta u} = \frac{\partial F}{\partial u} - \frac{d}{dx}\left(\frac{\partial F}{\partial u_x}\right). \tag{4}$$

Consider a Hamiltonian dynamical system represented in the *symplectic* (i.e., nondegenerate and skew-symmetric) form

$$u_t = J\frac{\delta \mathcal{H}}{\delta u} \tag{5}$$

where $u(\mathbf{x}, t)$ is the dynamical variable, $\mathcal{H}(u)$ is the Hamiltonian functional, and J is a sympletic operator, which is a transformation: $\{u\} \Rightarrow \{u\}$ satisfying the skew-symmetric property:

$$(u, Jv) = -(Ju, v). \tag{6}$$

---

where

$$\mathbf{J} = \begin{pmatrix} 0 & I \\ -I & 0 \end{pmatrix}, \nabla = \left(\frac{\partial}{\partial \mathbf{q}_i}, \frac{\partial}{\partial \mathbf{p}_i}\right)$$

I being the $n \times n$ identity matrix. The Poisson brackets are bilinear, skew-symmetric and they satisfy the *Jacobi identity*,

$$[[f, g], h] + [[g, h], f] + [[h, f], g] = 0.$$

2 The symbol δ is used to denote an increment everywhere else in this book, but in this chapter it denotes the variational derivative in keeping with the standard notation in *variational calculus*.

An equivalent Poisson bracket statement is

$$\frac{d\mathscr{F}}{dt} \equiv \left(\frac{\delta \mathscr{F}}{\delta u}, u_t\right) = \left(\frac{\delta \mathscr{F}}{\delta u}, J\frac{\delta \mathscr{H}}{\delta u}\right) \equiv [\mathscr{F}, \mathscr{H}]. \tag{7}$$

Equation (7) immediately implies that the Hamiltonian $\mathscr{H}$ is an integral invariant of this system, because

$$\frac{d\mathscr{H}}{dt} = \left(\frac{\delta \mathscr{H}}{\delta u}, J\frac{\delta \mathscr{H}}{\delta u}\right) = -\left(J\frac{\delta \mathscr{H}}{\delta u}, \frac{\delta \mathscr{H}}{\delta u}\right) = 0. \tag{8}$$

Now, suppose that there exists some steady state $u = U$ of this system. If the Hamiltonian representation is *canonical* in the sense that J is invertible, it follows, from (5), that the steady state $u = U$ is a *conditional extremum* of $\mathscr{H}$, namely,

$$J\frac{\delta \mathscr{H}}{\delta u}\bigg|_{u=U} = 0 \tag{9}$$

which implies

$$\frac{\delta \mathscr{H}}{\delta u}\bigg|_{u=U} = 0. \tag{10}$$

If, on the other hand, the Hamiltonian representation is *non-canonical*, in the sense that J is singular and non-invertible, then we instead have

$$J\frac{\delta \mathscr{H}}{\delta u}\bigg|_{u=U} = 0 \tag{11}$$

implying that

$$\frac{\delta \mathscr{H}}{\delta u}\bigg|_{u=U} = -\frac{\delta \mathscr{C}}{\delta u}\bigg|_{u=U} \tag{12}$$

for some *Casimir functional* $\mathscr{C}(u)$, which satisfies

$$J\frac{\delta \mathscr{C}}{\delta u} = 0. \tag{13}$$

Thus, the steady state $u = U$ is a conditional extremum of the *combined* invariant $(\mathscr{H} + \mathscr{C})$.

Note that, unlike an ordinary invariant A which commutes only with the Hamiltonian $\mathscr{H}$ as given by,

$$\dot{A} = \left(\frac{\partial A}{\partial u},\ J\ \frac{\partial \mathscr{H}}{\partial u}\right) = [A, \mathscr{H}] = 0 \tag{14}$$

a Casimir invariant commutes with all functionals $\mathscr{F}$ of the dynamical variables, because (13) implies that

$$[\mathscr{F}, \mathscr{C}] = \left(\frac{\delta \mathscr{F}}{\delta u}, J \frac{\delta \mathscr{C}}{\delta u}\right) = 0, \ \ \forall \mathscr{F} \tag{15}$$

and in particular,

$$[\mathscr{C}, \mathscr{H}] = 0. \tag{16}$$

It follows immediately from (16) that Casimir functionals are integral invariants of the dynamical system because

$$\frac{d\mathscr{C}}{dt} = [\mathscr{C}, \mathscr{H}] = 0. \tag{17}$$

The usual invariants are associated with explicit symmetries of the Hamiltonian itself. When these symmetries are continuous ones (like the translations in time and space), *Noether's Theorem* provides the connection between the two.

Theorem 3.1 *(Noether)* If $\mathscr{H}$ is invariant under translations in x, generated by the functional $\mathscr{M}$, i.e.,

$$J \frac{\delta \mathscr{M}}{\delta u} = -\frac{\partial u}{\partial x} \tag{18}$$

then, $\mathscr{M}$ is invariant, i.e.,

$$\frac{d\mathscr{M}}{dt} = 0. \tag{19}$$

Proof: We have

$$\frac{d\mathscr{M}}{dt} = \left(\frac{\delta \mathscr{M}}{\delta u}, J \frac{\delta \mathscr{H}}{\delta u}\right) = -\left(J \frac{\delta \mathscr{M}}{\delta u}, \frac{\delta \mathscr{H}}{\delta u}\right)$$

$$= \left(\frac{\partial u}{\partial x}, \frac{\partial \mathscr{H}}{\partial u}\right) = \frac{\partial \mathscr{H}}{\partial x} = 0.$$

□

Corollary: If $\mathscr{H}$ is invariant under translations in t, then $\mathscr{H}$ is an invariant, i.e.,

$$\frac{d\mathscr{H}}{dt} = 0. \tag{20}$$

Casimir invariants, which arise for non-canonical Hamiltonian systems cannot be obtained from *Noether's Theorem*. Casimir invariants are not associated with any property of the Hamiltonian function itself. Rather they arise from the *degenerate* nature of the symplectic operator J.[3] Note that a canonical system has no nontrivial Casimir invariants because, when J is invertible, the condition

$$J\frac{\delta \mathscr{C}}{\delta u} = 0$$

implies

$$\frac{\delta \mathscr{C}}{\delta u} = 0$$

which, in turn, implies that $\mathscr{C}$ is simply a constant.

3.2 Three-Dimensional Incompressible Flows

Consider a three-dimensional incompressible flow in a domain $V \subseteq \mathbf{R}^3$. The governing equations are (see equations (27) and (28) in Chapter 2),

$$\nabla \cdot \mathbf{v} = 0 \tag{21}$$

$$\frac{\partial \mathbf{v}}{\partial t} - \mathbf{v} \times \mathbf{\Omega} = -\nabla b \tag{22}$$

where

$$b \equiv \frac{p}{\rho} + \frac{1}{2}\mathbf{v}^2, \ \mathbf{\Omega} \equiv \nabla \times \mathbf{v}.$$

Upon taking the curl, we obtain the vorticity equation, from equation (22),

$$\frac{\partial \mathbf{\Omega}}{\partial t} - \nabla \times (\mathbf{v} \times \mathbf{\Omega}) = \mathbf{0}. \tag{23}$$

The Hamiltonian for this system is

$$\mathscr{H} = \frac{1}{2}\int_V \boldsymbol{\psi} \cdot \mathbf{\Omega}\, d\mathbf{x} \tag{24}$$

where,

$$\mathbf{v} = \nabla \times \boldsymbol{\psi}.$$

In deriving (24), we have set $|\boldsymbol{\psi}| = 0$ on the boundary $\partial \mathcal{V}$ (alternatively, one may impose the preservation of circulation Γ, $\Gamma \equiv \oint \mathbf{v} \cdot d\mathbf{x}$, i.e., $\delta\Gamma = 0$, on $\partial\mathcal{V}$), and $\boldsymbol{\psi}$ is made unique by imposing the *gauge condition*,

---

3 Casimir invariants are annihilators (with respect to any pairing functional) of the Poisson brackets which become degenerate when expressed in terms of these natural quantities.

$$\nabla \cdot \boldsymbol{\psi} = 0. \tag{25}$$

If one chooses $\boldsymbol{\Omega}$ to be the canonical or *Darboux* variable and takes the skew-symmetric operator J [4] to be (Olver, 1982)

$$J = -\nabla \times (\boldsymbol{\Omega} \times \nabla \times (\cdot)) \tag{26}$$

---

4 The operator J may be seen to induce the Poisson bracket (Kuznetsov and Mikhailov, 1980),

$$[F, G] = \left(\frac{\delta F}{\delta \boldsymbol{\Omega}}, J\frac{\delta G}{\delta \boldsymbol{\Omega}}\right)$$

$$= -\int_V \frac{\delta F}{\delta \boldsymbol{\Omega}} \cdot \nabla \times \left[\boldsymbol{\Omega} \times \left(\nabla \times \frac{\delta G}{\delta \boldsymbol{\Omega}}\right)\right] dV$$

$$= -\int_V \left(\nabla \times \frac{\delta F}{\delta \boldsymbol{\Omega}}\right) \cdot \left[\boldsymbol{\Omega} \times \left(\nabla \times \frac{\delta G}{\delta \boldsymbol{\Omega}}\right)\right] dV$$

$$= -\int_V \boldsymbol{\Omega} \cdot \left[\left(\nabla \times \frac{\delta F}{\delta \boldsymbol{\Omega}}\right) \times \left(\nabla \times \frac{\delta G}{\delta \boldsymbol{\Omega}}\right)\right] dV$$

which is a bilinear function defined on admissible functionals $F[\boldsymbol{\Omega}]$ and $G[\boldsymbol{\Omega}]$ satisfying,

$$\nabla \cdot \begin{pmatrix} \dfrac{\delta F}{\delta \boldsymbol{\Omega}} \\ \dfrac{\delta G}{\delta \boldsymbol{\Omega}} \end{pmatrix} = 0 \text{ in } V \text{ and } \left|\frac{\delta F}{\delta \boldsymbol{\Omega}}\right|, \left|\frac{\delta G}{\delta \boldsymbol{\Omega}}\right| = 0 \text{ on } \partial V.$$

The Poisson Bracket above satisfies the antisymmetry property,

$$[F, G] = -[G, F]$$

and the Jacobi identity,

$$[[F, G], K] + [[G, K], F] + [[K, F], G] = 0$$

and other required algebraic properties of Poisson brackets.

The dynamics underlying equations (21) and (23) can then be represented symplectically by

$$\frac{\partial F}{\partial t} = [F, H]$$

for admissible functionals $F[\boldsymbol{\Omega}]$.

Hamilton's equation is

$$\frac{\partial \boldsymbol{\Omega}}{\partial t} = J\frac{\delta \mathcal{H}}{\delta \boldsymbol{\Omega}} = -\nabla \times (\boldsymbol{\Omega} \times (\nabla \times \boldsymbol{\psi})) = \nabla \times (\mathbf{v} \times \boldsymbol{\Omega}) \tag{27}$$

which is just equation (23). Thus, in the Hamiltonian formulation, the Euler equations characterize the *geodesic flow* on an infinite-dimensional group of volume-preserving diffeomorphisms.

The Casimir invariants[5] for this problem are the solutions of

$$J\frac{\delta \mathscr{C}}{\delta \boldsymbol{\Omega}} = -\nabla \times \left[\boldsymbol{\Omega} \times \left(\nabla \times \frac{\delta \mathscr{C}}{\delta \boldsymbol{\Omega}}\right)\right] = \mathbf{0} \tag{28}$$

which implies that

$$\frac{\delta \mathscr{C}}{\delta \boldsymbol{\Omega}} = \mathbf{v}. \tag{29}$$

Then, a Casimir invariant is the *helicity*,

$$\mathscr{C} = \int_V \mathbf{v} \cdot \boldsymbol{\Omega} d\mathbf{x} \tag{30}$$

which describes the extent to which the vortex filaments wrap around one another and can be shown to characterize the *degree of knottedness* of the vortex line topology (Moffatt, 1969). The total helicity is not positive definite, so one cannot imagine a development of the minimum helicity state. Actually, one obtains a *Beltrami flow* (which has velocity parallel to vorticity) by minimizing energy while conserving the total helicity (see below).

Extremization of the Hamiltonian (24), while keeping the Casimir invariant (30) fixed leads to

$$\frac{\delta \mathcal{H}}{\delta \boldsymbol{\Omega}} = \mu \frac{\delta \mathscr{C}}{\delta \boldsymbol{\Omega}} \tag{31}$$

or

$$\boldsymbol{\psi} = \mu \mathbf{v}$$

or

$$\mathbf{v} = \mu \boldsymbol{\Omega} \tag{32}$$

where μ is the *Lagrange multiplier*. The flows, for which (32) is true, are called *Beltrami flows*.

Beltrami flows are actually known to correlate well with real fluid behavior; for example, the *Larichev-Reznik* (1976) *nonlinear dipole-vortex*-localized

5 Casimir invariants are traceable to the hidden symmetry in Euler equations associated with relabeling of fluid particles and they possess the same value when evaluated on functions that are related by particle rearrangement.

structure. *Beltramization* turns out to provide the means by which the underlying system can accomplish *ergodicity* of the streamlines. This follows by noting that Beltramization corresponds to relaxation of the constraint

$$(\mathbf{v} \cdot \nabla)\left(p/\rho + \mathbf{v}^2/2\right) = 0 \tag{33}$$

associated with the equation for steady flows,

$$\mathbf{v} \times \boldsymbol{\Omega} = \nabla\left(p/\rho + \mathbf{v}^2/2\right). \tag{34}$$

So, the streamlines are no longer confined to the *iso-baric* surfaces given by

$$p/\rho + \mathbf{v}^2/2 = const \tag{35}$$

and become ergodic (Moffatt, 1969). This raises the question: Do fluids have an intrinsic tendency toward Beltramization? A definitive answer does not seem to be at hand yet.

3.3 Two-Dimensional Incompressible Flows

Consider a two-dimensional incompressible flow in $D \subseteq R^2$. The governing equations are

$$\frac{\partial \zeta}{\partial t} + \partial(\psi, \zeta) = 0 \tag{36}$$

where

$$\left.\begin{array}{c} \mathbf{v} = \hat{\mathbf{i}}_z \times \nabla\psi,\ \zeta = \hat{\mathbf{i}}_z \cdot \nabla \times \mathbf{v} = \nabla^2\psi \\ \partial(\psi, \zeta) \equiv \psi_x \zeta_y - \psi_y \zeta_x \end{array}\right\}.$$

The Hamiltonian for this system is[6]

$$\mathcal{H} = -\frac{1}{2}\iint_D \psi\zeta\, dxdy \tag{37}$$

assuming that $\psi = 0$ on the boundary ∂D (D is taken to be *simply connected*).

---

6 Using the result, upon choosing appropriate boundary conditions,

$$\iint_D \nabla \cdot (\psi\nabla\psi)\, dx\, dy = \iint_D (\nabla\psi)^2\, dx\, dy + \iint_D (\psi\nabla^2\psi)\, dx\, dy = 0$$

we obtain

$$\mathcal{H} = \frac{1}{2}\iint_D (\nabla\psi)^2\, dxdy$$

as one would expect.

If one chooses ζ to be the canonical variable and takes the skew-symmetric operator J[7] to be

$$J = -\partial\left(\zeta, \cdot\right), \tag{38}$$

then Hamilton's equation is

$$\frac{\partial \zeta}{\partial t} = J\frac{\delta \mathscr{H}}{\delta \zeta} = -\partial\left(\zeta, (-\psi)\right) = -\partial\left(\psi, \zeta\right) \tag{39}$$

which is just equation (36)!

The Casimir invariants for this problem are the solutions to

$$J\frac{\delta \mathscr{C}}{\delta \zeta} = -\partial\left(\zeta, \frac{\partial \mathscr{C}}{\partial \zeta}\right) = 0 \tag{40}$$

which implies that, for some function $F(\zeta)$,

$$\frac{\delta \mathscr{C}}{\delta \zeta} = F'(\zeta). \tag{41}$$

Thus, the Casimir invariant is

$$\mathscr{C} = \iint\limits_{D} F(\zeta)\, dxdy. \tag{42}$$

One may understand Equation (42) by noting from equation (36) that material fluid elements carry their values of vorticity with them so that vorticity is actually a *Lagrangian* invariant. Consequently, any integral function of vorticity would be conserved. This introduces a strong restriction on the dynamics underlying two-dimensional flows.

---

7 The Poisson bracket for this problem is given by

$$[\mathscr{F}, \mathscr{G}] = \iint\limits_{D} \left(\frac{\delta \mathscr{F}}{\delta \zeta}, J\frac{\delta \mathscr{G}}{\delta \zeta}\right) dxdy$$

$$= -\iint\limits_{D} \frac{\delta \mathscr{F}}{\delta \zeta}\partial\left(\zeta, \frac{\delta \mathscr{G}}{\delta \zeta}\right) dxdy$$

$$= \iint\limits_{D} \zeta\partial\left(\frac{\delta \mathscr{F}}{\delta \zeta}, \frac{\delta \mathscr{G}}{\delta \zeta}\right) dxdy.$$

Note that

$$\frac{\partial \mathcal{H}}{\partial t} = \left(\frac{\delta \mathcal{H}}{\delta \zeta}, J\frac{\delta \mathcal{H}}{\delta \zeta}\right) = -\left(J\frac{\delta \mathcal{H}}{\delta \zeta}, \frac{\delta \mathcal{H}}{\delta \zeta}\right) = 0. \tag{43}$$

In fact, translational symmetry with respect to time implies the existence of an invariant $\mathcal{M}$, satisfying

$$J\frac{\partial \mathcal{M}}{\partial \zeta} = -\partial\left(\zeta, \frac{\delta \mathcal{M}}{\delta \zeta}\right) = -\zeta_t = \partial\left(\psi, \zeta\right) \tag{44}$$

which implies

$$\frac{\delta \mathcal{M}}{\delta \zeta} = \psi. \tag{45}$$

Thus,

$$\mathcal{M} = \iint_D \psi\zeta dxdy = -2\mathcal{H}. \tag{46}$$

So, $\mathcal{M}$ is indeed proportional to $\mathcal{H}$!

If this system has translational symmetry along the x-direction, according to *Noether's Theorem* (which provides conservation laws associated with the symmetry groups), there exists an invariant $\mathcal{M}$ given by

$$J\frac{\delta \mathcal{M}}{\delta \zeta} = -\frac{\partial \zeta}{\partial x} \tag{47}$$

which implies that,

$$\frac{\delta \mathcal{M}}{\delta \zeta} = y$$

or

$$\mathcal{M} = \iint y\zeta dxdy. \tag{48}$$

Note that, $\mathcal{M}$ Equation (48) is simply the x-component of *Kelvin's* (1868) *impulse.*

On the other hand, translational symmetry along the x-direction also implies that

$$\zeta(x, y, t) = \zeta(x - ct, y) \tag{49}$$

which in turn, gives

$$-c\frac{\partial \zeta}{\partial x} = \frac{\partial \zeta}{\partial t}. \tag{50}$$

Thus,

$$c\, J\frac{\delta \mathscr{M}}{\delta \zeta} = J\frac{\delta \mathscr{H}}{\delta \zeta} \tag{51}$$

or

$$J\frac{\delta(\mathscr{H} - c\mathscr{M})}{\delta \zeta} = 0. \tag{52}$$

Therefore, translationally symmetric states correspond to the extremization of $(\mathscr{H} - c\mathscr{M})$ under variations that preserve the Casimir invariants.

4

Surface Tension Effects

Surface tension is a cause of the fact that small water drops in air and small gas bubbles in water take up a spherical form. Surface tension is a consequence of intermolecular cohesive forces. When one of two media in contact is a liquid phase, work must be done on a molecule approaching the interface from the interior of the liquid because this molecule experiences an unbalanced cohesive force directed away from the interface. This results in a higher potential energy for the molecules at the interface and a tendency for all molecules of the liquid near the interface to move inward. The interface therefore tends to contract as if it were in a state of tension like a stretched membrane. Since, in a state of equilibrium, the interface energy must be a minimum, for a given volume, the sphere is the shape with the least surface area. A water drop in air and an air bubble in water are therefore spherical.

4.1 Shape of the Interface between Two Fluids

Consider the interface between two stationary fluids. In order to determine the shape of the interface corresponding to a mechanical equilibrium, we observe that a curved surface under tension exerts a normal stress across the surface. If the surface is described by

$$z - \xi(x, y) = 0 \tag{1}$$

then, the normal to the interface at the origin O is given by

$$\hat{\mathbf{n}} = \left(-\frac{\partial \xi}{\partial x}, -\frac{\partial \xi}{\partial y}, 1 \right), \tag{2}$$

Introduction to Theoretical and Mathematical Fluid Dynamics, Third Edition.
Bhimsen K. Shivamoggi.

The resultant of the tensile forces acting on a portion S of the interface containing O is

$$-T\oint_C \hat{\mathbf{n}} \times \mathbf{dx}, \tag{3}$$

where, T is the surface tension, and $\mathbf{dx}$ is a line element of the closed curve C enclosing S. The component of the resultant force along the z-axis is given by

$$-T\oint_C \left(-\frac{\partial \xi}{\partial x}dy + \frac{\partial \xi}{\partial y}dx\right)$$

or, on using Green's theorem, this becomes

$$T\iint_S \left(\frac{\partial^2 \xi}{\partial x^2} + \frac{\partial^2 \xi}{\partial y^2}\right) dxdy. \tag{4}$$

Now, let the interface deviate only slightly from the plane $z = 0$ so that ξ is small everywhere. The area A of the interface portion S is given by

$$A = \iint_S \sqrt{1 + \left(\frac{\partial \xi}{\partial x}\right)^2 + \left(\frac{\partial \xi}{\partial y}\right)^2}\, dxdy$$

$$\approx \iint_S \left[1 + \frac{1}{2}\left(\frac{\partial \xi}{\partial x}\right)^2 + \frac{1}{2}\left(\frac{\partial \xi}{\partial y}\right)^2\right] dxdy \tag{5}$$

from which, the change in the interface area on deformation is given by

$$\delta A = \iint_S \left[\frac{\partial \xi}{\partial x}\frac{\partial \delta\xi}{\partial x} + \frac{\partial \xi}{\partial y}\frac{\partial \delta\xi}{\partial y}\right] dxdy = -\iint_S \left(\frac{\partial^2 \xi}{\partial x^2} + \frac{\partial^2 \xi}{\partial y^2}\right) dxdy\delta\xi. \tag{6}$$

Let R_1 and R_2 be the principal radii of curvature at a given point on the surface. Let dl_1 and dl_2 be two arc length elements on S, which are also elements of circumference of circles with radii R_1 and R_2. Hence

$$A = \iint dl_1\left(1 + \frac{\delta\xi}{R_1}\right) dl_2\left(1 + \frac{\delta\xi}{R_2}\right) \approx \iint dl_1 dl_2\left(1 + \frac{\delta\xi}{R_1} + \frac{\delta\xi}{R_2}\right) \tag{7}$$

from which, the change in the interface area on deformation is given by

$$\delta A \approx \iint \delta\xi\left(\frac{1}{R_1} + \frac{1}{R_2}\right) dxdy. \tag{8}$$

On comparing (6) and (8), one obtains

$$\frac{1}{R_1} + \frac{1}{R_2} = -\left(\frac{\partial^2 \xi}{\partial x^2} + \frac{\partial^2 \xi}{\partial y^2}\right). \tag{9}$$

Thus, the tension acting on a curve enclosing the interface element is dynamically equivalent to a normal stress on the interface of magnitude given by

$$T\left(\frac{\partial^2 \xi}{\partial x^2} + \frac{\partial^2 \xi}{\partial y^2}\right) = T\left(\frac{1}{R_1} + \frac{1}{R_2}\right). \tag{10}$$

Note that it is necessary to recognize R_1 and R_2 as having appropriate signs, since the contribution to the equivalent pressure on the interface is directed toward the center of curvature.

Since the interface has negligible mass, a curved interface can be in equilibrium only if the normal stress due to surface tension is balanced by the jump in the pressure,

$$\Delta p = T\left(\frac{1}{R_1} + \frac{1}{R_2}\right) \tag{11}$$

which occurs on passing toward the center of curvature side of the interface.[1]

4.2 Capillary Rises in Liquids

Consider a free liquid meeting a plane vertical rigid wall (Figure 4.1). Let us determine the interface shape $z = \xi(y)$. Note that the principal curvatures of the interface are

$$\frac{1}{R_1} = 0, \ \frac{1}{R_2} = \frac{\xi''}{(1 + \xi'^2)^{3/2}} \tag{12}$$

where the primes denote differentiation with respect to y. One has from equation (11),

$$\Delta p = p_g - [p_g - \rho g \xi] = \rho g \xi = \frac{T\xi''}{(1 + \xi'^2)^{3/2}} \tag{13}$$

where we have used the boundary conditions,

$$y \Rightarrow \infty : \xi, \xi', \xi'' \Rightarrow 0. \tag{14}$$

and p_g refers to pressure in the gas.

1 Equation (11) implies that $\Delta p > 0$ ($\Delta p < 0$) across a convex (concave) interface (for which R_1 and $R_2 > 0$).

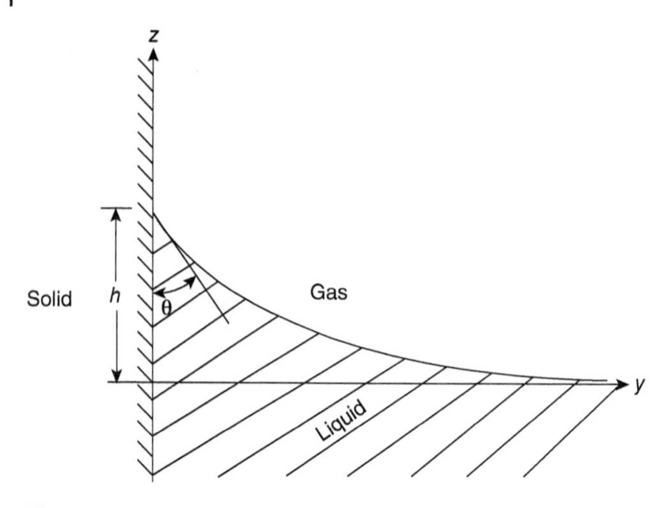

Figure 4.1 Free liquid meeting a plane vertical rigid wall.

Integrating equation (13), and using (14) again gives,

$$\frac{1}{2}\rho g \xi^2 + \frac{T}{(1+\xi'^2)^{1/2}} = T \tag{15}$$

from which, on noting that

$$\xi = h \, : \, \xi' = -\cot\theta \tag{16}$$

the rise of the liquid near the wall is given by

$$h^2 = \frac{2T}{\rho g}(1 - \sin\theta), \tag{17}$$

θ being the contact angle at the wall (see Figure 4.1).

The phenomenon of the rise or fall of the free surface of a liquid when it meets a rigid wall is called capillarity, which manifests in small tubes.

Exercises

1. If every fluid particle is moving on the surface of a sphere, show that the equation of mass conservation is

$$\frac{\partial \rho}{\partial t} + \frac{1}{\cos\theta}\frac{\partial}{\partial \theta}(\rho\omega\cos\theta) + \frac{1}{\cos\theta}\frac{\partial}{\partial \phi}(\rho\omega'\cos\theta) = 0$$

   ρ being the density, θ the latitude, ϕ the longitude of a fluid element, and ω, ω' being the latitudinal and longitudinal angular velocity components, respectively.

2. Consider a vessel containing a fluid of nonuniform density, subject to gravity and rotating steadily about the vertical z-axis with the fluid reaching the same steady rotation state. Show that the surfaces of constant pressure are paraboloids of revolution about the vertical axis.
3. Consider a capillary rise of a liquid in a circular tube of small radius a. Find the height of the column of liquid in the tube supported by the surface tension against gravity.
4. The equations governing two-dimensional magnetohydrodynamical flows are (Holm et al., 1985)

$$\left.\begin{aligned} &\frac{\partial A}{\partial t} + \partial(\psi, A) = 0 \\ &\frac{\partial \zeta}{\partial t} + \partial(\psi, \zeta) = \partial(A, j) \end{aligned}\right\}$$

where ψ is the stream function, A is the *magnetic potential*, ζ is the vorticity, and j is the *current density* defined by

$$\left.\begin{aligned} &\mathbf{v} = \hat{\mathbf{i}}_z \times \nabla\psi,\ \mathbf{B} = \hat{\mathbf{i}}_z \times \nabla A \\ &\nabla \times \mathbf{v} = \zeta \hat{\mathbf{i}}_z,\ \nabla \times \mathbf{B} = j\, \hat{\mathbf{i}}_z \end{aligned}\right\}$$

$\mathbf{v}$ being the velocity and $\mathbf{B}$ the magnetic field. Here,

$$\partial(a, b) \equiv \frac{\partial a}{\partial x}\frac{\partial b}{\partial y} - \frac{\partial a}{\partial y}\frac{\partial b}{\partial x}.$$

(i) Determine the Hamiltonian $\mathscr{H}$ such that Hamilton's equations

$$\left[\frac{\partial A}{\partial t}, \frac{\partial \zeta}{\partial t}\right]^T = \mathscr{J}\left[\frac{\partial \mathscr{H}}{\partial A}, \frac{\partial \mathscr{H}}{\partial \zeta}\right]^T$$

reduce to the above equations and exhibit a proper form for the symplectic matrix $\mathscr{J}$.

(ii) Show that the functional

$$\mathscr{C} = \iint \left[F(A) + \zeta G(A)\right] dx\, dy$$

for any functions F and G is conserved, and that it is indeed a Casimir invariant, i.e.,

$$\mathcal{J}\left[\frac{\partial \mathcal{C}}{\partial A}, \frac{\partial \mathcal{C}}{\partial \zeta}\right]^T = [0, 0]^T .$$

Part II

Dynamics of Incompressible Fluid Flows

5

Fluid Kinematics and Dynamics

Streamlines and streamtubes are theoretical devices which are of great help in visualizing fluid flow fields. Streamlines show the instantaneous direction of flow at every point on them. The fluid inside a streamtube flows like that inside a pipe with solid walls because there can be no flow across streamtube walls.

In practice, many fluid flows may be considered to be nearly vorticity-free or irrotational. Such flows allow the application of the full machinery of potential theory which considerably facilitates the formulation and analysis of these flows. The Laplace equation for the velocity potential governing irrotational flows is linear. This enables one to build up complex flow patterns by superposing special singular solutions of the Laplace equation.

5.1 Stream Function

Streamlines have the property that the instantaneous fluid velocity at any point is the tangent to the streamline through that point. A surface made up entirely of streamlines at any instant is called a *stream surface*. When the streamlines pass through a given closed curve in the fluid, the stream surface assumes the shape of a stream tube. The motion of a given fluid particle in space describes a *pathline* in space–time. In steady flow, the pathlines coincide with the instantaneous streamlines.

Streamlines are given by intersections of stream surfaces given by

$$\left.\begin{aligned} f(x,y,z) &= a \\ g(x,y,z) &= b \end{aligned}\right\} \tag{1}$$

Introduction to Theoretical and Mathematical Fluid Dynamics, Third Edition.
Bhimsen K. Shivamoggi.

or if $\mathbf{v} = (u, v, w)$ denotes the fluid velocity, (1) can be rewritten as

$$\left.\begin{aligned} \mathbf{v} \cdot \nabla f = u f_x + v f_y + w f_z = 0 \\ \mathbf{v} \cdot \nabla g = u g_x + v g_y + w g_z = 0 \end{aligned}\right\}. \tag{2}$$

From (2), one has

$$\mathbf{v} = \nabla f \times \nabla g \tag{3a}$$

or

$$\left.\begin{aligned} u &= (f_y g_z - f_z g_y) \\ v &= (f_z g_x - f_x g_z) \\ w &= (f_x g_y - f_y g_x) \end{aligned}\right\}. \tag{3b}$$

For two-dimensional flows, $g = z, f = \Psi(x, y, t)$, so (3b) leads to

$$u = \Psi_y, \quad v = -\Psi_x \tag{4a}$$

Ψ is called the *stream function*[1] because, the streamlines for the flow are, in fact, the set of level curves given by $\Psi(x, y, t) = f(t)$, where $f(t)$ is arbitrary. This may be seen by noting that the differential equation of the streamline is

$$\frac{dx}{u} = \frac{dy}{v}. \tag{5a}$$

Using (4a), equation (5a) leads to

$$\frac{\partial \Psi}{\partial x} dx + \frac{\partial \Psi}{\partial y} dy = d\Psi = 0. \tag{6a}$$

So, along a streamline, we have

$$\Psi(x, y, t) = f(t). \tag{7}$$

Similarly, the equation of a streamline in plane polar coordinates is given by,

$$\frac{dr}{u_r} = \frac{r d\theta}{u_\theta} \tag{8a}$$

1 The equations for particle trajectories in a two-dimensional incompressible flow, from equation (5a), are

$$\dot{x} = \frac{\partial \Psi}{\partial y}, \; \dot{y} = -\frac{\partial \Psi}{\partial x}$$

which constitute a one-degree-of-freedom *Hamiltonian* system, with *Hamiltonian* $\Psi(x, y, t)$, and the configuration space is the phase space in question. However, this representation is merely kinematic in content and does not contain any dynamical aspects of fluid motion.

or

$$u_r r d\theta - u_\theta dr = 0. \tag{8b}$$

Comparing (8b) with the alternate representation of the streamline given by,

$$d\Psi = \frac{\partial \Psi}{\partial r} dr + \frac{\partial \Psi}{\partial \theta} d\theta = 0 \tag{6b}$$

we obtain

$$u_r = \frac{1}{r}\frac{\partial \Psi}{\partial \theta}, \; u_\theta = -\frac{\partial \Psi}{\partial r}. \tag{4b}$$

In order to see further significance of the stream function, introduce, in the stream surface, a set of orthogonal coordinates (α, β). The velocity component normal to an area element $dS = hk\, d\alpha\, d\beta$ is

$$v_n = \frac{f_\alpha g_\beta - f_\beta g_\alpha}{hk} \tag{9}$$

where h and k are the scale factors for (α, β). Then, the discharge through a portion S of the surface bounded by the traces of stream surfaces,

$$\left.\begin{aligned} f(x,y,z) = f_1, \; f(x,y,z) = f_2 \\ g(x,y,z) = g_1, \; g(x,y,z) = g_2 \end{aligned}\right\} \tag{10}$$

is given by

$$Q = \iint_S v_n dS = \iint_S (f_\alpha g_\beta - f_\beta g_\alpha)\, d\alpha d\beta = \int_{g_1}^{g_2}\int_{f_1}^{f_2} df dg = (f_2 - f_1)(g_2 - g_1). \tag{11}$$

Note that Q is independent of the choice of the paths connecting f_1 to f_2 and g_1 to g_2.

In two-dimensional flows, the discharge per unit length across an arc AB is $(\Psi_B - \Psi_A)$. In axisymmetric flows, the discharge across the surface generated by a revolving arc AB about the axis of symmetry is $2\pi\,(\Psi_B - \Psi_A)$. For the latter case, taking φ to be the azimuthal angle, $g = -\varphi$, and $f = \Psi(r, z, t)$, the radial and axial velocity components, in cylindrical coordinates, are given by

$$u_r = \frac{1}{r}\Psi_z, \; u_z = -\frac{1}{r}\Psi_r. \tag{4c}$$

On the other hand, taking $g = -\varphi$ and $f = \Psi(R, \theta, t)$, the radial and meridional velocity components, in spherical coordinates, are given by

$$u_R = \frac{1}{R^2 \sin\theta}\frac{\partial \Psi}{\partial \theta}, \quad u_\theta = -\frac{1}{R\sin\theta}\frac{\partial \Psi}{\partial R}. \tag{4d}$$

Note that, a stream tube cannot end in the interior of the fluid; it must either be closed, or end on the boundary of the fluid, or extend to infinity.

Finally, one should note the possibility that Ψ is a multivalued function of position whenever sources of mass exist in the flow field.

5.2 Equations of Motion

Referring to Section 2.7, the equation of conservation of mass is

$$\nabla \cdot \mathbf{v} = 0. \tag{12}$$

The equation of motion (also called *Euler's equation*) is

$$\rho\left[\frac{\partial \mathbf{v}}{\partial t} + (\mathbf{v}\cdot\nabla)\,\mathbf{v}\right] = \rho\mathbf{F} - \nabla p, \tag{13}$$

$\mathbf{F}$ being the body force per unit mass, p is the pressure, and ρ is the mass density of the fluid.

5.3 Integrals of Motion

Equation (13) can be written as

$$\rho\left[\frac{\partial \mathbf{v}}{\partial t} + \nabla\left(\frac{\mathbf{v}^2}{2}\right) - \mathbf{v}\times(\nabla\times\mathbf{v})\right] = \rho\mathbf{F} - \nabla p. \tag{14}$$

When the body force $\mathbf{F}$ is *conservative*, i.e.,

$$\mathbf{F} = -\nabla U \tag{15}$$

and the motion is steady, the integration of equation (14) is possible along a streamline, and one obtains[2]

$$\frac{\mathbf{v}^2}{2} + \frac{p}{\rho} + U = const = C. \tag{16}$$

On the other hand, when the body force $\mathbf{F}$ is conservative, and the flow is *irrotational*, i.e.,

---

2 The constant C on the right-hand side in (16) may vary from streamline to streamline. Indeed, equation (14) shows that the gradient ∇C, called the *Lamb vector* is given by

$$\mathbf{v}\times(\nabla\times\mathbf{v}) = \nabla C.$$

$$\nabla \times \mathbf{v} = \mathbf{0} \text{ or } \mathbf{v} = \nabla\Phi \tag{17}$$

(where the *velocity potential* Φ may be *single-valued* or *multivalued* depending on whether or not the region is *simply-connected*, see Section 5.8) the integration of equation (14) for an unsteady flow is possible in any direction, and one obtains *Bernoulli's integral*,

$$\frac{\partial \Phi}{\partial t} + \frac{\mathbf{v}^2}{2} + \frac{p}{\rho} + U = F(t). \tag{18}$$

In general, for a steady flow and conservative body forces, equation (14) becomes

$$\nabla\left[\frac{\mathbf{v}^2}{2} + U + \frac{p}{\rho}\right] = \mathbf{v} \times (\nabla \times \mathbf{v}) \tag{19}$$

which, for a two-dimensional flow, leads to

$$\left.\begin{aligned} \frac{\partial}{\partial x}\left(\frac{\mathbf{v}^2}{2} + U + \frac{p}{\rho}\right) &= v\Omega \\ \frac{\partial}{\partial y}\left(\frac{\mathbf{v}^2}{2} + U + \frac{p}{\rho}\right) &= -u\Omega \end{aligned}\right\} \tag{20a}$$

where,

$$\boldsymbol{\Omega} \equiv \nabla \times \mathbf{v} = \Omega\, \hat{\mathbf{i}}_z,\ \Omega = const.$$

Using the stream function Ψ, given by (4a), equations (20a) lead to

$$\frac{\mathbf{v}^2}{2} + U + \frac{p}{\rho} + \Omega\Psi = const. \tag{20b}$$

5.4 Capillary Waves on a Spherical Drop

As an illustration of the use of the Bernoulli integral (18), consider small oscillations of a spherical drop of an incompressible fluid under the action of capillary forces. The surface area given in spherical polar coordinates (r, θ, φ) is

$$S = \int_0^{2\pi}\int_0^{\pi} \sqrt{1 + \frac{1}{r^2}\left(\frac{\partial r}{\partial \theta}\right)^2 + \frac{1}{r^2 \sin^2\theta}\left(\frac{\partial r}{\partial \varphi}\right)^2}\; r^2 \sin\theta\, d\theta\, d\varphi. \tag{21}$$

For small deviations ξ from a spherical surface of radius R, one may write $r = R + \xi, |\xi| \ll R$, so that (21) may be approximated by

$$S \approx \int_0^{2\pi}\int_0^{\pi} \left[(R+\xi)^2 + \frac{1}{2}\left\{ \left(\frac{\partial \xi}{\partial \theta}\right)^2 + \frac{1}{\sin^2\theta}\left(\frac{\partial \xi}{\partial \varphi}\right)^2 \right\} \right] \sin\theta \, d\theta \, d\varphi \tag{22}$$

from which, the small change in the surface area is given by

$$\delta S \approx \int_0^{2\pi}\int_0^{\pi} \left[2(R+\xi)\,\delta\xi + \frac{\partial \xi}{\partial \theta}\frac{\partial \delta\xi}{\partial \theta} + \frac{1}{\sin^2\theta}\frac{\partial \xi}{\partial \varphi}\frac{\partial \delta\xi}{\partial \varphi} \right] \sin\theta \, d\theta \, d\varphi. \tag{23a}$$

On integrating by parts, we obtain

$$\delta S \approx \int_0^{2\pi}\int_0^{\pi} \left[2(R+\xi) - \frac{1}{\sin\theta}\frac{\partial}{\partial \theta}\left(\sin\theta \frac{\partial \xi}{\partial \theta}\right) - \frac{1}{\sin^2\theta}\frac{\partial^2 \xi}{\partial \varphi^2} \right] \delta\xi \, \sin\theta \, d\theta \, d\varphi. \tag{23b}$$

If R_1 and R_2 are the principal radii of curvature, we have (see equation (8a) in Section 4.1],

$$\delta S \approx \iint \delta\xi \left(\frac{1}{R_1} + \frac{1}{R_2}\right)(R+\xi)^2 \sin\theta \, d\theta \, d\varphi. \tag{24}$$

We then obtain, on comparing (23b) and (24),

$$\frac{1}{R_1} + \frac{1}{R_2} = \frac{2}{R} - \frac{2\xi}{R^2} - \frac{1}{R^2}\left[\frac{1}{\sin\theta}\frac{\partial}{\partial \theta}\left(\sin\theta \frac{\partial \xi}{\partial \theta}\right) + \frac{1}{\sin^2\theta}\frac{\partial^2 \xi}{\partial \varphi^2} \right]. \tag{25}$$

The Bernoulli integral (18), then gives

$$\rho\frac{\partial \Phi}{\partial t} + T\left[\frac{2}{R} - \frac{2\xi}{R^2} - \frac{1}{R^2}\left\{ \frac{1}{\sin\theta}\frac{\partial}{\partial \theta}\left(\sin\theta \frac{\partial \xi}{\partial \theta}\right) + \frac{1}{\sin^2\theta}\frac{\partial^2 \xi}{\partial \varphi^2} \right\} \right] = const \tag{26}$$

where T is the surface tension.

Differentiating (26) with respect to t, and using the *kinematic condition* at the surface of the drop (implying that a fluid particle on the surface of the drop always remains there),

$$r = R : \frac{\partial \xi}{\partial t} = \frac{\partial \Phi}{\partial r} \tag{27}$$

one obtains

$$r = R : \rho \frac{\partial^2 \Phi}{\partial t^2} - \frac{T}{R^2} \left[2\frac{\partial \Phi}{\partial r} + \frac{\partial}{\partial r} \left\{ \frac{1}{\sin\theta} \frac{\partial}{\partial \theta} \left(\sin\theta \frac{\partial \Phi}{\partial \theta} \right) + \frac{1}{\sin^2\theta} \frac{\partial^2 \Phi}{\partial \varphi^2} \right\} \right] = 0. \tag{28}$$

We may now look for solutions of the form,

$$\Phi = e^{-i\omega t} f(r, \theta, \varphi) \tag{29}$$

with

$$f(r, \theta, \varphi) = r^l Y_{lm}(\theta, \varphi) \tag{30}$$

where $Y_{lm}(\theta, \varphi)$ are the *spherical harmonics*, satisfying the equation

$$\frac{1}{\sin\theta} \frac{\partial}{\partial \theta} \left(\sin\theta \frac{\partial Y_{lm}}{\partial \theta} \right) + \frac{1}{\sin^2\theta} \frac{\partial^2 Y_{lm}}{\partial \varphi^2} + \ell(\ell+1) Y_{lm} = 0. \tag{31}$$

The spherical harmonics are given by,

$$Y_{lm}(\theta, \varphi) = P_l^m(\cos\theta) e^{im\varphi}$$

where,

$$P_l^m(\cos\theta) = \sin^m\theta \frac{d^m P_l(\cos\theta)}{d(\cos\theta)^m}; m = 0, 1, 2, \dots, \ell; \ell = 0, 1, 2, \dots \tag{32}$$

are the *associated Legendre polynomials*, and $P_l(\cos\theta)$ are the *Legendre polynomials*.[3]

3

$$P_0(\cos\theta) = 1, P_1(\cos\theta) = \cos\theta, P_2(\cos\theta) = \frac{1}{2}(3\cos^2\theta - 1), \text{etc.}$$

Equation (28) then gives

$$\omega^2 = \frac{T(\ell - 1)(\ell + 2)\ell}{\rho R^3}. \tag{33}$$

For a given ℓ, there are $(2\ell + 1)$ different eigenfunctions, implying that these frequencies have a *degeneracy* of $(2\ell + 1)$. Note that,

$$\ell = 0, 1 \, : \, \omega = 0. \tag{34}$$

The case $\ell = 0$ corresponds to radial oscillations, and in an incompressible fluid such oscillations are impermissible. The case $\ell = 1$ corresponds to a translation of the drop as a whole, which does not lead to any deformation of the drop.

5.5 Cavitation

For an incompressible fluid, and for $\mathbf{F} = \mathbf{0}$, one obtains, from equation (14),

$$\nabla^2 p = -\frac{\rho}{2}\nabla^2 v^2 + \rho \nabla \cdot \{\mathbf{v} \times (\nabla \times \mathbf{v})\} = -\rho \frac{\partial v_i}{\partial x_j}\frac{\partial v_j}{\partial x_i}. \tag{35}$$

Integrating equation (35) over a volume $\mathcal{V}$ enclosed by a surface S, one obtains

$$\int_S \hat{\mathbf{n}} \cdot \nabla p dS = -\rho \int_{\mathcal{V}} \frac{\partial v_i}{\partial x_j}\frac{\partial v_j}{\partial x_i} d\mathbf{x} < 0 \tag{36}$$

from which, it is obvious that p can have a minimum value only at the boundary and not in the interior of the fluid. (Note, however, that the pressure can have maximum value in the interior of the fluid.) Now, *cavitation* or bubble formation occurs in a fluid when the absolute pressure falls below a critical value say, zero. This is the case near the tips of rapidly rotating turbine blades and ship propellers. Since a liquid cannot withstand tension, the cavitation will occur first at some point on the boundary as the pressure everywhere is decreased. However, when the pressure in the neighborhood of a cavity rises above the vapor pressure, the cavity collapses. The continual collapse of cavities leads to a deterioration and erosion of surfaces of solid bodies immersed in the liquid. The damage is believed to be caused by the impact of high-speed jets that form as the cavity collapses.

5.6 Rates of Change of Material Integrals

Consider the line integral of a scalar field F $(\mathbf{x}, t)$,

$$\int_P^Q F\mathbf{dx}$$

taken along a material curve joining the points P and Q. Since this curve moves with the fluid and consists of the same fluid particles, the integral above is a function of only t. In order to determine its rate of change with time, suppose that the position of the material curve at time t is PQ and its position at time $t + \delta t$ is $P'Q'$ (see Figure 5.1).

Note that

$$\frac{d}{dt}\int_P^Q F(\mathbf{x}, t)\,\mathbf{dx} = \lim_{\delta t \Rightarrow 0}\frac{1}{\delta t}\left[\int_{P'}^{Q'} F(\mathbf{x}', t + \delta t)\,\mathbf{dx}' - \int_P^Q F(\mathbf{x}, t)\,\mathbf{dx}\right]. \tag{37}$$

Dividing the curves PQ and $P'Q'$ into n segments by the points P_i and P'_i, $i = 0, \ldots, n$, respectively, such that P_iP_{i+1} of PQ moves to become $P'_iP'_{i+1}$ of $P'Q'$ over the time interval $[t, t + \delta t]$, one has

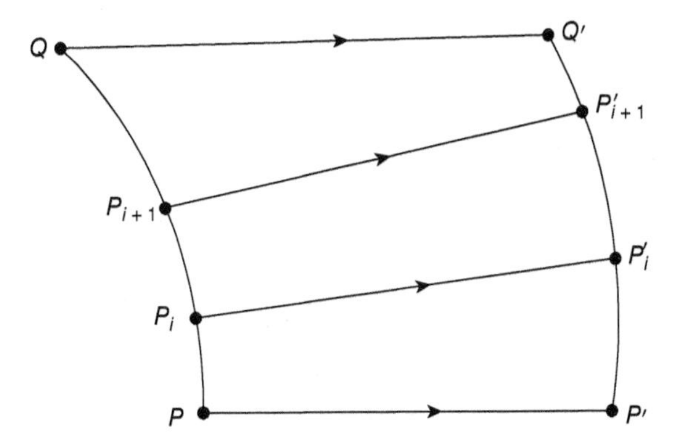

Figure 5.1 Displacement of a material curve.

$$\left.\begin{aligned}\int_P^Q F(\mathbf{x},t)\,\mathbf{dx}' &= \lim_{n\Rightarrow\infty}\sum_{i=0}^{n-1} F(\mathbf{x}_i,t)\,\overrightarrow{P_iP_{i+1}}\\ \int_{P'}^{Q'} F(\mathbf{x}',t+\delta t)\,\mathbf{dx}' &= \lim_{n\Rightarrow\infty}\sum_{i=0}^{n-1} F\left(\mathbf{x}_i',t+\delta t\right)\overrightarrow{P_i'P_{i+1}'}\end{aligned}\right\} \tag{38}$$

where $\mathbf{x}_i$ and $\mathbf{x}_i'$ are the position vectors of P_i and P_i', respectively.

Now, writing

$$F\left(\mathbf{x}_i',t+\delta t\right) \approx F(\mathbf{x}_i,t+\delta t) + \left(\overrightarrow{P_iP_i'}\cdot\nabla\right)F(\mathbf{x}_i,t+\delta t) + O\left(|\overrightarrow{P_iP_i'}|^2\right)$$

or

$$F\left(\mathbf{x}_i',t+\delta t\right) \approx F(\mathbf{x}_i,t) + \frac{\partial F(\mathbf{x}_i,t)}{\partial t}\delta t + \delta t\,\mathbf{v}(\mathbf{x}_i,t)\cdot\nabla F(\mathbf{x}_i,t) + O\left(\delta t^2\right)$$

and

$$\overrightarrow{P_i'P_{i+1}'} \approx \overrightarrow{P_iP_{i+1}} + \left[\mathbf{v}(\mathbf{x}_{i+1},t) - \mathbf{v}(\mathbf{x}_i,t)\right]\delta t$$

or

$$\overrightarrow{P_i'P_{i+1}'} \approx \overrightarrow{P_iP_{i+1}} + \left(\overrightarrow{P_iP_{i+1}}\cdot\nabla\right)\mathbf{v}(\mathbf{x}_i,t)\,\delta t, \tag{39}$$

one has

$$\begin{aligned}&F\left(\mathbf{x}_i',t+\delta t\right)\overrightarrow{P_i'P_{i+1}'} - F(\mathbf{x}_i,t)\,\overrightarrow{P_iP_{i+1}}\\ &= \left\{\left[\frac{\partial F(\mathbf{x}_i,t)}{\partial t} + \mathbf{v}(\mathbf{x}_i,t)\cdot\nabla F(\mathbf{x}_i,t)\right]\overrightarrow{P_iP_{i+1}}\right.\\ &\left.+F(\mathbf{x}_i,t)\left(\overrightarrow{P_iP_{i+1}}\cdot\nabla\right)\mathbf{v}(\mathbf{x}_i,t)\right\}\delta t + O\left(\delta t^2\right).\end{aligned} \tag{40}$$

Using (38) and (40), (37) leads to

$$\frac{d}{dt}\int_P^Q F\mathbf{dx} = \int_P^Q \frac{DF}{Dt}\mathbf{dx} + \int_P^Q F(\mathbf{dx}\cdot\nabla)\,\mathbf{v}. \tag{41a}$$

If $\mathbf{F}(\mathbf{x},t)$ is a vector field,

$$\mathbf{F}(\mathbf{x},t) = F_1(\mathbf{x},t)\,\hat{\mathbf{i}}_x + F_2(\mathbf{x},t)\,\hat{\mathbf{i}}_y + F_3(\mathbf{x},t)\,\hat{\mathbf{i}}_z \tag{42}$$

then, on applying (41a) to F_1, F_2 and F_3, we obtain,

$$\frac{d}{dt}\int_P^Q \mathbf{F}\cdot\mathbf{dx} = \int_P^Q \frac{D\mathbf{F}}{Dt}\cdot\mathbf{dx} + \int_P^Q \mathbf{F}\cdot(\mathbf{dx}\cdot\nabla)\,\mathbf{v}. \tag{41b}$$

5.7 The Kelvin Circulation Theorem

The *circulation* is defined by

$$\Gamma \equiv \oint_C \mathbf{v}\cdot\mathbf{ds} \tag{43}$$

where C is a closed material curve. We assume the region to be *simply-connected* so that all closed curves in it are reducible (they can be shrunk continuously down to a point without leaving the region). Using *Stokes' theorem*, (43) becomes

$$\Gamma = \iint_S (\nabla\times\mathbf{v})\cdot\hat{\mathbf{n}}\,dS = \iint_S \boldsymbol{\Omega}\cdot\hat{\mathbf{n}}\,dS \tag{44}$$

where, S is a surface bounded by the closed curve C, and $\hat{\mathbf{n}}$ is the outward drawn unit normal to the surface element dS. Thus, the circulation around any reducible closed curve is equal to the integral of vorticity over an open surface bounded by the curve.

Using equation (41b), one has

$$\frac{d\Gamma}{dt} = \oint_C \frac{D\mathbf{v}}{Dt}\cdot\mathbf{ds} + \oint_C \mathbf{v}\cdot(\mathbf{ds}\cdot\nabla\mathbf{v}). \tag{45}$$

Using equation (13), equation (45) becomes

$$\frac{d\Gamma}{dt} = \oint_C \mathbf{F}\cdot\mathbf{ds} - \oint_C \left(\frac{\nabla p}{\rho} - \nabla\left(\frac{1}{2}\mathbf{v}^2\right)\right)\cdot\mathbf{ds}. \tag{46}$$

If the body forces are conservative, (46) gives[4] *the Kelvin circulation theorem*:

$$\frac{d\Gamma}{dt} = \frac{d}{dt}\oint_S \boldsymbol{\Omega}\cdot\hat{\mathbf{n}}\,dS = 0 \tag{47}$$

4 Equation (47) holds even if the fluid is not incompressible but is *barotropic*. So, the enthalpy $\nabla p/\rho$ can be written as a perfect differential.

This implies that the circulation around any closed material curve and the strength of the vortex tube looped by it are invariants. Thus, under the action of conservative body forces, all motions of an inviscid, incompressible fluid set up from a state of rest or uniform motion (initial circulation is zero in both cases) are permanently irrotational.

5.8 The Irrotational Flow

The boundary-value problems for irrotational flows are governed by the *Laplace equation* for the velocity potential Φ,

$$\nabla^2\Phi = 0. \tag{48}$$

It is remarkable that equation (13), (which is nonlinear) can be avoided in the study of kinematics of irrotational flows of an ideal fluid, which are governed by the Laplace equation (which is linear). Once the velocity field is determined, the pressure can be found from the Bernoulli equation (18). It is noteworthy that the nonlinearity of the Euler equation is manifested only in the Bernoulli equation.

In a *simply-connected* region, one has for irrotational flows,

$$\Gamma = \oint_C \mathbf{v} \cdot \mathbf{ds} = \oint_C d\Phi = 0 \tag{49}$$

implying that Φ is *single-valued*. In a *doubly-connected* region, on the other hand, some closed curves C are not reducible, and Φ may be *multivalued* for such curves. The various values of Φ differ from one another by multiples of Γ.

Being the potential equation, (48) provides standard results from potential theory that can be readily proved.

Theorem 5.1 **(i)** The potential Φ can have neither a maximum nor a minimum in the *interior* of the fluid.

Proof: Consider a closed surface S enclosing a region $\mathcal{V}$. We have by the *divergence theorem*,

$$\oiint_S \frac{\partial \Phi}{\partial n}\, dS = \iiint_{\mathcal{V}} \nabla^2\Phi \, d\mathbf{x}.$$

If Φ satisfies the Laplace equation,

$$\nabla^2\Phi = 0 \text{ in } \mathcal{V},$$

then, we have

$$\oiint_S \frac{\partial \Phi}{\partial n}\, dS = 0$$

which rules out a maximum or a minimum for Φ at any point P inside $\mathcal{V}$, because otherwise $\partial\Phi/\partial n > 0$ or < 0 everywhere in the neighborhood of P.

□

(ii) The solution to the *Neumann exterior problem* (wherein the derivatives of Φ are prescribed on the boundary) in a *simply-connected* region is unique up to an additive constant.

Lemma If S is a closed surface enclosing a region $\mathcal{V}$, we have

$$\iiint_{\mathcal{V}} (\nabla\Phi)^2\, \mathbf{dx} = \oiint_S \Phi\nabla\Phi \cdot \hat{\mathbf{n}} dS$$

where $\hat{\mathbf{n}}$ is the appropriate outward drawn normal vector to S.

Proof: Consider the *Neumann problem* in the *simply connected* region $\mathcal{V}$ exterior to a bounded volume R whose boundary surface is given by S.

If Φ_1, Φ_2 are two solutions satisfying

$$\left.\begin{array}{l} \nabla^2\Phi = 0 \text{ in region } \mathcal{V} \text{ exterior to } S \\ \nabla\Phi \cdot \hat{\mathbf{n}} = f(\mathbf{x}, t) \text{ on } S \end{array}\right\}$$

then we have

$$\left.\begin{array}{l} \nabla^2(\Phi_1 - \Phi_2) = 0 \; in \; \mathcal{V} \\ \nabla(\Phi_1 - \Phi_2) \cdot \hat{\mathbf{n}} = 0 \; on \; S \end{array}\right\}.$$

If $\sum$ is a surface at infinity enclosing S, we have by the above **Lemma**,

$$\begin{aligned} \iiint_{\mathcal{V}} [\nabla(\Phi_1 - \Phi_2)]^2\, \mathbf{dx} &= \oiint_{\Sigma} (\Phi_1 - \Phi_2)\nabla(\Phi_1 - \Phi_2) \cdot \hat{\mathbf{n}}\, dS \\ &\quad - \oiint_S (\Phi_1 - \Phi_2)\nabla(\Phi_1 - \Phi_2) \cdot \hat{\mathbf{n}}\, dS = 0 \end{aligned}$$

from which,

$$\nabla(\Phi_1 - \Phi_2) = 0 \text{ everywhere in } \mathcal{V}$$

or

$$\Phi_1 - \Phi_2 = const. \qquad \square$$

(iii) The solution of the *Neumann exterior problem* in a *doubly-connected* region is uniquely determined (up to an additive constant) only when the circulation is specified.

Proof: Consider the *Neumann exterior problem* in the *doubly-connected* region S exterior to an infinite cylinder whose boundary is given by a closed curve $\mathscr{C}_0$ (see Figure 5.2).

If Φ_1, Φ_2 are two solutions satisfying

$$\left.\begin{aligned} \nabla^2\Phi &= 0 \text{ in } S \\ \nabla\Phi \cdot \hat{\mathbf{n}} &= f(\mathbf{x}, t) \text{ on } \mathscr{C}_0 \end{aligned}\right\}$$

then we have

$$\left.\begin{aligned} \nabla^2(\Phi_1 - \Phi_2) &= 0 \text{ in } S \\ \nabla(\Phi_1 - \Phi_2) \cdot \hat{\mathbf{n}} &= 0 \text{ on } \mathscr{C}_0 \end{aligned}\right\}.$$

Suppose $\mathscr{C}$ is a closed curve at infinity enclosing $\mathscr{C}_0$. Since S is *doubly-connected*, it needs to be rendered *simply-connected* by inserting a barrier, so that *Green's theorem* in the plane can apply.

Consider two adjacent points P_+ and P_- on either side of the barrier. Then the integral of $d\Phi$ along any irreducible path $\mathscr{C}$ connecting P_+ to P_- leads to

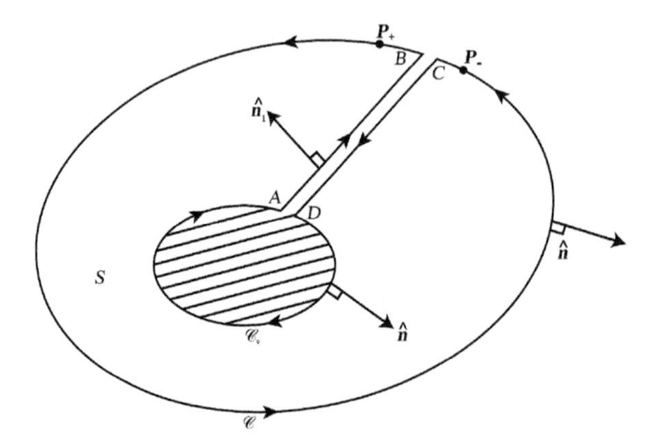

Figure 5.2 Doubly - connected region.

$$\lim_{P_+ \to P_-} \int_{P_+}^{P_-} d\Phi = \lim_{P_+ \to P_-} [\Phi(P_-) - \Phi(P_+)] = \Gamma$$

where Γ is the circulation around $\mathscr{C}$.

Green's theorem in the plane then gives

$$\iint_S [\nabla(\Phi_1 - \Phi_2)]^2 \, dS = \int_{AB} (\Phi_1 - \Phi_2) \frac{\partial(\Phi_1 - \Phi_2)}{\partial n_1} ds$$

$$+ \int_{\mathscr{C}} (\Phi_1 - \Phi_2) \frac{\partial(\Phi_1 - \Phi_2)}{\partial n} ds - \int_{CD} (\Phi_1 - \Phi_2) \frac{\partial(\Phi_1 - \Phi_2)}{\partial n_1} ds$$

$$- \int_{\mathscr{C}_0} (\Phi_1 - \Phi_2) \frac{\partial(\Phi_1 - \Phi_2)}{\partial n} ds$$

$$= (\Gamma_1 - \Gamma_2) \int_{barrier} \frac{\partial(\Phi_1 - \Phi_2)}{\partial n_1} ds$$

where Γ_1 and Γ_2 are the circulations corresponding to Φ_1 and Φ_2, respectively, and $\hat{n}_1$ is the appropriate normal vector to the segment AB. If Γ is specified, i.e., $\Gamma_1 = \Gamma_2$, then we obtain

$$\nabla(\Phi_1 - \Phi_2) = 0 \text{ everywhere in } S$$

or

$$\Phi_1 - \Phi_2 = const.$$

□

Thus, for the *Neumann exterior problem* in a *doubly-connected* region, with the same boundary and infinity conditions, different values of the circulation yield different solutions.

The lack of explicit appearance of time variation in equation (48) implies that the instantaneous flow pattern depends only on the instantaneous boundary conditions. The history of the flow becomes irrelevant here since the flow has no memory.[5]

Thus, when a rigid body moves through an initially stationary fluid, the flow field is uniquely determined by the instantaneous velocity of the body (together with its shape) independent of its acceleration and history of motion. Of course,

5 We can interpret that the irrotational flow of an ideal fluid at any instant is generated impulsively from a state of rest by applying suitable impulsive conservative forces. The velocity potential of this flow is the potential of the impulse.

this is valid because equation (48) ignores compressibility of the fluid making speed of sound infinity!

5.9 Simple-Flow Patterns

Since the Laplace equation (48) is linear, it is permissible to build up some flow patterns by superposing its *singular* solutions. Let us first consider a few two-dimensional simple flow patterns described by such singular solutions.

(i) The Source Flow

Consider a flow directed radially from a *point source* (see Figure 5.3). The incompressibility condition

$$\frac{1}{r}\frac{\partial}{\partial r}(ru_r) = 0 \tag{50}$$

gives

$$u_r = \frac{A}{r} \tag{51}$$

where A is an arbitrary constant.

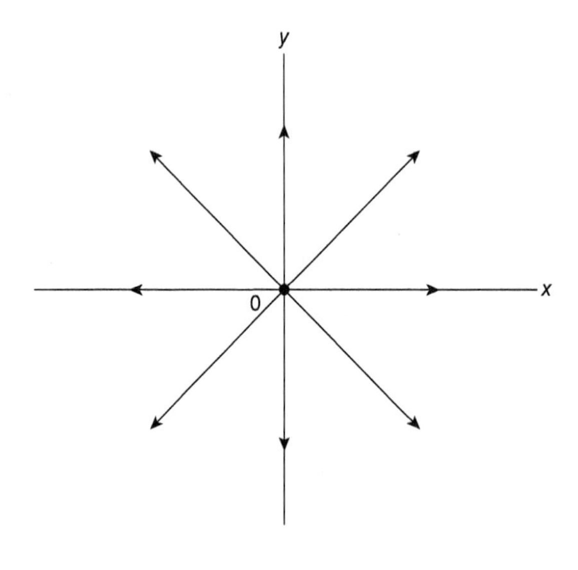

Figure 5.3 Source flow.

The conservation of mass, expressed by,

$$q = \int_0^{2\pi} u_r \cdot r d\theta = const, \tag{52}$$

q being the strength of the source. Here, q is equal to the volume of fluid flowing out of a closed curve enclosing the source per unit time. Using (51) gives,

$$A = \frac{q}{2\pi}. \tag{53}$$

Using (53), (51) becomes

$$u_r = \frac{q}{2\pi r}. \tag{54}$$

Then, from the relations (see (4b) and (17)),

$$u_r = \frac{\partial \Psi}{r \partial \theta} = \frac{\partial \Phi}{\partial r} \tag{55}$$

one obtains, for the stream function Ψ and the velocity potential Φ,

$$\Psi = \frac{q\theta}{2\pi}, \quad \Phi = \frac{q}{2\pi} \ell n r. \tag{56}$$

Note that the stream function is a multivalued function of position, owing to the existence of a nonzero volume flux across a closed curve around the origin.

(ii) The Doublet Flow

Consider the superposition of a sink (strength $-q$) at $(-a, 0)$ and a source (strength q) at $(a, 0)$ (see Figure 5.4). One has, from (56), for the velocity potential of the combined flow,

$$\Phi = \frac{q}{2\pi} \ell n \sqrt{r^2 + a^2 - 2ra\cos\theta} - \frac{q}{2\pi} \ell n \sqrt{r^2 + a^2 + 2ra\cos\theta} \tag{57}$$

or

$$\Phi = \frac{q}{4\pi} \ell n \left[1 - \frac{4ra\cos\theta}{r^2 + a^2 + 2ra\cos\theta}\right].$$

If $a/r \ll 1$, (57) may be approximated by[6]

$$\Phi \approx \frac{q}{4\pi}\left(-\frac{4a\cos\theta}{r}\right). \tag{58}$$

Let us choose q and a such that we have the following limit.

$$\lim_{a\to 0 q\to\infty} q\cdot 2a = const \equiv \mu \tag{59}$$

In this limit, (58) becomes,[7]

$$\Phi = -\frac{\mu\cos\theta}{2\pi r} \tag{60a}$$

and describes the flow due to a *doublet*. Using the relations,

$$\left.\begin{aligned} \frac{1}{r}\frac{\partial\Psi}{\partial\theta} &= \frac{\partial\Phi}{\partial r} = \frac{\mu\cos\theta}{2\pi r^2} \\ -\frac{\partial\Psi}{\partial r} &= \frac{1}{r}\frac{\partial\Phi}{\partial\theta} = \frac{\mu\sin\theta}{2\pi r^2} \end{aligned}\right\}$$

6

$$ln(1+x) = x - \frac{x^2}{2} + \frac{x^3}{3} - \ldots, \ |x| < 1$$

7 Alternatively, note that one may generate the doublet flow from the source flow by simply differentiating the velocity potential corresponding to the source flow in a direction opposite to the doublet axis. This follows from

$$\lim_{dx\to 0} -\phi'(x)\,dx = \lim_{dx\to 0}\left[\phi(x-0) - \phi(x-(-dx))\right].$$

Thus,

$$\Phi = -\frac{\partial}{\partial x}\left[\frac{q}{2\pi} ln\sqrt{x^2+y^2}\right]dx$$

$$= -\frac{(qdx)}{2\pi}\frac{x}{x^2+y^2} = -\frac{\mu}{2\pi}\frac{\cos\theta}{r},$$

where,

$$\mu \equiv \lim_{\substack{dx\to 0\\ q\to\infty}} qdx.$$

the stream function for this flow is, then given by

$$\Psi = \frac{\mu \sin\theta}{2\pi r}. \tag{61a}$$

In cartesian coordinates, (60a) and (61a) become

$$\Phi = -\frac{\mu x}{2\pi\,(x^2 + y^2)},\ \Psi = \frac{\mu y}{2\pi\,(x^2 + y^2)}. \tag{60b,61b}$$

Thus, the equipotentials given by $\Phi = const$ are the coaxial circles,

$$x^2 + y^2 = 2k_1 x \tag{62a}$$

and the streamlines given by $\Psi = const$ are the coaxial circles,

$$x^2 + y^2 = 2k_2 y. \tag{62b}$$

The circles in the first family have centers $(k_1, 0)$ and radii k_1; and those in the second family have centers $(0, k_2)$ and radii k_2 (see Figure 5.4). The two families are mutually orthogonal.

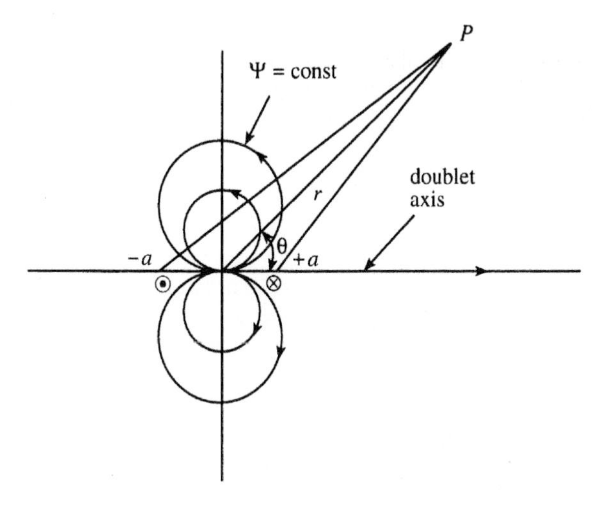

Figure 5.4 Doublet flow.

(iii) The Vortex Flow

Consider a *vortex* flow given by the streamlines in concentric circles. The velocity potential and the stream function for this flow are given by

$$\Phi = \frac{\Gamma\theta}{2\pi}, \ \Psi = -\frac{\Gamma}{2\pi}\ell nr \tag{63}$$

Γ being the counterclockwise circulation of the flow.

Let us now build up some simple flow patterns using the singular solutions (i)-(iii).

(iv) Doublet in a Uniform Stream

The velocity potential and the stream function for a uniform stream with velocity (U, V) are given by

$$\left.\begin{aligned} \Phi &= Ux + Vy \\ \Psi &= -Vx + Uy \end{aligned}\right\} \tag{64}$$

Consider a doublet in a uniform stream $(U, 0)$ such that the doublet axis is in a direction opposite to that of the uniform stream. One then obtains for the stream function the following expression,

$$\Psi = Ur\sin\theta - \frac{\mu}{2\pi}\frac{\sin\theta}{r} = U\sin\theta\left(r - \frac{\mu}{2\pi U}\frac{1}{r}\right). \tag{65}$$

The velocity components are then given by

$$\left.\begin{aligned} u_r &= \frac{1}{r}\frac{\partial\Psi}{\partial\theta} = U\left(1 - \frac{\mu}{2\pi U}\frac{1}{r^2}\right)\cos\theta \\ u_\theta &= -\frac{\partial\Psi}{\partial r} = -U\left(1 + \frac{\mu}{2\pi U}\frac{1}{r^2}\right)\sin\theta \end{aligned}\right\} \tag{66}$$

which shows that the stagnation points $(u_r = 0,\ u_\theta = 0)$ are at $r = \sqrt{\mu/2\pi U}$, and $\theta = 0, \pi$. Note that, (66) represents the flow past a circular cylinder of radius,

$$a \equiv \sqrt{\frac{\mu}{2\pi U}} \tag{67}$$

because, the kinematic condition

$$r = a : u_r = 0 \tag{68}$$

is automatically satisfied on such a body.

The stream function (65) may then be rewritten as

$$\Psi = U\left(1 - \frac{a^2}{r^2}\right) r \sin\theta. \tag{69}$$

Using (69) in (16), one obtains, for the pressure on the cylinder,

$$r = a : C_p \equiv \frac{p - p_\infty}{\frac{1}{2}\rho U^2} = 1 - \left(\frac{u_\theta}{U}\right)^2 = 1 - 4\sin^2\theta \tag{70}$$

which is symmetric about the directions parallel and perpendicular to direction of streaming so that in such a flow there is neither a *drag* D nor a *lift* L on the cylinder. This is shown below,

$$D = a\int_0^{2\pi}\left(1 - 4\sin^2\theta\right)\cos\theta\, d\theta = 2a\int_0^{\pi}\left(1 - 4\sin^2\theta\right)\cos\theta d\theta = 0$$

$$L = a\int_0^{2\pi}\left(1 - 4\sin^2\theta\right)\sin\theta\, d\theta = 0.$$

(v) Uniform Flow Past a Circular Cylinder with Circulation

The stream function for a uniform flow past a circular cylinder with clockwise circulation (see Figure 5.5) is, (using (63) and (69)), given by[8].

$$\Psi = Ur\sin\theta + \frac{\Gamma}{2\pi} ln\left(\frac{r}{a}\right) - \frac{Ua^2}{r}\sin\theta. \tag{71}$$

From this, we obtain the velocity components as below,

$$\left.\begin{aligned} u_r &= \frac{1}{r}\frac{\partial\Psi}{\partial\theta} = U\left(1 - \frac{a^2}{r^2}\right)\cos\theta \\ u_\theta &= -\frac{\partial\Psi}{\partial r} = -U\left(1 + \frac{a^2}{r^2}\right)\sin\theta - \frac{\Gamma}{2\pi r} \end{aligned}\right\}. \tag{72a}$$

---

8 From (71), we see that $r = a$ represents the streamline $\Psi = 0$, as required. This is also confirmed by (72a), which shows that when $r = a, u_r = 0$

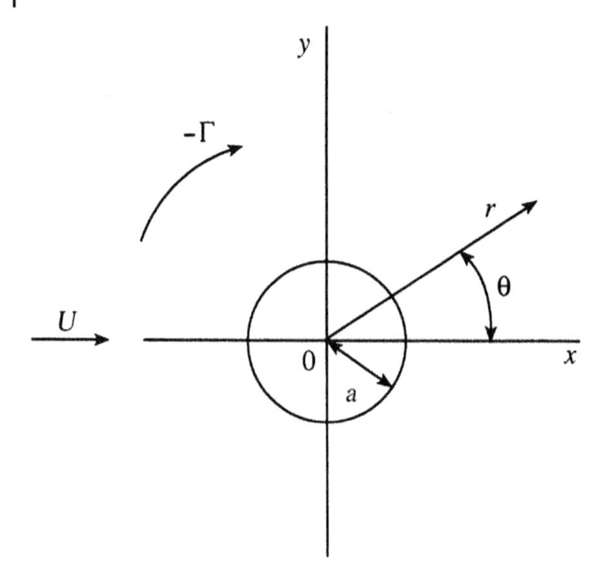

Figure 5.5 Uniform flow past a circular cylinder with circulation.

This shows that the stagnation points ($u_r = 0, u_\theta = 0$) are given by

$$r = a : \theta = \sin^{-1}(-\Gamma/4\pi Ua),\ \Gamma \leq 4\pi Ua$$

$$r \geq a : \cos\theta = 0 \text{ or } \theta = -\frac{\pi}{2} \tag{73a}$$

and

$$U\left(1 + \frac{a^2}{r^2}\right) = \frac{\Gamma}{2\pi r},\ \Gamma > 4\pi Ua. \tag{73b}$$

Equation (73b) shows that there are two stagnation points $r = r_1, r_2$ given by,

$$r_{1,2} = \frac{1}{2}\left[\frac{\Gamma}{2\pi U} \pm \sqrt{\frac{\Gamma^2}{4\pi^2 U^2} - 4a^2}\right],\ r_1 r_2 = a^2. \tag{74}$$

If $\Gamma > 4\pi Ua$, then one stagnation point lies outside $r = a$ and the other inside. However, if $\Gamma = 4\pi Ua$, then the two points coincide with $r = a$.

From (72a), the velocity on the cylinder is given by

$$r = a : u_r = 0,\ u_\theta = -\left(2U \sin\theta + \frac{\Gamma}{2\pi a}\right). \tag{72b}$$

For $r = a$, from (72b), the stagnation points (for $\Gamma \leq 4\pi U a$) are given by

$$\Gamma = -4\pi U a \sin\theta. \tag{75}$$

If $\Gamma \equiv 4\pi U a \sin\alpha$, then the stagnation points are given by

$$-\sin\theta = \sin\alpha \text{ or } \theta = -\alpha \text{ and } \pi + \alpha. \tag{76}$$

The flow pattern described by (71) for various values of the circulation is shown in Figure 5.6.

Using (72a) in (16), one finds, for the pressure on the cylinder,

$$\begin{aligned} r = a : p &= \left(p_\infty + \frac{1}{2}\rho U^2\right) - \frac{\rho u_\theta^2}{2} \\ &= \left(p_\infty + \tfrac{1}{2}\rho U^2\right) - 2\rho U^2 \sin^2\theta - \frac{\rho \Gamma^2}{8\pi a^2} - \frac{\rho \Gamma U}{\pi a} \sin\theta. \end{aligned} \tag{77}$$

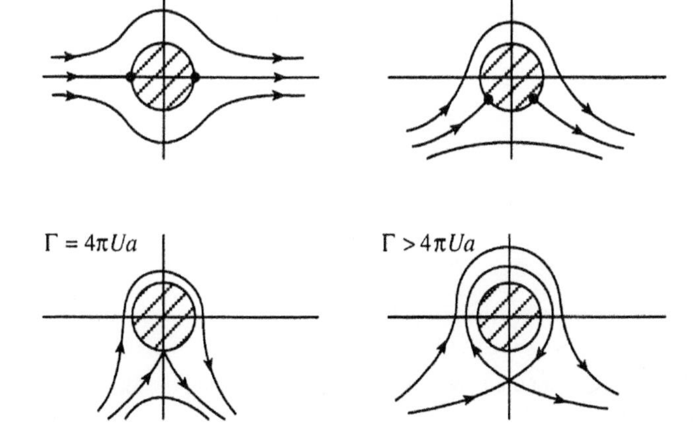

Figure 5.6 Uniform flows past a circular cylinder with various values of the circulation.

The net force on the cylinder is directed along the y-axis[9]. This force is called the *lift* (the *Magnus* (1853) *effect*), and is given by

$$L = -\int_0^{2\pi} p \sin\theta \cdot a d\theta = \rho U \Gamma. \tag{78}$$

This result is, in fact, *independent* of the shape of the cross section of the cylinder (*Kutta-Joukowski Theorem*, see Section 11.1).

Exercises

1. Consider a vortex-flow, and a doublet flow between two circular cylinders of radii R_1, R_2. Calculate the kinetic energy of the fluid between the two cylinders by a volume integral and an integral over the boundaries separately.
2. Show that the irrotational flow of a fluid is impossible if the boundaries are fixed. [10]

---

9 Thanks to the *fore-aft* symmetry of the flow with respect to the cylinder, the force in the x-direction (the *drag*) is zero.
10 This result implies that when the solid boundaries in motion are instantaneously brought to rest, the flow of the fluid will instantaneously cease to be irrotational but will not come to a stop!

6

The Complex-Variable Method

The velocity potential and the stream function of a two-dimensional irrotational flow of an inviscid fluid satisfy relations identical to the Cauchy-Riemann conditions for an analytic function of a complex variable. This allows the application of the full machinery of complex variable theory to formulate and analyze two dimensional irrotational flows of an inviscid fluid. More specifically, the velocity potential and the stream function can be combined to form an analytic function of a complex variable $z = x + iy$, in the x, y-plane occupied by the flow. Furthermore, one may use the idea of conformal mapping in complex variable theory to relate the solution to a two-dimensional potential flow for one geometry with that for a simpler geometry. In cases where the flow separates from a rigid boundary and free streamlines appear, one may use the Schwarz-Christoffel transformation to formulate and analyze the flow.

6.1 The Complex Potential

The velocity potential and stream function of a two-dimensional irrotational flow of an inviscid fluid satisfy the following properties.

$$\left.\begin{aligned} &\frac{\partial^2 \Phi}{\partial x^2} + \frac{\partial^2 \Phi}{\partial y^2} = 0, \ \frac{\partial^2 \Psi}{\partial x^2} + \frac{\partial^2 \Psi}{\partial y^2} = 0 \\ &\frac{\partial \Phi}{\partial x} = \frac{\partial \Psi}{\partial y}, \ \frac{\partial \Phi}{\partial y} = -\frac{\partial \Psi}{\partial x} \\ &\frac{\partial \Phi}{\partial x} \cdot \frac{\partial \Psi}{\partial x} + \frac{\partial \Phi}{\partial y} \cdot \frac{\partial \Psi}{\partial y} = 0 \end{aligned}\right\} \tag{1}$$

These properties are identical to the *Cauchy-Riemann conditions* exhibited by the *real* and *imaginary* parts of an *analytic function of a complex variable.*

Introduction to Theoretical and Mathematical Fluid Dynamics, Third Edition.
Bhimsen K. Shivamoggi.

Therefore, it is natural to combine Φ and Ψ into an analytic function F of a complex variable $z = x + iy$ in the region of the z-plane occupied by the flow.

Thus, one may consider the *complex potential*,

$$F(z) \equiv \Phi(x, y) + i\Psi(x, y). \tag{2}$$

All the resources of the *theory of analytic function of a complex variable* now become available for the description of two-dimensional irrotational flows. More specifically, the problem reduces to the determination of an analytic function whose real and imaginary parts satisfy the given boundary conditions. The complex potential $F(z)$, as defined by (2), has a unique derivative with respect to z at all points in the region of the flow. Thus, one obtains for the *"complex" velocity*,

$$W(z) \equiv \frac{dF}{dz} = \frac{\partial \Phi}{\partial x} + i\frac{\partial \Psi}{\partial x} = \frac{\partial \Psi}{\partial y} - i\frac{\partial \Phi}{\partial y} = u - iv. \tag{3}$$

Example 1: For a uniform flow, one has

$$F(z) = Az. \tag{4}$$

Example 2: For a source flow or a vortex flow, one has

$$F(z) = A\ell nz \tag{5}$$

depending on whether A is real or imaginary, respectively.

Example 3: For a flow in a corner of angle α, one has

$$F(z) = Az^{\pi/\alpha} \tag{6}$$

which is valid only in the neighborhood of the corner. From (6) we obtain

$$\Phi = r^{\pi/\alpha} \cos \pi\theta/\alpha, \Psi = r^{\pi/\alpha} \sin \pi\theta/\alpha$$

so, $\Psi = 0$ corresponds to the walls given by $\theta = 0$ and $\theta = \alpha$.

For illustration, consider the case $\alpha = \pi/2$, for which (6) becomes

$$F(z) = Az^2 \tag{7}$$

which describes the *stagnation-point* flow at a plane (by symmetry).

The velocity potential and the stream function are then given by,

$$\Phi = A\left(x^2 - y^2\right), \Psi = 2Axy. \tag{8}$$

Thus, the equipotentials are the rectangular hyperbolae (Figure 6.1)

$$x^2 - y^2 = const$$

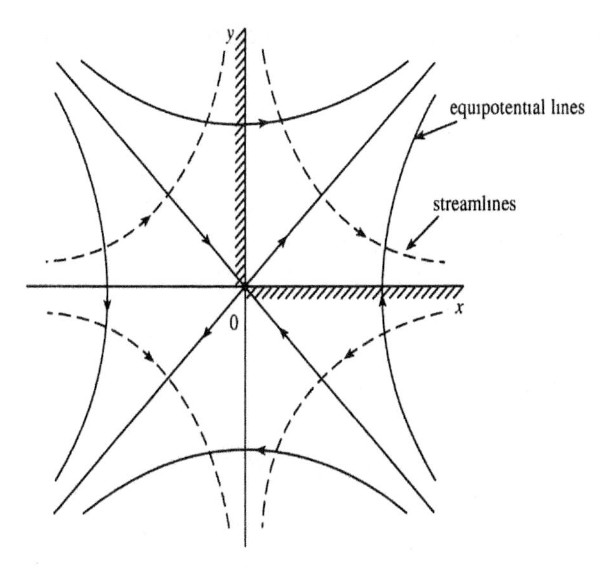

Figure 6.1 Flow in a corner of angle /2.

having the asymptotes given by $y = \pm x$. The streamlines are also the rectangular hyperbolae (Figure 6.1),

$$xy = const$$

having the axes $x = 0,\ y = 0$ as asymptotes. These two families of hyperbolae are orthogonal to each other. Furthermore, the only stagnation point in the flow occurs at the origin.

Now, the fact that there is no flow across a streamline implies that the flow is unaltered if any one of the streamlines is replaced by a rigid barrier. The present case, namely, $\alpha = \pi/2$, corresponds to replacing the axes $x = 0, y = 0$ by rigid boundaries.[1]

Example 4: For a doublet flow, one has upon superposing a source at $(a, 0)$ and a sink at $(-a, 0)$,

$$F(z) = \frac{q}{2\pi}\ell n(z - a) - \frac{q}{2\pi}\ell n(z + a) = \frac{q}{2\pi}\ell n\left(1 - \frac{2a}{z + a}\right) \approx -\frac{q}{2\pi}\frac{2a}{z}. \tag{9}$$

1 This implies an appropriate system of images is prescribed in the barrier for the flow on the other side (see Section 6.3).

Let us choose q and a such that we have the following limit.

$$\lim_{\substack{a\to 0\\ q\to\infty}} 2qa = const \equiv \mu. \tag{10}$$

In this limit, (9) becomes,

$$F(z) = -\frac{\mu}{2\pi z} \tag{11}$$

which recovers (51) and (52) given in Section 5.9.

6.2 Conformal Mapping of Flows

Another important aspect of the *theory of analytic function of a complex variable* becomes apparent while relating the solution for the two-dimensional potential flow for one geometry with that for another geometry *via a conformal mapping*. Consider a two-dimensional flow, for which the velocity potential Φ satisfies the *Laplace equation* (i.e., it is a *harmonic* function),

$$\frac{\partial^2\Phi}{\partial x^2} + \frac{\partial^2\Phi}{\partial y^2} = 0. \tag{12}$$

Consider a transformation given by

$$\left.\begin{aligned} (x,y) &\to [\xi(x,y), \eta(x,y)] \\ \Phi(x,y) &\to \Phi(\xi,\eta) \end{aligned}\right\}. \tag{13}$$

Then, equation (12) leads to

$$\begin{aligned} &\left[\left(\frac{\partial\xi}{\partial x}\right)^2 + \left(\frac{\partial\xi}{\partial y}\right)^2\right]\frac{\partial^2\Phi}{\partial\xi^2} + \left[\left(\frac{\partial\eta}{\partial x}\right)^2 + \left(\frac{\partial\eta}{\partial y}\right)^2\right]\frac{\partial^2\Phi}{\partial\eta^2} \\ &\quad + \left[\frac{\partial^2\xi}{\partial x^2} + \frac{\partial^2\xi}{\partial y^2}\right]\frac{\partial\Phi}{\partial\xi} + \left[\frac{\partial^2\eta}{\partial x^2} + \frac{\partial^2\eta}{\partial y^2}\right]\frac{\partial\Phi}{\partial\eta} \\ &\quad + 2\left[\frac{\partial\xi}{\partial x}\frac{\partial\eta}{\partial x} + \frac{\partial\xi}{\partial y}\frac{\partial\eta}{\partial y}\right]\frac{\partial^2\Phi}{\partial\xi\partial\eta} = 0. \end{aligned} \tag{14}$$

If we want the coordinate transformation (13) to relate the solutions of a two dimensional potential flow for two different geometries, i.e.

$$\frac{\partial^2\Phi}{\partial\xi^2} + \frac{\partial^2\Phi}{\partial\eta^2} = 0 \tag{15}$$

then, the coordinate transformation (13) must satisfy

$$\left(\frac{\partial \xi}{\partial x}\right)^2+\left(\frac{\partial \xi}{\partial y}\right)^2=\left(\frac{\partial \eta}{\partial x}\right)^2+\left(\frac{\partial \eta}{\partial y}\right)^2 \neq 0$$

$$\frac{\partial^2 \xi}{\partial x^2}+\frac{\partial^2 \xi}{\partial y^2}=\frac{\partial^2 \eta}{\partial x^2}+\frac{\partial^2 \eta}{\partial y^2}=0 \tag{16}$$

$$\frac{\partial \xi}{\partial x}\frac{\partial \eta}{\partial x}+\frac{\partial \xi}{\partial y}\frac{\partial \eta}{\partial y}=0.$$

These are the *Cauchy-Riemann conditions* governing the real and imaginary parts of an analytic function of a complex variable, making $\zeta = \xi + i\eta$ an analytic function of $z = x + iy$.

Theorem 6.1 Suppose that an analytic function, $\zeta = \zeta(z) = \xi(x,y) + i\eta(x,y)$ maps a domain D in the z-plane onto a domain D' in the ζ-plane. If $\Phi(x,y)$ is a harmonic function defined on D, then the function $\Phi(\xi,\eta) = \Phi[x(\xi,\eta), y(\xi,\eta)]$ is harmonic in D' as well. If $\zeta = \zeta(z)$ describes a mapping of the z-plane onto the ζ-plane, then both the magnification ratio $|\zeta'(z)|$ and the angular rotation $\arg \zeta'(z)$ at any point z are the same for all curves C passing through z.

This theorem implies that the corresponding infinitesimal figures in the z- and ζ-planes are similar. ($\zeta'(z)$ may be viewed as the *local scale factor* of the transformation (13)). Note that the mapping generated by an analytic function $\zeta(z)$ is *conformal* in the neighborhood of a point z, except when it is a *critical point*, i.e., $\zeta'(z_0) = 0$.[2]

Example 5: Consider the flow past a parabolic cylinder (see Figure 6.2). The mapping

$$\zeta(z) = \sqrt{z} \text{ or } z = \zeta^2$$

yields

$$x = \xi^2 - \eta^2, \;\; y = 2\xi\eta.$$

2 At a critical point of mapping (where the mapping is not conformal), the angles between two curves intersecting at this point are magnified n times, when $\frac{d^n\zeta}{dz^n}$ is the lowest derivative that does not vanish at this point.

This maps the paraboloid in the z-plane,

$$x - \xi_0^2 = -\frac{y^2}{4\xi_0^2}$$

into the plane $\xi = \xi_0$ in the ζ-plane (see Figure 6.2).

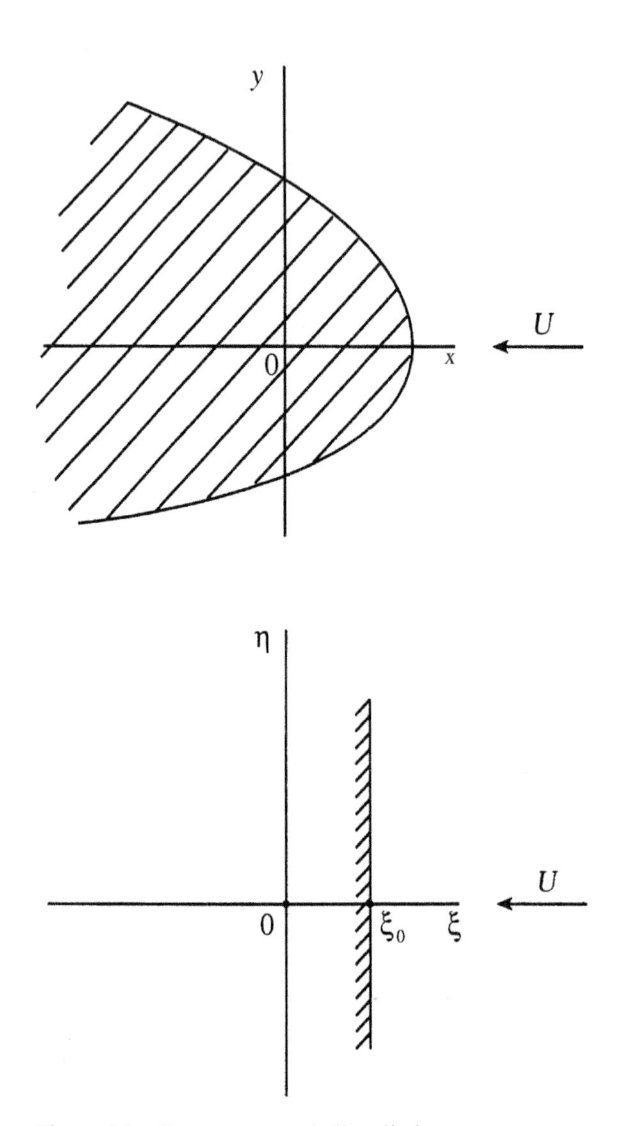

Figure 6.2 Flow past a parabolic cylinder.

A flow impinging perpendicularly on a plane $\xi = \xi_0$ in the ζ-plane is given by

$$\tilde{F}(\zeta) = A(\zeta - \xi_0)^2$$

so that one has, for the complex velocity in the z-plane,

$$W = \frac{d\tilde{F}/d\zeta}{dz/d\zeta} = \frac{2A(\zeta - \xi_0)}{2\zeta}.$$

Using the boundary condition dictating that the two flows at infinity be the same, i.e.,

$$|\zeta| \to \infty : W \to -U$$

one obtains

$$A = -U.$$

Thus, the complex velocity in the z-plane is given by

$$W = -\frac{U\left(\sqrt{z} - \xi_0\right)}{\sqrt{z}}.$$

Example 6: Consider the transformation to *elliptic* coordinates,

$$z = C\cosh\zeta$$

or

$$x = C\cosh\ \xi \cdot \cos\eta,\ y = C\sinh\ \xi \cdot \sin\eta$$

which leads to the following relations,

$$\left.\begin{aligned} \frac{x^2}{C^2\cosh^2\xi} + \frac{y^2}{C^2\sinh^2\xi} = 1 \\ \frac{x^2}{C^2\cos^2\eta} - \frac{y^2}{C^2\sin^2\eta} = 1 \end{aligned}\right\}.$$

Thus, the line $\xi = \xi_0$ gives an ellipse in the x, y-plane, with major and minor radii, respectively,

$$a \equiv C\cosh\xi_0,\ \ b \equiv C\sinh\xi_0$$

from which,

$$\left.\begin{aligned} a^2 - b^2 &= C^2 \\ a \pm b = Ce^{\pm\xi_0} \text{ or } \xi_0 &= \frac{1}{2}\ell n \frac{a+b}{a-b} \end{aligned}\right\}.$$

Now, the kinematic boundary condition at the body requires the flow to have no relative velocity component normal to the body. This gives (see Figure 6.3)

$$\frac{\partial \Psi}{\partial s} = U\cos\theta = -U\frac{\partial y}{\partial s}$$

from which,

$$\Psi = -Uy + const, \text{ at the body.}$$

Note that this prescription satisfies the boundary condition identically for *any* shape of the body translating in the x-direction.

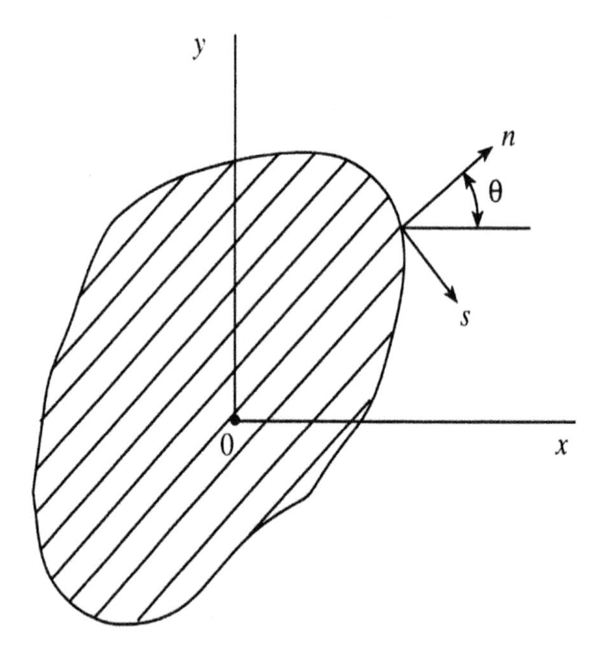

Figure 6.3 Body-fixed coordinates.

Consider, for illustration, the flow in the ζ-plane to be given by,

$$F = -\tilde{C}e^{-\zeta}$$

which leads to the following expressions for the velocity potential and the stream function,

$$\Phi = -\tilde{C}e^{-\xi}\cos\eta, \quad \Psi = \tilde{C}e^{-\xi}\sin\eta.$$

The boundary condition at the body, then gives

$$\tilde{C}e^{-\xi_0}\sin\eta = -UC\sinh\xi_0 \cdot \sin\eta = -Ub\sin\eta$$

from which,

$$\tilde{C} = -Ube^{\xi_0} = -Ub\sqrt{\frac{a+b}{a-b}}.$$

The stream function then becomes

$$\Psi = -Ub\sqrt{\frac{a+b}{a-b}}e^{-\xi}\sin\eta$$

which gives the flow produced by an elliptic cylinder translating along the major axis with velocity U.

On the other hand, for a steady flow past an elliptic cylinder, the complex potential is given by

$$\tilde{F} = -Ub\sqrt{\frac{a+b}{a-b}}e^{-\zeta} - UC\cosh\zeta.$$

Example 7: Consider a rigid cylinder rotating about an axis through the origin parallel to its length. One has from the boundary condition at the body, (see Figure 6.4),

$$\frac{\partial\Psi}{\partial s} = \omega r\cos\theta = \omega r\frac{dr}{ds}$$

which implies that,

$$\Psi = \frac{1}{2}\omega r^2 + D$$

D being an arbitrary constant.

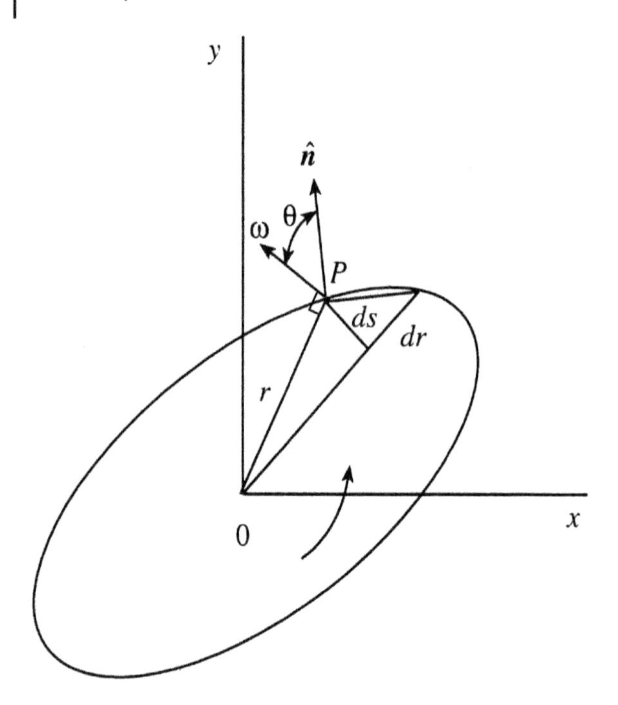

Figure 6.4 Rigid body rotating in a fluid.

As an example, consider the flow to be given by the complex potential,

$$F = iAz^2$$

so that the velocity potential and the stream function are given by

$$\Phi = -2Axy, \;\; \Psi = A\left(x^2 - y^2\right).$$

The boundary condition at the body gives

$$A\left(x^2 - y^2\right) = \frac{1}{2}\omega\left(x^2 + y^2\right) - C$$

C being an arbitrary constant. This relation may be rewritten as

$$\frac{x^2}{\left[C/\left(\frac{\omega}{2} - A\right)\right]} + \frac{y^2}{\left[C/\left(\frac{\omega}{2} + A\right)\right]} = 1$$

which is an ellipse, if $A < \omega/2$. If a and b are the major and minor radii of this ellipse, we have

$$a^2 = \frac{C}{\omega/2 - A}, \quad b^2 = \frac{C}{\omega/2 + A}.$$

Eliminating C, we obtain

$$A = \frac{\omega}{2}\frac{a^2 - b^2}{a^2 + b^2}.$$

Thus, the stream function becomes

$$\Psi = \frac{\omega}{2}\frac{a^2 - b^2}{a^2 + b^2}\left(x^2 - y^2\right)$$

which represents the flow within a rotating hollow elliptic cylinder.

Consider next the transformation to *elliptic* coordinates,

$$z = C\cosh\zeta$$

and a flow in the ζ-plane given by the complex potential,

$$\tilde{F} = i\tilde{C}e^{-2\zeta}$$

so that the velocity potential and the stream function are given by

$$\Phi = \tilde{C}e^{-2\xi}\sin 2\eta, \quad \Psi = \tilde{C}e^{-2\xi}\cos 2\eta.$$

The boundary condition at the body gives

$$\tilde{C}e^{-2\xi_0}\cos 2\eta = \frac{1}{4}C^2\omega\left(\cosh 2\xi_0 + \cos 2\eta\right) + D$$

from which,

$$\tilde{C}e^{-2\xi_0} = \frac{1}{4}C^2\omega, \quad \frac{1}{4}C^2\omega\cosh 2\xi_0 + D = 0$$

If a and b are the major and minor radii of the elliptic cylinder given by $\xi = \xi_0$, one then obtains

$$a = C\cosh\xi_0, \quad b = C\sinh\xi_0$$

from which,

$$e^{\xi_0} = \sinh\xi_0 + \cosh\xi_0 = \frac{a+b}{C}.$$

Thus,

$$\tilde{C} = (a+b)^2\frac{\omega}{4}$$

and the stream function becomes

$$\Psi = \frac{\omega}{4}(a+b)^2 e^{-2\xi}\cos 2\eta$$

which represents the flow due to rotation of an elliptic cylinder about its axis with angular velocity ω.

6.3 Hydrodynamic Images

The hydrodynamic images are distributions of sources and vortices placed inside a body B, which produce the disturbance flow generated outside B when it is placed in a flow field. Thus, the image of a uniform stream $(U,0)$ in the circular cylinder $r = a$ is a doublet of moment $\left(-2\pi U a^2, 0\right)$ placed at the center $(0,0)$.

Consider a source outside a plane wall (Figure 6.5). The complex potential is given by

$$F = \frac{q}{2\pi}\left[\ell n\left(z - z_0\right) + \ell n\left(z + \bar{z}_0\right)\right]. \tag{17}$$

The presence of the image source ensures that, on superposition with the given source, the plane wall remains a streamline of the flow.

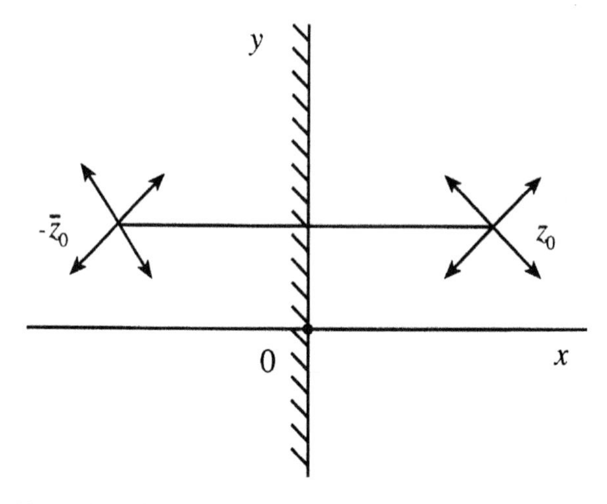

Figure 6.5 Source outside a plane wall.

On the wall $z = iy$, (17) leads to

$$F = \Phi + i\Psi = \frac{q}{2\pi}\ell n\left[|(z - z_0)(z + \overline{z}_0)|\right]$$
$$= \frac{q}{2\pi}\ell n\left[x_0^2 + (y - y_0)^2\right] \tag{18}$$

from which we have,

$$\Psi = 0 \tag{19}$$

as required.

Let us next determine the complex potential representing the flow of a fluid in the presence of a circular interior boundary. Let the complex potential of a given flow be given by

$$F = f(z)$$

which is free from singularities in the region $|z| \leq a$. Suppose that a stationary circular cylinder of radius a is now placed with its center at the origin. Then, for every singularity z_0 of $f(z)$ (outside $|z| = a$), we place an image in the circular boundary such that the two together will render the circle $|z| = a$, a streamline. One thus obtains, for the complex potential for the flow in the presence of a circular cylinder the following expression (*Milne-Thompson* (1940) *Circle Theorem*),

$$f(z) + \overline{f}(a^2/z)$$

which is purely real on $|z| = a$, and therefore, gives a streamline described by $\Psi = 0$.

Example 8: Consider the flow due to a point source of strength q placed at $z = z_0$. One then has the following complex potential for this flow,

$$f(z) = \frac{q}{2\pi} ln(z - z_0).$$

In the presence of a circular cylinder centered at the origin with radius $a < |z_0|$, the complex potential becomes

$$F = \frac{q}{2\pi}\, ln(z - z_0) + \frac{q}{2\pi}\, ln\left(\frac{a^2}{z} - \overline{z}_0\right)$$

$$= \frac{q}{2\pi}\, ln(z - z_0) + \frac{q}{2\pi} ln\left(z - \frac{a^2}{\overline{z}_0}\right) - \frac{q}{2\pi}\, ln\, z + \frac{q}{2\pi}\ell n\overline{z}_0 \tag{19}$$

which shows that, for a source outside the cylinder, the image system consists of a source q at the inverse point $z = a^2/\overline{z}_0$, and a sink $-q$ at the origin.

6.4 Principles of Free-Streamline Flow

The boundary conditions imposed on the flow cases in the preceding examples do not permit *separation of flow* from a boundary. However, separation of the flow is inevitable whenever sudden changes occur in the direction of boundaries, which otherwise cause infinite velocities. At separation points in steady flow of a fluid around a body, the streamlines leave the body. These dividing streamlines are called *free streamlines* in two-dimensional flow. The fluid in contact with the body downstream from the separation points and separated from the main flow (assumed to be inviscid and irrotational) by the free streamlines is known as the *wake*. The fluid in the wake is assumed to be at rest in steady-flow problems. Since this is not quite so in reality (the fluid in the wake is in a relatively slow recirculating flow), the theory yields better results if the wake contains another fluid of much less density.

If the effects of gravity are ignored, the pressure in the wake is constant, and the pressure along the free streamline must also be constant. From Bernoulli's equation (18) in Section 5.3, the velocity of the free streamline must also be constant. This somewhat alleviates the difficulty that the location of the free streamlines is not known beforehand.

The method of solution involves introducing a *hodograph transformation*,

$$Q \equiv ln\,(U/W) = ln\;q + i\theta,$$

where, $W \equiv Ve^{-i\theta}$ and $q \equiv U/V$. Note that the real part of Q is constant on each free streamline and the imaginary part of Q is constant on each straight portion of the rigid boundary. Since both parts of this boundary are streamlines, the stream function ψ can be prescribed. Consequently, if all the solid boundaries are straight, the flow region is mapped onto a straight-sided figure in the Q-plane. One may then use a *Schwarz-Christoffel transformation* to map the interior of the straight-sided figure onto the upper half of another plane.

(i) Schwarz-Christoffel Transformation

This transformation maps the inside of a polygon in the z-plane onto the upper half of the ζ-plane.

Theorem 6.2 Consider the transformation given by

$$\frac{dz}{d\zeta} = C\prod_{i=1}^{n} (\zeta - a_i)^{\beta_i} \tag{20}$$

where the a_i's are real and are ordered as follows

$$a_l < a_2 < \cdots < a_n.$$

This transformation maps the real axis in the ζ-plane into the boundary of a polygon in the z-plane in such a way that the vertices of the polygon correspond to the points $a_1, a_2, \cdots$ and the interior angles of the polygon are $\alpha_1, \alpha_2, \cdots$ where

$$\alpha_i \equiv \pi\,(1 + \beta_i)\,,\ \forall i.$$

Proof: From equation (20), we obtain

$$arg\left(\frac{dz}{d\zeta}\right) = arg\,C + \beta_1\,arg\,(\zeta - a_1) + \beta_2\,arg\,(\zeta - a_2) + \cdots + \beta_n\,arg\,(\zeta - a_n)\,. \tag{21}$$

Now, since ζ is real, one has

$$arg\left(\frac{dz}{d\zeta}\right) = arg\,(dz) = \theta_i, \text{ in the } i\text{th interval.} \tag{22}$$

Furthermore, when ζ lies on the real axis in the ith interval between a_i and a_{i+1}, note that

$$\left.\begin{aligned} arg\ (\zeta - a_k) &= 0,\ \ k < i \\ arg\ (\zeta - a_k) &= \pi,\ \ k > i \end{aligned}\right\}. \tag{23}$$

Using (22) and (23), (21) gives

$$\theta_i = arg\,C + \pi\,(\beta_{i+1} + \beta_{i+2} + \cdots + \beta_n)$$

Thus, all points on the real axis segment $(a_{i+1} - a_i)$ in the ζ-plane are mappings of a line segment with slope θ_i in the z-plane (see Figure 6.6). Moreover,

$$\theta_{i+1} - \theta_i = -\pi\beta_{i+1}. \tag{24}$$

From Figure 6.6, we note that

$$\theta_{i+1} - \theta_i = \pi - \alpha_{i+1} \tag{25}$$

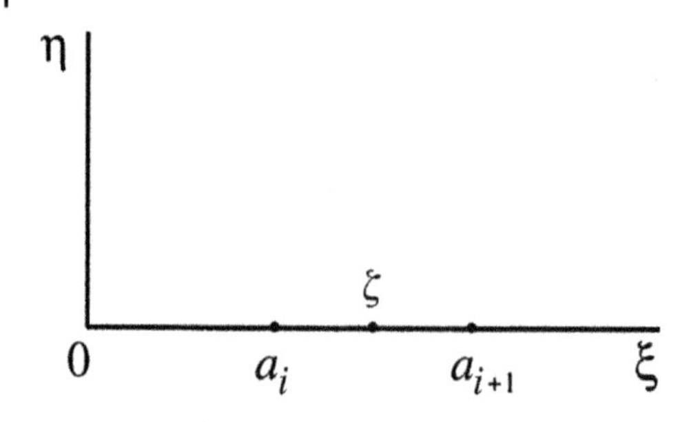

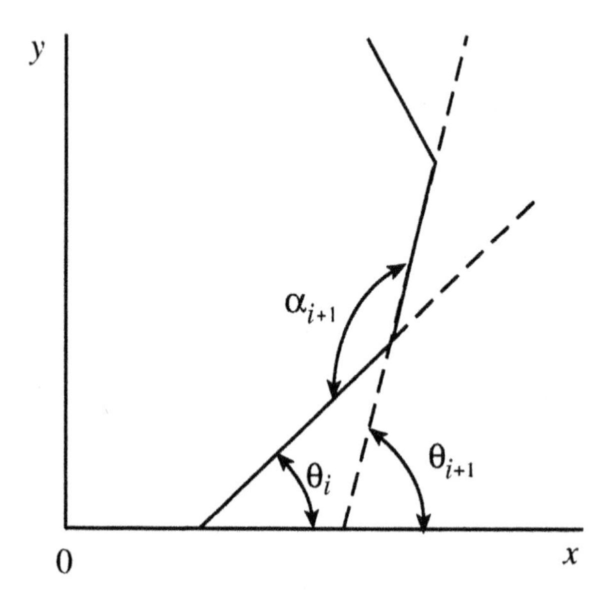

Figure 6.6 Schwarz - Christoffel mapping.

α_{i+1} being the interior angle at the corner in question, one obtains

$$\alpha_{i+1} = \pi + \pi\beta_{i+1}$$

or

$$\beta_i = \frac{\alpha_i}{\pi} - 1. \tag{26}$$

Thus, (20) becomes

$$\frac{dz}{d\zeta} = C\prod_{i=1}^{n}(\zeta - a_i)^{(\alpha_i/\pi - 1)}. \tag{27}$$

The *scale factor* C prescribes both the relative scale and the relative angular orientation of the two geometries. Note that the applicability of the Schwarz-Christoffel transformation depends crucially on the representation of the flow region into a polygon in the Q-plane. This is possible only if all the solid boundaries are straight. □

Example 9: Consider a source of strength q in a channel (see Figure 6.7). One then has, from (27), that the Schwarz-Christoffel transformation,

$$\frac{dz}{d\zeta} = \frac{C}{\zeta}$$

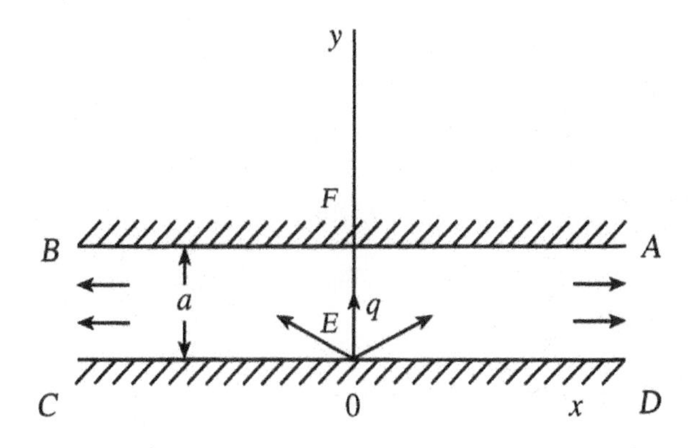

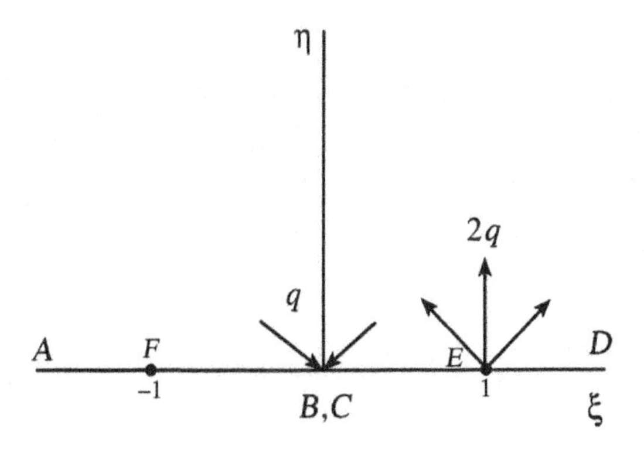

Figure 6.7 Conformal mapping for a source in a channel.

on imposing the boundary conditions

$$\left.\begin{array}{l} \zeta = 1 : \; z = 0 \\ \\ \zeta = -1 : \; z = ia \end{array}\right\}$$

leads to

$$z = \frac{a}{\pi}\,\ell n\,\zeta.$$

This transformation maps the given flow in the z-plane onto the one in the ζ-plane, as shown in Figure 6.7. One has for the latter,

$$\tilde{F}(\zeta) = \frac{2q}{2\pi}\ell n\,(\zeta - 1) - \frac{q}{2\pi}\ell n\zeta = \frac{q}{\pi}\ell n \sin h\frac{\pi z}{2a}.$$

The sources at $\zeta = 0, 1$ have strengths $q, 2q$, respectively. Note that

$$W(z) = \frac{d\tilde{F}}{dz} = \frac{q}{2a}\cot h\frac{\pi z}{2a}$$

which vanishes at $z = ia$, confirming the fact that the dividing streamline from O (or E) goes straight up to F on the opposite wall.

Example 10: Consider the flow past a corner (see Figure 6.8). Making first the *hodograph* transformation given by

$$Q \equiv ln\left(\frac{U}{W}\right) = ln\left(\frac{U}{V}\right) + i\theta$$

U being the uniform flow speed away from the corner, the given flow maps onto a straight-sided figure shown in Figure 6.8. Next, the Schwarz-Christoffel transformation,

$$\frac{dQ}{d\zeta} = C\,(\zeta + 1)^{-1/2}\,(\zeta - 1)^{-1/2}$$

on imposing the boundary conditions,

$$\left.\begin{array}{l} \zeta = 1 : Q = O \\ \\ \zeta = -1 : Q = i\gamma \end{array}\right\}.$$

leads to

$$Q = \frac{\gamma}{\pi}\cosh^{-1}\zeta.$$

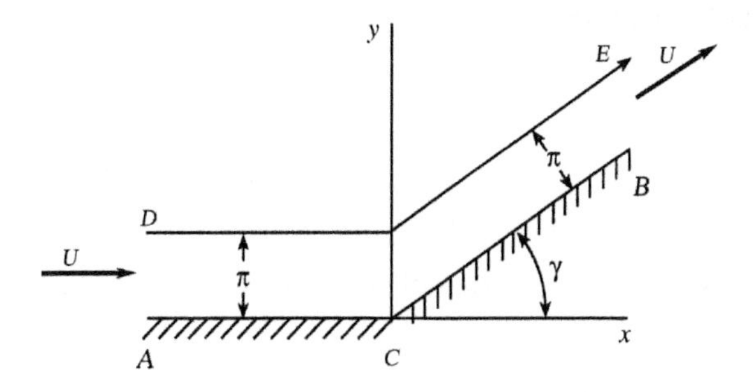

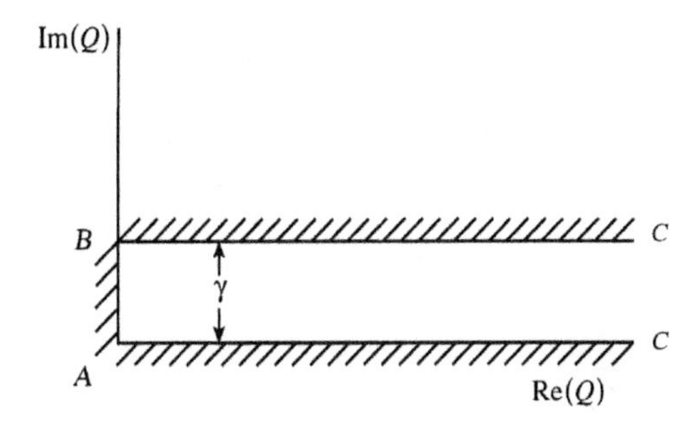

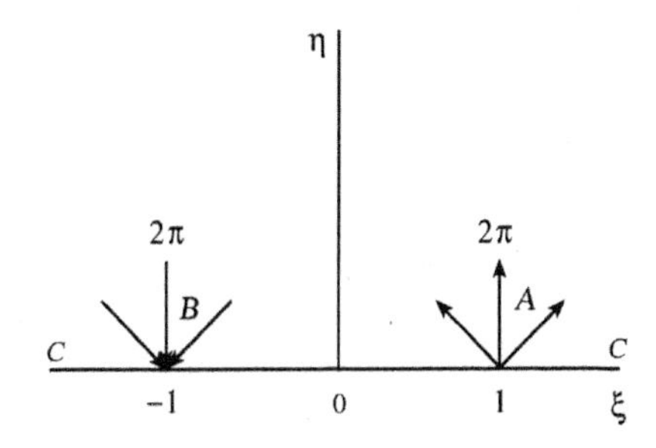

Figure 6.8 Conformal mapping for a flow past a corner.

This transformation maps the given flow onto the one in the ζ-plane shown in Figure 6.8. The complex potential for the latter is given by

$$\tilde{F}(\zeta) = \ell n(\zeta - 1) - \ell n(\zeta + 1) = \ell n\left(\frac{\zeta - 1}{\zeta + 1}\right).$$

The sources at $\zeta = \underline{+}1$ have a strength of 2π.

On the free streamline DE given by $d\Psi = 0$, we have

$$ds = \frac{1}{U}(d\Phi + id\Psi) = \frac{1}{U}d\tilde{F} = \frac{1}{U}\left(\frac{d\zeta}{\zeta - 1} - \frac{d\zeta}{\zeta + 1}\right) = \frac{2}{U}\frac{d\zeta}{\zeta^2 - 1}$$

from which,

$$\frac{Us}{2} + C = -\tanh^{-1}\zeta, |\zeta| < 1 \text{ (see expression for } \zeta \text{ below)}$$

C being the constant of integration.

On the other hand, we have from the hodograph transformation, on the free streamline DE,

$$i\theta = \frac{\gamma}{\pi}\cosh^{-1}\zeta$$

or

$$\zeta = \cos\left(\frac{\pi\theta}{\gamma}\right).$$

Thus,

$$\frac{Us}{2} + C = -\tanh^{-1}\left(\cos\frac{\pi\theta}{\gamma}\right).$$

Upon using the boundary conditions

$$\left.\begin{array}{r} s \to -\infty : \theta \to 0 \\ s \to \infty : \theta \to \gamma \end{array}\right\}$$

we obtain

$$C = 0.$$

So,

$$\cos\left(\frac{\pi\theta}{\gamma}\right) = -\tanh\left(\frac{Us}{2}\right)$$

which shows, as required, that θ varies from 0 to γ, as s varies from $-\infty$ to $+\infty$.

Example 11: Consider a jet emerging from an orifice (see Figure 6.9). Once again, we first make the hodograph transformation, given by

$$Q = ln\left(\frac{U}{W}\right) = lnq + i\theta, q \equiv \frac{U}{V}$$

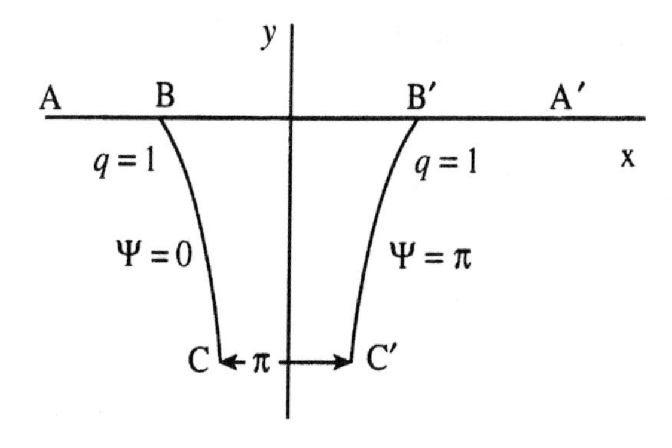

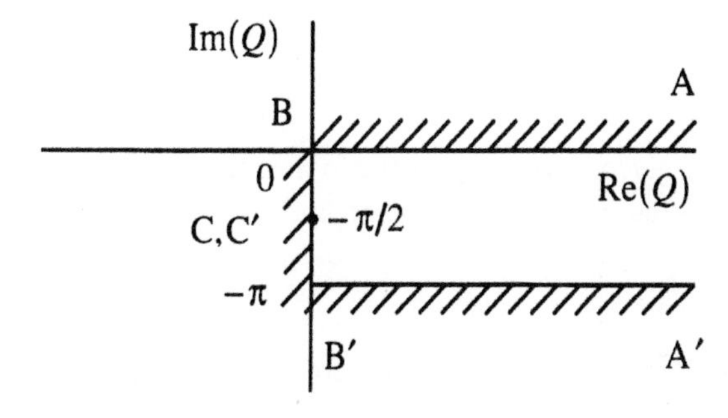

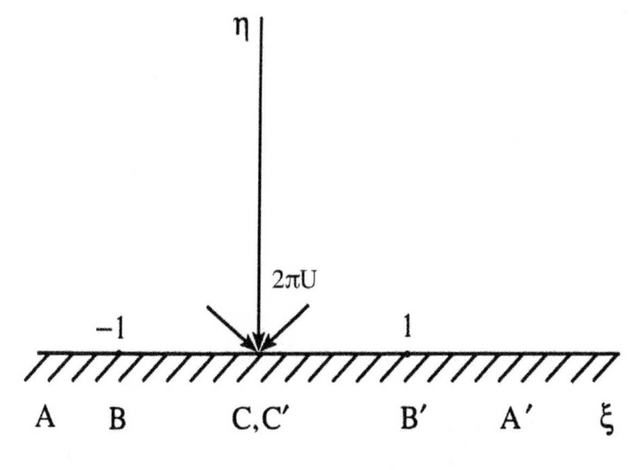

Figure 6.9 Conformal mapping for a jet emerging an orifice.

where U is the uniform speed of the fluid on the free streamlines separating from the edges of the orifice. Moreover, this is the speed of the fluid in the interior of the jet far downstream from the orifice where the streamlines are straight and parallel.

The Schwarz-Christoffel transformation,

$$\frac{dQ}{d\zeta} = C\left(\zeta + 1\right)^{-1/2}\left(\zeta - 1\right)^{-1/2}$$

on imposing the boundary conditions,

$$\left.\begin{aligned} \zeta = -1 &: \; Q = 0 \\ \zeta = 1 &: \; Q = -i\pi \end{aligned}\right\}$$

leads to

$$Q = \cosh^{-1}\zeta - i\pi.$$

This transformation maps the given flow onto the one in the ζ-plane shown in Figure 6.9. The complex potential for the latter is given by

$$\tilde{F}\left(\zeta\right) = -U\ell n\,\zeta.$$

The sink at $\zeta = 0$ has strength $2\pi U$. Note that, on the free streamline BC (see Figure 6.9), given by $d\Psi = 0$, one has

$$ds = \frac{1}{U}d\Phi = -\frac{d\zeta}{\zeta} = -\frac{d\left[\cosh\left(i\theta + i\pi\right)\right]}{\cosh\left(i\theta + i\pi\right)} = -\frac{d\left(\cos\theta\right)}{\cos\theta} = \tan\theta\; d\theta$$

from which

$$dx = \sin\theta\; d\theta,\;\; dy = \frac{\sin^2\theta}{\cos\theta}d\theta$$

with the boundary conditions,

$$\theta = 0 : x = 0, y = 0.$$

Thus, the free streamline BC is given by

$$x = 1 - \cos\theta,\;\; y = -\sin\theta + ln\left[\tan\left(\frac{\theta}{2} + \frac{\pi}{4}\right)\right].$$

The coefficient of contraction of the jet is then given by (Kirchhoff, 1869)

$$\frac{CC'}{BB'} = \frac{\pi}{\pi + 2}.$$

(ii) Hodograph Method

This method affords an alternate approach to find solutions to problems of free streamline flows. First, noting that

$$\left.\begin{aligned} F(z) &= \Phi + i\Psi \\ \frac{dF}{dz} &= qe^{-i\theta} \end{aligned}\right\} \tag{28}$$

one obtains

$$dz = \frac{e^{i\theta}}{q} dF = \frac{e^{i\theta}}{q} \left(\frac{\partial F}{\partial q} dq + \frac{\partial F}{\partial \theta} d\theta \right). \tag{29}$$

Since dz is a perfect differential, (29) implies

$$\frac{\partial}{\partial \theta} \left(\frac{e^{i\theta}}{q} \frac{\partial F}{\partial q} \right) = \frac{\partial}{\partial q} \left(\frac{e^{i\theta}}{q} \frac{\partial F}{\partial \theta} \right) \tag{30}$$

from which, one obtains

$$i \frac{\partial F}{\partial q} = -\frac{1}{q} \frac{\partial F}{\partial \theta}. \tag{31}$$

Equating the real and imaginary parts of (31), one obtains

$$\frac{\partial \Phi}{\partial q} = -\frac{1}{q} \frac{\partial \Psi}{\partial \theta}, \quad \frac{\partial \Psi}{\partial q} = \frac{1}{q} \frac{\partial \Phi}{\partial \theta} \tag{32}$$

from which, one derives

$$\frac{\partial^2 \Psi}{\partial q^2} + \frac{1}{q} \frac{\partial \Psi}{\partial q} + \frac{1}{q^2} \frac{\partial^2 \Psi}{\partial \theta^2} = 0. \tag{33}$$

If it is possible to specify the boundary conditions in terms of q and θ, equation (33) may be used to solve for $\Psi(q, \theta)$. Using (32), one then solves for $\partial F/\partial q$ and $\partial F/\partial \theta$. These are used to determine $z = z(q, \theta)$, and in turn to find $x = x(q, \theta)$ and $y = y(q, \theta)$.

Looking for a free-streamline flow type solution to the problem in which the flow is divided into two regions (the fluid is at rest in one and flows irrotationally in the other) can be cumbersome. Alternatively, we consider a somewhat simpler problem of two-dimensional jet flow impinging on a flat plate. The jet is supposed to originate as a uniform stream U of width $2K/U$ as $x \to -\infty$ and is separated from the stagnant fluid by the free streamlines $\Psi = \pm K$. The x-axis is taken to be an axis of symmetry so that $\Psi = 0$ is the dividing streamline composed of the negative half of the x-axis together with the y-axis. It is sufficient

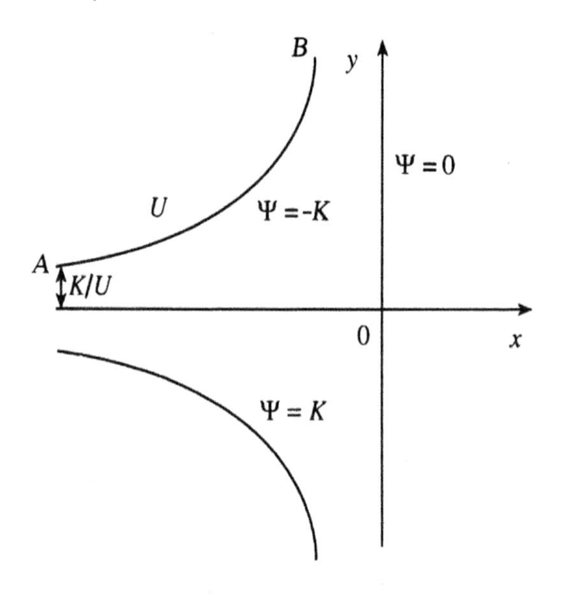

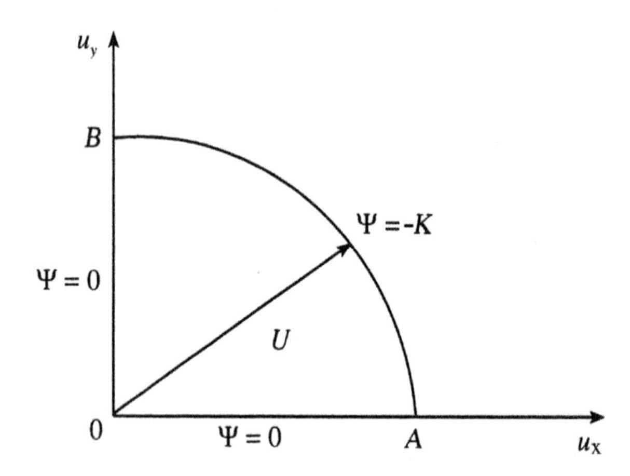

Figure 6.10 Flow enjoining on a vertical wall (from Rutherford, 1959).

to consider only the region $y \geq 0$. Referring to Figure 6.10, the points A and B represent points at infinity.

The hodograph of the free streamline AB where $\Psi = -K$ is a quadrant of the circle $q = U$, and those of the dividing streamlines OX and OY, where $\Psi = 0$

are the axes OA and OB. Noting that

$$\theta = 0, \pi/2 : \Psi = 0 \tag{34}$$

one obtains, from equation (33),

$$\Psi = \sum_s a_s q^{2s} \sin 2s\theta. \tag{35}$$

The boundary condition

$$\Psi(U, \theta) = -K, \;\; 0 \leq \theta \leq \frac{\pi}{2} \tag{36}$$

then gives

$$-K = \sum_s a_s U^{2s} \sin 2s\theta, \;\; 0 \leq \theta \leq \frac{\pi}{2} \tag{37}$$

from which,

$$-\frac{a_s U^{2s}}{K} = \frac{4}{\pi} \int_0^{\pi/2} \sin 2st dt = \frac{4}{\pi} \left[\frac{1 - (-1)^s}{2s} \right]. \tag{38}$$

Using (38), (35) becomes

$$\Psi = -\sum_{p=0} \frac{4K}{\pi (2p+1)} \left(\frac{q}{U} \right)^{2(2p+1)} \sin 2(2p+1)\theta. \tag{39}$$

Thus, the complex potential is given by

$$F = \frac{8K}{\pi} \sum_{p=0}^{\infty} \frac{1}{4p+2} \left(\frac{qe^{-i\theta}}{U} \right)^{4p+2} = \frac{8K}{\pi} \sum_{p=0}^{\infty} \frac{Z^{4p+2}}{4p+2}, \tag{40}$$

where,

$$Z \equiv \frac{qe^{-i\theta}}{U}.$$

Noting, from (28) and (40), that

$$Udz = \frac{dF}{Z} = \frac{8K}{\pi} \sum_{p=0}^{\infty} Z^{4p} dZ = \frac{8K}{\pi} \frac{dZ}{1 - Z^4}, \tag{41}$$

one obtains, on integration,

$$z = \frac{2K}{\pi U} \left[\ell n \left(\frac{1+Z}{1-Z} \right) + i \, \ell n \left(\frac{1 - iZ}{1 + iZ} \right) \right]. \tag{42}$$

On the bounding streamline $\Psi = -K$, one has $q = U$ or $Z = e^{-i\theta}$.

Noting then, that

$$\left.\begin{aligned} \ell n\left(\frac{1+Z}{1-Z}\right) &= -\ell n\left(i \tan\frac{\theta}{2}\right) = -\frac{i\pi}{2} - \ell n \tan\frac{\theta}{2} \\ \ell n\left(\frac{1-iZ}{1+iZ}\right) &= -\frac{i\pi}{2} - \ell n\left(\frac{1+\tan\theta/2}{1-\tan\theta/2}\right) \end{aligned}\right\}$$

one obtains finally,

$$z = x + iy = \frac{2K}{\pi U}\left[-\frac{i\pi}{2} - \ell n t + \frac{\pi}{2} + i\,\ell n\frac{1-t}{1+t}\right] \tag{43}$$

where $t \equiv \tan\theta/2$. Thus, one has, for the free streamlines,

$$x = \frac{-2K}{\pi U}\left(\ell n t - \frac{\pi}{2}\right),\ y = \frac{-2K}{\pi U}\left(\frac{\pi}{2} - \ell n\frac{1-t}{1+t}\right). \tag{44}$$

Exercises

1. Find the complex potential for the flow past a wedge of angle 2α.
2. Find the complex potential due to a doublet-flow outside a circular cylinder.
3. Find the complex potential for the orifice flow shown in Figure 6.11a. Determine the coefficient of contraction. Use the Schwarz-Christoffel transformation.
4. Find the complex potential for the flow past a flat plate with a cavity of constant ambient pressure, shown in Figure 6.11b. Calculate the drag on the plate. Use the Schwarz-Christoffel transformation.

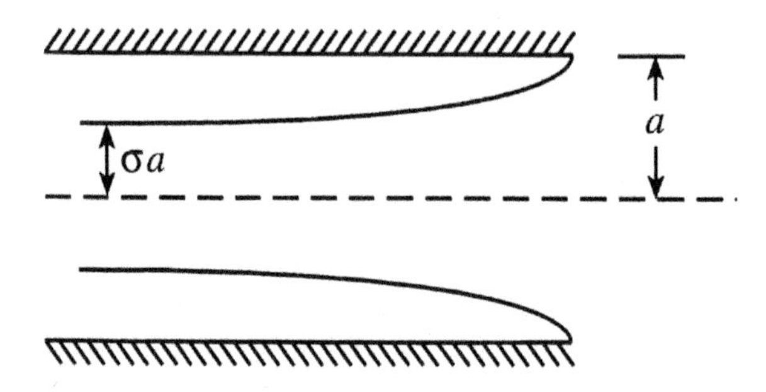

Figure 6.11a Borda Orifice flow.

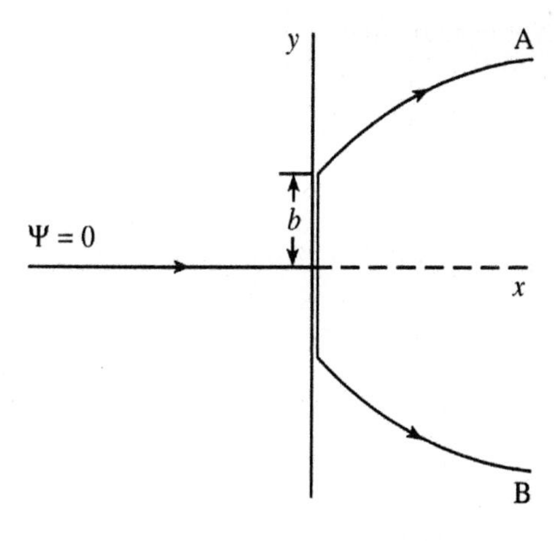

Figure 6.11b Flow past a flat plate placed perpendicular to the flow.

7

Three-Dimensional Irrotational Flows

The region occupied by the fluid is necessarily *doubly-connected* in a two-dimensional flow field and *simply-connected* in a three-dimensional flow field. Therefore, differences between the properties of the two flow fields can be expected to arise.

7.1 Special Singular Solutions

As in the case of two-dimensional flows, one may build up complicated three-dimensional flow fields by superposing certain singular solutions of the Laplace equation.

(i) The Source Flow

Using the spherical polar coordinates (R, θ, φ), for the velocity potential of the source flow, one has

$$\Phi(R) = \frac{A}{R}. \tag{1}$$

Using the condition of mass conservation,

$$q = \oiint_S \nabla\Phi \cdot \hat{\mathbf{n}} dS = const \tag{2}$$

q being the source strength, (1) gives

$$A = -\frac{q}{4\pi}. \tag{3}$$

Introduction to Theoretical and Mathematical Fluid Dynamics, Third Edition.
Bhimsen K. Shivamoggi.

Thus, (1) becomes

$$\Phi = -\frac{q}{4\pi R}. \tag{4}$$

From the relations,

$$u_r = \frac{\partial \Phi}{\partial R} = \frac{1}{R^2 \sin\theta}\frac{\partial \Psi}{\partial \theta},\ u_\theta = \frac{1}{R}\frac{\partial \Phi}{\partial \theta} = \frac{-1}{R\sin\theta}\frac{\partial \Psi}{\partial R} \tag{5}$$

one obtains, for the *axisymmetric stream function*,

$$\Psi = -\frac{q\cos\theta}{4\pi}. \tag{6}$$

Example 1: Consider a source in a uniform flow. On taking the polar axis along the flow direction, one obtains for the stream function describing this flow,

$$\Psi = \frac{1}{2}UR^2\sin^2\theta - \frac{q\cos\theta}{4\pi}.$$

The stagnation streamline, on noting that the stagnation point would be on the line $\theta = \pi$, is then given by,

$$\Psi_0 = \frac{1}{2}UR^2\sin^2\theta - \frac{q\cos\theta}{4\pi} - \frac{q}{4\pi}$$

which represents the flow past a body of revolution of asymptotic radius (see Figure 7.1),

$$\theta \Rightarrow 0 : R\sin\theta = \sqrt{\frac{q}{\pi U}}.$$

The velocity potential for this flow is

$$\Phi = UR\cos\theta - \frac{q}{4\pi R}$$

from which, the velocity components are given by

$$\left.\begin{aligned} u_R &= \frac{\partial \Phi}{\partial R} = U\cos\theta + \frac{q}{4\pi R^2} \\ u_\theta &= \frac{1}{R}\frac{\partial \Phi}{\partial \theta} = -U\sin\theta \end{aligned}\right\}.$$

The stagnation point S is therefore at

$$\theta = \pi,\ R = \sqrt{\frac{q}{4\pi U}}.$$

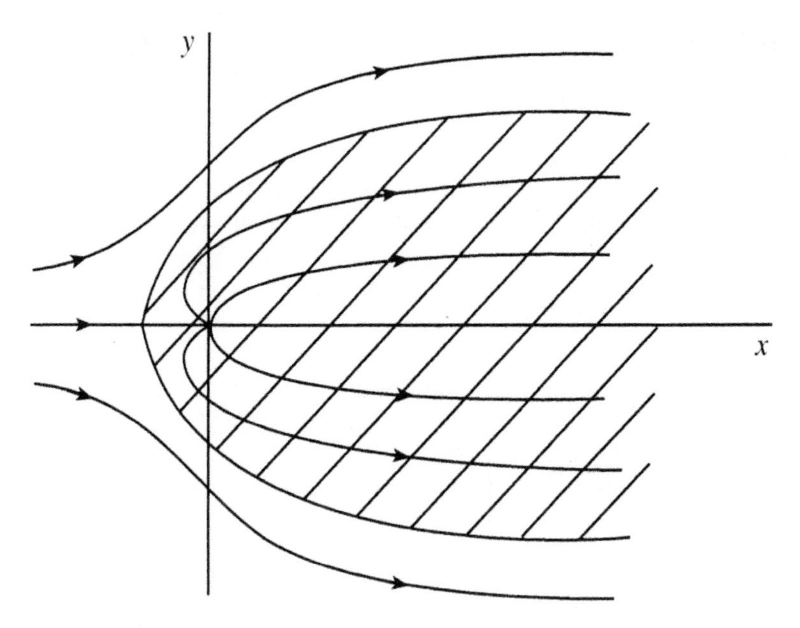

Figure 7.1 Flow past a body of revolution.

(ii) The Doublet Flow

Note that one can generate the doublet flow from the source flow by simply differentiating the velocity potential corresponding to the source flow in a direction opposite to the doublet axis along the x-axis. Thus, one obtains

$$\Phi = -\frac{d}{dx}\left(-\frac{q/4\pi}{\sqrt{x^2+r^2}}\right)dx = -\frac{\mu\cos\theta}{4\pi R^2} \tag{7}$$

where, let us choose q and dx such that we have the following limit.

$$\mu \equiv \lim_{\substack{dx\to 0\\ q\to\infty}} q\,dx. \tag{8}$$

Using the relations,

$$\left.\begin{aligned} u_R &= \frac{\partial\Phi}{\partial R} = \frac{1}{R^2\sin\theta}\frac{\partial\Psi}{\partial\theta} \\ u_\theta &= \frac{1}{R}\frac{\partial\Phi}{\partial\theta} = \frac{-1}{R\sin\theta}\frac{\partial\Psi}{\partial R} \end{aligned}\right\}. \tag{9}$$

one obtains, for the axisymmetric stream function,

$$\Psi = \frac{\mu \sin^2 \theta}{4\pi R}. \tag{10}$$

Example 2: Consider a doublet in a uniform stream, with the doublet axis opposite to the stream direction. One then has, for the velocity potential describing the flow,

$$\phi = UR \cos\theta + \frac{\mu \cos\theta}{4\pi R^2}$$

from which, the velocity components are given by

$$\left.\begin{aligned} u_R &= \left(U - \frac{\mu}{2\pi R^3}\right)\cos\theta \\ u_\theta &= -\left(U + \frac{\mu}{4\pi R^3}\right)\sin\theta \end{aligned}\right\}.$$

Then, the axisymmetric stream function for this flow is,

$$\Psi = \left(\frac{UR^2}{2} - \frac{\mu}{4\pi R}\right)\sin^2\theta$$

from which, we find that the stagnation streamline is

$$R = R_0 \equiv \left(\frac{\mu}{2\pi U}\right)^{1/3}.$$

This therefore represents the flow past a sphere of radius $a = R_0$. Thus, the velocity potential for this flow is given by

$$\Phi = U\left(1 + \frac{a^3}{2R^3}\right)R \cos\theta$$

with the velocity components,

$$\left.\begin{aligned} u_R &= U\left(1 - \frac{a^3}{R^3}\right)\cos\theta \\ u_\theta &= -U\left(1 + \frac{a^3}{2R^3}\right)\sin\theta \end{aligned}\right\}.$$

The pressure on the surface of the sphere is then given by

$$R = a : C_p \equiv \frac{p - p_\infty}{\frac{1}{2}\rho U^2} = 1 - \frac{u_\theta^2}{U^2} = 1 - \frac{9}{4}\sin^2\theta$$

which is symmetric about $\theta = 0$, and $\pi/2$, so that there is no net force exerted by the flow on the sphere - *d'Alembert's paradox.*

It is of interest to compare the above result with the corresponding result for the case of flow past an infinite cylinder (equation (70) in Section 5.9), namely,

$$r = a : C_p = 1 - 4\sin^2\theta$$

(see Figure 7.2). The shallower *pressure well* in the former case is due to the three-dimensional *flow-spreading* effect (see also Section 18.2).

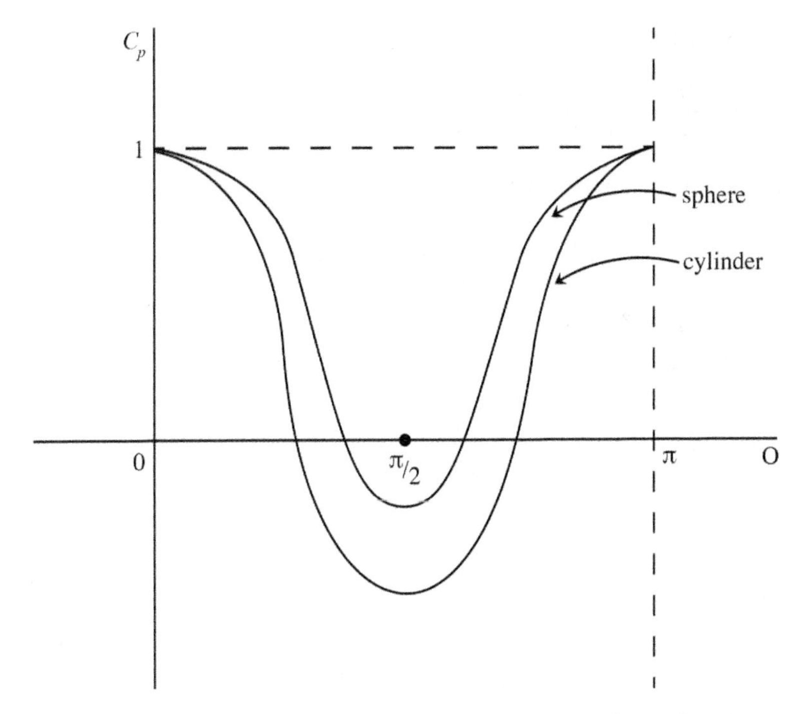

Figure 7.2 Comparison of the pressure coefficient on the surfaces of a sphere and a cylinder.

7.2 d'Alembert's Paradox

Example 2 seems to imply that the zero resultant force experienced by a sphere when it moves with uniform translational velocity in a quiescent unbounded ideal fluid depends on the particular symmetric flow geometry. However, this result has greater generality as seen in the following, The resultant force exerted by the fluid on a finite body of arbitrary shape translating as in the case of the sphere above vanishes if no vorticity is generated in the fluid during this process.

Consider a finite body bounded by a surface S translating with constant velocity $\mathbf{U}$ in an unbounded fluid which occupies a region $\mathcal{V}$ and is at rest at infinity. The boundary-value problem for the velocity potential Φ, then consists of

$$\nabla^2\Phi = 0 \; in \; \mathcal{V} \tag{11}$$

along with boundary conditions,

$$\left.\begin{aligned} &\frac{\partial \Phi}{\partial n} = \mathbf{U}\cdot\hat{\mathbf{n}} \; on \; S \\ &|\nabla\Phi| \Rightarrow 0 \text{ at infinity} \end{aligned}\right\}, \tag{12}$$

$\hat{\mathbf{n}}$ being the outward drawn unit normal vector to S.

The force exerted by the fluid on the body is given by

$$\mathbf{F} = -\iint_S \hat{\mathbf{n}}\, p\, dS. \tag{13}$$

Using the Bernoulli integral (18) in Section 5.3, (13) becomes

$$F_j = -p_\infty \iint_S n_j\, dS + \rho \iint_S n_j \frac{\partial \Phi}{\partial t} dS + \frac{1}{2}\rho \iint_S n_j (\nabla\Phi)^2 dS \tag{14}$$

p_∞ being the constant pressure at infinity.

Now, one may note the following results,

$$\iint_S n_j dS = 0$$

$$\text{on } S: \; \frac{\partial \Phi}{\partial t} = \frac{\partial \Phi}{\partial x_i}\frac{dx_i}{dt} = -\nabla\Phi\cdot\mathbf{U} = -U_i u_i$$

$$\frac{\partial}{\partial x_j}(\nabla\Phi)^2 = 2u_i \frac{\partial u_i}{\partial x_j} = 2u_i \frac{\partial u_j}{\partial x_i} = 2\frac{\partial}{\partial x_i}(u_i u_j)$$

$$\iint_S n_j(\nabla\Phi)^2 = -\iiint_{\mathcal{V}} \frac{\partial}{\partial x_j}(\nabla\Phi)^2 dv + \iint_\Sigma n_j(\nabla\Phi)^2 d\Sigma$$

$$= 2\iint_S n_i u_i u_j dS + 2\iint_\Sigma n_i u_i u_j d\Sigma + \iint_\Sigma n_j (\nabla\Phi)^2 d\Sigma$$

$$= 2U_i \iint_S n_i u_j dS \tag{15}$$

where Σ is the surface at infinity enclosing the region $\mathcal{V}$, and the boundary conditions (12) have been used in deriving the last result.

Using (15), and the boundary conditions (12) again, (14) becomes,

$$F_j = \rho U_i \iint_S (n_i u_j - n_j u_i)\, dS$$

$$= -\rho U_i \iiint_{\mathcal{V}} \left(\frac{\partial u_j}{\partial x_i} - \frac{\partial u_i}{\partial x_j}\right) dv + \rho U_i \iint_\Sigma (n_i u_j - n_j u_i)\, d\Sigma \tag{16}$$

$$= 0$$

Thus, a finite body of arbitrary shape translating steadily through a quiescent unbounded ideal fluid experiences no resultant force exerted by the fluid. This result is called *d'Alembert's paradox*, because it is contrary to experience. It is to be noted that the fallacy has little to do with the neglect of viscous effects. On the contrary, it has more to do with the assumption that there is no vorticity in the fluid outside the body. Thus, the above result is applicable to a streamlined body for which there is only a thin wake containing vorticity.

7.3 Image of a Source in a Sphere

Let us find the image of a source of strength $4\pi m$ at $(0, 0, a)$ (in cylindrical polar coordinates) in the sphere of radius $b(b < a)$ centered at the origin. In the absence of the sphere, the potential for the source is (see Figure 7.3) $-m/\sqrt{r^2 + a^2 - 2ar\cos\theta}$. In the presence of the sphere, one may then write for the flow

$$\Phi = \Phi_1 - \frac{m}{\sqrt{r^2 + a^2 - 2ar\cos\theta}} \tag{17}$$

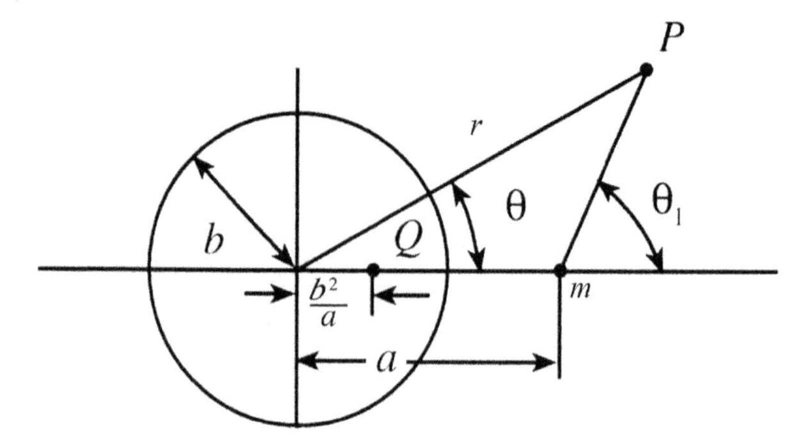

Figure 7.3 Source outside a sphere.

where

$$\left.\begin{aligned} r > b &: \nabla^2\Phi_1 = 0 \\ r \Rightarrow \infty &: \nabla\Phi_1 \Rightarrow 0 \end{aligned}\right\}. \tag{18}$$

One may write

$$\Phi_1 = \sum_{n=0}^{\infty} \frac{B_n}{r^{n+1}} P_n(\cos\theta) \tag{19}$$

where $P_n(z)$ are the *Legendre polynomials*. Thus, one has, from (17) and (19), for $r < a$,

$$\Phi = \sum_{n=0}^{\infty} \left[\frac{B_n}{r^{n+1}} P_n(\cos\theta) - \frac{mr^n}{a^{n+1}} P_n(\cos\theta) \right]. \tag{20}$$

The boundary condition on the sphere given by,

$$r = b : \frac{\partial\Phi}{\partial r} = 0 \tag{21}$$

then, gives

$$B_n = -\frac{mn}{b(n+1)} \left(\frac{b^2}{a} \right)^{n+1}. \tag{22}$$

Using (22), (19) becomes

$$\Phi = -\sum_{n=0}^{\infty} \frac{m}{b}\left(1 - \frac{1}{n+1}\right)\left(\frac{b^2}{a}\right)^{n+1} \frac{1}{r^{n+1}} P_n(\cos\,\theta). \tag{23}$$

Note that the first term on the right in (23) can be written as

$$-\sum_{n=0}^{\infty} \frac{m}{b}\left(\frac{b^2}{a}\right)^{n+1} \frac{1}{r^{n+1}} P_n(\cos\theta) = -\frac{mb}{a}\sum_{n=0}^{\infty}\left(\frac{b^2}{a}\right)^{n} \frac{1}{r^{n+1}} P_n(\cos\theta)$$

$$= -\frac{(mb/a)}{\sqrt{r^2 + (b^2/a)^2 - 2r(b^2/a)\cos\theta}},$$

$$\left(\frac{b^2}{a}\right) < r$$

which corresponds to a source of strength $4\pi mb/a$ located at $(0, 0, b^2/a)$. Next, the second term on the right in (23) can be written as

$$-\sum_{n=0}^{\infty} \frac{m}{b(n+1)}\left(\frac{b^2}{a}\right)^{n+1} \frac{1}{r^{n+1}} P_n(\cos\theta) = \frac{m}{b}\sum_{n=0}^{\infty}\int_0^{b^2/a} \frac{z^n P_n(\cos\theta)dz}{r^{n+1}}$$

$$= \frac{m}{b}\int_0^{b^2/a} \frac{dz}{\sqrt{r^2 + z^2 - 2rz\cos\theta}},$$

$$z < r.$$

which corresponds to a uniform *line sink* of strength $4\pi mb/a$ per unit length stretching from $(0, 0, 0)$ to $(0, 0, b^2/a)$. The image system therefore consists of a source $4\pi mb/a$ at $(0, 0, b^2/a)$ and this line sink.

7.4 Flow Past an Arbitrary Body

Let us use *Green's theorem* for the region $\mathcal{V}$ between the body and an infinite sphere enclosing the body. It turns out that the nonvanishing contribution to the velocity potential from the surface integral on the infinite sphere is a constant. Ignoring this constant contribution, we then have,

$$-\oiint_{S_0} (\Phi_1 \nabla'\Phi_2 - \Phi_2 \nabla'\Phi_1)\cdot \hat{\mathbf{n}} dS' = \iiint_{\mathcal{V}} (\Phi_1 \nabla'^2\Phi_2 - \Phi_2 \nabla'^2\Phi_1) d\mathbf{x}' \tag{24}$$

where S_0 is the surface of the body, and $\hat{\mathbf{n}}$ points into the fluid.

Taking

$$\Phi_1 = \frac{1}{R}, \ \Phi_2 = \Phi, \ \nabla^2\Phi = 0 \tag{25}$$

where $R = |\mathbf{x} - \mathbf{x}'|$ is the distance between the field point $P(\mathbf{x})$ and an element of area $dS(\mathbf{x}')$ on the body, (24) becomes

$$-\oiint\limits_{S_0} \left[\frac{1}{R}\nabla'\Phi - \Phi\nabla'\left(\frac{1}{R}\right)\right]\cdot\hat{\mathbf{n}}dS = \iiint\limits_{\mathcal{V}} \left[\frac{1}{R}\nabla'^2\Phi - \Phi\nabla'^2\left(\frac{1}{R}\right)\right]d\mathbf{x}'. \tag{26}$$

If the field point $P(\mathbf{x})$ is not in $\mathcal{V}$, then (26) gives

$$-\oiint\limits_{S_0} \left[\frac{1}{R}\nabla'\Phi - \Phi\nabla'\left(\frac{1}{R}\right)\right]\cdot\hat{\mathbf{n}}dS = 0. \tag{27}$$

If the field point $P(\mathbf{x})$ is within $\mathcal{V}$, enclose it by a small sphere S_ε of radius ε, then (26) gives

$$\oiint\limits_{S_\varepsilon} \left(\frac{1}{R}\frac{\partial\Phi}{\partial R} + \frac{\Phi}{R^2}\right)dS + \oiint\limits_{S_0} \left[\frac{1}{R}\nabla'\Phi - \Phi\nabla'\left(\frac{1}{R}\right)\right]\cdot\hat{\mathbf{n}}dS = 0$$

As $\varepsilon \Rightarrow 0$, this gives

$$\Phi(P) = -\frac{1}{4\pi}\oiint\limits_{S_0} \left[\frac{1}{R}\nabla'\Phi + \Phi\nabla\left(\frac{1}{R}\right)\right]\cdot\hat{\mathbf{n}}dS \tag{28}$$

which shows explicitly how Φ is determined throughout the fluid by the conditions on the body.

One may write the *Taylor expansion*,

$$\frac{1}{R} = \frac{1}{r} - x_i'\frac{\partial}{\partial x_i}\left(\frac{1}{r}\right) + \frac{1}{2}x_i'x_j'\frac{\partial^2}{\partial x_i \partial x_j}\left(\frac{1}{r}\right) + \cdots \tag{29}$$

which is convergent for $r' < r$; here $r = |\mathbf{x}|, r' = |\mathbf{x}'|$.

Using (29), (28) gives the *multipole expansion*,

$$\Phi(P) = \frac{c}{r} + c_i\frac{\partial}{\partial x_i}\left(\frac{1}{r}\right) + c_{ij}\frac{\partial^2}{\partial x_i \partial x_j}\left(\frac{1}{r}\right) + \cdots \tag{30}$$

where

$$c \equiv -\frac{1}{4\pi}\oiint_{S_0} \hat{\mathbf{n}} \cdot \nabla'\Phi dS$$

$$c_i \equiv \frac{1}{4\pi}\oiint_{S_0} (x_i'\hat{\mathbf{n}} \cdot \nabla'\Phi - \hat{n}_i\Phi)\, dS$$

$$c_{ij} \equiv \frac{1}{4\pi}\oiint_{S_0} \left(-\frac{1}{2}x_i'x_j'\hat{\mathbf{n}} \cdot \nabla'\Phi + x_i'\hat{n}_j\Phi\right) dS$$

etc.

Thus, the flow field at large distances from a moving body can be expressed as a sum of *multipole flow-fields* where the coefficients are certain integrals on the given body. At such large distances the dominant action of the moving body on the fluid is equivalent to the action of a point source. The detailed distribution of the surface sources on the moving body are of lesser importance.

7.5 Unsteady Flows

Consider an infinite mass of fluid of density ρ, initially at rest, containing an expanding spherical cavity of radius $\hat{R}_0$ produced by an underwater explosion. Pressure applied uniformly over the surface of the cavity forces the fluid to move outwards. There is no pressure at infinity and no body forces act.[1] Since the fluid motion is spherically symmetric, Laplace equation for the velocity potential takes the form

$$\frac{\partial}{\partial R}\left(R^2\frac{\partial\Phi}{\partial R}\right) = 0 \tag{31}$$

1 The cavity dynamics displays different scenarios depending on whether the cavity consists of a permanent noncondensable gas or the vapor of the surrounding fluid. In the former case, the inertia of the surrounding fluid plays a dominant role, while in the latter case, the latent heat flow associated with boiling phenomena plays a dominant role.

while the kinematic condition at the surface of the cavity is

$$R = \hat{R}(t) : \frac{\partial \Phi}{\partial R} = \frac{d\hat{R}}{dt}. \tag{32}$$

This condition implies that the radius of the cavity increases at a rate equal to the local fluid velocity so that no fluid then ever crosses the cavity surface, which always consists of the same fluid particles.

One has, from (31) and (32),

$$\Phi (R, t) = -\frac{\hat{R}^2(t)}{R}\hat{R}' \tag{33}$$

where the primes denote differentiation with respect to t.

Using (33), the Bernoulli integral (18) in Section 5.3, then gives

$$\frac{1}{\rho}p + \frac{1}{2}\left(\frac{\hat{R}^2\hat{R}'}{R^2}\right)^2 - \frac{\hat{R}^2\hat{R}'' + 2\hat{R}\hat{R}'^2}{R} = 0. \tag{34}$$

Using the *adiabatic equation of state* (see Chapter 12) for the expanding gas in the cavity,

$$\frac{p}{p_0} = \left(\frac{\hat{R}_0^3}{\hat{R}^3}\right)^\gamma \tag{35}$$

where γ is the ratio of specific heats of the gas in the cavity, and p_0 is the initial pressure when $\hat{R} = \hat{R}_0$, one obtains from (34), for $R = \hat{R}$,

$$\hat{R}\hat{R}'' + \frac{3}{2}\hat{R}'^2 = \frac{p_0}{\rho}\left(\frac{\hat{R}_0}{\hat{R}}\right)^{3\gamma}. \tag{36}$$

Upon integrating (36), and using

$$\hat{R} = \hat{R}_0 : \hat{R}' = 0 \tag{37}$$

one obtains

$$\hat{R}'^2 = \frac{2a_0^2}{3(\gamma - 1)}\left[\left(\frac{\hat{R}_0}{\hat{R}}\right)^3 - \left(\frac{\hat{R}_0}{\hat{R}}\right)^{3\gamma}\right] \tag{38}$$

where

$$a_0 \equiv \sqrt{\frac{p_0}{\rho}}.$$

Equation (38) shows that the cavity expansion process is *monotonic* (see Figure 7.4).

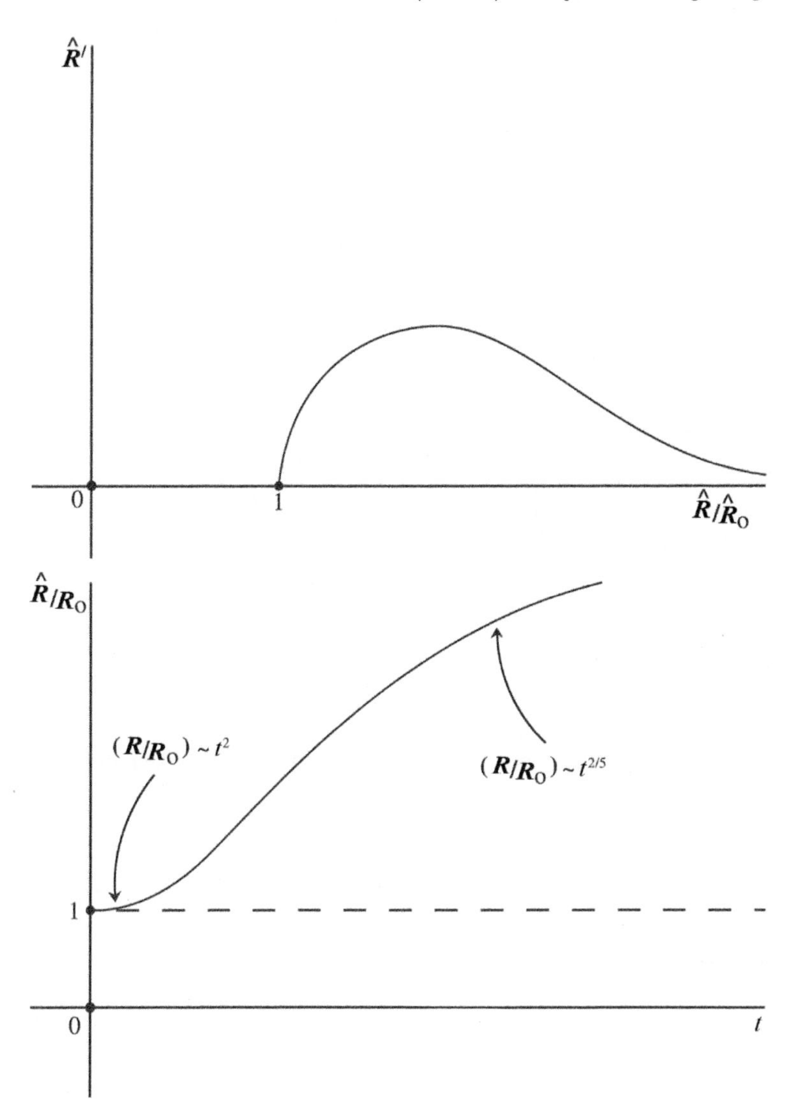

Figure 7.4 Expansion of a spherical gas cavity in a fluid.

7.6 Renormalized (or Added) Mass of Bodies Moving through a Fluid

The reaction of the fluid to a body translating through it is to change the inertia of the body and hence *renormalize* its mass. Consider the example of a sphere of radius a translating unsteadily through a fluid with velocity $\mathbf{U} = -U\hat{\imath}_x$.

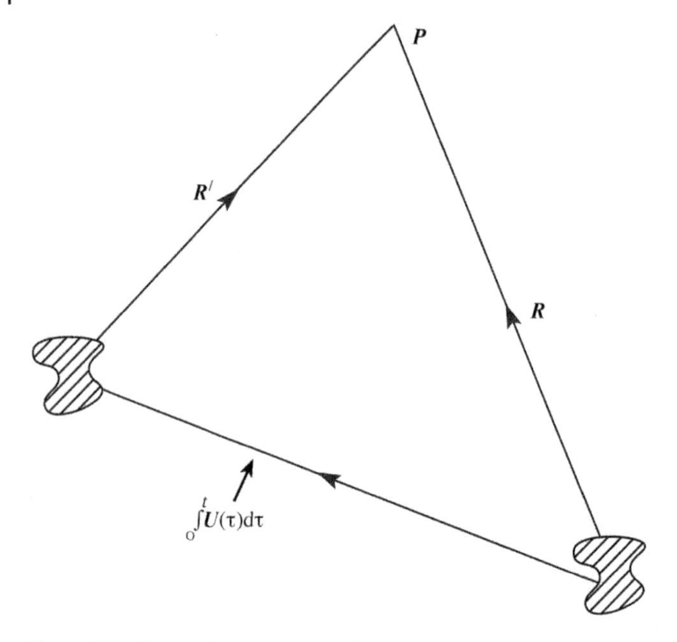

Figure 7.5 Space-fixed and body-fixed reference frames.

One then has, for the velocity potential for this flow in the body-fixed reference frame (R', θ', t')[2]

$$\Phi = \frac{U(t)a^3}{2}\frac{\cos\theta'}{R'^2}. \tag{39}$$

2 Since the body-fixed reference frame (R', θ', t') is translating with a velocity $\mathbf{U}(t)$ with respect to the space-fixed reference frame (R, θ, t), assuming that the two frames coincide at $t = 0$, we have (see Figure 7.5)

$$\mathbf{R}' = \mathbf{R} - \int_0^t \mathbf{U}(\tau)d\tau,\ t' = t$$

from which,

$$\nabla = \nabla',\ \frac{\partial}{\partial t} = \frac{\partial}{\partial t'} - \mathbf{U}(t)\cdot\nabla'.$$

The kinetic energy of the fluid set into motion by this translating sphere is given by

$$T = \frac{1}{2}\rho \iiint_{\mathcal{V}} (\nabla\Phi)^2 d\mathbf{x} = -\frac{1}{2}\rho \oiint_S \Phi \frac{\partial \Phi}{\partial n} dS \tag{40}$$

where S denotes the surface of the sphere, and the normal vector $\hat{\mathbf{n}}$ points into the fluid.

Using (39), (40) leads to

$$T = -\frac{\rho}{2} \oiint_S \frac{Ua\cos\theta}{2}(-U\cos\theta)\, dS = \frac{1}{2}\left(\frac{2\pi a^3 \rho}{3}\right) U^2. \tag{41}$$

It is obvious from (41) that one may view the kinetic energy of the fluid as an *"added" mass*,[3]

$$m_a = \frac{2\pi \rho a^3}{3} \tag{42}$$

for the body moving through the fluid. Note that this *added mass* is half of the mass of the fluid displaced by the sphere.

In order to see the dynamical significance of the *added mass* further, let us calculate the force exerted by the fluid on the sphere.

Using (39), the Bernoulli integral (18) in Section 5.3 gives for the pressure at a point on the sphere,

$$\begin{gathered} p\Big|_{R'=a} = p_\infty - \rho \left[\frac{\partial \Phi}{\partial t'} - \mathbf{U}(t') \cdot \nabla'\Phi + \frac{1}{2}(\nabla'\Phi)^2\right]\Big|_{R'=a} \\ \text{or} \\ p\Big|_{R'=a} = p_\infty + \frac{1}{2}\rho U^2 \left(1 - \frac{9}{4}\sin^2\theta'\right) - \frac{1}{2}\rho a \dot{U}\cos\theta' \end{gathered} \tag{43}$$

p_∞ being the constant pressure at infinity.

---

3 The concept of *"added"* mass becomes viable for an incompressible fluid because the disturbances propagate in the fluid with infinite speed. So, the effects of accelerating the fluid around a moving body are determined completely by the acceleration of the body at that instant. However, in a compressible fluid, the disturbances propagate with a *finite* speed. So, the entire history of the motion of the body is required to determine the acceleration effect at any one instant. So, the concept of *added* mass of a moving body becomes cumbersome to implement in the case of a compressible fluid (Tsien, 1958).

The force exerted by the fluid on the sphere is, then, given by

$$\mathbf{F} = \oiint_S \hat{\mathbf{n}} p \Big|_{R'=a} dS = -\hat{\mathbf{i}}_x \oiint_S p \Big|_{R'=a} \cos\theta' dS - \hat{\mathbf{i}}_y \oiint_S p \Big|_{R'=a} \sin\theta' dS \tag{44}$$

and on using (43),

$$F = \left(\frac{2\pi\rho a^3}{3}\right)\dot{U}\hat{\mathbf{i}}_x \tag{45}$$

which is simply the inertia force of the accelerating fluid matter of mass equal to the *added mass* given in (42), and is in the direction opposing the motion of the sphere.

Exercises

1. Find the velocity potential for a doublet-flow outside a sphere when the axis of the doublet passes through the center of the sphere.
2. A spherical gas bubble of radius $\hat{R}_0$ and pressure p_0 starts to expand in an infinite mass of fluid of density ρ with zero pressure at infinity. Suppose that the gas is initially at rest and that its pressure and volume $\mathcal{V}$ are related by the equation of state $p\mathcal{V}^{4/3} = const.$ Show that the bubble radius becomes double the original value $\hat{R}_0$ in time $\frac{28\hat{R}_0}{15}\sqrt{2\rho_0/p_o}$. Show also that (see Figure 7.4)

$$\frac{\hat{R}(t)}{\hat{R}_0} \sim \begin{cases} t^2, & \text{for small } t \\ t^{2/5}, & \text{for large } t. \end{cases}$$

3. For the special case $\gamma = 4/3$ (Lamb, 1945), show that equation (38) admits an *exact* solution. (Hint: Set $\hat{R}/\hat{R}_0 = 1 + z$ and solve for z.)
4. Calculate the added mass of a plate (length 4ℓ) moving normal to its plane in a fluid. Use the *Joukowski transformation* (see Section 11.1).
5. Calculate the added mass of a sphere moving through a fluid inside a bigger hollow concentric sphere.
6. Determine the flow past a nearly spherical surface given by

$$R = a\left(1 - \varepsilon \sin^2\theta\right), \quad \varepsilon \ll 1,$$

the flow at infinity being uniform with a velocity U in the x-direction.

8

Vortex Flows

The vorticity vector is the curl of fluid velocity (Section 2.6) and is twice the effective local angular velocity of the fluid (Section 1.4). Vorticity plays a central role in understanding fluid flow phenomena, in particular, at high Reynolds numbers (Chapter 21). A case in point is the vortex stretching process underlying the energy transport phenomena in turbulent flows.

Moreover, vorticity provides a vital tool while discussing the analytic structure underlying the theoretical formulations of fluid dynamics. It turns out that the vorticity vector is the appropriate *canonical* variable in the Hamiltonian formulations of fluid flows (Section 3.2). These formulations afford interesting alternative geometric perspectives (Section 3.2) on the Euler equations of fluid flows (Section 5.2).

8.1 Vortex Tubes

The Euler equations of motions for an incompressible fluid are

$$\nabla \cdot \mathbf{v} = 0 \tag{1}$$

$$\frac{\partial \mathbf{v}}{\partial t} + (\mathbf{v} \cdot \nabla)\,\mathbf{v} = -\frac{1}{\rho}\nabla p. \tag{2}$$

The vorticity is given by

$$\boldsymbol{\Omega} \equiv \nabla \times \mathbf{v}. \tag{3}$$

The *vortex lines* are those which are tangential everywhere to the local vorticity vector. The surface in the fluid formed by all the vortex lines passing through a given reducible closed curve C drawn in the fluid is a *vortex tube*. Consider the integral of vorticity over an open surface S bounded by C lying entirely in the fluid,

Introduction to Theoretical and Mathematical Fluid Dynamics, Third Edition.
Bhimsen K. Shivamoggi.

$$\iint_S \boldsymbol{\Omega} \cdot \hat{\mathbf{n}} dS = \oint_C \mathbf{v} \cdot d\mathbf{s} = \Gamma$$

where $\hat{\mathbf{n}}dS$ is a surface area element, and Γ is the flow circulation around C. We obtain from (3),

$$\nabla \cdot \boldsymbol{\Omega} = 0 \tag{4}$$

from which, it follows that the above integral has the same value for any open surface S in the fluid bounded by the closed curve C lying on the vortex tube and passing around the tube once. The strength of a vortex tube is the flux of vorticity through its cross section or the circulation Γ about a closed loop C lying on its surface. Thus, the vortex tubes have constant strength and are either closed tubes or do not end in the interior of the fluid.

Equation (2) gives for the vorticity,

$$\frac{\partial \boldsymbol{\Omega}}{\partial t} = \nabla \times (\mathbf{v} \times \boldsymbol{\Omega}) \tag{5a}$$

which, on using equation (1), becomes

$$\frac{\partial \boldsymbol{\Omega}}{\partial t} + (\mathbf{v} \cdot \nabla)\boldsymbol{\Omega} = (\boldsymbol{\Omega} \cdot \nabla)\mathbf{v}. \tag{5b}$$

Now, if $\mathbf{d}\boldsymbol{\ell}$ is an infinitesimal material line element, one has the following *kinematical equation* governing the stretching and reorientation of this element,

$$\left[\frac{\partial}{\partial t} + (\mathbf{v} \cdot \nabla)\right]\mathbf{d}\boldsymbol{\ell} = (\mathbf{d}\boldsymbol{\ell} \cdot \nabla)\mathbf{v}. \tag{6}$$

Comparison of equation (5b) with equation (6) shows that $\boldsymbol{\Omega}$ evolves like the vector representing a material line element which coincides instantaneously with an element of the local vortex line. Consequently, vortex lines move with the fluid. This implies that a vortex tube consists always of the same fluid particles, so that its volume is conserved. Any stretching of the vortex tube would therefore intensify the local vorticity. The term on the right-hand side in equation (5b) indeed represents the vortex-line stretching. This leads to a concentration and intensification of vorticity, no matter how dispersed the vorticity may be initially. An example is the bath-plug vortex wherein the extension due to the draining motion of the vortex lines produces a concentration of the vorticity.

For a two-dimensional flow, equation (5a) gives

$$\frac{\partial \Omega}{\partial t} + (\mathbf{v} \cdot \nabla)\Omega = 0 \tag{7}$$

where Ω is the component of vorticity normal to the plane of the flow. Equation (7) leads to the *Lagrange invariant*,

$$\Omega = const. \tag{8}$$

8.2 Induced Velocity Field

Note that equation (1), which indicates that the velocity field is solenoidal implies the existence of a *vector potential* **A**,

$$\mathbf{v} = \nabla \times \mathbf{A}. \tag{9}$$

Then, the vorticity is given by

$$\boldsymbol{\Omega} = \nabla \times \mathbf{v} = \nabla \times (\nabla \times \mathbf{A}) = \nabla (\nabla \cdot \mathbf{A}) - \nabla^2 \mathbf{A}. \tag{10}$$

Now, note that the vector potential **A** is not uniquely determined by (9) when the velocity **v** is known. In fact,

$$\mathbf{A}' \equiv \mathbf{A} + \nabla \chi \tag{11}$$

represents the same velocity field. But, it is always possible to choose the *gauge* χ such that a *gauge condition*,

$$\nabla \cdot \mathbf{A} = \mathbf{0} \tag{12}$$

is in place.

Consequently, (10) gives

$$\nabla^2 \mathbf{A} = -\boldsymbol{\Omega} \tag{13}$$

from which, one has

$$\mathbf{A}(\mathbf{r}, t) = \frac{1}{4\pi} \iiint_{\mathscr{V}} \frac{\boldsymbol{\Omega}(\mathbf{r}', t)}{|\mathbf{r} - \mathbf{r}'|} d\mathscr{V}(\mathbf{r}'). \tag{14}$$

It may in fact be verified that (14) leads to $\nabla \cdot \mathbf{A} = 0$, in accordance with equation (12), if the vorticity vanishes at infinity. Finally, using (14) in (9), one has, for the velocity

$$\mathbf{v}(\mathbf{r}, t) = \frac{1}{4\pi} \nabla \times \iiint_{\mathscr{V}} \frac{\boldsymbol{\Omega}(\mathbf{r}', t)}{|\mathbf{r} - \mathbf{r}'|} d\mathscr{V}(\mathbf{r}'). \tag{15}$$

8.3 Biot-Savart's Law

From (14), note that, for a vortex tube shown in Figure 8.1, one has (momentarily suspending the argument t),

$$\delta \mathbf{A}(\mathbf{r}) = \frac{1}{4\pi} \frac{\Omega(\mathbf{r}')}{|\mathbf{r} - \mathbf{r}'|} (\hat{\mathbf{n}} \delta S \cdot \mathbf{d\ell}). \tag{16}$$

Noting that

$$\mathbf{d\ell} = \frac{\boldsymbol{\Omega}}{|\boldsymbol{\Omega}|} d\ell, \; \Gamma = \hat{\mathbf{n}} \cdot \boldsymbol{\Omega} dS \tag{17}$$

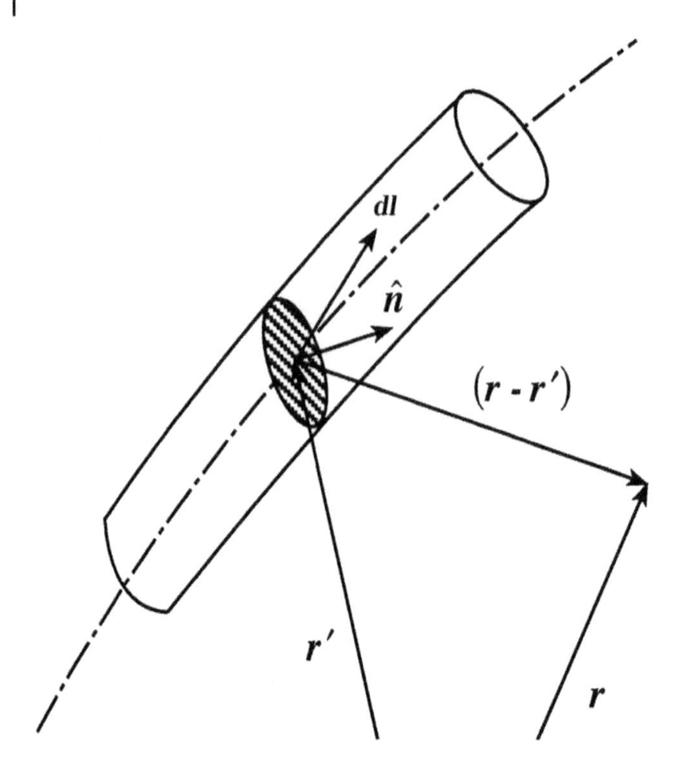

Figure 8.1 Vortex tube in a fluid.

Γ being the circulation around the vortex tube, (16) becomes

$$\delta \mathbf{A}(\mathbf{r}) = \frac{\Gamma}{4\pi} \frac{\mathbf{d}\ell}{|\mathbf{r} - \mathbf{r}'|}. \tag{18}$$

Using (18), (9) gives

$$\delta \mathbf{v}(\mathbf{r}) = \nabla \times \frac{\Gamma}{4\pi} \frac{\mathbf{d}\ell}{|\mathbf{r} - \mathbf{r}'|} \tag{19a}$$

from which, the velocity induced by the vortex tube is (Biot and Savart, 1820)

$$\mathbf{v}(\mathbf{r}) = \frac{\Gamma}{4\pi} \iiint_{\mathscr{V}} \frac{\mathbf{d}\ell \times (\mathbf{r} - \mathbf{r}')}{|\mathbf{r} - \mathbf{r}'|^3}. \tag{19b}$$

Example 1: Consider a line vortex of finite length ℓ (Figure 8.2). Tornadoes and whirlpools are some examples of such a concentration of vorticity.

Let

$$\mathbf{r}_1 \equiv \mathbf{r} - \mathbf{r}'.$$

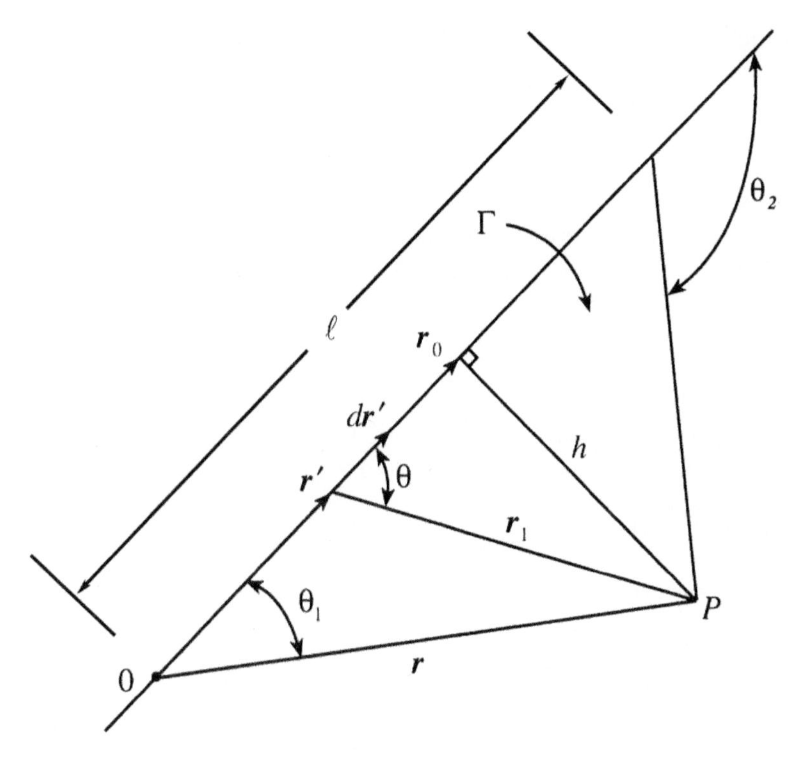

Figure 8.2 A line vortex.

Then, one has (see Figure 8.2)

$$d\mathbf{r}' \times \mathbf{r}_1 = r_1 dr' \sin\theta\ \hat{\mathbf{e}}$$

where $\hat{\mathbf{e}}$ is the unit vector pointing into the plane of the paper.

Then, (19b) gives

$$\mathbf{v}(\mathbf{r}) = \hat{\mathbf{e}}\frac{\Gamma}{4\pi}\int_0^{\ell}\frac{\sin\theta}{r_1^2}d\mathbf{r}'.$$

Noting, from Figure 8.2, that

$$r_1 = h\ \text{cosec}\theta,\ \ r_0 - r' = h\cot\theta,$$

we obtain

$$\mathbf{v}(\mathbf{r}) = \hat{\mathbf{e}}\frac{\Gamma}{4\pi h}\int_{\theta_1}^{\theta_2}\sin\theta d\theta = \hat{\mathbf{e}}\frac{\Gamma}{4\pi h}(\cos\theta_1 - \cos\theta_2).$$

For an infinitely long vortex line ($\theta_1 \Rightarrow 0, \theta_2 \Rightarrow \pi$), we have

$$\mathbf{v}(\mathbf{r}) = \hat{\mathbf{e}}\frac{\Gamma}{2\pi h}.$$

Example 2: Consider a vortex between two perpendicular planes (see Figure 8.3). The image system corresponding to the given vortex at A consists of vortices at A_1, A_2, A_3 as shown in Figure 8.3. The velocity induced at A by the image system is (using the results of example 1)

$$\left.\begin{aligned} \dot{r} &= \frac{\Gamma}{2\pi}\left(\frac{\cos\theta}{AA_3} - \frac{\sin\theta}{AA_1}\right) = \frac{\Gamma\cot 2\theta}{2\pi r} \\ r\dot{\theta} &= \frac{\Gamma}{2\pi}\left(\frac{1}{AA_2} - \frac{\cos\theta}{AA_1} - \frac{\sin\theta}{AA_3}\right) = -\frac{\Gamma}{4\pi r} \end{aligned}\right\}$$

from which

$$\frac{1}{r}\frac{dr}{d\theta} = -2\cot 2\theta$$

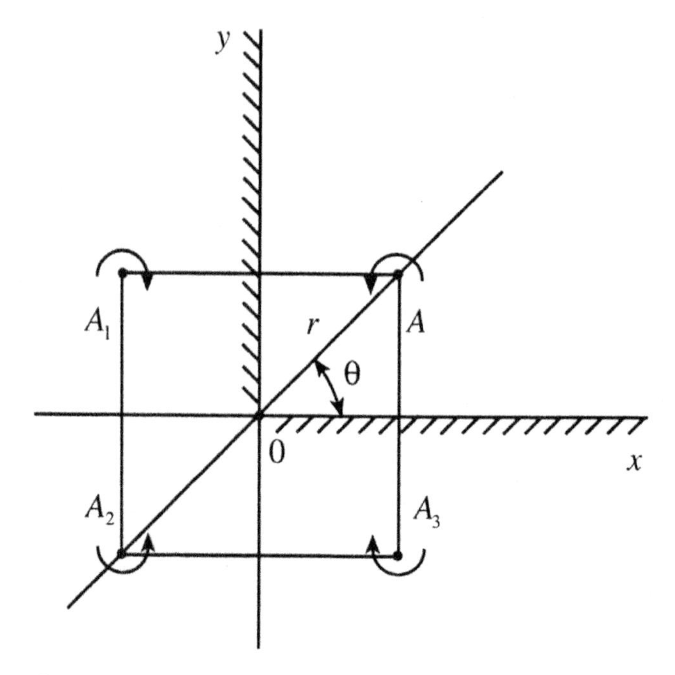

Figure 8.3 A line vortex between two perpendicular planes.

and hence,

$$r = C\operatorname{cosec} 2\theta$$

C being an arbitrary constant.

On the other hand, putting,

$$u \equiv \frac{1}{r}$$

one obtains

$$\frac{d^2u}{d\theta^2} + u = -3u$$

which implies that the radial acceleration of the vortex A is $3\Gamma^2/16\pi^2 r^3$, and that the transverse acceleration is zero. Therefore, vortex A moves as if it were under a *repulsion* from the origin, inversely proportional to r^3.

Example 3: Consider a closed vortex filament. Then, (19a) gives, on using *Stokes' Theorem*,

$$\left.\begin{aligned} \mathbf{v}(\mathbf{r}) &= -\frac{\Gamma}{4\pi}\nabla \times \iint_S \left[\nabla'\left(\frac{1}{|\mathbf{r}-\mathbf{r}'|}\right)\right] \times \hat{\mathbf{n}}\,dS(\mathbf{r}') \\ &\text{or} \\ \mathbf{v}(\mathbf{r}) &= -\frac{\Gamma}{4\pi}\iint_S \hat{\mathbf{n}}\cdot\nabla\left[\nabla'\left(\frac{1}{|\mathbf{r}-\mathbf{r}'|}\right)\right] dS(\mathbf{r}') = -\frac{\Gamma}{4\pi}\nabla\omega \end{aligned}\right\}$$

where

$$\omega \equiv \iint_S \frac{(\mathbf{r}-\mathbf{r}')\cdot\hat{\mathbf{n}}}{|\mathbf{r}-\mathbf{r}'|^3}\,dS(\mathbf{r}')$$

which is the solid angle subtended by the closed line vortex at $\mathbf{r}$.

8.4 von Kármán Vortex Street

Consider an infinite system of equal parallel rectilinear vortices each of strength Γ arranged along the x-axis at $x = 0, \underline{+}a, \underline{+}2a \cdots$ Then, the complex potential for the resulting flow is given by (see Section 8.1)

$$F(z) = -\frac{i\Gamma}{2\pi}\sum_{n=-\infty}^{\infty}\ell n(z-na) = -\frac{i\Gamma}{2\pi}\ell n\left[(z)\cdot(z^2-a^2)\cdot(z^2-4a^2)\cdots\right]$$

$$= -\frac{i\Gamma}{2\pi}\ell n\left[\frac{\pi z}{a}\left(1-\frac{z^2}{a^2}\right)\left(1-\frac{z^2}{4a^2}\right)\cdots\right]$$

$$-\frac{i\Gamma}{2\pi}\ell n\left[\frac{a}{\pi}(-a^2)(-4a^2)\cdots\right]$$

$$= -\frac{i\Gamma}{2\pi}\ell n\left(\sin\frac{\pi z}{a}\right)$$

where we have used the result

$$\frac{\sin z}{z} = \prod_{n=1}^{\infty}\left(1-\frac{z^2}{n^2\pi^2}\right).$$

Now, recall that any vortex will move with the fluid. Therefore, the motion of any particular line vortex in the above row cannot take place in isolation and has to be induced by the other line vortices in the row. The complex potential at the mth line vortex due to all the others is

$$F_{(m)}(z) = -\frac{i\Gamma}{2\pi}\sum_{\substack{n=-\infty \\ n\neq m}}^{\infty}\ell n(z-na)\Big|_{z=ma}.$$

The corresponding complex velocity at the mth line vortex is, then given by

$$W_{(m)}(z) = -\frac{i\Gamma}{2\pi}\sum_{\substack{n=-\infty \\ n\neq m}}^{\infty}\frac{d}{dz}\ell n(z-na)\Big|_{z=ma} = -\frac{i\Gamma}{2\pi a}\sum_{\substack{n=-\infty \\ n\neq m}}^{\infty}\frac{1}{(m-n)} = 0$$

so that all line vortices in the row are at rest.

The stream function for this flow is given by

$$\Psi = \frac{1}{2i}\left[F(z)-\overline{F}(z)\right] = -\frac{\Gamma}{4\pi}\ell n\left(\sin\frac{\pi z}{a}\cdot\sin\frac{\pi\overline{z}}{a}\right)$$

$$= -\frac{\Gamma}{4\pi}\ell n\left[\cos\frac{\pi(\overline{z}-z)}{a} - \cos\frac{\pi(\overline{z}+z)}{a}\right] + \frac{\Gamma}{4\pi}\ell n 2$$

$$= -\frac{\Gamma}{4\pi}\ell n\left[\cosh\frac{2\pi y}{a} - \cos\frac{2\pi x}{a}\right] + \frac{\Gamma}{4\pi}\ell n 2.$$

One then obtains, for the velocity components,

$$\left.\begin{aligned} u &= \frac{\partial \Psi}{\partial y} = -\frac{\Gamma}{2a}\frac{\sinh\dfrac{2\pi y}{a}}{\cosh\dfrac{2\pi y}{a}-\cos\dfrac{2\pi x}{a}} \\ v &= -\frac{\partial \Psi}{\partial x} = \frac{\Gamma}{2a}\frac{\sin\dfrac{2\pi x}{a}}{\cosh\dfrac{2\pi y}{a}-\cos\dfrac{2\pi x}{a}} \end{aligned}\right\}.$$

Note that

$$\left.\begin{aligned} y \Rightarrow \pm\infty : u \approx \mp\frac{\Gamma}{2a} \\ x = \pm\frac{1}{2}a,\ \pm a,\ \pm\frac{3}{2}a,\cdots : v = 0 \end{aligned}\right\}.$$

Thus, an infinite row of line vortices simulates a *vortex sheet* which produces a uniform flow parallel to the sheet on both sides, but in opposite directions (see Section 8.7).

Now, if one places another row of vortices, which are similar, but of opposite sign, at the points

$$\left(\pm\frac{1}{2}a, b\right), \left(\pm\frac{3}{2}a, b\right), \cdots$$

a typical vortex of the new set has a velocity, induced by the first set, given by

$$\left.\begin{aligned} u &= -\frac{\Gamma}{2a}\frac{\sinh\dfrac{2\pi b}{a}}{\cosh\dfrac{2\pi b}{a}+1} = -\frac{\Gamma}{2a}\tanh\frac{\pi b}{a} \\ v &= 0 \end{aligned}\right\}.$$

From symmetry considerations, the first set must have exactly the same velocity, so that the whole double row of vortices, called a *von Kármán vortex street*, moves with speed $(\Gamma/2a)\tanh(\pi b/a)$ in the negative x-direction (Figure 8.4). Such a vortex street can be used to represent the wake behind a body moving slowly through a fluid. Each vortex in this street is generated by direct transfer of vorticity to a fluid mass adjacent to a body *via* viscous diffusion which then *separates* from the body.[1]

---

1 For the flow past a body such as a cylinder, when the flow speed is small, the streamline pattern is essentially symmetrical both fore and aft of the cylinder. However, as the flow speed increases, asymmetry of the flow develops. Eddies are observed to form in regions of separated flow attached to the downstream side of the cylinder. With further increase in the flow speed, these eddies detach from the cylinder and are carried downstream. The eddies are shed alternately from the top and bottom of the cylinder on its downstream side, which then give a chase

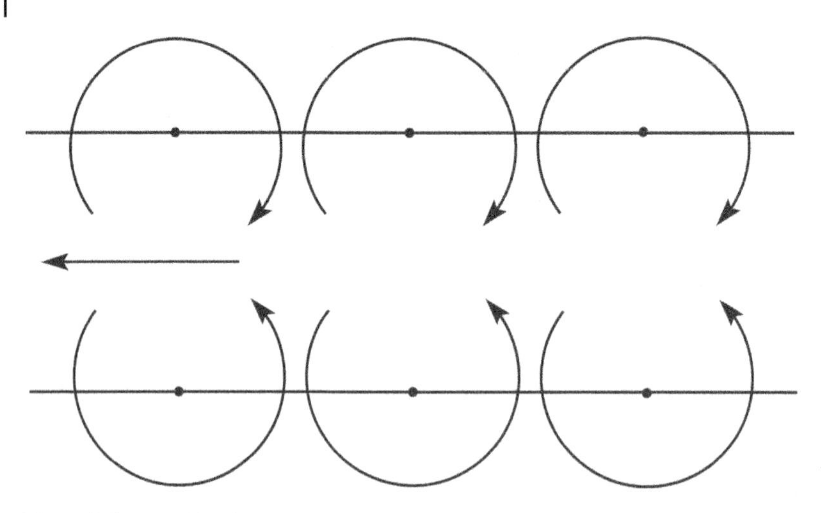

Figure 8.4 von Kármán vortex street.

8.5 Vortex Ring

A common example of a vortex ring (i.e., a flow with circular vortex lines) is the *smoke ring* which is easy to generate by ejecting sudden bursts of smoke blobs through a circular hole on a smoke-filled container.

In an axisymmetric flow with cylindrical polar coordinates (r, φ, z), the velocity field is given by

$$\mathbf{v} = u_r(r, z, t)\,\hat{\mathbf{i}}_r + u_z(r, z, t)\,\hat{\mathbf{i}}_z. \tag{20}$$

The streamlines then lie in meridional planes $\varphi = const$, and the vorticity is given by $\mathbf{\Omega} = \Omega\hat{\mathbf{i}}_\varphi$, where

$$\Omega = \frac{\partial u_r}{\partial z} - \frac{\partial u_z}{\partial r}. \tag{21}$$

In an axisymmetric flow, the vortex tubes are, therefore, ring-shaped, around the axis of symmetry. During their motion with the fluid, the vortex rings will expand and contract about the axis of symmetry. However, since the fluid is incompressible, the volume of these rings must remain constant so that the

---

to the cylinder and their pattern is closely simulated by the *von Kármán vortex street*. Observations of *von Kármán vortex street* date back to several centuries - they are clearly visible in Leonardo da Vinci's sketches.

vorticity Ω will be proportional to the circumference of the ring $2\pi r$.[2] This may also be seen by noting that equation (5a), for the present situation, becomes

$$\frac{\partial}{\partial t}\left(\frac{\Omega}{r}\right) + u_r \frac{\partial}{\partial r}\left(\frac{\Omega}{r}\right) + u_z \frac{\partial}{\partial z}\left(\frac{\Omega}{r}\right) = 0. \tag{22}$$

Consider now an arbitrarily thin circular vortex filament. Upon introducing the axisymmetric stream function Ψ according to

$$u_r = \frac{1}{r}\frac{\partial \Psi}{\partial z},\ u_z = -\frac{1}{r}\frac{\partial \Psi}{\partial r} \tag{23}$$

one obtains, from (22),

$$\frac{\partial^2 \Psi}{\partial z^2} + \frac{\partial^2 \Psi}{\partial r^2} - \frac{1}{r}\frac{\partial \Psi}{\partial r} - r\Omega = 0. \tag{24}$$

Outside the ring, one has an irrotational flow so that equation (24) leads to

$$\frac{\partial^2 \Psi'}{\partial z^2} + \frac{\partial^2 \Psi'}{\partial r^2} - \frac{1}{r}\frac{\partial \Psi'}{\partial r} = 0. \tag{25}$$

Putting

$$\Psi = \chi r,\ \Psi' = \chi' r \tag{26}$$

equations (24) and (25) lead to

$$\frac{\partial^2 \chi}{\partial z^2} + \frac{\partial^2 \chi}{\partial r^2} + \frac{1}{r}\frac{\partial \chi}{\partial r} - \frac{\chi}{r^2} - \Omega = 0 \tag{27}$$

$$\frac{\partial^2 \chi'}{\partial z^2} + \frac{\partial^2 \chi'}{\partial r^2} + \frac{1}{r}\frac{\partial \chi'}{\partial r} - \frac{\chi'}{r^2} = 0 \tag{28}$$

which imply that $\chi \cos\varphi$ may be viewed as the gravitational potential of a distribution of matter of density $\Omega \cos\varphi/4\pi$ occupying the same region of space as the vortex ring.

Referring to Figure 8.5, one obtains

$$\chi \cos\varphi = \frac{r_0 \Omega \sigma}{4\pi} \int_{\varphi}^{2\pi+\varphi} \frac{\cos\varphi' d\varphi'}{\left[(z-z')^2 + r^2 + r_0^2 - 2rr_0\cos(\varphi-\varphi')\right]^{1/2}} \tag{29}$$

where σ is the area of cross section of the core of the ring, and r_0 is the radius of the ring.

2 If sigma is the area of cross section of the core of the ring, then the constancy of the volume of the vortex ring implies $\sigma \sim 1/r$. On the other hand, the constancy of the vorticity flux through the ring implies $\Omega \sim 1/S$. So, $\Omega \sim r$.

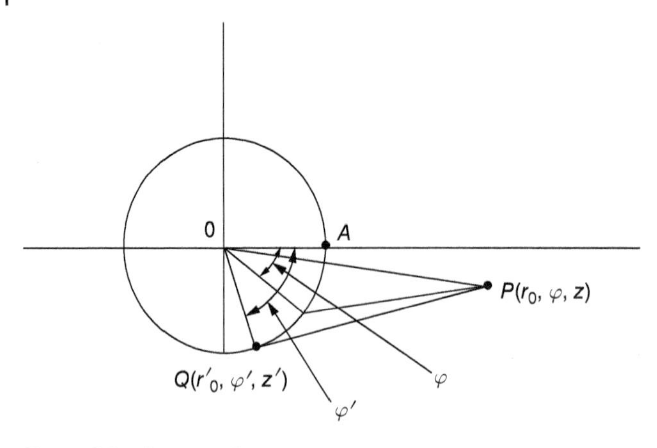

Figure 8.5 A vortex ring.

Putting

$$\varepsilon \equiv \varphi' - \varphi \tag{30}$$

equation (29) becomes

$$\chi = \frac{\sigma\Omega r_0}{2\pi}\int_0^{\pi}\frac{\cos\varepsilon d\varepsilon}{\left[(z-z')^2 + r^2 + r_0^2 - 2rr_0\cos\varepsilon\right]^{1/2}}. \tag{31a}$$

So, from (26), one obtains

$$\Psi = \frac{\sigma\Omega rr_0}{2\pi}\int_0^{\pi}\frac{\cos\varepsilon d\varepsilon}{\left[(z-z')^2 + r^2 + r_0^2 - 2rr_0\cos\varepsilon\right]^{1/2}}. \tag{31b}$$

Introducing,

$$\left.\begin{aligned} k &\equiv \frac{2(rr_0)^{1/2}}{\left[(z-z')^2 + (r+r_0)^2\right]^{1/2}} \\ \eta &\equiv \varepsilon/2 \end{aligned}\right\} \tag{32}$$

we may write (31b) as,

$$\Psi = \sigma\Omega\frac{(rr_0)^{1/2}k}{2\pi}\int_0^{\pi/2}\frac{(2\cos^2\eta - 1)}{(1-k^2\cos^2\eta)^{1/2}}d\eta \tag{33}$$

or

$$\Psi = \sigma\Omega\frac{(rr_0)^{1/2}}{2\pi}\left[\left(\frac{2}{k} - k\right)K(k) - \frac{2}{k}E(k)\right]. \tag{34}$$

Here,

$$K(k) \equiv \int_0^{\pi/2} \frac{d\eta}{\left(1 - k^2\cos^2\eta\right)^{1/2}}, \quad E(k) \equiv \int_0^{\pi/2} \left(1 - k^2\cos^2\eta\right)^{1/2} d\eta$$

are the *complete elliptic integrals of first and second kinds*, k being the modulus, $0 < k < 1$. The streamline pattern for this flow is sketched in Figure 8.6.

Observing that, on the surface of the ring, $z \approx z'$, $r \approx r_0$, $k \approx 1$,[3] using the asymptotic results,

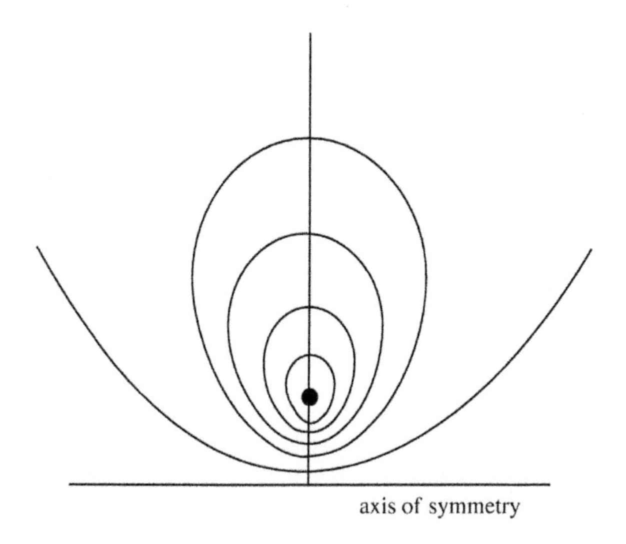

Figure 8.6 Streamline pattern near vortex ring.

---

3 More precisely

$$\sqrt{1-k^2} \approx \frac{|r - r_o|}{2r_o} \approx \frac{a}{2r_o},$$

a being the radius of the core of the vortex ring.

$$\left.\begin{aligned}K(k) &\sim ln\frac{4}{(1-k^2)^{1/2}}+\frac{1}{2}\left(1-k^2\right)^{1/2}\left[ln\frac{4}{(1-k^2)^{1/2}}-1\right]\\E(k) &\sim 1+\frac{1}{2}\left(1-k^2\right)\left[ln\frac{4}{(1-k^2)^{1/2}}-\frac{1}{2}\right]\end{aligned}\right\}\tag{35}$$

one obtains, from (34), for the velocity components

$$u_r \approx \frac{\Gamma}{\pi}\left[\frac{2r_0}{a}-\frac{3a}{4r_0}\left(ln\frac{8r_0}{a}-\frac{3}{2}\right)\right]\frac{z-z'}{2r_0a}\tag{36}$$

$$u_z \approx \frac{\Gamma}{4\pi r_0}\left(ln\frac{8r_0}{a}-\frac{1}{4}\right)\tag{37}$$

where a is the radius of the core of the vortex ring, and $\Gamma \equiv \sigma\Omega$.[4] Note that at $z = z', u_r = 0$, so that the radius of the ring remains unchanged.[5] Thus, an isolated vortex ring in an unbounded ideal fluid will move without noticeable change of size in a direction perpendicular to its plane with a constant velocity. From (37) we infer that the axial speed is greater for small values of r_0. In order to illustrate the consequences of this aspect, consider two similar vortex rings at some distance apart on a common axis of symmetry. The velocity field induced by the vortex ring at the back has a radially outward component at the position of the vortex ring in the front so that the radius of the latter gradually increases. According to (37), this leads to a decrease in the speed of motion of the vortex ring in the front. Conversely, there is a corresponding increase in the speed of motion of the vortex ring at the back, which subsequently passes through the larger vortex ring in the front and becomes now, the front vortex ring. This *"leap-frogging"* (or trailer/leader vortex) sequence keeps repeating itself. This process has been observed in laboratory experiments (Yamada and Matsui, 1978) although it is limited to only a few cycles because of viscous effects.

---

4 Kelvin originally gave (37) and proposed to explain the behavior of atoms by considering them as knots of ether imparted with vortex motion. It is of interest to note that an arbitrary vortex filament experiences a self-induced motion which may be approximated locally as that of an *osculating vortex ring* of radius same as the local radius of curvature of the vortex filament (Da Rios, 1906).

5 This may also be seen from the Biot-Savart's law (19b) which shows that, for points in the plane of the ring, the induced velocity has no radial component.

8.6 Hill's Spherical Vortex

Hill's spherical vortex corresponds to a flow configuration in which the vorticity is confined to the interior of a sphere. Let us now consider a sphere of radius a. Here, the vortex lines are circles about an axis passing through the center of the sphere. So, $\mathbf{\Omega} = \Omega\hat{\mathbf{i}}_\varphi$, and the streamlines lie in meridional planes. In addition, let the flow outside the sphere be irrotational.

The equations of continuity and vorticity transport for a steady axisymmetric flow in cylindrical polar coordinates are

$$\frac{\partial (u_r r)}{\partial r} + \frac{\partial (u_z r)}{\partial z} = 0 \tag{38}$$

$$\frac{\partial (u_r \Omega)}{\partial r} + \frac{\partial (u_z \Omega)}{\partial z} = 0. \tag{39}$$

From these, one derives the steady version of equation (22),

$$u_r \frac{\partial}{\partial r}\left(\frac{\Omega}{r}\right) + u_z \frac{\partial}{\partial z}\left(\frac{\Omega}{r}\right) = 0 \tag{22}$$

or, in terms of the axisymmetric streamfunction Ψ,

$$\frac{1}{r}\frac{\partial \Psi}{\partial z}\frac{\partial}{\partial r}\left(\frac{\Omega}{r}\right) - \frac{1}{r}\frac{\partial \Psi}{\partial r}\frac{\partial}{\partial z}\left(\frac{\Omega}{r}\right) = 0. \tag{40}$$

Equation (40) implies that

$$\frac{\Omega}{r} = f(\Psi). \tag{41}$$

In spherical polar coordinates (R, θ, φ) note that,

$$\Omega = \frac{1}{R}\left[\frac{\partial}{\partial R}(R u_\theta) - \frac{\partial u_R}{\partial \theta}\right] \tag{42a}$$

where

$$u_R = \frac{1}{R^2 \sin\theta}\frac{\partial \Psi}{\partial \theta}, \quad u_\theta = -\frac{1}{R\sin\theta}\frac{\partial \Psi}{\partial R}. \tag{42b}$$

Substituting (42a) and (42b) in (41), we obtain,

$$\frac{\partial}{\partial R}\left(\frac{1}{\sin\theta}\frac{\partial \Psi}{\partial R}\right) + \frac{\partial}{\partial \theta}\left(\frac{1}{R^2\sin\theta}\frac{\partial \Psi}{\partial \theta}\right) = -R^2 \sin\theta \cdot f(\Psi). \tag{43a}$$

If $f(\Psi) = const = -A$, looking for a solution of the form,

$$\Psi = F(R)\sin^2\theta \tag{44}$$

equation (43a) leads to the *Cauchy-Euler equation*,

$$R^2 F'' - 2F = AR^4 \tag{43b}$$

where primes denote differentiation with respect to the argument. Thus,

$$F(R) = \frac{B}{R} + CR^2 + \frac{AR^4}{10}. \tag{45}$$

The motion within a fixed sphere of radius a is described by (44) and (45) implying that its surface is a material surface. Moreover, if Ψ is well-behaved for $R < a$ it requires that,

$$B = 0. \tag{46a}$$

The normal velocity of this flow vanishes at the boundary,

$$R = a : \left(\frac{1}{R \sin\theta} \frac{\partial \Psi}{R\partial\theta} \right) = 0. \tag{46b}$$

Using (44) and (45), (46a) and (46b) give

$$C = -\frac{Aa^2}{10}. \tag{47}$$

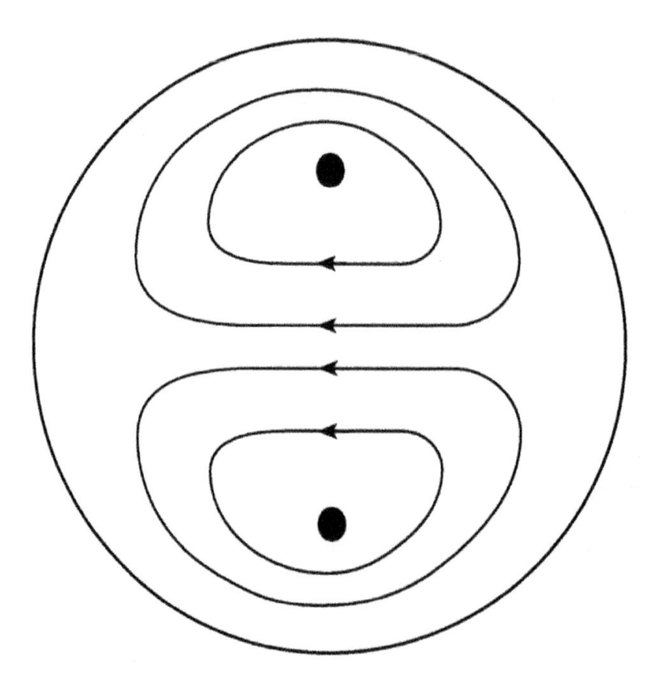

Figure 8.7 Streamline pattern in Hill's vortex.

Using (45), (46a), and (47), (44) leads to

$$R < a : \Psi = -\frac{A}{10}\left(a^2 - R^2\right) R^2 \sin^2\theta. \tag{48}$$

The streamlines in the meridional plane described by (48) are shown in Figure 8.7. It also shows that the vortex lines are circles perpendicular to the axis of symmetry.

The arbitrary constant A may be exploited to immerse such a spherical vortex in an irrotational flow of a fluid with a uniform speed U, in the positive x-direction. One then obtains a uniformly translating spherical vortex,

$$R < a : \Psi = -\frac{A}{10}\left(a^2 - R^2\right) R^2 \sin^2\theta \tag{49a}$$

while the flow outside the spherical vortex is given by (see Section 7.1, Example-2),

$$R > a : \Psi = \frac{1}{2} U R^2 \sin^2\theta \left(1 - \frac{a^3}{R^3}\right). \tag{49b}$$

The continuity of the tangential velocity $(1/R\sin\theta)\,\partial\Psi/\partial R$ at $R = a$ (which precludes a vortex sheet there) leads to

$$A = \frac{15}{2}\frac{U}{a^2}. \tag{50}$$

Hill's spherical vortex may be viewed as the opposite extreme of a circular line vortex in the family of vortex rings.

8.7 Vortex Sheet

Vortex sheets are surfaces on which the vorticity is infinite, and the tangential component of the velocity field is discontinuous. In addition to being kinematically possible, such surfaces can exist in the limit of vanishing viscosity as long as the discontinuities are consistent with the integral forms of the Euler equations. An example of a surface concentration of vorticity is the flow field behind an airplane wing (see Section 11.4). Let

$$\boldsymbol{\Gamma} \equiv \lim_{\varepsilon \Rightarrow 0^+} \int_{\varepsilon} \boldsymbol{\Omega}\, dn \tag{51}$$

where n is the distance normal to the sheet S and the integral is over some small range ε containing the surface. $\mathbf{\Gamma}$ may be taken to be the vortex strength per unit area. Using (52), (15) gives

$$\mathbf{v}(\mathbf{r}) = -\frac{1}{4\pi}\iint_S \frac{(\mathbf{r}-\mathbf{r}')\times\mathbf{\Gamma}(\mathbf{r}')}{|\mathbf{r}-\mathbf{r}'|^3}dS(\mathbf{r}'). \tag{52}$$

For a single plane sheet of uniform vorticity, (52) becomes

$$\mathbf{v}(\mathbf{r}) = \frac{\mathbf{\Gamma}}{4\pi}\times\iint_S \frac{(\mathbf{r}-\mathbf{r}')}{|\mathbf{r}-\mathbf{r}'|^3}dS(\mathbf{r}'). \tag{53a}$$

Setting

$$(\mathbf{r}-\mathbf{r}') = a\hat{\mathbf{n}} + \beta\hat{\mathbf{s}} + c\hat{\mathbf{t}} \tag{54}$$

where $\hat{\mathbf{n}}$ is the unit normal to the vortex sheet directed toward the side on which the point $\mathbf{r}$ lies, and $\hat{\mathbf{s}}$ and $\hat{\mathbf{t}}$ are unit vectors mutually orthogonal and lying in the vortex sheet. Noting that the integrals involving $\hat{\mathbf{s}}$ and $\hat{\mathbf{t}}$ vanish due to the antisymmetry of the integrand, we obtain

$$\mathbf{v}(\mathbf{r}) = \frac{\mathbf{\Gamma}}{4\pi}\times\hat{\mathbf{n}}\left[\iint_S \frac{\hat{\mathbf{n}}\cdot(\mathbf{r}-\mathbf{r}')}{|\mathbf{r}-\mathbf{r}'|^3}dS(\mathbf{r}')\right]. \tag{53b}$$

The integral in the bracket represents the solid angle ω subtended by the infinitely extended vortex sheet at $\mathbf{r}$, which is 2π. This gives us

$$\mathbf{v}(\mathbf{r}) = \frac{1}{2}\mathbf{\Gamma}\times\hat{\mathbf{n}}. \tag{55}$$

Thus, the fluid velocity produced by the vortex sheet is uniform and parallel to the sheet on each side, but in opposite directions on the two sides[6] (see Figure 8.8).

6 The application of the integral form of Euler's equation to an element of the vortex sheet shows that the pressure and the normal component of the velocity field are continuous across the vortex sheet.

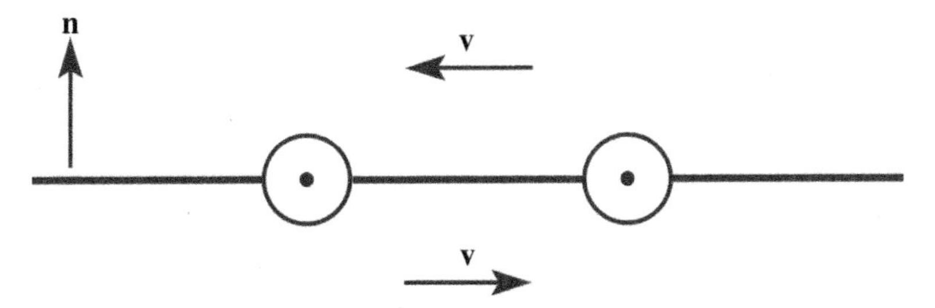

Figure 8.8 Vortex sheet.

In the case that $\mathbf{\Gamma}$ is not constant, and the sheet were not a plane,[7] one may nonetheless obtain from (55) for the local jump across the sheet in the induced flow velocity,

$$[\mathbf{v}] = \mathbf{\Gamma} \times \hat{\mathbf{n}}. \tag{56}$$

Consider next a vortex sheet in the form of a cylinder of arbitrary cross section, over which $\mathbf{\Gamma}$ defined by (51) is uniform and is *perpendicular* to the generators of the cylinder everywhere. This indicates that the vortex lines are plane curves, all of the same shape, passing around the cylinder. From (52) we obtain

$$\mathbf{v}(\mathbf{r}) = -\frac{\Gamma}{4\pi} \oint_C \int_{-\infty}^{\infty} \frac{(\mathbf{r} - \mathbf{r}') \times \mathbf{d\ell}(\mathbf{r}')}{|\mathbf{r} - \mathbf{r}'|^3} ds(\mathbf{r}') \tag{57}$$

where $ds(\mathbf{r}')$ is a length element of a generator and $\mathbf{d\ell}(\mathbf{r}')$ is a length element vector of a vortex line (Figure 8.9). Thanks to the antisymmetry of the integrand in (57), the component of $(\mathbf{r} - \mathbf{r}')$ parallel to the generators makes no contribution to the integral with respect to s. Consequently, we have

$$\mathbf{v}(\mathbf{r}) = -\frac{\Gamma}{4\pi} \oint_C \int_{-\infty}^{\infty} \frac{\mathbf{p} \times \mathbf{d\ell}(\mathbf{r}')}{(p^2 + s^2)^{3/2}} ds(\mathbf{r}') \tag{58a}$$

or

$$\mathbf{v}(\mathbf{r}) = -\frac{\Gamma}{2\pi} \oint_C \frac{\mathbf{p} \times \mathbf{d\ell}(\mathbf{r}')}{p^2} \tag{58b}$$

7 Indeed, vortex sheets tend to roll up into spirals that continuously tighten as the distance between neighboring turns becomes smaller and smaller, eventually developing into circular vortices.

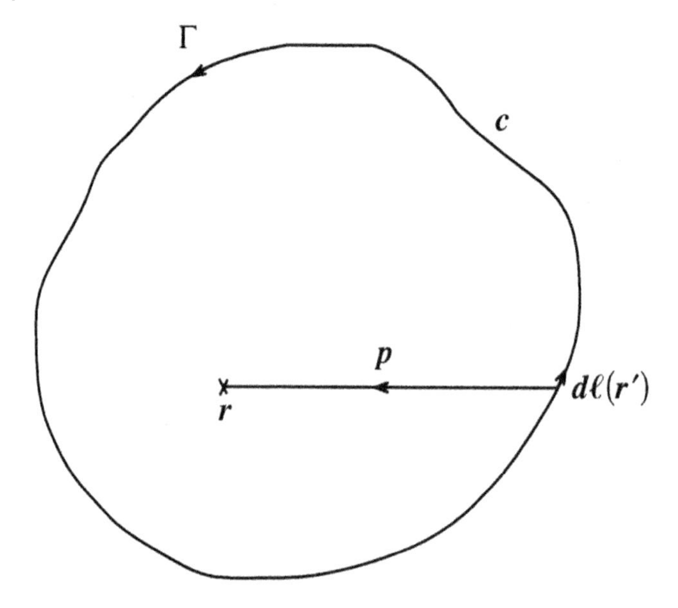

Figure 8.9 Sectional view of a cylindrical vortex sheet (from Batchelor, 1967).

where $\mathbf{p}$ is the projection of $(\mathbf{r} - \mathbf{r}')$ on a cross-sectional plane. The expression $|\mathbf{p} \times \mathbf{d\ell}\,(\mathbf{r}')|/p^2$ is the angle subtended at $\mathbf{r}$ by $\mathbf{d\ell}$ in this cross-sectional plane. So, at any point $\mathbf{r}$ within the cylinder, $\mathbf{v}$ is parallel to the generators and has a uniform magnitude Γ, while at any point $\mathbf{r}$ outside the cylinder $\mathbf{v}$ is zero.[8]

Example 4: Consider n equal rectilinear vortex filaments arranged symmetrically *along* the generators of a circular cylinder of radius a. One then has, for the complex potential of this flow,

$$F(z) = \frac{i\Gamma}{2\pi}\sum_{s=0}^{n-1} ln\left(z - ae^{2\pi si/n}\right) = \frac{i\Gamma}{2\pi} ln \prod_{s=0}^{n-1}\left(z - ae^{2\pi si/n}\right)$$
$$= \frac{i\Gamma}{2\pi} ln\,(z^n - a^n).$$

8 This result is similar to the one in electromagnetic theory, namely, that the magnetic field due to a steady electric current in a solenoid (which is a long wire in the form of a closely wound helix) is uniform and parallel to the axis within the solenoid but vanishes outside the solenoid.

Here, note that $e^{2\pi is/n}, s = 0, 1, \cdots, n-1$ are the n complex roots of unity. The stream function Ψ of this flow is then given by

$$2i\Psi = F - \overline{F}$$

or

$$e^{4\pi\Psi/\Gamma} = (z^n - a^n)\left(\overline{z}^n - a^n\right)$$
$$= z^n\,\overline{z}^n - a^n\left(z^n + \overline{z}^n\right) + a^{2n} = r^{2n} - 2a^n r^n \cos n\theta + a^{2n}.$$

The velocity components are, given by

$$\left.\begin{aligned} u_r &= -\frac{1}{r}\frac{\partial\Psi}{\partial\theta} = \frac{-\Gamma n a^n r^{n-l}\sin n\theta}{2\pi\left(r^{2n} - 2a^n r^n\cos n\theta + a^{2n}\right)} \\ u_\theta &= \frac{\partial\Psi}{\partial r} = \frac{\Gamma n\left(r^{2n-1} - a^n r^{n-l}\cos n\theta\right)}{2\pi\left(r^{2n} - 2a^n r^n\cos n\theta + a^{2n}\right)} \end{aligned}\right\}.$$

For $n \gg 1$, and $r \gg a$, these give

$$u_r \approx 0,\; u_\theta \approx \frac{\Gamma n}{2\pi r}.$$

On the other hand, for $n \gg 1$, and $r < a$, these give

$$u_r \approx 0,\; u_\theta \approx 0.$$

One obtains a cylindrical vortex sheet by letting $n \Rightarrow \infty$, and keeping $n\Gamma = const.$ Therefore, outside this sheet, the fluid moves as if all the vorticity were concentrated along the axis of the cylinder and inside the sheet, the fluid is stagnant.

8.8 Vortex Breakdown: Brooke Benjamin's Theory

Vortex breakdown refers to the major structural changes that a concentrated vortex core experiences while embedded in a decelerating irrotational flow. Typically, this involves the formation of an internal stagnation point on the vortex axis, followed by a localized reversed axial flow. Downstream of the breakdown region, a new vortex structure with an expanded core develops. As a model of the vortex breakdown process, consider an otherwise cylindrical vortex, embedded in a decelerating irrotational flow, passing through a region of noncylindrical flow. At some distance downstream, the cylindrical state (no variations in the flow properties along the axis) prevails again. One initially has, the following velocity profile,

$$r \leq a : u_z = U_1, \quad u_\theta = \omega r$$

$$r \geq a : u_z = U_1, \quad u_\theta = \frac{\omega a^2}{r} \tag{59a}$$

and the final velocity profile is

$$r \geq b : u_z = U_2, \quad u_\theta = \frac{\omega b^2}{r}. \tag{59b}$$

In the theoretical models, the principal mechanisms underlying the vortex breakdown process are assumed to exist in axisymmetric flow conditions. However, this is now known to be inaccurate. Note that, for an axisymmetric flow, the vorticity components are given by

$$\Omega_z = \frac{1}{r}\frac{\partial (r u_\theta)}{\partial r}, \quad \Omega_r = -\frac{\partial u_\theta}{\partial z}, \quad \Omega_\theta = \frac{\partial u_r}{\partial z} - \frac{\partial u_z}{\partial r} \tag{60}$$

while the velocity components are given by

$$u_r = \frac{1}{r}\frac{\partial \Psi}{\partial z}, \quad u_z = -\frac{1}{r}\frac{\partial \Psi}{\partial r}. \tag{23}$$

Using (23), (60) leads to

$$\Omega_\theta = -\frac{1}{r}\left(\frac{\partial^2 \Psi}{\partial z^2} + \frac{\partial^2 \Psi}{\partial r^2} - \frac{1}{r}\frac{\partial \Psi}{\partial r}\right). \tag{61}$$

Next, the equations of motion give, for a steady axisymmetric flow,

$$u_r \Omega_\theta - u_\theta \Omega_r = \frac{\partial H}{\partial z} \tag{62a}$$

$$u_\theta \Omega_z - u_z \Omega_\theta = \frac{\partial H}{\partial r} \tag{62b}$$

$$u_z \Omega_r - u_r \Omega_z = 0 \tag{62c}$$

where H is the total energy per unit mass,

$$H \equiv \frac{1}{2}\left(u_r^2 + u_\theta^2 + u_z^2\right) + \frac{p}{\rho}.$$

One obtains, from equations (60) and (62c),

$$\frac{D}{Dt}(r u_\theta) = 0 \tag{63}$$

which implies that

$$ru_\theta = C(\Psi). \tag{64}$$

Using (64), (60) gives

$$\Omega_z = u_z \frac{dC}{d\Psi}, \Omega_r = u_r \frac{dC}{d\Psi}. \tag{65}$$

Substituting (65) in equation (62a) we obtain,

$$\frac{\Omega_\theta}{r^2} = \frac{C}{r^2}\frac{dC}{d\Psi} - \frac{dH}{d\Psi}. \tag{66}$$

Combining (66) and (61) results in

$$\frac{\partial^2 \Psi}{\partial z^2} + \frac{\partial^2 \Psi}{\partial r^2} - \frac{1}{r}\frac{\partial \Psi}{\partial r} = r^2 \frac{dH}{d\Psi} - C\frac{dC}{d\Psi}. \tag{67}$$

The functions H and C are arbitrary and must be prescribed to suit a given flow. They can be determined uniquely by the flow properties on open streamlines extending to upstream infinity where the flow properties are uniform along the axis. The functional forms so determined will then apply to all regions of the flow permeated by these streamlines. At upstream infinity, one has, on using (64),

$$\frac{1}{\rho}\frac{dp}{dr} = \frac{u_\theta^2}{r} = \frac{C^2}{r^3}. \tag{68}$$

Substituting (68) in the expression for H gives,

$$H(\Psi) = \frac{1}{2}\left(u_r^2 + u_\theta^2 + u_z^2\right) + \int \frac{C^2}{r^3} dr$$

or

$$H(\Psi) = \frac{1}{2}\left(u_r^2 + u_z^2\right) + \int \frac{C}{r^2}\frac{dC}{dr} dr. \tag{69}$$

If the flow far upstream has an axial velocity U_1 and rotates as a rigid body with angular velocity ω, then we have

$$\Psi = \frac{1}{2}U_1 r^2, \ C = \omega r^2. \tag{70}$$

Substituting (70) in (69) gives,

$$H(\Psi) = \frac{1}{2}U_1^2 + \omega^2 r^2. \tag{71}$$

Alternately, the upstream conditions may be written using (70) and (71) as,

$$C(\Psi) = \frac{2\omega}{U_1}\Psi, \ H(\Psi) = \frac{1}{2}U_1^2 + \frac{2\omega^2}{U_1}\Psi. \tag{72}$$

Using (72), equation (67) becomes

$$\frac{\partial^2 \Psi}{\partial z^2} + \frac{\partial^2 \Psi}{\partial r^2} - \frac{1}{r}\frac{\partial \Psi}{\partial r} = \frac{2\omega^2}{U_1} r^2 - \frac{4\omega^2}{U_1^2} \Psi \tag{73}$$

which is linear!

Looking for a solution of the form,

$$\Psi(z,r) = \frac{1}{2} U_1 r^2 + rF(z,r) \tag{74}$$

equation (73) leads to

$$\frac{\partial^2 F}{\partial z^2} + \frac{\partial^2 F}{\partial r^2} + \frac{1}{r}\frac{\partial F}{\partial r} + \left(k^2 - \frac{1}{r^2}\right) F = 0 \tag{75}$$

where

$$k \equiv \frac{2\omega}{U_1}.$$

Using the boundary condition (59b), (74) leads to

$$r = b : F = \frac{1}{2} U_1 \left(\frac{a^2 - b^2}{b}\right). \tag{76}$$

From equations (75) and (76), one obtains

$$F = \frac{1}{2} U_1 \left(\frac{a^2}{b^2} - 1\right) \frac{bJ_1(kr)}{J_1(kb)} \tag{77}$$

where $J_n(x)$ is *Bessel's function* of order n.

Using (74) and (77), (23) gives

$$u_z = U_1 + \frac{1}{2} U_1 \left(\frac{a^2}{b^2} - 1\right) \frac{kbJ_0(kr)}{J_1(kb)} \tag{78}$$

from which we have,

$$\left(\frac{u_z}{U_2}\right)_{r=0} = \frac{U_1}{U_2}\left[1 + \left(\frac{a^2}{b^2} - 1\right)\frac{\frac{1}{2}kb}{J_1(kb)}\right] \tag{79}$$

and

$$\left(\frac{U_2}{U_1}\right) = 1 + \left(\frac{a^2}{b^2} - 1\right)\frac{\frac{1}{2}kb}{J_1(kb)} J_0(kb). \tag{80}$$

It can be seen from equation (79) shows that a change in the axial velocity in the irrotational flow surrounding a vortex produces significant changes (see

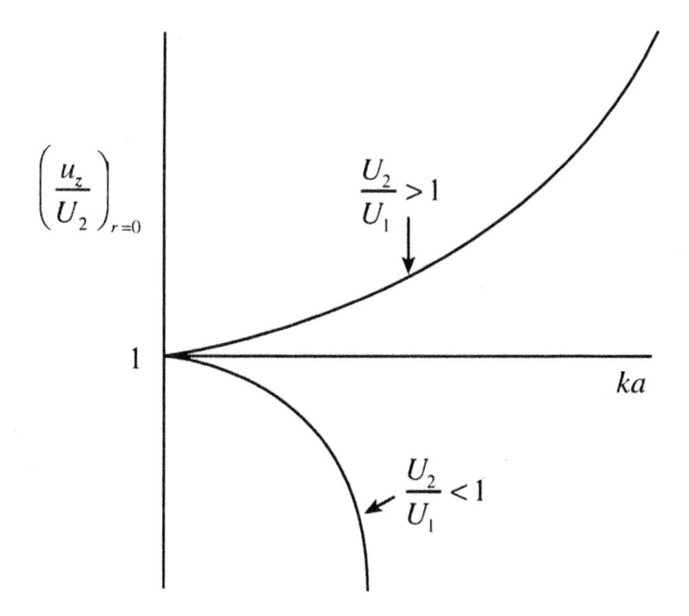

Figure 8.10 Variation of the axial velocity in the vortex core with the radius of the vortex filament (from Batchelor, 1967).

Figure 8.10) in the structure of the vortex (particularly, when the external fluid is *decelerated*).

Figure 8.11 shows the way in which kb varies with $U_1/U_2\,(> 1)$ for a given value ka ($0 \leq ka \leq 2,4$, the first zero of $J_0\,(x)$) according to (80). One observes that the increased thickening of the vortex due to deceleration of the external stream becomes catastrophic at a critical value $(U_2/U_1)^*$ - an event identified to be the *vortex breakdown*.

Figure 8.11 predicts the possibility of a finite transition from a state on the lower branch to a state on the upper branch. Indeed, Brooke Benjamin (1962) proposed that the vortex breakdown be viewed as such a finite transition between *dynamically conjugate* states of axisymmetric flow. The transition is from a *supercritical flow*, which cannot support standing waves, to a *subcritical flow*, which can (like that associated with a *hydraulic jump*, see Section 10.10). In support of Brooke Benjamin's proposal were Harvey's (1962) experimental results showing that

- the breakdown can be made axisymmetric, whereas the original flow is highly stable to axisymmetric disturbances;
- the breakdown can then comprise an abrupt expansion of the stream surfaces near the axis; and
- the above configuration can be made approximately steady.

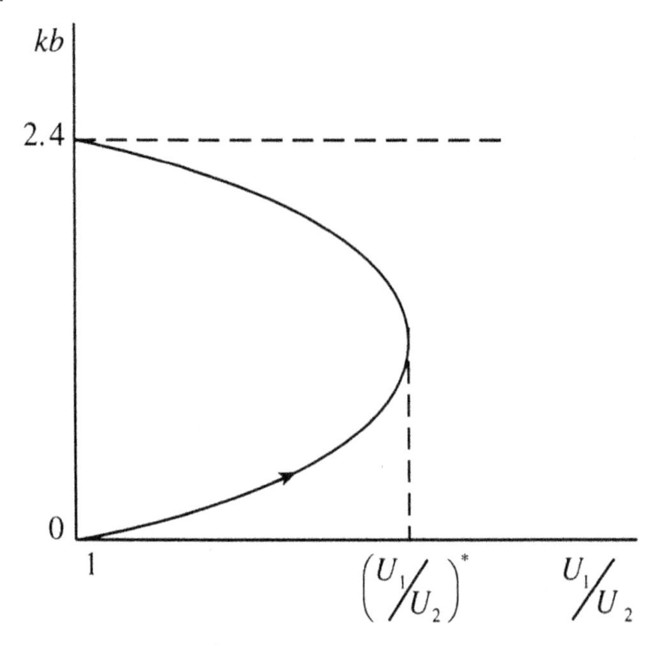

Figure 8.11 Variation of the size of the vortex with the axial flow (from Batchelor, 1967).

Typically, the transitional breakdown region occupies a small length so that the oncoming and outgoing regions may be thought of as joined by a sudden jump. The jump conditions connecting these two regions follow from the continuity of the fluxes of mass and momentum:

$$U_1 a^2 = U_2 b^2 \tag{81}$$

$$2\pi \int_0^a (p_1 + \rho U_1^2 + \rho u_\theta^2)\, r dr = 2\pi \int_0^b (p_2 + \rho U_2^2 + \rho u_\theta^2)\, r dr \tag{82}$$

from which, one obtains

$$\left.\begin{aligned} U_1^2 &= \frac{1}{2}\Omega^2 \frac{b^3}{a} \\ U_2^2 &= \frac{1}{2}\Omega^2 \frac{a^3}{b} \end{aligned}\right\}. \tag{83}$$

The energy fluxes E on both sides of the transition are not the same and their difference is the amount of energy dissipated in the transition. Thus, if

$$\triangle E \equiv 2\pi\rho \int_0^a U_1 H_1 r dr - 2\pi\rho \int_0^b U_2 H_2 r dr \tag{84a}$$

we then have

$$\triangle E = \frac{2Q\Omega^2}{4a^2b^2}\left(b^4 + b^3 a - \frac{b^2a^2}{3} + ba^3 + a^4\right)(b-a) \tag{84b}$$

where Q is the mass flux through the cylindrical vortex,

$$Q \equiv \pi a^2 \rho U_1$$

the total energy per unit mass is

$$H \equiv \frac{p}{\rho} + \frac{1}{2}\left(U^2 + u_\theta^2\right).$$

Noting that

$$b^4 + b^3 a - \frac{b^2a^2}{3} + ba^3 + a^4 = (a-b)^4 + ab\left[5(a-b)^2 + \frac{11}{3}ab\right] > 0,$$

$\triangle E > 0$ implies

$$b > a. \tag{85}$$

Therefore, in such a transition the vortex always *expands* (Shivamoggi and Uberoi, 1981).

Next, from equation (78), it is seen that there is a maximum value for b/a only below which it is possible to preserve axisymmetry in the flow downstream of the transition. From equation (75), we observe that the presence of a wave-motion downstream of the transition increases this maximum value of b/a. In other words, a wavemotion promotes the possibility of axisymmetric flow downstream of the transition.

Exercises

1. Find the velocity induced on the axis of a vortex ring.
2. Find the motion of the line vortices of equal and opposite strengths $\pm\Gamma$ located at $z = \pm a$.
3. Find the flow generated in a doubly infinite strip between rigid barriers along the lines $x = \pm\frac{\pi}{2}$ by a line vortex of strength Γ placed at the origin. Determine the streamlines of this flow.

9

Rotating Flows

The dominance of *Coriolis forces* in a rotating flow can lead to interesting consequences. One such example is when the flow is steady relative to the rotating axes. The flow then shows a strong tendency towards two dimensionality. toward two-dimensionality. In general, rotation imparts some kind of rigidity to the flow. It also confers a certain elasticity on the fluid that supports the propagation of inertial waves.

9.1 Governing Equations and Elementary Results

When the motion of a uniformly rotating fluid is referred to a frame of reference that rotates with the fluid, the equation of motion of the fluid is changed only by the addition of an apparent body force,

$$\frac{\partial \mathbf{v}}{\partial t} + \frac{1}{2}\nabla(\mathbf{v}\cdot\mathbf{v}) + (\nabla\times\mathbf{v})\times\mathbf{v} + 2\boldsymbol{\Omega}\times\mathbf{v} + \boldsymbol{\Omega}\times(\boldsymbol{\Omega}\times\mathbf{r}) = -\frac{1}{\rho}\nabla p \quad (1)$$

where $\boldsymbol{\Omega}$ is the angular velocity of the fluid.

The *centrifugal force* $\boldsymbol{\Omega}\times(\boldsymbol{\Omega}\times\mathbf{r})$ plays a significant role only when the density ρ is nonuniform. If ρ is uniform, the centrifugal force is *conservative* and is equivalent to an effective radial pressure gradient. It may be transformed away by incorporating this effective pressure in the actual pressure.

Thus, introducing

$$P \equiv \frac{1}{\rho}\left[p - \frac{\rho}{2}|\boldsymbol{\Omega}\times\mathbf{r}|^2\right] \quad (2)$$

equation (1) becomes

$$\frac{\partial \mathbf{v}}{\partial t} + \frac{1}{2}\nabla(\mathbf{v}\cdot\mathbf{v}) + (\nabla\times\mathbf{v})\times\mathbf{v} + 2\boldsymbol{\Omega}\times\mathbf{v} = -\nabla P \quad (3)$$

Introduction to Theoretical and Mathematical Fluid Dynamics, Third Edition.
Bhimsen K. Shivamoggi.

so that the effects of rotation are contained only in the term $2\boldsymbol{\Omega} \times \mathbf{v}$, called the *Coriolis force*. The Coriolis force does no work on the fluid but serves only to change the direction of its velocity.

9.2 Taylor-Proudman Theorem

If equation (3) is written in a dimensionless form by introducing reference length and velocity scales L and U, one finds that the following dimensionless parameter, called the *Rossby number*,

$$R_0 = \frac{U}{\Omega L} \tag{4}$$

occurs. The Rossby number measures the importance of the nonlinear convective acceleration term relative to the Coriolis term. For small Rossby number flows (like the steady large-scale circulations of the oceans and atmospheres) the inertia forces are negligible. Equation (3) then reduces to a balance between the Coriolis term and the pressure-gradient term - the *geostrophic*[1] balance,

$$2\boldsymbol{\Omega} \times \mathbf{v} = -\nabla P. \tag{5}$$

Taking the curl of equation (5), one obtains

$$-(2\boldsymbol{\Omega} \cdot \nabla)\mathbf{v} = \mathbf{0}. \tag{6}$$

Thus, the velocity components of such a flow do not vary in the direction of the angular velocity. This leads to the *Taylor-Proudman Theorem*.

Theorem 9.1 All steady, slow motions in a rotating inviscid fluid are very nearly two-dimensional.

Consider an experiment where a solid cylinder of finite height and horizontal ends is steadily towed horizontally in a fluid rotating about the vertical axis. Then, a fluid column coaxial with the solid cylinder is found to move along with it. This is due to the fact that when the lateral velocity is made to vanish at a point along the direction of the angular velocity by a small obstruction, the

1 *Geostrophic* flows are governed by a reduced equation that is mathematically degenerate, being of lower order than the complete equation of motion. Consequently, the reduced equation is incapable of providing a solution that meets all the necessary boundary and initial conditions. So, the regions of highly non-geostrophic flows (occurring say, on the boundaries of the system) are necessary concomitants of geostrophic motion.

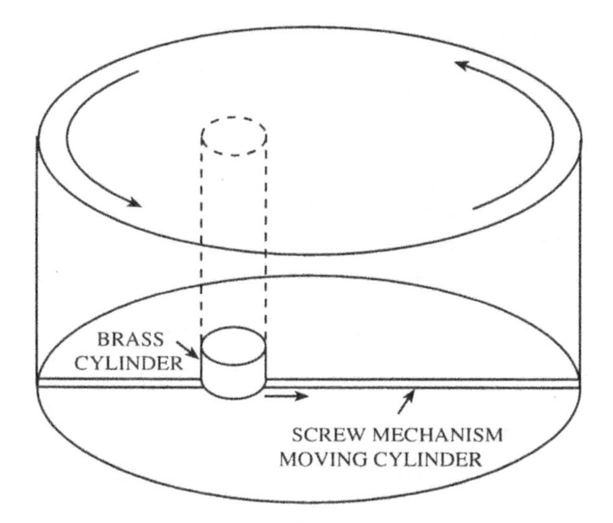

Figure 9.1 "Taylor column" experiment.

former will vanish along the direction of the latter. The fluid will, then, flow two-dimensionally around this so-called *Taylor column*.[2]

If we take the angular velocity $\mathbf{\Omega}$ to be along the z-direction, say $\mathbf{\Omega} = (f/2)\,\hat{\mathbf{i}}_z$, and include a body force $-\nabla\Phi\,(\Phi = gz)$, due to gravity, then, equation (5) gives

$$\frac{1}{\rho}\frac{\partial p}{\partial x} - fv = 0 \tag{7}$$

$$\frac{1}{\rho}\frac{\partial p}{\partial y} + fu = 0 \tag{8}$$

$$\frac{1}{\rho}\frac{\partial p}{\partial z} + g = 0. \tag{9}$$

Equation (9) is simply the hydrostatic approximation, while equations (7) and (8) can be combined to give

2 To demonstrate this, Taylor devised an ingenious experiment. He slowly moved a short cylinder standing on end, across the bottom of a rotating slab of water. He then introduced coloring matter into the water in order to be able to observe the streamlines. These streamlines were then observed to flow discretely around the column of water directly above the cylinder - apparently almost as if this column is impenetrable (see Figure 9.1).

$$u\frac{\partial p}{\partial x} + v\frac{\partial p}{\partial y} = 0. \tag{10}$$

Equation (10) shows that the horizontal streamlines of a steady geostrophic flow are perpendicular to the (horizontal) pressure gradient, i.e. they are along the *isobars*.

On the other hand, elimination of p from equations (7) and (8) gives

$$\frac{\partial u}{\partial x} + \frac{\partial v}{\partial y} = 0 \tag{11}$$

which shows that the strong Coriolis forces compel the flow in a lateral plane to become solenoidal.

9.3 Propagation of Inertial Waves in a Rotating Fluid

Rotation confers a certain elasticity on a fluid. Consequently, a restoring mechanism becomes available in the fluid to sustain propagation of inertial waves involving perturbations on rigid body rotations. The restoring force arises from the tendency of the Coriolis forces to oppose any deviation from solenoidal flow in the lateral plane (as shown by equation (11)).

The equations governing this flow are given by,

$$\frac{\partial v_i}{\partial t} + v_j\frac{\partial v_i}{\partial x_j} = -\frac{\partial P}{\partial x_i} + 2\varepsilon_{ijk}v_j\Omega_k \tag{12}$$

$$\frac{\partial v_i}{\partial x_i} = 0. \tag{13}$$

Seeking solutions of the form

$$q(\mathbf{x}, t) \sim e^{i(k_j x_j + nt)} \tag{14}$$

and linearizing, equations (12) and (13) give

$$inv_i = -ik_iP + 2\varepsilon_{ijk}v_j\Omega_k \tag{15}$$

$$k_j v_j = 0. \tag{16}$$

Multiplying equation (15) by v_i, k_i, Ω_i, one obtains

$$nv_i^2 = 0 \tag{17}$$

$$k^2P + 2i\varepsilon_{ijk}k_iv_j\Omega_k = 0 \tag{18}$$

$$nv_i\Omega_i + Pk_i\Omega_i = 0. \tag{19}$$

Letting,

$$\boldsymbol{\Omega} = (\Omega_x, 0, \Omega_z), \qquad \mathbf{k} = (0, 0, k) \tag{20}$$

equation (16) gives

$$kv_z = 0 \quad \text{or} \quad v_z \equiv 0. \tag{21}$$

Using (20) and (21), equations (17)-(19) become

$$n\left(v_x^2 + v_y^2\right) = 0 \tag{22}$$

$$2i\Omega_x v_y = kP \tag{23}$$

$$n\Omega_x v_x = -kP\Omega_z \tag{24}$$

from which

$$v_x = \pm i v_y \tag{25}$$

and

$$\left(1 - \frac{n^2}{4\Omega_z^2}\right) v_x^2 = 0 \tag{26a}$$

implying

$$n = \pm 2\Omega_z. \tag{26b}$$

We infer from (20) and (25) that the inertial waves are *transverse* and *circularly polarized.*

9.4 Plane Inertial Waves

Having just seen that small disturbances in a uniformly rotating, incompressible fluid can propagate as wavemotions, let us now proceed to explore certain special features of wave propagation in a rotating fluid. One has, for the linearized motion in a rotating fluid,

$$\nabla \cdot \mathbf{v} = 0 \tag{27}$$

$$\frac{\partial \mathbf{v}}{\partial t} + 2\boldsymbol{\Omega} \times \mathbf{v} = -\nabla P. \tag{28}$$

Looking for plane wave solutions of the form

$$\left.\begin{aligned} \mathbf{v} &= \mathbf{V}e^{i(\mathbf{k}\cdot\mathbf{x}-\sigma t)} \\ P &= \mathscr{P}e^{i(\mathbf{k}\cdot\mathbf{x}-\sigma t)} \end{aligned}\right\} \tag{29}$$

equation (27) gives

$$\mathbf{k}\cdot\mathbf{V} = 0 \tag{30}$$

i.e., the velocity is perpendicular to the propagation direction and the waves are *transverse*. This is the only possible type of motion in an incompressible fluid.

Equation (28) gives

$$-i\sigma\mathbf{V} + 2\boldsymbol{\Omega}\times\mathbf{V} = ip\mathbf{k} \tag{31}$$

$$-i\sigma\mathbf{V}\cdot\mathbf{V} = 0. \tag{32}$$

Setting

$$\mathbf{V} = \mathbf{a} + i\mathbf{b} \tag{33a}$$

equation (32), in turn gives

$$a^2 = b^2 \qquad \text{and} \qquad \mathbf{a}\cdot\mathbf{b} = 0 \tag{33b}$$

which implies again that the waves are *circularly polarized*. Furthermore, using equation (30) we see that $\mathbf{a}, \mathbf{b}$, and $\mathbf{k}$ form an orthogonal triad.

Equations (30) and (31) also lead to

$$\left[\sigma^2 - \frac{4}{k^2}(\boldsymbol{\Omega}\cdot\mathbf{k})^2\right]\mathbf{v} = \mathbf{0} \tag{34a}$$

from which, we obtain the dispersion relation

$$\sigma = \pm 2\boldsymbol{\Omega}\cdot\hat{\mathbf{k}} \tag{34b}$$

where

$$\hat{\mathbf{k}} \equiv \frac{\mathbf{k}}{|\mathbf{k}|}.$$

The *phase velocity* of the wave moving in the direction $\hat{\mathbf{k}}$ is then given by

$$\mathbf{C}_p = \frac{\sigma}{|\mathbf{k}|}\hat{\mathbf{k}} = 2\frac{\boldsymbol{\Omega}\cdot\hat{\mathbf{k}}}{|\mathbf{k}|}\hat{\mathbf{k}}. \tag{35}$$

We note from equation (35) that the phase speed is inversely proportional to the magnitude of the wave vector and that the waves are in general, *dispersive* and *anisotropic*. The long waves (small $|\mathbf{k}|$) travel faster and the short waves slower, like the *surface waves* on water (see Section 10.3).

The *group velocity* of the waves is the velocity of energy propagation, and is given, by using (34b)

$$\mathbf{C}_g \equiv \nabla_k \sigma(\mathbf{k}) = \frac{2}{|\mathbf{k}|}\boldsymbol{\Omega} - 2\frac{\boldsymbol{\Omega}\cdot\hat{\mathbf{k}}}{|\mathbf{k}|}\hat{\mathbf{k}} \tag{36a}$$

or

$$\mathbf{C}_g = \frac{2}{|\mathbf{k}|}\hat{\mathbf{k}} \times (\boldsymbol{\Omega}\times\hat{\mathbf{k}}) \tag{36b}$$

which shows that the energy transport is at right angles to the phase velocity.

When $\sigma \sim \boldsymbol{\Omega}\cdot\hat{\mathbf{k}} \approx 0$, the phase velocity $\mathbf{C}_p$ is zero. However, there is a steady propagation of wave energy along the axis of rotation with a group velocity $\mathbf{C}_g = 2\boldsymbol{\Omega}/|\mathbf{k}|$. This leads to the formation of Taylor columns in a rotating fluid.

Consider a flow field which has a rigid rotation with angular velocity $\boldsymbol{\Omega}$ and a uniform streaming motion with velocity $\mathbf{U}$. Now introduce waves propagating in this flow field. One has, for linearized wave motion, in this flow field,

$$\nabla\cdot\mathbf{v} = 0 \tag{27}$$

$$\frac{\partial\mathbf{v}}{\partial t} + \mathbf{U}\cdot\nabla\mathbf{v} + 2\boldsymbol{\Omega}\times\mathbf{v} = -\nabla P. \tag{37}$$

Looking for plane wave solution of the form (29), equation (37) gives

$$\sigma = \underline{+}2\boldsymbol{\Omega}\cdot\hat{\mathbf{k}} + \mathbf{U}\cdot\mathbf{k}. \tag{38}$$

The phase velocity is

$$\mathbf{C}_p = \left(\underline{+}2\frac{\boldsymbol{\Omega}\cdot\hat{\mathbf{k}}}{|\mathbf{k}|} + \mathbf{U}\cdot\hat{\mathbf{k}}\right)\hat{\mathbf{k}} \tag{39}$$

and the group velocity is

$$\mathbf{C}_g = \underline{+}\frac{2}{|\mathbf{k}|}\hat{\mathbf{k}}\times(\boldsymbol{\Omega}\times\hat{\mathbf{k}}) + \mathbf{U}. \tag{40}$$

Note that the frequency is corrected for a *Doppler shift*, and the wave speeds are augmented by the free-stream convection.

Suppose $\mathbf{U}$ were in a direction opposite to that of $\boldsymbol{\Omega}$. Then, energy can be transmitted *upstream*, against the streaming flow if $\mathbf{C}_g\cdot\boldsymbol{\Omega} > 0$, i.e., if

$$\frac{2}{k\Omega}\left[\Omega^2 - (\boldsymbol{\Omega}\cdot\hat{\mathbf{k}})^2\right] > U.$$

Thus, waves with wavenumber k and $\boldsymbol{\Omega}\cdot\hat{\mathbf{k}} = 0$ can propagate against the streaming flow if $U < 2\Omega/k$.

9.5 Forced Wavemotion in a Rotating Fluid

Consider now, forced wavemotion in a rotating fluid. The equation governing the waves turns out to be *elliptic, parabolic*, or *hyperbolic*. This depends on whether the frequency of forcing oscillation ω is greater than, equal to, or less than twice the angular velocity Ω of the fluid. Futhermore, in the hyperbolic case, there exist real characteristic surfaces in the flow (similar to the *Mach cones* in gas dynamics, see Ch. 18,) across which the disturbances become discontinuous. Experiments have confirmed the predicted dependence of the apex angle of the cone on the forcing frequency. We will give here a linearized calculation of motion induced by an oscillatory point source in a rotating incompressible, inviscid fluid (Rao, 1967, Shivamoggi, 1986). Since the flow velocities near the source are large, the linearization of the equations governing the flow is valid only at distances far away from the source. The solutions in the elliptic and hyperbolic cases are obtained using the *Fourier-transform* method. In the hyperbolic case, an appropriate *radiation condition* has to be imposed to find the correct solution. We will find that in the elliptic case, the flow is continuous everywhere. In the hyperbolic case, a cone with vertex at the source and axis along the axis of rotation becomes a surface of discontinuity dividing the flow into three separate regions with different characteristic features.

Consider the motion induced by an oscillatory point source of strength $\rho_0 q e^{i\omega t}$, placed at the origin in an inviscid, incompressible fluid rotating with a constant angular velocity Ω about the x-direction through the origin. The linearized equations for this flow, when referred to a frame rotating with the fluid are

$$\frac{\partial u'}{\partial x} + \frac{\partial v'}{\partial y} + \frac{\partial w'}{\partial z} = q e^{i\omega t} \delta(x) \delta(y) \delta(z) \tag{41}$$

$$\frac{\partial u'}{\partial t} = \frac{\partial P'}{\partial x} \tag{42}$$

$$\frac{\partial v'}{\partial t} - 2\Omega w' = \frac{\partial P'}{\partial y} \tag{43}$$

$$\frac{\partial w'}{\partial t} + 2\Omega v' = \frac{\partial P'}{\partial z} \tag{44}$$

where

$$P' \equiv -\frac{p'}{\rho}.$$

The primes denote the disturbances. The basic state of the fluid (denoted by subscript 0) is governed by,

$$\left.\begin{aligned} -2\Omega^2 y &= -\frac{\partial p_0}{\partial y} \\ -2\Omega^2 z &= -\frac{\partial p_0}{\partial z}. \end{aligned}\right\}. \tag{45}$$

Using equations (41)-(44), one obtains

$$\frac{\partial^2}{\partial t^2}\nabla^2 P' + 2\Omega\frac{\partial^2}{\partial t^2}\left(\frac{\partial w'}{\partial y} - \frac{\partial v'}{\partial z}\right) = -i\omega^3 q e^{i\omega t}\delta(x)\delta(y)\delta(z). \tag{46}$$

Equations (43) and (44) give,

$$\frac{\partial^2}{\partial t^2}\left(\frac{\partial w'}{\partial y} - \frac{\partial v'}{\partial z}\right) = -2\Omega\frac{\partial}{\partial t}\left(\frac{\partial v'}{\partial y} + \frac{\partial w'}{\partial z}\right) \tag{47a}$$

which, may be rewritten using equations (41) and (42), as,

$$\frac{\partial^2}{\partial t^2}\left(\frac{\partial w'}{\partial y} - \frac{\partial v'}{\partial z}\right) = -2\Omega i\omega q e^{i\omega t}\delta(x)\delta(y)\delta(z) + 2\Omega\frac{\partial^2 P'}{\partial x^2}. \tag{47b}$$

Using (47a) and (47b), equation (46) becomes

$$\frac{\partial^2}{\partial t^2}\nabla^2 P' + 4\Omega^2\frac{\partial^2 P'}{\partial x^2} = i\omega\left(4\Omega^2 - \omega^2\right) q e^{i\omega t}\delta(x)\delta(y)\delta(z). \tag{48}$$

Applying Fourier transform according to

$$F(\ell, m, n)\, e^{i\omega t} = \frac{1}{(2\pi)^{3/2}}\int_{-\infty}^{\infty}\int_{-\infty}^{\infty}\int_{-\infty}^{\infty} e^{i(\ell x+my+nz)} f(x,y,z,t)\, dxdydz \tag{49}$$

equation (48) gives, on inversion,

$$P' = \frac{e^{i\omega t}}{8\pi^3}\int_{-\infty}^{\infty}\int_{-\infty}^{\infty}\int_{-\infty}^{\infty}\frac{iq\omega\left(4\Omega^2 - \omega^2\right)}{-\ell^2\left(4\Omega^2 - \omega^2\right) + \omega^2\left(m^2 + n^2\right)} e^{i(\ell x+my+nz)} d\ell\, dmdn. \tag{50}$$

We shall evaluate the integral in (50) by first performing the ℓ-integration by the *method of residues*, and then, the m, n integrations by changing the variables. We have two different cases to consider here.[3]

$$\begin{aligned} &(i)\ B_1^2 \equiv \omega^2 - 4\Omega^2 > 0: \quad \text{elliptic case} \\ &(ii)\ B_2^2 \equiv 4\Omega^2 - \omega^2 > 0: \quad \text{hyperbolic case.} \end{aligned} \tag{51}$$

3 For the case with no oscillatory point source ($q = 0$), equation (48) gives, assuming $P' \sim e^{-i\omega_o t}$,

$$\frac{\partial^2 P'}{\partial y^2} + \frac{\partial^2 P'}{\partial z^2} + \left(1 - \frac{4\Omega^2}{\omega_o^2}\right)\frac{\partial^2 P'}{\partial x^2} = 0$$

which is elliptic/hyperbolic for $\omega_o \gtrless 2\Omega$. Assuming further that,

$$P' \sim e^{i(k_y y + k_z z)}$$

we obtain the dispersion relation,

$$\omega^2 k^2 - 4\Omega^2 k_x^2 = 0$$

from which,

$$\left.\begin{aligned} C_{g_x} &= \pm\frac{2\Omega}{k^3}\left(k_y^2 + k_z^2\right) \\ C_{g_y} &= \mp\frac{2\Omega}{k^3}k_x k_y \\ C_{g_z} &= \mp\frac{2\Omega}{k^3}k_x k_z \end{aligned}\right\}.$$

For a point source of constant frequency ω_o on the x-axis, we have for the hyperbolic case,

$$\frac{x}{\sqrt{y^2+z^2}} = \frac{C_{g_x}}{\sqrt{C_{g_y}^2 + C_{g_z}^2}} = \pm\sqrt{\frac{k^2}{k_x^2} - 1} = \pm\sqrt{\frac{4\Omega^2}{\omega^2} - 1}.$$

So, the disturbance is found on the characteristic cone of semivortex angle $\tan^{-1}\left(\frac{1}{\sqrt{4\Omega^2/\omega^2 - 1}}\right)$ around the x-axis.

(i) The Elliptic Case

The poles of the integrand in (50) now occur at,

$$\ell = \pm \frac{i\omega s}{B_1} \tag{52}$$

where

$$s^2 \equiv m^2 + n^2.$$

For $x > 0$, the path of ℓ-integration along the $Re(\ell)$-axis is closed by a large semicircle in the upper-half of the ℓ-plane. The contribution to the integral in (50), then comes from the positive values in (52), and one obtains

$$P' = -\frac{iqB_1 e^{i\omega t}}{8\pi^2} \int_{-\infty}^{\infty} \int_{-\infty}^{\infty} \frac{e^{(-\omega s/B_1)x + imy + inz}}{s} dmdn. \tag{53}$$

Changing the variables, according to

$$\left.\begin{aligned} &x = r\cos\theta, \quad y = r\sin\theta\cos\phi, \quad z = r\sin\theta\sin\phi \\ &m = s\cos\psi, \quad n = s\sin\psi \end{aligned}\right\} \tag{54}$$

one obtains from (53)

$$P' = -\frac{iqB_1 e^{i\omega t}}{8\pi^2} \int_0^{\infty} \int_0^{2\pi} \frac{e^{-(s\omega r\cos\theta)/B_1}}{s} e^{ir\sin\theta \cdot s\cos(\psi - \phi)} s ds d\psi$$

or[4]

$$P' = -\frac{iqB_1 e^{i\omega t}}{4\pi} \int_0^{\infty} \frac{e^{-(s\omega r\cos\theta)/B_1}}{s} J_0(r\sin\theta \cdot s) s ds \tag{55}$$

4 Note the relation,

$$\frac{1}{2\pi} \int_0^{2\pi} e^{i\lambda\cos(\theta - \phi)} d\theta = J_o(\lambda).$$

where $J_{\upsilon}(x)$ is the *Bessel function* of order υ. Evaluating the integral in (55), one obtains[5]

$$P' = -\frac{iqB_1^2 e^{i\omega t}}{4\pi r\sqrt{\omega^2 - 4\Omega^2 \sin^2\theta}}. \tag{56}$$

Observe that, according to equation (56), the flow is *continuous* everywhere.

(ii) The Hyperbolic Case

The poles of the integrand in (50) now occur at

$$\ell = \pm\frac{\omega s}{B_2}. \tag{57}$$

For $x > 0$, the path of ℓ-integration along the $Re(\ell)$-axis is closed again by a large semicircle in the upper-half of the ℓ-plane. In order to determine which pole in (57) contributes to the integral in (50), one imposes an appropriate *radiation condition* (Lighthill, 1960). This is done by introducing in ω a small negative imaginary part $-i\varepsilon$ ($\varepsilon \Rightarrow 0^+$). This is equivalent to posing, a more realistic *initial-value problem*, where the applied steady-source is "*switched-on*" at time $t = 0$. The contribution to the integral in (50), then, comes from the negative value in (57), and one obtains, upon changing the variables, according to (54),

$$P' = -\frac{qB_2 e^{i\omega t}}{8\pi^2}\int_0^{\infty}\int_0^{2\pi}\frac{e^{-(is\omega r\cos\theta)/B_2}}{s}e^{ir\sin\theta\cdot s\cos(\psi-\phi)}s\,ds\,d\psi$$

or

$$P' = -\frac{qB_2 e^{i\omega t}}{4\pi}\int_0^{\infty}\frac{e^{-(is\omega r\cos\theta)/B_2}}{s}J_0(r\sin\theta\cdot s)\,s\,ds. \tag{58}$$

---

5 Note the relation,

$$\int_0^{\infty} e^{-ax}J_0(bx)\,dx = \frac{1}{\sqrt{a^2+b^2}}.$$

Carrying out the integration in (58) one obtains

$$P' = \begin{cases} \dfrac{iqB_2^2 e^{i\omega t}}{4\pi r\sqrt{\omega^2 - 4\Omega^2 \sin^2\theta}}, & \dfrac{\omega}{2\Omega} > \sin\theta \\[2ex] \dfrac{qB_2^2 e^{i\omega t}}{4\pi r\sqrt{\Omega^2 \sin^2\theta - \omega^2}}, & \dfrac{\omega}{2\Omega} < \sin\theta. \end{cases} \tag{59}$$

This shows that the pressure becomes infinity on the cone $\omega/2\Omega = \sin\theta$ with vertex at the source and axis along the axis of rotation. This is a surface of discontinuity in the flow dividing it into three separate regions with different characteristic features. Observe that the solution inside the cone resembles that of the elliptic case, as it should. The adjustment of the flow from inside the cone to outside is made through thin viscous shear layers about the surface of the cone.

Note that, when $\omega = 2\Omega$, the pressure vanishes everywhere except on the cone of discontinuity, which now becomes a plane through the source and perpendicular to the axis of rotation.

9.6 Slow Motion along the Axis of Rotation

Let us consider a simple case of transient evolution of inertial waves to illustrate the process by which these waves organize to form a Taylor column ahead of the body producing the waves. Specifically, consider a finite disk of radius r_0 that is initially rotating rigidly about an axis perpendicular to it within an infinite body of fluid. The disk is then moved slowly along the axis of rotation. As far as steady longitudinal motion of the body is concerned, only its projected shape on a plane perpendicular to the axis of rotation is important. It appears that a body of any shape in slow forward motion is related to a disk of the same projected cross-sectional area.

The linearized problem for the slow forward motion of the disk in a rotating frame of cylindrical coordinates with the z-axis along the axis of rotation is governed by

$$\nabla \cdot \mathbf{v} = 0 \tag{60}$$

$$\frac{\partial \mathbf{v}}{\partial t} + 2\boldsymbol{\Omega} \times \mathbf{v} = -\frac{1}{\rho}\nabla p \tag{61}$$

with boundary conditions,

$$\left.\begin{array}{l} t \geq 0: \ \mathbf{v}\cdot\hat{\mathbf{i}}_z = U \quad \text{on} \quad r \leq r_0 \\ \mathbf{v} \Rightarrow 0 \quad \text{as} \quad |r| \Rightarrow \infty \end{array}\right\} \tag{62}$$

$$t \leq 0: \ \mathbf{v} \equiv 0. \tag{63}$$

We shall apply the *Laplace* and *Hankel transforms* consecutively to reduce the above partial differential equation to an ordinary differential equation with z as the independent variable. Let u_r, u_θ, u_z denote the velocity components along the r, θ, z-directions. We introduce a stream function ψ, through

$$u_r = \frac{1}{r}\frac{\partial \psi}{\partial z}, \quad u_z = -\frac{1}{r}\frac{\partial \psi}{\partial r} \tag{64}$$

so that equation (60) is identically satisfied. Applying Laplace transform, according to

$$\Psi = \int_0^\infty e^{-st}\psi dt \tag{65}$$

and nondimensionalizing r and z using r_0, equation (61) leads to

$$r\frac{\partial}{\partial r}\frac{1}{r}\frac{\partial \Psi}{\partial r} + \left(1 + \frac{4\Omega^2}{s^2}\right)\frac{\partial^2 \Psi}{\partial z^2} = 0 \tag{66a}$$

with, (62) giving

$$\left.\begin{array}{l} z = 0, r \leq 1: \ -\frac{1}{r}\frac{\partial \Psi}{\partial r} = \frac{U}{s} \\ \qquad\quad r \geq 1: \ -\frac{1}{r}\frac{\partial \Psi}{\partial z} = 0 \\ z \Rightarrow \infty: \ \Psi \Rightarrow 0 \end{array}\right\}. \tag{67}$$

Setting

$$\Psi(r,z) = r\chi(r,z) \tag{68}$$

equation (66a) becomes

$$\frac{\partial^2 \chi}{\partial r^2} + \frac{1}{r}\frac{\partial \chi}{\partial r} - \frac{1}{r^2}\chi + \left(1 + \frac{4\Omega^2}{s^2}\right)\chi = 0. \tag{66b}$$

Applying Hankel transform, according to

$$\overline{\chi}(k,z) = \int_{o}^{\infty} r\chi(r,z) J_1(kr)\, dr \tag{69}$$

equation (66b) leads to

$$-k^2\overline{\chi} + \left(1 + \frac{4\Omega^2}{s^2}\right)\frac{d^2\overline{\chi}}{dz^2} = 0. \tag{70}$$

Equation (70) yields

$$\overline{\chi}(k,z) = \frac{1}{k}A(k)\,e^{-\dfrac{|k|z}{\sqrt{1+4\Omega^2/s^2}}} \tag{71}$$

and (68) leads to

$$\Psi = r\int_0^{\infty} A(k) J_1(kr)\, e^{-k|z|/\sqrt{1+4\Omega^2/s^2}}\, dk. \tag{72}$$

Then, (67) gives

$$\left.\begin{aligned} &r \le 1 : \int_0^{\infty} kA(k) J_0(kr)\, dk = -\frac{U}{s} \\ &r > 1 : \int_0^{\infty} kA(k) J_1(kr)\, dk = 0 \\ &\quad \text{or} \\ &\quad \frac{1}{r}\int_o^{\infty} (kA(k))' J_o(kr)\, dk = 0 \end{aligned}\right\} \tag{73a,b}$$

from which, one has[6]

$$kA(k) = \frac{2U}{\pi s}\left(\cos k - \frac{\sin k}{k}\right). \tag{74}$$

Using (74), (72) leads to

$$\Psi = \frac{2Ur}{\pi}\int_0^\infty \frac{J_1(kr)}{k}\left(\cos k - \frac{\sin k}{k}\right)\cdot\frac{1}{s}\, e^{-(kz)/\sqrt{1+4\Omega^2/s^2}}dk \tag{75}$$

and equations (61) and (64) give

$$U_\theta = \frac{4U\Omega}{\pi}\int_0^\infty J_1(kr)\cdot\left(\cos k - \frac{\sin k}{k}\right)\frac{e^{-(kz)/\sqrt{1+4\Omega^2/s^2}}}{s^2\sqrt{1+4\Omega^2/s^2}}dk \tag{76}$$

U_θ being the Laplace transform of u_θ. While inverting the Laplace transform, note that there is a simple pole at $s = 0$ and two branch points at $s = \underline{+}2i\Omega$, and a branch cut is introduced between the two. Let us now find an asymptotic approximation for large t. The condition that a certain minimum time must

6 The *dual integral equations* (73a,b) may be solved as follows. Noting the results,

$$\left.\begin{aligned} 0<r<1 &: \int_0^\infty J_0(kr)\frac{\sin k}{k}dk = \frac{\pi}{2} \\ r>1 &: \int_0^\infty J_0(kr)\sin k dk = 0 \end{aligned}\right\}$$

we obtain from equations (73a,b)

$$kA(k) = C\cos k - \frac{2U}{\pi s}\frac{\sin k}{k}$$

where C is an arbitrary constant. Note that,

$$\lim_{k\Rightarrow 0} kA(k) = 0$$

which implies

$$C = \frac{2U}{\pi s}.$$

elapse before waves of a given wavenumber can arrive at a given location on the axis of rotation implies that

$$t > \frac{z}{C_g}. \tag{77}$$

Here,

$$C_g = \frac{2\Omega}{k}$$

is the group velocity of plane waves of wavenumber k whose phase velocity is in a direction perpendicular to the axis of rotation. The group velocity is, therefore, along the axis of rotation. Thus, the Laplace inversion of (75) and (76) leads to

$$\psi \sim \frac{2Ur}{\pi} \int_0^{2\Omega t/z} \frac{J_1(kr)}{k} \left(\cos k - \frac{\sin k}{k} \right) dk \tag{78}$$

$$u_\theta \sim \frac{2U}{\pi} \int_0^{2\Omega t/z} J_1(kr) \left(\cos k - \frac{\sin k}{k} \right) dk. \tag{79}$$

In the frame of reference moving with the disk, the stream function is given by,

$$\tilde{\psi} = \frac{1}{2} Ur^2 + \psi \tag{80}$$

and u_θ is unchanged. Figure 9.2 shows the instantaneous streamlines $\tilde{\psi}$ about the disk. Observe the appearance of a stagnation point in the flow, the broad bluff front, and the reverse cellular flow behind it. This is similar to the flow about an imaginary obstacle with projected cross-sectional area same as that of the disk, but of a constantly increasing dimension along the axis of rotation. This represents, of course, the formation and development of a Taylor column. Recall that this column cannot form if the waves are convected downstream by a stream flowing with a speed exceeding their group velocities.

Figure 9.3 shows the variation of the azimuthal velocity u_θ with respect to r for different values of $2\Omega t r_0/z$. Observe the formation of a velocity discontinuity across the cylinder circumscribing the disk, with the generators parallel to the axis of rotation. This implies simply, the existence of a thin viscous shear layer there. In this layer, the Taylor-Proudman Theorem does not apply, and there is, in fact, some interchange of fluid between the interior and the exterior of the Taylor column.

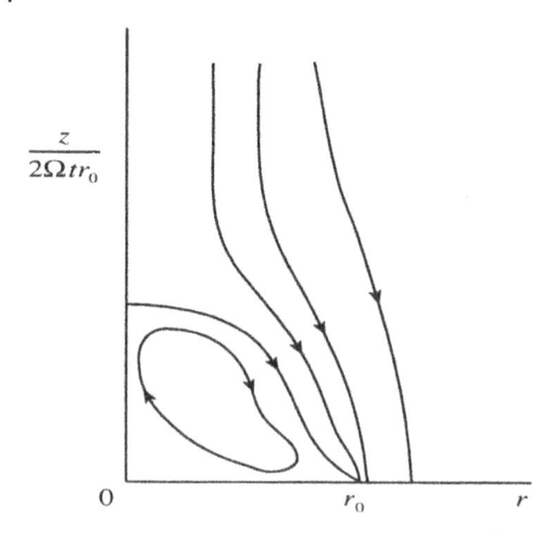

Figure 9.2 Instantaneous streamline pattern about a rotating disk moving slowly perpendicular to itself along the axis of rotation (from Greenspan, 1968).

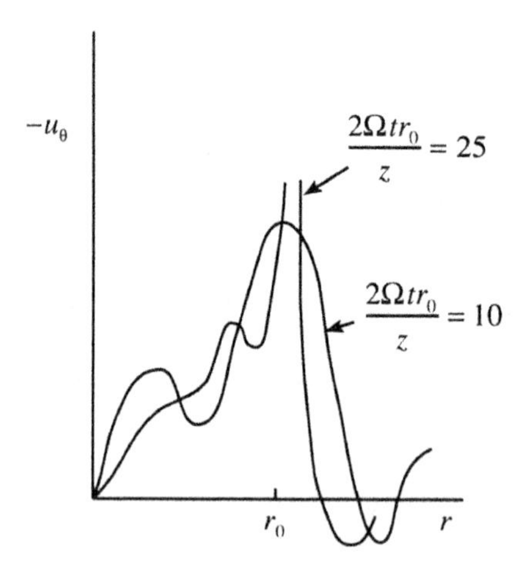

Figure 9.3 Variation of the azimuthal velocity with radius for different values of $2\Omega t r_0/z$ (from Greenspan, 1968).

9.7 Rossby Waves

Rossby waves, occuring in oceans and atmospheres are transverse waves propagating in a plane perpendicular to the earth's angular velocity $\mathbf{\Omega}$. They arise due to the variation of the vertical component of $\mathbf{\Omega}$ with latitude, i.e., $\partial\Omega_z/\partial y \neq 0$. If this variation is assumed to be constant, then it supplies a steady gradient of planetary vorticity Ω_z, called the Coriolis parameter along a meridian. Furthermore, it simulates the first order effect of the planet's curvature in the tangent plane (called the β-plane).[7] The law of vorticity conservation then implies that a displaced fluid element is subjected to a restoring torque.

In order to study large-scale motions in a fluid layer on a rotating globe, one further makes the following assumptions,

- the upper boundary of the layer of fluid remains spherical;
- the bulk motion of the fluid is nearly horizontal and is uniform across this fluid layer.[8]

The equations of motion in spherical polar coordinates then are

$$\frac{Dv_\theta}{Dt} - \frac{v_\theta^2 \cot\theta}{R} - f v_\varphi = -\frac{1}{\rho R}\frac{\partial P}{\partial \theta} \tag{81}$$

$$\frac{Dv_\varphi}{Dt} + \frac{v_\theta v_\varphi \cot\theta}{R} + f v_\theta = -\frac{1}{\rho R \sin\theta}\frac{\partial P}{\partial \varphi} \tag{82}$$

where f is twice the vertical component of the angular velocity $\boldsymbol{\Omega}$,

$$f \equiv 2\Omega\cos\theta \tag{83}$$

the complementary latitude angle θ is measured from the north pole, and

$$\frac{D}{Dt} \equiv \frac{\partial}{\partial t} + \frac{v_\theta}{R}\frac{\partial}{\partial \theta} + \frac{v_\varphi}{R\sin\theta}\frac{\partial}{\partial \varphi}.$$

Here, P is the effective pressure that includes the centrifugal force arising from the rotation of the coordinate system (see (2)). We expressed in (83), the fact that the angular velocity $\boldsymbol{\Omega}$ of the rotating sphere makes different angles with the vertical at different latitudes.

The radial component of the vorticity of the fluid motion relative to the globe is

$$\omega = \frac{1}{R\sin\theta}\left\{\frac{\partial\left(v_\varphi \sin\theta\right)}{\partial\theta} - \frac{\partial v_\theta}{\partial\varphi}\right\} \tag{84}$$

and one obtains, from equations (81) and (82),

$$\frac{D\omega}{Dt} = -\frac{v_\theta}{R}\frac{df}{d\theta} - \frac{1}{R\sin\theta}\left\{\frac{\partial\left(v_\theta \sin\theta\right)}{\partial\theta} + \frac{\partial v_\varphi}{\partial\varphi}\right\}(f+\omega). \tag{85}$$

---

7 So, in the β-plane approximation, one ignores the geometrical effects of the curvature of the earth and yet retains its dynamical effects via the latitudinal variation of the Coriolis parameter.

8 Thanks to the density stratification, no matter how small, if the fluid layer is thin, then the geophysical flows typically have a horizontal scale larger than the vertical scale. Furthermore, the velocity vectors are nearly horizontal. In such a situation, the fluid layer motion is primarily influenced by the vertical component of the angular velocity Ω.

Now, from the conservation of mass, one has

$$\frac{1}{R\sin\theta}\left[\frac{\partial\left(v_\theta\sin\theta\right)}{\partial\theta}+\frac{\partial v_\varphi}{\partial\varphi}\right]=-\frac{1}{H}\frac{DH}{Dt} \tag{86}$$

where H is the thickness of the fluid layer. We consider here, flows with a horizontal characteristic length L larger than the layer thickness H.

Using (86), equation (85) gives the *Ertel Theorem:*

$$\frac{D}{Dt}\left(\frac{f+\omega}{H}\right)=0. \tag{87}$$

According to equation (87), the absolute vorticity of each individual column of fluid changes whenever this column moves to a place where the height of the fluid column is different. The cyclonic vertical component of the earth's angular velocity is stronger (weaker) as one moves towards higher (lower) latitudes. Equation (87) therefore implies that a fluid column of constant height will experience decreasing (increasing) cyclonic rotation when displaced toward higher (lower) latitudes.

Consider localized motions, near the latitude $\theta=\theta_0$. If these motions have length scales relatively small compared to the radius of the earth, then they maybe modeled by a plane layer flow. Thus, we have a plane horizontal layer of fluid rotating about a vertical axis with angular velocity equal to the vertical component $f=2\Omega\cos\theta_0$ of the earth's angular velocity. Introducing

$$\left.\begin{array}{ll} x=R\sin\theta_0\cdot\varphi, & \text{the eastward coordinate} \\ y=R\left(\theta_0-\theta\right), & \text{the northward coordinate} \end{array}\right\} \tag{88}$$

one has, for the material derivative,

$$\frac{D}{Dt}\equiv\frac{\partial}{\partial t}+u\frac{\partial}{\partial x}+v\frac{\partial}{\partial y}.$$

One may improve upon this approximation by including the variation of the vertical component f of the earth's angular velocity $\mathbf{\Omega}$ with latitude. Thus, one may write

$$f=f_0+\beta y \tag{89}$$

where

$$\beta\equiv\frac{2\Omega\sin\theta_0}{R}.$$

Thus, the flow field may now be regarded as occurring in a plane layer with a normal rotation vector f whose magnitude varies linearly in the y-direction. This is called the *β-plane approximation.*

Let us demonstrate the existence of Rossby waves in a plane layer of fluid with uniform thickness and linearly varying f. For such a wave in a fluid otherwise at rest, the stream function is assumed to be,

$$\psi \sim e^{i(kx+\ell y-\sigma t)}. \tag{90}$$

The concomitant relative vorticity is

$$\omega = -\nabla^2\psi = \left(k^2 + \ell^2\right)\psi. \tag{91}$$

Then, the invariance of the absolute vorticity of a material element set forth by equation (87) leads to

$$-\frac{\partial\psi}{\partial x}\frac{df}{dy} + \left(k^2 + \ell^2\right)\frac{\partial\psi}{\partial t} = i\psi\left[-\beta k - \sigma\left(k^2 + \ell^2\right)\right] = 0 \tag{92}$$

from which,

$$\sigma = -\frac{\beta k}{k^2 + \ell^2}. \tag{93}$$

These are transverse waves, for which the fluid velocity is everywhere at right angles to the wave vector (k, ℓ). Rossby waves propagate toward *west*, opposite to the direction of earth's rotation, and have far *lower* frequency than that of the inertial waves for which $\sigma > 2\Omega$ (see (38)).

Let us now consider the Rossby wave generation in a situation that includes in the basic state, a horizontal flow perpendicular to the planetary vorticity gradient. Such a situation arises in the atmosphere when *zonal winds* move (east-west) past a topographical obstale. Thus, we consider the steady flow over a ridge in the form of a step along a north-south line at $x = 0$ (Figure 9.4). Let the oncoming stream have a uniform velocity U_0 along the x-direction and zero relative vorticity. One then has over the ridge, from the conservation of mass and absolute vorticity,

$$U_1 = \frac{U_0 H_0}{H_1}, \quad \omega = \left(f_0 + \beta y\right)\frac{H_1 - H_0}{H_0}. \tag{94}$$

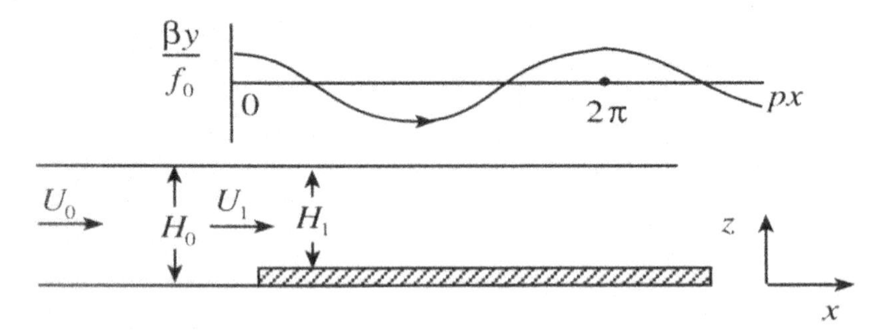

Figure 9.4 Streamline pattern for the steady flow over a ridge in the form of a step along the north-south line at $x = 0$ (from Batchelor, 1967).

Now, at $x = 0$, one has for the stream function $\psi = U_1 y$. So, (94) yields[9]

$$f + \omega = \frac{H_1}{H_0}\left(f_0 + \frac{\beta}{U_1}\psi\right). \tag{95}$$

Thus, in the region $x > 0$, one has

$$\nabla^2\psi = -\omega = f - \frac{H_1}{H_0}\left(f_0 + \frac{\beta}{U_1}\psi\right) \tag{96}$$

or

$$\nabla^2\psi = f_0\left(\frac{H_0 - H_1}{H_0}\right) + \beta y - p^2\psi \tag{97}$$

where

$$p^2 \equiv \frac{\beta H_1}{U_1 H_0}.$$

---

9 For steady flows, with $H = const$,

$$[\psi, (f + \omega)] = 0$$

where

$$[A, B] \equiv \frac{\partial A}{\partial x}\frac{\partial B}{\partial y} - \frac{\partial A}{\partial y}\frac{\partial B}{\partial x}.$$

Thus,

$$(f + \omega) = g(\psi).$$

The initial conditions for this flow are,

$$x = 0 : \psi = U_1 y, \frac{\partial \psi}{\partial x} = 0. \tag{98}$$

Thus,

$$\psi = (y + A) F(x) + \frac{1}{p^2}\left[f_0\left(1 - \frac{H_1}{H_0}\right) + \beta y\right]. \tag{99}$$

Using (99), the initial value problem (97), (98) becomes,

$$\frac{d^2F}{dx^2} + p^2 F = 0 \tag{100}$$

$$x = 0 : \frac{dF}{dx} = 0. \tag{101}$$

Thus,

$$F(x) = B\cos px. \tag{102}$$

Using (102), (99) becomes

$$\psi = (y + A) B\cos px + \frac{1}{p^2}\left[f_0\left(1 - \frac{H_1}{H_0}\right) + \beta y\right]. \tag{103}$$

The initial conditions (98) then give

$$x = 0 : \psi = (y + A) B + \frac{1}{p^2}\left[f_o\left(1 - \frac{H_1}{H_0}\right) + \beta y\right] = U_1 y \tag{104}$$

from which we obtain

$$\left.\begin{aligned} B + \frac{\beta}{p^2} &= U_1 \\ AB + \frac{f_0}{p^2}\left(1 - \frac{H_1}{H_0}\right) &= 0 \end{aligned}\right\}. \tag{105}$$

This yields,

$$B = U_1\left(1 - \frac{H_0}{H_1}\right), A = \frac{f_0}{\beta}. \tag{106}$$

Using (106), (103) becomes

$$\psi = \left(y + \frac{f_0}{\beta}\right) U_1\left(1 - \frac{H_0}{H_1}\right)\cos px + U_1\left[\frac{f_0}{\beta}\left(\frac{H_0}{H_1} - 1\right) + \frac{H_0}{H_1} y\right] \tag{107a}$$

which may be rewritten as

$$\psi = U_1 y + \frac{U_1}{\beta}\left(\frac{H_0}{H_1} - 1\right)(f_0 + \beta y)(1 - \cos px) \tag{107b}$$

which is sketched in Figure 9.4.

So, the Rossby waves are driven by the nonuniformity of the Coriolis parameter. Furthermore, the effect of the bottom topography is to generate a relative vorticity at $x = 0$ and turn the flow toward the south.

Exercises

1. For the steady case, find an integral of motion for a rotating flow of a fluid.
2. Consider linear Rossby waves in a plane layer of fluid of variable thickness $H(x, y, t)$ in the β-plane approximation. Assuming the perturbations to vary like $e^{i(kx+\ell y-\sigma t)}$ show that the dispersion relation for linear Rossby waves is given by

$$\sigma = -\frac{\beta k}{k^2 + \ell^2 + (f_o^2/gH_o)}$$

where g is the acceleration due to gravity and the subscript o denotes the mean values.

10

Water Waves

The phenomena of water waves historically provided a great deal of impetus and framework for the development of the theory of dispersive waves. The purpose of this chapter is to give a brief account of the mathematical theory of wavemotion in fluids with a free surface subjected to gravitational and other forces.

There are two types of surface-wave motions. *Shallow-water waves* arise when the wavelength of the oscillations is much greater than the depth of the fluid. Here, the vertical acceleration of the fluid is small in comparison to the horizontal acceleration. *Surface waves* correspond to disturbances that do not extend far below the surface. The wavelength is much less than the depth of the fluid, and the vertical acceleration is, then, no longer negligible.

The features that make an analysis of water waves difficult are

- the presence of nonlinearities;
- the free surface being unknown *a priori*, besides being variant with time.

In order to make progress with the theory of water waves, it is, in general, necessary to simplify the model by making special hypotheses of one kind or another. These suggest themselves on the basis of general physical circumstances contemplated in a given class of problems. Thus, two approximate theories result when

- the amplitude of the wave is considered to be small (surface waves);
- the depth of the water is considered to be small with respect to the wavelength (shallow-water waves).

The first hypothesis leads to a linear theory for boundary-value problems of nearly classical type. The second leads to a nonlinear theory for initial-value problems, which, in the lowest order is of the type corresponding to the wave propagation in compressible fluids (see Chapter 15).

Introduction to Theoretical and Mathematical Fluid Dynamics, Third Edition.
Bhimsen K. Shivamoggi.

10.1 Governing Equations

The equilibrium configuration of a fluid in a container of finite size is one of rest with the plane surface. One may produce a wavemotion on the surface which is due to the action of gravity that takes place in the direction of restoring the undisturbed state of rest. If the wavemotion is assumed to have started from rest relative to the undisturbed state of flow, which is itself assumed irrotational, then the wavemotions will be irrotational. The wave propagation is taken to occur along the x-direction, and the gravity acts opposite to the z-direction. One has for the velocity potential Φ,

$$\frac{\partial^2 \Phi}{\partial x^2} + \frac{\partial^2 \Phi}{\partial z^2} = 0 \tag{1}$$

and at a rigid stationary boundary, the condition signifying the fluid impenetrability at the boundary,

$$\frac{\partial \Phi}{\partial n} = 0. \tag{2a}$$

Let $z = \xi(x, t)$ denote the displacement of the free surface from its mean position (or equilibrium position). Since a fluid particle on that surface will remain there, we have the following *kinematic* condition expressing the fact that the free surface remains a material surface,

$$\frac{D}{Dt}(z - \xi) = 0. \tag{3}$$

On linearizing, (3) gives

$$\frac{\partial \Phi}{\partial z} = \frac{\partial \xi}{\partial t}. \tag{4}$$

The *dynamic* condition at the free surface is (see Section 4.1),

$$-p = T\left(\frac{1}{R_1} + \frac{1}{R_2}\right) \tag{5}$$

where R_1, R_2 are the two principal radii at the free surface, counted positive when the center of curvature is above the surface (see Section 4.1), and T is the surface tension. Since the interface is taken to be planar, we have,

$$\frac{1}{R_1} = 0, \frac{1}{R_2} = \frac{(\partial^2 \xi / \partial x^2)}{\left[1 + (\partial \xi / \partial x)^2\right]^{3/2}}. \tag{6}$$

On linearizing, (6) gives

$$p \approx -T \frac{\partial^2 \xi}{\partial x^2}. \tag{7}$$

The *Bernoulli integral*,

$$\frac{p}{\rho} = -\frac{\partial \Phi}{\partial t} - gz - \frac{u^2 + w^2}{2} + f(t) \tag{8}$$

upon linearization, and combining with (4) and (7) gives

$$z = 0: \frac{\partial^2 \Phi}{\partial t^2} = \left(\frac{T}{\rho}\frac{\partial^2}{\partial x^2} - g\right)\frac{\partial \Phi}{\partial z}. \tag{9}$$

Here, the fluid velocity is $\mathbf{v} = (u, v, w)$.

We will neglect the no-slip condition at the bottom wall and the shear stress of the free surface. This implies that there must be boundary layers (see Chapter 21) at the bottom wall and the free surface which are contaminated by vorticity. However, if the viscosity of the fluid is small, these boundary layers are thin and may be neglected.

10.2 A Variational Principle for Surface Waves

Bateman (1929) used an expression for the pressure as the *Lagrangian* and derived the equations of motion in an inviscid, incompressible fluid from a variational principle. Luke (1967) showed that a simple extension of this variational principle also provides the boundary conditions appropriate for fluid motion with a free surface.

Consider an inviscid fluid of density ρ subjected to a gravitational field $\mathbf{g} = -g\hat{\mathbf{i}}_z$, and confined in the region $0 < z < h$. Then, the variational principle is

$$\delta J \equiv \delta \int_{t_1}^{t_2} \int_{x_1}^{x_2} L \, dx \, dt = 0 \tag{10}$$

where

$$L = \int_0^{h(x,t)} \left(\frac{1}{2}\Phi_x^2 + \frac{1}{2}\Phi_z^2 + \Phi_t + gz\right) dz. \tag{11}$$

Here, $\Phi(x, z, t)$ is the velocity potential describing the motion of the fluid. $\Phi(x, z, t)$ and $h(x, t)$ are allowed to vary subject to the restrictions $\delta\Phi = 0$ and $\delta h = 0$ at the boundaries $x = x_1, x = x_2, t = t_1$ and $t = t_2$.

Following the usual procedure in calculus of variations, (10) gives, for small changes $\delta\Phi$ and δh in Φ and h, respectively,

$$\delta J = \int_{t_1}^{t_2} \int_{x_1}^{x_2} \left\{ \left[\frac{1}{2}\Phi_x^2 + \frac{1}{2}\Phi_z^2 + \Phi_t + gz \right]_{z=h} \delta h \right.$$

$$\left. + \int_0^{h(x,t)} (\Phi_x \delta\Phi_x + \Phi_z \delta\Phi_z + \delta\Phi_t)\, dz \right\} dxdt = 0. \tag{12}$$

Upon integrating by parts, (12) gives

$$\delta J = \int_{t_1}^{t_2} \int_{x_1}^{x_2} \left\{ \left[\frac{1}{2}\Phi_x^2 + \frac{1}{2}\Phi_z^2 + \Phi_t + gz \right]_{z=h} \delta h \right.$$

$$+ \left[(-h_x \Phi_x + \Phi_z - h_t)\, \delta\Phi \right]_{z=h} - \left[\Phi_z \delta\Phi \right]_{z=0}$$

$$\left. - \int_0^{h(x,t)} (\Phi_{xx} + \Phi_{zz})\, \delta\Phi dz \right\} dxdt = 0. \tag{13}$$

First let us choose

$$\delta h = 0, \quad [\delta\Phi]_{z=0,h} = 0.$$

Since $\delta\Phi$ is arbitrary otherwise, one deduces from (13) that,

$$0 < z < h: \quad \Phi_{xx} + \Phi_{zz} = 0. \tag{14}$$

Furthermore, $\delta h, [\delta\Phi]_{z=0}$, and $[\delta\Phi]_{z=h}$ may be given arbitrary independent values. So, (13) gives

$$z = h : -h_x \Phi_x + \Phi_z - h_t = 0 \tag{15}$$

$$z = h : \frac{1}{2}\Phi_x^2 + \frac{1}{2}\Phi_z^2 + \Phi_t + gz = 0 \tag{16}$$

$$z = 0 : \Phi_z = 0. \tag{17}$$

Equation (15) describes the kinematic condition on the velocity field at the interface. Equation (16) describes the force balance at the interface. Equation (17) describes the condition of the fluid impenetrability at the boundary $y = 0$.

10.3 Water Waves in a Semi-Infinite Fluid

If the fluid fills the semi-infinite space $-\infty < z \leq 0, -\infty < x < \infty$, with a free surface at $z = 0$, then, one requires

$$z \Rightarrow -\infty : \frac{\partial \Phi}{\partial z} \Rightarrow 0. \tag{18}$$

From (1) and (18), one finds

$$\Phi = Ce^{kz} \cos k\,(x - ct)\,. \tag{19}$$

Using (19) in the free-surface dynamic condition (9), we obtain the dispersion relation,

$$c^2 = \frac{g}{k} + \frac{kT}{\rho}. \tag{20}$$

This is sketched in Figure 10.1. Observe that according to (20), the phase velocity c takes a minimum value,

$$c_{min} = \left(4\frac{Tg}{\rho}\right)^{1/4}, \text{ for } k = k_* = \left(\frac{\rho g}{T}\right)^{1/2}. \tag{21}$$

For long waves, gravity effects dominate, and (20) yields the *gravity waves* in deep water,

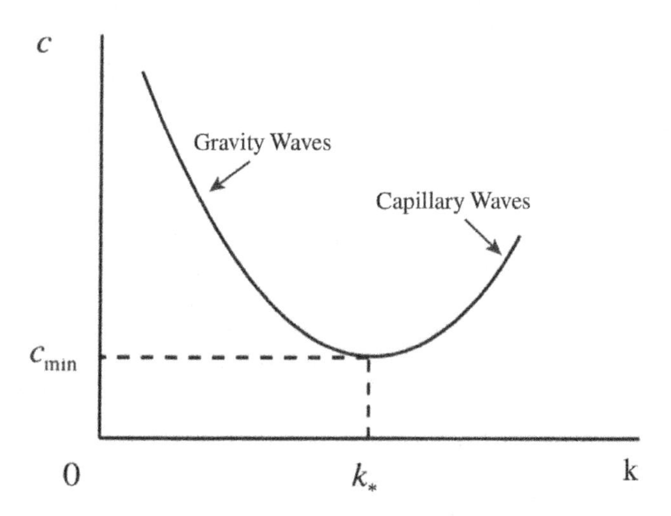

Figure 10.1 Dispersion curve for surfaces waves.

$$c^2 \approx \frac{g}{k}. \tag{22}$$

For short waves, capillary effects dominate, and (11) yields the *capillary waves*,

$$c^2 \approx \frac{kT}{\rho}. \tag{23}$$

Both (22) and (23) show that the water waves are dispersive, which is caused by the sharp density inhomogeneity represented by the free surface in this system.

10.4 Water Waves in a Fluid Layer of Finite Depth

For a fluid layer of finite depth, one needs to comply with the fluid impenetrability condition at the bottom surface which is taken to be rigid and stationary. This condition is given by,

$$z = -h_0 : \frac{\partial \phi}{\partial z} = 0. \tag{2b}$$

Any irrotational-flow solution so obtained involves a non-zero tangential velocity at the boundary. In a viscous fluid, this solution has to be reconciled with the exact boundary condition of zero fluid velocity at a stationary solid surface. This is accomplished through the intervention of a boundary layer (see Chapter 21) between the irrotational flow and the surface.

From (1) and (2b), one obtains

$$\Phi = C \cosh k (z + h_0) \cos k (x - ct) \tag{24}$$

C being an arbitrary constant.

Using (24) in the free-surface dynamic condition (9), we obtain the dispersion relation,

$$c^2 = \left(g + \frac{k^2 T}{\rho} \right) \frac{\tanh k h_0}{k} \tag{25}$$

which reduces to the semi-infinite fluid result (20), in the limit $h_0 \to \infty$.

For long waves, (25) gives the gravity waves in finite-depth water,

$$c^2 \approx g h_0. \tag{26}$$

The dispersion relation shows that short waves, on the other hand, do not feel the effect of finite depth of water, and resemble the capillary waves in a semi-infinite fluid,

$$c^2 \approx \frac{kT}{\rho}. \tag{27}$$

The fluid velocity components are given, from (24), by

$$u = \frac{\partial \Phi}{\partial x} = -kC \cosh k (z + h_0) \sin k (x - ct) \tag{28}$$

$$w = \frac{\partial \Phi}{\partial z} = -kC \sinh k (z + h_0) \cos k (x - ct). \tag{29}$$

For long waves, $\frac{\partial w}{\partial t} \approx 0$, so that the vertical acceleration is negligible, and the pressure distribution in the fluid is then nearly *hydrostatic.*

The locus of a material particle is given by

$$u \equiv \frac{dX}{dt}, \quad w \equiv \frac{dZ}{dt}. \tag{30}$$

So, on using (24) and (29), (30) leads to

$$\left.\begin{aligned} X &= -\frac{C}{c} \cosh k (z + h_0) \cos k (x - ct) \\ Z &= -\frac{C}{c} \sinh k (z + h_0) \sin k (x - ct) \end{aligned}\right\}. \tag{31}$$

This shows that the locus is an ellipse in a vertical plane,

$$\frac{X^2}{A^2} + \frac{Z^2}{B^2} = 1 \tag{32}$$

with the major and minor axes given by,

$$A \equiv \frac{C}{c} \cosh k (z + h_0), \quad B \equiv \frac{C}{c} \sinh k (z + h_0).$$

Note that, at $z = -h_0$, the ellipse degenerates into a horizontal line segment. Consequently, the particles on the bottom surface move only to and fro, and do not rise and fall. For infinite depth, we have

$$\lim_{h_0 \Rightarrow \infty} \frac{A}{B} = 1 \tag{33}$$

So, the locus is a circle.

10.5 Shallow-Water Waves

It is seen from (28) and (29) that in the shallow-water approximation (the wavelength is large compared with the depth), the pressure distribution is nearly *hydrostatic* due to very small vertical accelerations. The horizontal velocity at any section normal to the direction of propagation is very nearly constant. So, the fluid particles that are originally in a vertical plane remain there.

One has, for the two-dimensional flow in a layer of fluid over a rigid surface at the bottom, the following boundary-value problem,

$$u_x + w_z = 0 \tag{34}$$

$$z = \xi \, : \, \xi_t + u\xi_x - w = 0 \tag{35}$$

$$z = \xi \, : \, p = 0 \tag{36}$$

$$z = -h_0 \, : \, uh_{0_x} + w = 0 \tag{37}$$

where we have allowed for a spatial variation in the bottom topography.[1]

From (34), (35), and (37), one obtains

$$\int_{-h_0}^{\xi} u_x dz + \xi_t + u|_\xi \cdot \xi_x - u|_{-h_0} \cdot h_{0_x} = 0$$

or

$$\frac{\partial}{\partial x}\int_{-h_0}^{\xi} udz = -\xi_t. \tag{38}$$

Equation (38), in the shallow-water approximation, gives

$$[u(\xi + h_0)]_x = -\xi_t. \tag{39}$$

---

1 If the wall at the bottom is given by

$$f(x, z) = z + h_0(x) = 0$$

then the normal to the bottom is given by

$$\nabla f = < h_{0x}, 0, 1 > .$$

The fluid impenetrability condition at the bottom wall leads to

$$\mathbf{v} \cdot \nabla f = uh_{0_x} + w = 0.$$

The shallow-water wave theory is based on the assumption that the vertical acceleration of the fluid is negligible so that the pressure behaves *hydrostatically*. Using (36), the pressure is given by,

$$p = \rho g(\xi - z). \tag{40}$$

Using (40), the equation of motion becomes

$$u_t + uu_x \approx -g\xi_x. \tag{41}$$

If $h_0 = const$, (39) and (41) give upon linearization

$$u_{xx} - \frac{1}{gh_0}u_{tt} = 0. \tag{42}$$

Thus, the solution to a water wave problem of *elliptic* type is approximated, in the shallow-water model by a problem of *hyperbolic* type.

(i) Analogy with Gas Dynamics

The equations of shallow-water wave theory become analogous to those of one-dimensional gas flows, as shown below. Introducing

$$\bar{\rho} \equiv \rho(\xi + h_0) \tag{43}$$

$$\bar{p} \equiv \int_{-h_0}^{\xi} p dz \tag{44}$$

and using (40), one obtains

$$\bar{p} = \frac{\rho g}{2}(\xi + h_0)^2 = \frac{g}{2\rho}\bar{\rho}^2. \tag{45}$$

Using (45), equation (41) then leads to

$$\bar{\rho}(u_t + uu_x) = -\bar{p}_x \tag{46}$$

while (39) leads to the mass conservation equation for one-dimensional gas flows (See Chapter 15)

$$(\bar{\rho}u)_x = -\bar{\rho}_t. \tag{47}$$

If $h_0 = const$, the speed of the shallow-water waves, in analogy with gas dynamics (see Section 13.3), is given, from (45), by

$$c = \sqrt{\frac{d\bar{p}}{d\bar{\rho}}} = \sqrt{g(\xi + h_0)}. \tag{48}$$

(ii) Breaking of Waves

The nonlinear shallow-water equations which neglect dispersion altogether lead to *breaking* of the typical hyperbolic kind (see Chapter 15), with the development of a vertical tangent and a multivalued profile. In particular, (48) implies that a wave crest will travel faster than a trough. Any wave profile will gradually steepen until it ultimately falls over forward as is seen in the breaking of waves on the sea shore when they reach shallow water.

10.6 Water Waves Generated by an Initial Displacement over a Localized Region

Consider water waves generated by an initial displacement over a localized region near the origin, according to

$$t = 0 : \xi = f(x), \frac{\partial \xi}{\partial t} = 0. \tag{49}$$

If, the initial displacement is an even function of x, i.e.,

$$f(x) = f(-x) \tag{50}$$

then, it has the *Fourier cosine transform*,

$$f(x) = \sqrt{\frac{2}{\pi}} \int_0^\infty \phi(k) \cos kx dk \tag{51}$$

where

$$\phi(k) = \sqrt{\frac{2}{\pi}} \int_0^\infty f(x) \cos kx dx. \tag{52}$$

The solution for ξ, for $x > 0$, at any subsequent position and time, with the initial conditions (49), is

$$\xi(x,t) = \frac{1}{\sqrt{2\pi}} \int_{-\infty}^{\infty} \phi(k) \cos(kx - \omega t) dk. \tag{53}$$

Since $\omega/k = h(k)$, waves of different wavenumber propagate at different velocities. Consequently, the overall wave profile, represented by (53), changes its shape as it moves.

Example 1: Consider

$$f(x) = \begin{cases} \dfrac{A}{2b}, -b < x < b \\ 0 \quad \text{otherwise} \end{cases}.$$

Then, one has, from (52)

$$\phi(k) = \sqrt{\frac{2}{\pi}} \int_0^b \frac{A}{2b} \cos kx dx = \frac{1}{\sqrt{2\pi}} \frac{A \sin kb}{kb}.$$

Consider the asymptotic behavior of (53) as $t \Rightarrow \infty$, with x/t held fixed. Let us first write (53) in the form,

$$\xi(x,t) = \frac{1}{\sqrt{2\pi}} Re\left[\int \phi(k) e^{-i\chi(k)t} dk\right] \tag{54}$$

where

$$\chi(k) \equiv \omega(k) - k\frac{x}{t}$$

and x/t is presently a fixed parameter. This integral is evaluated by using the *method of stationary phase.* This is based on the fact that harmonic components with nearly same phase will reinforce, and those differing in phase will annul each other due to mutual interference. One, therefore, looks for positions and times at which a large number of harmonic components have nearly the same phase.

Thus, as $t \Rightarrow \infty$, the integral (54) takes on an asymptotic form determined entirely in terms of the values of k near the points where the phase $\chi(k)$ is stationary. Let $k = k_0$ be such a stationary point where,

$$\chi'(k_0) = \omega'(k_0) - \frac{x}{t} = 0. \tag{55}$$

This condition implies that the group of contributing waves moves with the *group velocity* $C(k_0) \equiv \omega'(k_0)$ (while the individual waves constituting the group move with the *phase velocity* $c(k) \equiv \omega/k$). One writes for $k \approx k_0$,

$$\left.\begin{aligned} \phi(k) &\approx \phi(k_0) \\ \chi(k) &\approx \chi(k_0) + \frac{1}{2}(k-k_0)^2 \chi''(k_0) \end{aligned}\right\} \tag{56}$$

provided, $\chi''(k_0) \neq 0$.

From (54), one then has

$$\xi(x,t) \approx \frac{1}{\sqrt{2\pi}} Re\left[\phi(k_0)\, e^{-i\chi(k_0)t} \int_{-\infty}^{\infty} e^{\frac{i}{2}(k-k_0)^2\chi''(k_0)t} dk\right]. \tag{57}$$

This approximation becomes valid and accurate, as $t \Rightarrow \infty$. The remaining integral is reduced to the real *error integral*,

$$\int_{-\infty}^{\infty} e^{-\alpha z^2} dz = \sqrt{\frac{\pi}{\alpha}} \tag{58}$$

by rotating the path of integration through $\pm\pi/4$ (the sign being that of $\chi''(k_0)$). This corresponds to changing the path of integration to the *path of steepest descents*. This path compresses the region $Im[\chi(k)] \gg 1$ into a short space so that the arcs at $k \Rightarrow \pm\infty$ contribute little to the integral. As a result, the error incurred in truncating the *Taylor series* (56) for $\chi(k)$ around $k = k_0$ is minimized.

Thus, the asymptotic form of ξ, given by (57), for $t \Rightarrow \infty$, is

$$\xi(x,t) \approx Re\left[\phi(k_0)\sqrt{\frac{1/2}{t|\chi''(k_0)|}}\, e^{-i[\chi(k_0)t+\pi/4\,sgn\,\chi''(k_0)]}\right]$$

or

$$\xi(x,t) \approx \phi(k_0)\sqrt{\frac{1/2}{t|\omega''(k_0)|}} \cos\left[k_0 x - \omega(k_0)t - \frac{\pi}{4} sgn\,\omega''(k_0)\right]. \tag{59}$$

From (59), one has, for the wave amplitude,

$$a \equiv \phi(k_0)\sqrt{\frac{1/2}{t|\omega''(k_0)|}}\, e^{-i(\pi/4)sgn\,\omega''(k_0)}. \tag{60}$$

We now introduce the quantity Q(t) given by,

$$Q(t) \equiv \int_{x_1}^{x_2} |a|^2 dx. \tag{61}$$

This may be rewritten using (60) as,

$$Q(t) = \int_{x_1}^{x_2} \frac{|\phi(k_0)|^2}{2t|\omega''(k_0)|} dx = \int_{k_{01}}^{k_{02}} \frac{1}{2}|\phi(k_0)|^2 dk_0. \tag{62}$$

If we let

$$x_1 \equiv \omega'(k_{01})t, \quad x_2 \equiv \omega'(k_{02})t \tag{63}$$

and hold k_{01} and k_{02} fixed, as t varies, then $Q(t)$ remains constant. Thus, the total amount of $|a|^2$ between points x_1 and x_2 moving with the group velocity remains unchanged. Thus

$$\frac{dQ}{dt} = \int_{x_1}^{x_2} \frac{\partial q}{\partial t} dx + C_2 q_2 - C_1 q_1 = \int_{x_1}^{x_2} \left[\frac{\partial q}{\partial t} + \frac{\partial}{\partial x}\{C(k)q\}\right] dx = 0 \tag{64}$$

where,

$$q \equiv |a|^2, \quad C_i \equiv \omega'(k_{0i}); i = 1, 2.$$

Equation (64) implies that

$$\frac{\partial q}{\partial t} + \frac{\partial}{\partial x}[C(k)q] = 0 \tag{65a}$$

or

$$\frac{\partial}{\partial t}(|a|^2) + \frac{\partial}{\partial x}\left[C(k)|a|^2\right] = 0. \tag{65b}$$

In this sense, $|a|^2$ (which may be proportional to energy in some simple cases) propagates with the group velocity.

Thus, in the special case of a wavepacket where the initial disturbance is localized around $x = 0$ containing appreciable amplitude only in wavenumbers near some $k_0 = K_0$, the resulting disturbance will evolve as a group around

K_0. The above wavepacket as a whole moves with the particular group velocity $\omega'(K_0)$.

Example 2: For gravity waves in deep water, one has (see (22))

$$\omega = \sqrt{gk}.$$

Recalling Example 1, (59) becomes

$$\xi(x,t) \approx \frac{A}{b}\sqrt{\frac{2x}{\pi g t^2}} \sin\frac{gt^2 b}{4x^2} \cos\left(\frac{gt^2}{4x} - \frac{\pi}{4}\right).$$

Now, consider the initial disturbance to be an infinite displacement concentrated at the origin, i.e., let $b \Rightarrow 0$. We then obtain

$$\xi(x,t) \approx \frac{At}{4}\sqrt{\frac{2g}{\pi x^3}} \cos\left(\frac{gt^2}{4x} - \frac{\pi}{4}\right).$$

On the other hand, for very large x, we obtain

$$\xi(x,t) \approx \frac{At}{4}\sqrt{\frac{2g}{\pi x^3}} \cos\left(\frac{gt^2}{4x} - \frac{\pi}{4}\right).$$

Observe the agreement between these two results. This implies that the asymptotic behavior of water waves far away from a finite localized region of their generation corresponds to that of those generated by an infinite line displacement in this region.

Example 3: Suppose the initial displacement is a wavepacket and consists of a limited train of harmonic waves $k \approx k'$.

Furthermore, let the initial displacement vary symmetrically with respect to the origin, according to

$$f(x) = \begin{cases} \cos k'x, \ |x| < \left(2\pi + \frac{1}{2}\right)\pi/k' \\ 0 \quad \text{otherwise} \end{cases}.$$

We then have from (52)

$$\phi(k) = \sqrt{\frac{2}{\pi}} \int_0^{\left(2n+\frac{1}{2}\right)\pi/k'} \cos k'x \cos kx dx = \sqrt{\frac{2}{\pi}} k' \frac{\cos\left[\left(2n + \frac{1}{2}\right)\frac{\pi k}{k'}\right]}{k'^2 - k^2}.$$

Using the above result in (53), the solution for ξ when $x > 0$ and $t > 0$ is then given by

$$\xi(x,t) = \frac{2k'}{\pi}\int_0^\infty \frac{\cos\left(2n+\frac{1}{2}\right)\frac{\pi k}{k'}}{k'^2-k^2}\cos(kx-\omega t)\,dk.$$

As $t \Rightarrow \infty$ keeping x/t constant, we obtain, by using the *method of steepest descent*,

$$\xi(x,t) \approx \frac{16\sqrt{\frac{2g}{\pi}}k'x^{5/2}t}{16k'^2x^4 - g^2t^4}\cos\left[\left(2n+\frac{1}{2}\right)\frac{\pi g t^2}{4k'x^2}\right]\cos\left(\frac{gt^2}{4x}-\frac{\pi}{4}\right)$$

$$\equiv A\cos\left(\frac{gt^2}{4x}-\frac{\pi}{4}\right), \text{ say.}$$

The variation of the amplitude A with k is shown in Figure 10.2.

The main undulatory disturbance appears as a simple group around $k = k'$, moving forward with $d\omega/dk = \frac{1}{2}\omega/k$. But, in advance of this main group of undulations there are two or three subsidiary groups of appreciable amplitude and larger wavelengths. Behind the main group, a series of alternating groups show up, following each other much more quickly. Their wavelengths and velocities are less separated out than those in advance of the main group.

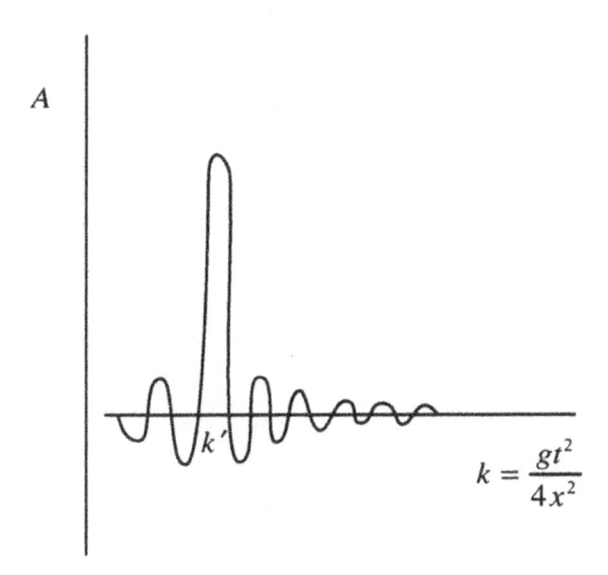

Figure 10.2 Variation of the amplitude with wave number for waves generated by an initial displacement.

Hence, the disturbance in the rear, close to the origin, may be expected to consist of small, irregular motion resulting from the superposition of this latter system of groups.

10.7 Waves on a Steady Stream

We now consider the waves which stand steady in a uniform stream of velocity U. We take the source to be an external steady pressure applied to the surface of the stream, rather than a prescribed displacement, since this represents the effect of a floating body more correctly. If the stream velocity is greater than a certain minimum value, two sets of standing waves can often be observed - one of rather long waves (*gravity waves*) behind the body, the other of short waves (*capillary waves*) ahead of the body (Figure 10.3).

While solving *stationary wave* problems by *Fourier transforms*, difficulties arise in ensuring uniqueness. These may be resolved by taking the applied pressure on $z = 0$ to be effectively zero in the distant past and then built up to a steady value at current time as given below,[2]

$$\frac{p - p_0}{\rho} = f(x, y)\, e^{\varepsilon t}, \quad \varepsilon > 0. \tag{66}$$

One then has the following boundary-value problem for the velocity potential Φ:

$$z < 0 : \nabla^2 \Phi = 0 \tag{67}$$

$$z = 0 : \xi_t + U \xi_x = \Phi_z \tag{68}$$

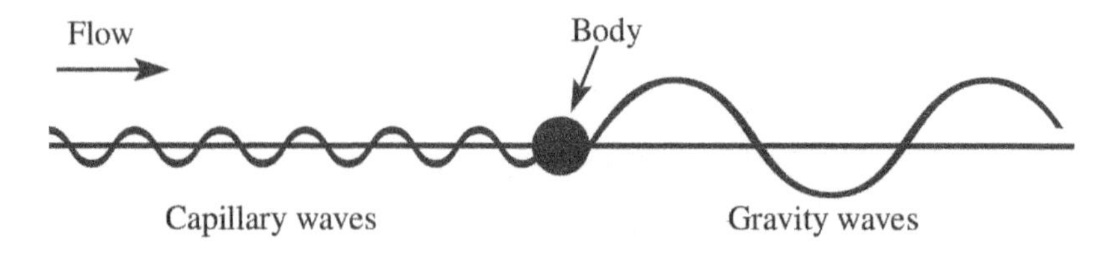

Figure 10.3 Capillary waves and gravity waves generated by a body plaved in a flow.

---

2 Alternatively, one may impose an appropriate *radiation condition* (Lighthill, 1960) or incorporate an infinitesimal amount of dissipation in the system.

$$z = 0 : \Phi_t + U\Phi_x + g\xi - \frac{T}{\rho}\left(\xi_{xx} + \xi_{yy}\right) = -f(x,y)\,e^{\varepsilon t} \tag{69}$$

$$z \Rightarrow \infty : \Phi \Rightarrow 0. \tag{70}$$

Applying the Fourier transform, according to

$$\left.\begin{aligned}
\Phi &= \frac{e^{\varepsilon t}}{2\pi}\iint_{-\infty}^{\infty} B(k)\,e^{i(k_x x + k_y y) + kz}\,dk_x dk_y \\
\xi &= \frac{e^{\varepsilon t}}{2\pi}\iint_{-\infty}^{\infty} A(k)\,e^{i(k_x x + k_y y)}\,dk_x dk_y \\
f &= \frac{1}{2\pi}\iint_{-\infty}^{\infty} F(k)\,e^{i(k_x x + k_y y)}\,dk_x dk_y
\end{aligned}\right\} \tag{71}$$

we obtain from (67)-(70)

$$\xi(x,t) = \frac{e^{\varepsilon t}}{2\pi}\iint_{-\infty}^{\infty} \frac{kF(k)\,e^{i(k_x x + k_y y)}\,dk_x dk_y}{\left(k_x U - i\varepsilon\right)^2 - \omega_0^2(k)} \tag{72}$$

where

$$\omega_0^2 \equiv gk + \frac{T}{\rho}k^3, \quad k \equiv \sqrt{k_x^2 + k_y^2}.$$

The poles of the integrand in (72) lie at

$$k_x^2 U^2 - \omega_0^2(k) \approx 0 \tag{73}$$

which is also the condition for the waves to *stand steady* in the stream.

(i) One-Dimensional Gravity Waves

One has, for the one-dimensional gravity waves, from (72)

$$\xi(x,t) = \frac{e^{\varepsilon t}}{\sqrt{2\pi}}\int_{-\infty}^{\infty} \frac{k_x F(k_x)\,e^{ik_x x}}{\left(k_x U - i\varepsilon\right)^2 - gk_x}\,dk_x. \tag{74}$$

If the applied pressure is localized around $x = 0$, i.e.,

$$f(x) = P\delta(x) \tag{75}$$

then, we have from (71)

$$F(k_x) = P/\sqrt{2\pi}. \tag{76}$$

Using (76), (74) becomes

$$\xi(x,t) = \frac{Pe^{\varepsilon t}}{2\pi}\int_{-\infty}^{\infty}\frac{k_x e^{ik_x x}}{(k_x U - i\varepsilon)^2 - gk_x}dk_x \tag{77a}$$

or

$$\frac{2\pi\xi(x,t)}{P} = \lim_{\varepsilon\Rightarrow 0}\int_{-\infty}^{\infty}\frac{e^{ik_x x}}{k_x U^2 - 2i\varepsilon U - g}dk_x. \tag{77b}$$

The poles of the integrand in (77a) lie at

$$k_x = \frac{g}{U^2} + \frac{2}{U}i\varepsilon \text{ or } U \approx c(k_x) \equiv \sqrt{\frac{g}{k_x}} \tag{78}$$

which is again the condition for the gravity waves to *stand steady* in the stream.

For $x > 0$, the path of k_x-integration along the $Re(k_x)$-axis is closed by a large semicircle in the upper half of the k_x-plane. We then obtain from (77b),

$$i\frac{U^2\xi(x,t)}{P} = -e^{iKx}, \quad x > 0 \tag{79}$$

where

$$K \equiv \frac{g}{U^2}.$$

For $x < 0$, the path of integration is closed the other way, and one then obtains

$$\frac{U^2\xi(x,t)}{P} = 0, \quad x < 0. \tag{80}$$

Therefore, *standing gravity waves* appear *downstream* ($x > 0$) of the applied pressure disturbance.

(ii) One-Dimensional Capillary-Gravity Waves

On including the surface tension, (77b) becomes

$$\frac{2\pi\xi(x,t)}{P} = \lim_{\varepsilon\Rightarrow 0^+}\int_{-\infty}^{\infty}\frac{e^{ik_x x}dk_x}{k_x U^2 - 2i\varepsilon U - g - \frac{T}{\rho}k_x^2}. \tag{81}$$

The poles of the above integrand lie at

$$k_x U^2 - 2i\varepsilon U - g - \frac{T}{\rho}k_x^2 = -\frac{T}{\rho}\left[(k_x - k_T)(k_x - k_g) + \frac{2\rho U}{T}i\varepsilon\right] = 0$$

or

$$U \approx c(k_x) = \left[\frac{g}{k_x} + \frac{k_x T}{\rho}\right]^{1/2} \tag{82}$$

which is again the condition for the *capillary-gravity* waves to *stand steady* in the stream. If $U < c_{\min} = \left(4\frac{Tg}{\rho}\right)^{1/4}$, as per (21), then there are no real roots for k_x. Consequently, there is no standing wave pattern. If $U > c_{\min}$ (see Figure 10.1), then there are two real roots, given by

$$k_x = k_{x_g}, k_{x_T} \equiv \frac{\rho U^2}{2T} \mp \sqrt{\left(\frac{\rho U^2}{2T}\right)^2 - \frac{\rho g}{T}} \tag{83}$$

where

$$k_{x_T} > k_{x_g}$$

and the poles will lie at

$$k_{x_1} = k_{x_g} + \frac{2\rho U}{\left(k_{x_T} - k_{x_g}\right)T} i\varepsilon, \quad k_{x_2} = k_{x_T} - \frac{2\rho U}{\left(k_{x_T} - k_{x_g}\right)T} i\varepsilon. \tag{84}$$

For $x > 0$, the path of integration for the integral in (81) may again be closed by a large semicircle in the upper half of the k_x-plane. Contributions then come from the poles at $k_x \approx k_{x_g}$. Consequently, *gravity waves* appear once again, *downstream* of the source (Figure 10.3). For $x < 0$, the paths of integration may be closed the other way.

Contributions now come from the pole at $k_x \approx k_{x_T}$. One then obtains *capillary waves upstream* of the source (Figure 10.3). Thus, we have

$$\xi(x,t) \sim \begin{cases} \dfrac{iP\rho}{\left(k_{x_T} - k_{x_g}\right)T} e^{ik_{x_g} x}, & x > 0 \\ \dfrac{-iP\rho}{\left(k_{x_T} - k_{x_g}\right)T} e^{ik_{x_T} x}, & x < 0. \end{cases} \tag{85}$$

(iii) Ship Waves

Consider the wave pattern produced by a ship traveling with uniform velocity U in the negative x-direction on the surface of deep water. We will make use of a reference frame in which the ship is fixed, so that there is a uniform stream with velocity U in the x-direction. In this frame, the ship is equivalent to an applied pressure localized at $x = 0, y = 0$, i.e.,

$$f(x, y) = P\delta(x, y) \tag{86}$$

so that one has, from (71),

$$F(k) = P/2\pi. \tag{87}$$

Using (87), and ignoring the capillary effects, (72) gives

$$\frac{4\pi^2 U^2 \xi(x,t)}{P} = e^{\varepsilon t} \int_{-\infty}^{\infty}\int_{-\infty}^{\infty} \frac{k e^{i(k_x x + k_y y)}}{\left(k_x - \frac{i\varepsilon}{U}\right)^2 - \frac{gk}{U^2}} dk_x dk_y. \tag{88}$$

Introducing

$$\left.\begin{array}{ll} x = r\cos\alpha, & y = r\sin\alpha \\ k_x = -k\cos\chi, & k_y = k\sin\chi \end{array}\right\} \tag{89}$$

and noting that the contribution from the range $\pi/2 < \chi < 3\pi/2$ is the complex conjugate of that from the range $-\pi/2 < \chi < \pi/2$, (88) becomes

$$\frac{2\pi^2 U^2 \xi(x,t)}{P} = Re\left[\lim_{\varepsilon \Rightarrow 0} \int_{-\pi/2}^{\pi/2-\alpha} \frac{d\chi}{\cos^2\chi} \int_0^{\infty} \frac{k e^{-ikr\cos(\alpha+\chi)}}{k - k_0} dk\right] \tag{90}$$

where,

$$k_0 = \frac{g}{U^2\cos^2\chi} - \frac{2i\varepsilon}{U\cos\chi}.$$

Since the pattern is symmetrical about the x-axis, it is sufficient to consider the range $0 < \alpha < \pi$. When $\cos(\alpha + \chi) > 0$, i.e., $-\pi/2 < \chi < \pi/2 - \alpha$, the path of integration in the k-plane may be closed by a large semicircle in the lower-half of the k-plane with contribution from the pole at $k = k_0$ (which corresponds to $\sqrt{g/k_0} = U\cos\chi$). Physically this implies that the ship can feed energy continuously to a wave only when its bow travels with the crest of the wave (which is like the "surf-riding" condition!). When $\cos(\alpha + \chi) < 0$, i.e.,

$-\pi/2 > \chi > \pi/2 - \alpha$, the path of integration is closed the other way, with no contribution from the pole at $k = k_0$. Thus

$$\xi(x,t) \approx \frac{gP}{\pi U^4} Im \left[\int_{-\pi/2}^{\pi/2-\alpha} \frac{e^{-irs(\chi)}}{\cos^4 \chi} d\chi \right] \tag{91}$$

where

$$s(\chi) \equiv k_0 \cos(\alpha + \chi) = \frac{g \cos(\alpha + \chi)}{U^2 \cos^2 \chi}.$$

$s(\chi)$ has a stationary point at $\chi = \Psi$, where

$$\tan(\alpha + \Psi) = 2 \tan \Psi. \tag{92}$$

So, one obtains by using the *method of stationary phase*,

$$\xi(x,t) \sim \frac{gP}{\pi U^4} Im \left[\sqrt{\frac{\pi}{r|s''(\Psi)|}} \frac{e^{-irs(\Psi) - i\pi/4 sgn s''(\Psi)}}{\cos^4 \psi} \right]$$

or

$$\xi(x,t) \sim -\sqrt{\frac{g}{\pi r}} \frac{P}{U^3 \cos^3 \Psi} \frac{\left(1 + 4\tan^2 \Psi\right)^{1/4}}{|1 - 2\tan^2 \Psi|^{1/2}} \times \sin\left[k_{0_*} r \cos(\alpha + \Psi) + \frac{\pi}{4} sgn s''(\Psi)\right] \tag{93}$$

where

$$k_{0_*} \equiv \frac{g}{U^2 \cos^2 \Psi}.$$

Note that the amplitude is singular at the maximum wedge angle $\alpha = \alpha_* = \tan^{-1}\left(1/2\sqrt{2}\right) = 19.5°$ where $\tan \Psi_c = 1/\sqrt{2}$. One may see from (93) that α_* also corresponds to the maximum wedge angle α_{max}.[3] This is where the lateral and transverse crests meet at a cusp (see Figure 10.4) and it corresponds to the confluence of the two stationary points of the exponent $s(\chi)$. Thus, the wave pattern is confined to a wedge-shaped region spreading out behind the ship, and the semivertex angle of the wedge is 19.5°. Note that this result is independent

---

3 This may be seen by noting from (92),

$$\frac{d\alpha}{d\Psi}\left(1 + 4\tan^2 \Psi\right) + 2\tan^2 \Psi = 1.$$

from which,

$$\Psi = \Psi_c : \frac{d\alpha}{d\Psi} = 0 ; \tan \Psi_c = \frac{1}{\sqrt{2}}.$$

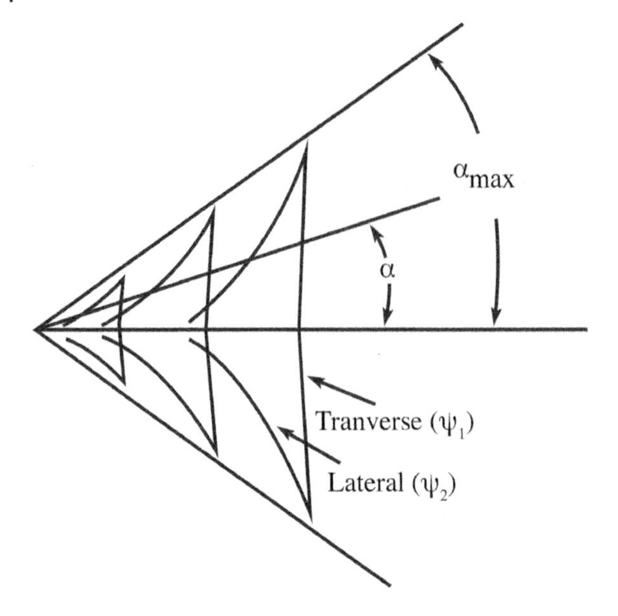

Figure 10.4 Ship-wave pattern.

of the ship velocity U, provided it is constant. The semi vertex angle depends only on the fact that the ratio of the group velocity and the phase velocity for deep water waves is $1/2$ (see (22))!

10.8 Gravity Waves in a Rotating Fluid

Consider wavemotions in a liquid of depth h_0, rotating with an angular velocity Ω about a vertical axis (along the z-direction)- a situation applicable to oceans on a rotating planet. Consider long waves, for which the vertical accelerations are negligible, and one may assign, for the pressure, the hydrostatic value corresponding to the distance from the free surface given by $z = h_0 + \xi$. One then has in the rotating frame, the following linearized equations governing this flow,

$$\frac{\partial u}{\partial t} - \Omega v = -g\frac{\partial \xi}{\partial x} \tag{94}$$

$$\frac{\partial v}{\partial t} + \Omega u = -g\frac{\partial \xi}{\partial y} \tag{95}$$

$$\frac{\partial \xi}{\partial t} = -h_0\frac{\partial u}{\partial x} - h_0\frac{\partial v}{\partial y}. \tag{96}$$

Considering one-dimensional waves, one obtains, from equations (94)-(96)

$$\frac{\partial^2 u}{\partial t^2} + \Omega^2 u = c^2 \frac{\partial^2 u}{\partial x^2} \tag{97}$$

where c is the phase velocity of long surface waves in water of depth h_0

$$c = \sqrt{g h_0}. \tag{26}$$

Let us prescribe the initial conditions as follows,

$$t = 0 : u = u_0, \frac{\partial u}{\partial t} = s_0. \tag{98}$$

Consider the traveling waves of the form $e^{i(kx \mp \omega t)}$, for which equation (97) yields the dispersion relation

$$\omega = c\sqrt{k^2 + \Omega^2/c^2}. \tag{99}$$

Applying Fourier transform, according to

$$\left.\begin{aligned} u_0(x) &= \frac{1}{\sqrt{2\pi}} \int_{-\infty}^{\infty} a(k) e^{ikx} dk \\ s_0(x) &= \frac{1}{\sqrt{2\pi}} \int_{-\infty}^{\infty} b(k) e^{ikx} dk \end{aligned}\right\} \tag{100}$$

one obtains, from equation (97),

$$u(x,t) = \frac{1}{\sqrt{2\pi}} \int_{-\infty}^{\infty} A(k) e^{i(kx-\omega t)} + \frac{1}{\sqrt{2\pi}} \int_{-\infty}^{\infty} B(k) e^{i(kx+\omega t)} dk \tag{101}$$

where

$$\left.\begin{aligned} A(k) &\equiv \frac{1}{2}\left[a(k) + \frac{i}{\omega(k)} b(k)\right] \\ B(k) &\equiv \frac{1}{2}\left[a(k) - \frac{i}{\omega(k)} b(k)\right] \end{aligned}\right\}.$$

Thus

$$u(x,t) = \frac{1}{\sqrt{2\pi}} \int_{-\infty}^{\infty} a(k) e^{ikx} \cos\left(\sqrt{k^2 + \Omega^2/c^2}ct\right) dk +$$

$$+ \frac{1}{c}\frac{1}{\sqrt{2\pi}} \int_{-\infty}^{\infty} b(k) e^{ikx} \frac{\sin\left(\sqrt{k^2 + \Omega^2/c^2}ct\right)}{\sqrt{k^2 + \Omega^2/c^2}} dk$$

or

$$u(x,t) = \frac{1}{\pi} \int_{-\infty}^{\infty} u_0(\xi)\, d\xi \int_0^{\infty} \cos k(x-\xi) \cos\left(\sqrt{k^2 + \Omega^2/c^2}ct\right) dk +$$

$$+ \frac{1}{\pi c} \int_{-\infty}^{\infty} s_0(\xi)\, d\xi \int_0^{\infty} \cos k(x-\xi) \frac{\sin\left(\sqrt{k^2 + \Omega^2/c^2}ct\right)}{\sqrt{k^2 + \Omega^2/c^2}} dk. \tag{102}$$

This may be written as

$$u(x,t) = \frac{1}{\pi c}\frac{\partial}{\partial t} \int_{-\infty}^{\infty} G(x-\xi, ct)\, u_0(\xi)\, d\xi + \frac{1}{\pi c} \int_{-\infty}^{\infty} G(x-\xi, ct)\, s_0(\xi)\, d\xi \tag{103}$$

where

$$G(x,\tau) \equiv \int_0^{\infty} \cos \zeta x \frac{\sin\left(\sqrt{\zeta^2 + \Omega^2/c^2}\tau\right)}{\sqrt{\zeta^2 + \Omega^2/c^2}} d\zeta \tag{104}$$

which can, in turn, be written as

$$G(x,\tau) = \frac{1}{2} \int_{-\infty}^{\infty} e^{i\zeta x} \frac{\sin\left(\sqrt{\zeta^2 + \Omega^2/c^2}\tau\right)}{\sqrt{\zeta^2 + \Omega^2/c^2}} d\zeta. \tag{105}$$

Setting

$$\zeta = \frac{\Omega}{c} \sin h\varphi \tag{106}$$

we obtain

$$G(x,\tau) = \frac{1}{2} \int_{-\infty}^{\infty} e^{i(\Omega/c)x \sin h\varphi} \sin\left(\frac{\Omega\tau}{c} \cos h\varphi\right) d\varphi.$$

Thus,

$$G(x,\tau) = \begin{cases} \dfrac{\pi}{2} J_0\left(\dfrac{\Omega}{c}\sqrt{\tau^2 - x^2}\right), & \tau > x \\ 0, & \tau < x \end{cases} \tag{107}$$

which implies that $G(x,\tau)$ assumes the role as the *propagator of causal effects* from the source region. Here, $J_n(x)$ is Bessel's function of first kind of order n. Using (107), (103) gives

$$u(x,t) = \frac{1}{2c}\frac{\partial}{\partial t}\int_{x-ct}^{x+ct} J_0\left[\frac{\Omega}{c}\sqrt{c^2t^2 - (x-\xi)^2}\right] u_0(\xi)\, d\xi +$$

$$+ \frac{1}{2c}\int_{x-ct}^{x+ct} J_0\left[\frac{\Omega}{c}\sqrt{c^2t^2 - (x-\xi)^2}\right] s_0(\xi)\, d\xi$$

or

$$u(x,t) = \frac{1}{2}\left[u_0(x-ct) + u_0(x+ct)\right] +$$

$$- \frac{1}{2}\Omega t\int_{x-ct}^{x+ct} \frac{J_1\left[\frac{\Omega}{c}\sqrt{c^2t^2 - (x-\xi)^2}\right]}{\sqrt{c^2t^2 - (x-\xi)^2}} u_0(\xi)\, d\xi + \tag{108}$$

$$+ \frac{1}{2c}\int_{x-ct}^{x+ct} J_0\left[\frac{\Omega}{c}\sqrt{c^2t^2 - (x-\xi)^2}\right] s_0(\xi)\, d\xi.$$

Let us now assume an excitation moving with a speed c, i.e.,

$$s_0(x) = -c\frac{du_0}{dx}. \tag{109}$$

Then (108) becomes

$$u(x,t) = u_0(x-ct) +$$

$$- \frac{\Omega}{2c}\int_{x-ct}^{x+ct} \frac{J_1\left[\frac{\Omega}{c}\sqrt{c^2t^2 - (x-\xi)^2}\right]}{\sqrt{c^2t^2 - (x-\xi)^2}} \left[ct + (x-\xi)\right] u_0(\xi)\, d\xi. \tag{110}$$

If the excitation is localized around $x = 0$, i.e.,

$$u_0(x) = A\delta(x) \tag{111}$$

then, (110) becomes

$$u(x,t) = A\delta(x - ct) - \frac{A\Omega}{2c}\frac{ct + x}{\sqrt{c^2t^2 - x^2}}J_1\left(\frac{\Omega}{c}\sqrt{c^2t^2 - x^2}\right). \tag{112}$$

The excitation represented by this expression comprises a sharp front advancing in the x-direction at the speed c. Behind this, a rotation-induced wavetrain extends over the steadily widening range $|x| < ct$, with an amplitude that diminishes as the trailing edge $x = -ct$ is approached. As $t \Rightarrow \infty$, this wavetrain is described by

$$u(x,t) \sim A\sqrt{\frac{\Omega}{2\pi c}}\frac{(ct + x)}{(c^2t^2 - x^2)^{3/4}}\cos\left[\frac{\Omega}{c}\sqrt{c^2t^2 - x^2} + \frac{\pi}{4}\right].[4] \tag{113}$$

4 Alternatively, this may be deduced by noting that

$$u(x,t) = u_+(x,t) + u_-(x,t)$$

where

$$u_{\pm}(x,t) = \frac{A}{4\pi}\int_{-\infty}^{\infty}\left[1 \pm \frac{kc}{\omega(k)}\right]e^{i[kx \mp \omega(k)t]}dk.$$

Using the method of stationary phase, we obtain

$$u_{\pm}(x,t) \sim \frac{A}{2}\sqrt{\frac{\Omega}{2\pi c}}\frac{(ct + x)}{(c^2t^2 - x^2)^{3/4}}e^{\mp\frac{i\Omega}{c}\sqrt{c^2t^2 - x^2} \mp \frac{i\pi}{4}}.$$

Thus,

$$u(x,t) \sim A\sqrt{\frac{\Omega}{2\pi c}}\frac{ct + x}{(c^2t^2 - x^2)^{3/4}}\cos\left[\frac{\Omega}{c}\sqrt{c^2t^2 - x^2} + \frac{\pi}{4}\right], t \Rightarrow \infty$$

in agreement with (113)!

10.9 Theory of Tides

Tides are the periodic rise and fall of the free surface of water in the oceans caused by the *lunar gravitational attraction.*[5] Let us consider here the equatorial tides, in the interest of simplicity.

In the *long-wave* approximation, one has the following equations for the flow,

$$\rho \frac{\partial u}{\partial t} = -\frac{\partial p}{\partial x} + \rho F_x \tag{114}$$

$$0 = -\frac{\partial p}{\partial z} - \rho g \ \text{ or } \ p = p_0 + \rho g (h + \xi - z) \tag{115}$$

$$\frac{\partial u}{\partial x} + \frac{\partial w}{\partial z} = 0 \ \text{ or } \ w = -h \frac{\partial u}{\partial x} \tag{116}$$

$$z = 0 : \ \frac{\partial \xi}{\partial t} = w \tag{117}$$

where h is the depth of water, and F_x is the gravitational force component (see Figure 10.5). One obtains from (114)-(117),

$$\frac{\partial^2 \xi}{\partial t^2} = gh \frac{\partial^2 \xi}{\partial x^2} - h \frac{\partial F_x}{\partial x}. \tag{118}$$

Let us neglect the dynamical effects of the earth's rotation (i.e., neglect centrifugal and Coriolis forces). Let the only effect of this rotation be an apparent revolution of the moon (as seen by an observer) around the earth once each day. Let us assume further that the earth is a uniform sphere covered with a static ocean of uniform depth, and ignore the presence of land masses.

Due to of the finite size of the earth, different points on its surface will generally experience different gravitational attractions by the moon. The differences between local values of attraction leads to the tidal force. The gravitational forces per unit mass at a point P on the surface of the earth and the center of mass C of the earth are

$$\left. \begin{aligned} \mathbf{F}_P &= \frac{GM_m}{r^2} \cos(\alpha + \phi) \, \hat{\mathbf{i}}_z + \frac{GM_m}{r^2} \sin(\alpha + \phi) \, \hat{\mathbf{i}}_x \\ \mathbf{F}_C &= \frac{GM_m}{D^2} \cos\phi \ \hat{\mathbf{i}}_z + \frac{GM_m}{D^2} \sin\phi \ \hat{\mathbf{i}}_x. \end{aligned} \right\} \tag{119}$$

The tidal force $\mathbf{F}$ is then

$$\mathbf{F} = \mathbf{F}_P - \mathbf{F}_C. \tag{120}$$

5 Tidal forces arise on any body of finite size moving in a spatially variable force field.

Noting (see Figure 10.5)

$$\left.\begin{aligned} r^2 &= D^2 + a^2 - 2aD\cos\phi \\ \frac{\sin\alpha}{\sin\phi} &\approx \frac{a}{D} \ll 1 \end{aligned}\right\}$$

we have

$$\left.\begin{aligned} F_x &= \frac{GM_m}{D^2}\left[\frac{D^2}{r^2}\sin(\alpha+\phi) - \sin\phi\right] \\ &\approx \frac{GM_m}{D^2}\left[\left(\sin\phi\cos\phi\cdot\frac{a}{D} + \sin\phi\right)\left(1 + 2\frac{a}{D}\cos\phi\right) - \sin\phi\right] \\ &= \frac{3GM_m a}{2D^3}\sin 2\phi. \end{aligned}\right\} \tag{121}$$

Using (121), and noting $x \approx a\phi$, equation (118) becomes

$$\frac{\partial^2\xi}{\partial t^2} - \frac{gh}{a^2}\frac{\partial^2\xi}{\partial\phi^2} = \frac{3GM_m h}{D^3}\cos 2\phi. \tag{122}$$

Let ω be the frequency of the moon about the earth, as observed from the point P, then

$$\phi = \omega t. \tag{123}$$

So, equation (122) gives

$$\frac{\partial^2\xi}{\partial t^2} = \frac{3GM_m h}{D^3\left(1 - \frac{gh}{\omega^2 a^2}\right)}\cos 2\omega t \tag{124}$$

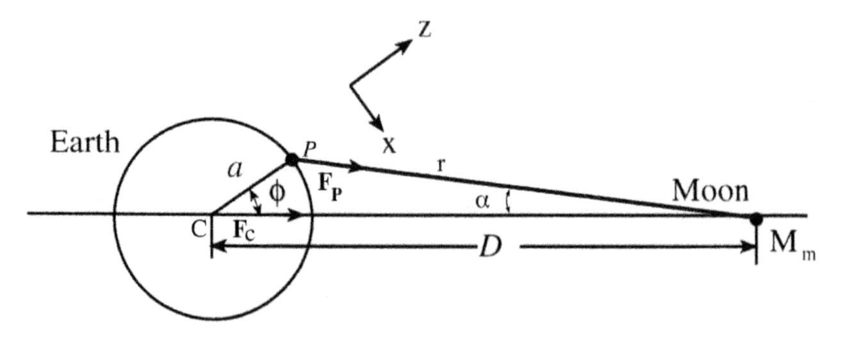

Figure 10.5 The earth-moon system.

from which the solution is,

$$\xi = \frac{3GM_m a^2 h}{4D^3 \left(gh - \omega^2 a^2\right)} \cos 2\omega t \tag{125}$$

Note, from (125), that

- the tides are *semidiurnal*, i.e., the water level at a particular point will reach its maximum value twice a day - even though the moon traverses its path only once in the same period of time;
- since $gh \ll \omega^2 a^2$, the tide is *inverted*, i.e., one has a low tide when the moon is directly overhead.

10.10 Hydraulic Jump

(i) Tidal Bores

Tidal bores occur on rivers when an unusually high tide enters a narrowing estuary.

Consider a semi-infinite body of shallow water lying at rest in the region $x > 0$, which is held back by a wall at $x = 0$. At $t = 0$, the wall starts to move into the water with velocity V which creates a *hydraulic jump* that propagates away from the wall into the water at constant speed U like a tidal bore (Figure 10.6).

Flow in a channel downstream of a sluice gate experiences a sudden transition from a *supercritical* ($V > \sqrt{gh}$, h being the water depth) flow to a *subcritical* $\left(V < \sqrt{gh}\right)$ flow. The free surface rises sharply and is usually covered with a mass of foam beneath which there is a violent turbulent motion - this is called a *hydraulic jump*. This is analogous to a *shock wave* in supersonic gas flow (see Chapter 16). Under these conditions, infinitesimal disturbances cannot travel upstream. However, finite disturbances can travel upstream with a speed equal to V to make a stationary jump possible. The relations between the values of the flow variables on either side of the discontinuity can be obtained from the conditions of continuity of fluxes of mass and momentum. We assume the following,

- the cross section of the channel is uniform,
- the velocity over each cross section is uniform,
- the depth is uniform across the width.

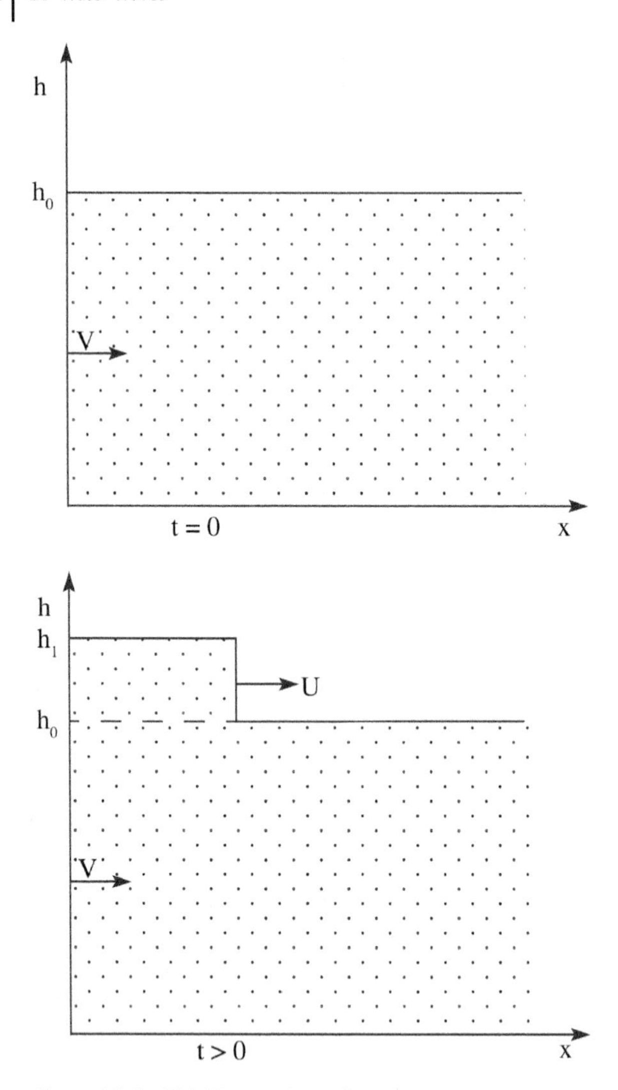

Figure 10.6 Tidal bore schemetic.

The mass flux density is $\rho V h$. The momentum flux density is

$$\int_{0}^{n} (p + \rho V^2)\, dz \equiv \rho V^2 h + \frac{1}{2}\rho g h^2 \tag{126}$$

where we have noted that in the shallow-water limit, the pressure behaves hydrostatically, so $p = \rho g z$. If the states upstream and downstream of the discontinuity are denoted by subscripts 1 and 2, then one has the following conditions,[6]

$$V_1 h_1 = V_2 h_2 \equiv Q \tag{127}$$

$$V_1^2 h_1 + \frac{1}{2} g h_1^2 = V_2^2 h_2 + \frac{1}{2} g h_2^2 \tag{128}$$

Q being the mass flux.

The total energy per unit mass E of a fluid particle on the two sides of the discontinuity is not the same, and their difference accounts for the energy dissipated in the discontinuity. Note that

$$E = \frac{V^2}{2} + gh = \frac{Q^2}{2h^2} + gh. \tag{129}$$

---

6 The equations governing shallow-water flows are

$$\left.\begin{aligned} h_t + uh_x + hu_x &= 0 \\ u_t + uu_x + gh_x &= 0 \end{aligned}\right\}$$

or in the *conservation* form,

$$\left.\begin{aligned} h_t + (uh)_x &= 0 \\ (uh)_t + \left(u^2 h + g\frac{h^2}{2}\right)_x &= 0 \end{aligned}\right\}.$$

If u and h are discontinuous across a smooth curve $\sum\ :\ x = x(t)$ moving with speed $dx/dt = U$, the jump conditions (denoted by rectangular brackets) are then given by

$$\left.\begin{aligned} U[h] &= [uh] \\ U[uh] &= \left[u^2 h + g\frac{h^2}{2}\right]. \end{aligned}\right\}$$

Transforming to a frame of reference moving with the shock, i.e., putting

$$V \equiv U - u$$

we obtain for the jump conditions

$$\left.\begin{aligned} [Vh] &= 0 \\ \left[V^2 h + \frac{1}{2} g h^2\right] &= 0, \end{aligned}\right\}$$

which are the same as (127) and (128)!

From equation (128), we have

$$Q^2 = (g/2)\, h_1 h_2 \,(h_1 + h_2)\,. \tag{130}$$

Using (129) and (130), we obtain

$$E_1 - E_2 = (g/4h_1 h_2)\,(h_2 - h_1)^3\,. \tag{131}$$

Thus, $h_2 > h_1$ or the liquid attains a higher surface level across such a jump.

On using equations (127) and (128), the *Froude numbers* $F \equiv \dfrac{V}{\sqrt{gh}}$ upstream and downstream of the jump, are given by

$$\left.\begin{aligned} F_1 &\equiv \frac{V_1}{\sqrt{gh_1}} = \frac{1}{\sqrt{2}\,h_1}\sqrt{(h_1+h_2)\,h_2} \\ F_2 &\equiv \frac{V_2}{\sqrt{gh_2}} = \frac{1}{\sqrt{2}\,h_2}\sqrt{(h_1+h_2)\,h_1} \end{aligned}\right\} \tag{132}$$

from which

$$\left.\begin{aligned} F_1^2 - 1 &= \frac{(H+2)(H-1)}{2} \\ 1 - F_2^2 &= \frac{(H+1/2)(H-1)}{H^2} \end{aligned}\right\} \tag{133}$$

where

$$H \equiv h_2/h_1.$$

So, if $H > 1$, (133) yields

$$F_1 > 1 \text{ and } F_2 < 1$$

as with the *Mach numbers* upstream and downstream of the shock wave in supersonic gas flows (see Chapter 16).

(ii) The Dam-Break Problem

Consider water of depth h_0 retained in the region $x < 0$ by a dam, which is removed suddenly at $t = 0$. This initial-value problem is then governed by

$$t > 0 : h_t + uh_x + hu_x = 0 \tag{134}$$

$$u_t + uu_x = -gh_x \tag{135}$$

$$t = 0 : \quad u = 0, \quad -\infty < x < \infty$$

$$h = \begin{cases} h_0, & -\infty < x < 0 \\ 0, & 0 < x < \infty. \end{cases} \tag{136}$$

Let us rewrite the equations (134) and (135) as

$$2\frac{\partial c}{\partial t} + 2u\frac{\partial c}{\partial x} + c\frac{\partial u}{\partial x} = 0 \tag{137}$$

$$\frac{\partial u}{\partial t} + u\frac{\partial u}{\partial x} + 2c\frac{\partial c}{\partial x} = 0 \tag{138}$$

where

$$c \equiv \sqrt{gh}.$$

Rearranging equations (137) and (138) as

$$\frac{\partial}{\partial t}(u \pm 2c) + (u \pm c)\frac{\partial}{\partial x}(u \pm 2c) = 0 \tag{139}$$

we obtain the *Riemann invariants* (see Section 15.1) for this problem,

$$C_\pm : \frac{dx}{dt} = u \pm c : \quad u \pm 2c = const. \tag{140}$$

The characteristics C_+ come from the *uniform* state prevalent in the region $x < 0$, so,

$$u + 2c = const = 2c_0, \forall x > 0.$$

This represents a *simple-wave* region (see Section 15.2). So, the characteristics C_- are the rays emerging from the origin given by,

$$C_- : \frac{x}{t} = u - c = const. \tag{141}$$

We have in the disturbed region,

$$\left.\begin{aligned} u &= \frac{2}{3}\left(\frac{x}{t} + c_0\right) \\ c &= \frac{1}{3}\left(2c_0 - \frac{x}{t}\right) \end{aligned}\right\} \tag{142}$$

which is bounded on the left by (see Figure 10.7)

$$C_-^L : \frac{x}{t} = -c_0 \tag{143a}$$

and on the right by

$$C_-^R : \frac{x}{t} = 2c_0. \tag{143b}$$

The solution to the dam-break problem is, therefore,

$$h = \begin{cases} h_0, & x \leq -c_0 t \\ \frac{1}{9g}\left(2c_0 - \frac{x}{t}\right)^2, & -c_0 t \leq x \leq 2c_0 t \\ 0, & x \geq 2c_0 t \end{cases} \tag{144}$$

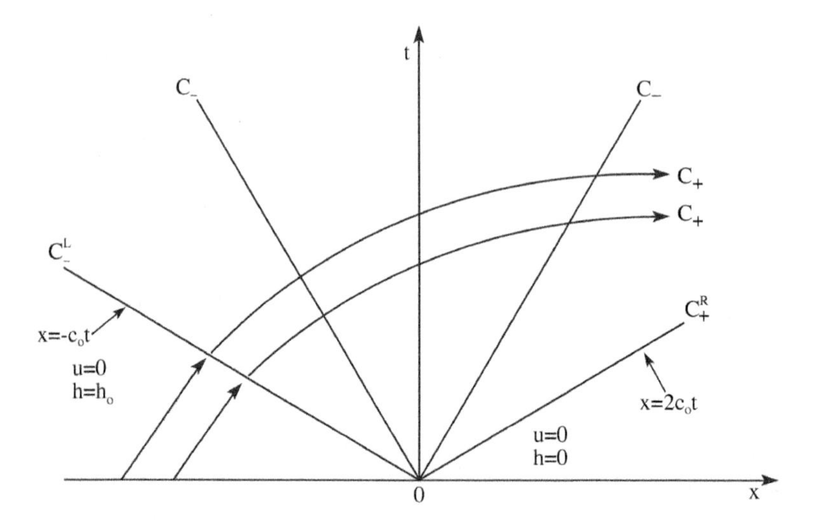

Figure 10.7 Characteristics for the dam-break problem.

and

$$u = \begin{cases} 0, & x \leq -c_0 t \\ \frac{2}{3}\left(\frac{x}{t} + c_0\right), & -c_0 t \leq x \leq 2c_0 t \\ 0, & x \geq 2c_0 t. \end{cases} \tag{145}$$

Observe that u is discontinuous across C_-^R while h is continuous there (see Figure 10.8).

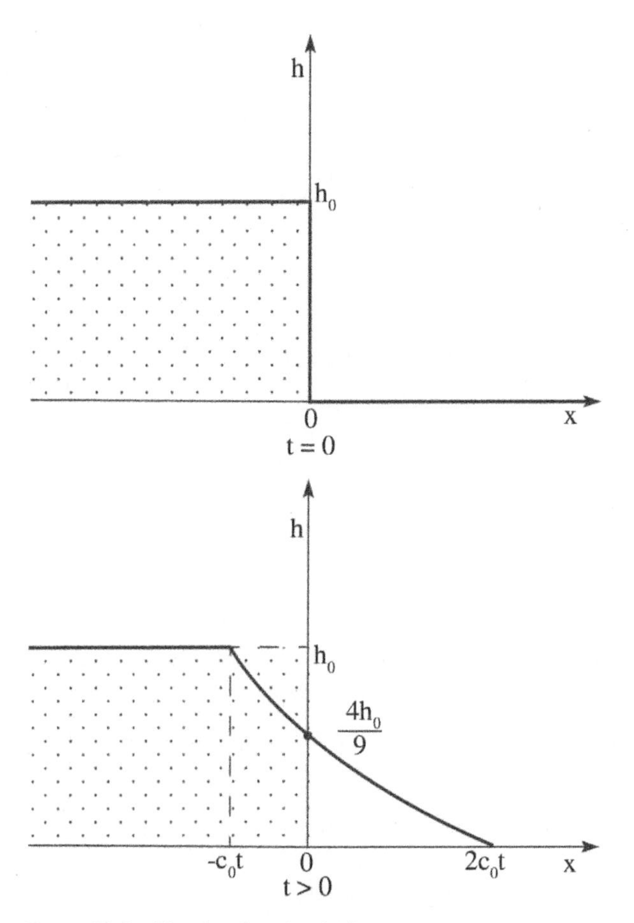

Figure 10.8 The dam break solution.

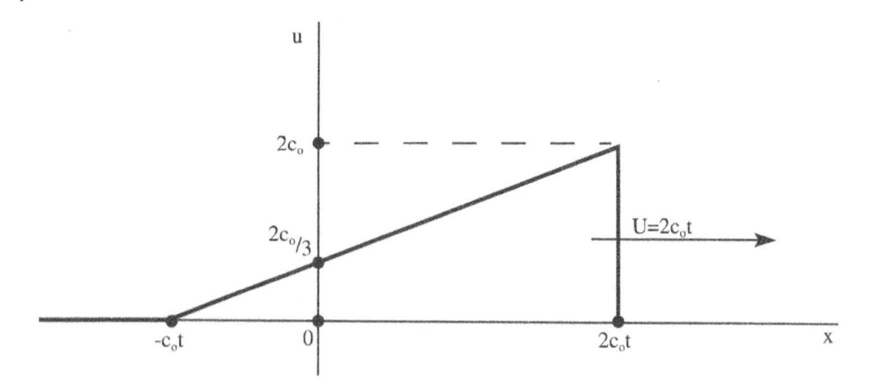

Figure 10.9 Burgers shock front.

The free surface is a parabola between the front $h = 0, x = 2c_0t$, traveling with speed $2c_0$, and the undisturbed dam level h_0 at $x = -c_0t$. Note that $h = 4h_0/9$ and $u = 2c_0/3$ remain constant at the dam position, $\forall t > 0$!

This solution represents a shock at $x = 2c_0t$ moving with velocity $U = 2c_0$ (Figure 10.9).

10.11 Nonlinear Shallow-Water Waves

Nonlinear effects in a train of surface waves lead to many frequently observed phenomena. In deep water, the wave profile becomes distorted, with the crests slightly sharper than the troughs and the phase speed increases slightly with increasing wave slope. In shallow water, the nonlinear effects are stronger and hence more easily observed. Consequently, the linearized model here demands much more stringent conditions on the wave amplitude than those in deep water.

Consider small (but finite)-amplitude long waves in shallow water of depth h. Suppose that these waves have amplitude a and wavelength λ such that,

$$\alpha \equiv \frac{a}{h} \ll 1 \text{ and } \beta \equiv \frac{h}{\lambda} \ll 1. \tag{146}$$

Two limiting cases arise for such waves depending on the parameter $S \equiv \frac{\alpha}{\beta^2} = \frac{a\lambda^2}{h^3}$. The first case corresponds to $S >> 1$ called the *Stokes approximation*. The other corresponds to $S \approx 1$ called the *Boussinesq approximation*. The first case leads to nonlinear periodic waves while the second removing affords a balance between nonlinearity and dispersion, leading to *solitary waves*.

Consider a homogeneous, incompressible fluid whose undisturbed depth above a rigid horizontal boundary is h_0. Let the disturbed free surface be at a height $h(x,t)$ (see Figure 10.10). We will assume that the flow is two-dimensional, and that there is no dependence on the transverse horizontal coordinate.

One then has the following boundary-value problem for the velocity potential ϕ:

$$\frac{\partial^2 \phi}{\partial x^2} + \frac{\partial^2 \phi}{\partial z^2} = 0 \tag{147}$$

$$z = 0 : \frac{\partial \phi}{\partial z} = 0 \tag{148}$$

$$z = h : \frac{\partial \phi}{\partial z} = \frac{\partial h}{\partial t} + \frac{\partial \phi}{\partial x}\frac{\partial h}{\partial x} \tag{149}$$

$$\frac{\partial \phi}{\partial t} + \frac{1}{2}\left[\left(\frac{\partial \phi}{\partial x}\right)^2 + \left(\frac{\partial \phi}{\partial z}\right)^2\right] + g\,(h - h_0) = 0. \tag{150}$$

Consider waves moving only to the right, and so introduce

$$\phi\,(x,z,t) = \varepsilon^{-\frac{1}{2}}\psi\,(\xi,z,\tau)$$

where

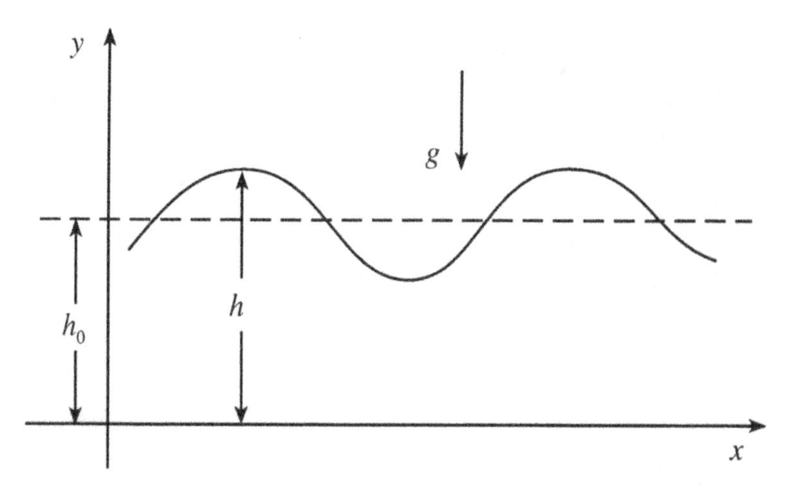

Figure 10.10 Surface waves on water of mean depth h_0.

$$\xi = \sqrt{\varepsilon}(x - c_0 t), \quad \tau = \varepsilon^{3/2} t,$$

$$c_0 = \sqrt{g h_0}, \quad \varepsilon \ll 1. \tag{151}$$

The parameter ε therefore characterizes the smallness of the amplitude of the waves in question. Thus, (147)-(150) become

$$\varepsilon \frac{\partial^2 \psi}{\partial \xi^2} + \frac{\partial^2 \psi}{\partial z^2} = 0 \tag{152}$$

$$z = 0 : \frac{\partial \psi}{\partial z} = 0 \tag{153}$$

$$z = h : \frac{\partial \psi}{\partial z} = \varepsilon^2 \frac{\partial h}{\partial \tau} + \varepsilon \left(\frac{\partial \psi}{\partial \xi} - c_0 \right) \frac{\partial h}{\partial \xi} \tag{154}$$

$$\varepsilon^2 \frac{\partial \psi}{\partial \tau} - \varepsilon c_0 \frac{\partial \psi}{\partial \xi} + \frac{1}{2} \left[\varepsilon \left(\frac{\partial \psi}{\partial \xi} \right)^2 + \left(\frac{\partial \psi}{\partial z} \right)^2 \right] + \varepsilon g (h - h_0) = 0. \tag{155}$$

Looking for solutions of the form,

$$\left. \begin{aligned} h &= h_0 + \varepsilon h_1 + \varepsilon h_2 + \cdots \\ \psi &= \varepsilon \psi_1 + \varepsilon^2 \psi_2 + \cdots \end{aligned} \right\} \tag{156}$$

equation (152) leads to

$$\frac{\partial^2 \psi_1}{\partial z^2} = 0 \tag{157}$$

$$\frac{\partial^2 \psi_n}{\partial z^2} + \frac{\partial^2 \psi_{n-1}}{\partial \xi^2} = 0, \quad n > 1. \tag{158}$$

Solving equations (157) and (158) using (153) gives,

$$\begin{aligned} \psi_1 &= \psi_1 (\xi, \tau) \\ \psi_2 &= -\frac{1}{2} z^2 \frac{\partial^2 \psi_1}{\partial \xi^2} \\ \psi_3 &= \frac{1}{24} z^4 \frac{\partial^4 \psi_1}{\partial \xi^4} \end{aligned} \tag{159}$$

etc.

The boundary conditions (154) and (155) are imposed at a boundary which is unknown without the solution and varies slightly with the perturbation parameter ε. To deal with this difficulty, the solution is assumed to be *analytic* in spatial coordinates. The transfer of the boundary conditions to the basic configuration of the boundary (corresponding to $\varepsilon = 0$) is then effected by expanding the solution in a *Taylor series* about the values at the basic configuration. Thus, (154) and (155) give

$$\begin{aligned} z = h_0 : \frac{\partial \psi_2}{\partial z} &= -c_0 \frac{\partial h_1}{\partial \xi} \\ \frac{\partial \psi_3}{\partial z} + h_1 \frac{\partial^2 \psi_2}{\partial z^2} &= \frac{\partial h_1}{\partial \tau} - c_0 \frac{\partial h_2}{\partial \xi} + \frac{\partial \psi_1}{\partial \xi} \frac{\partial h_1}{\partial \xi}, \\ &\text{etc.} \end{aligned} \tag{160}$$

and

$$\begin{aligned} z = h_0 : -c_0 \frac{\partial \psi_1}{\partial \xi} + g h_1 = 0 \\ \frac{\partial \psi_1}{\partial \tau} - c_0 \frac{\partial \psi_2}{\partial \xi} + \frac{1}{2} \left(\frac{\partial \psi_1}{\partial \xi} \right)^2 + g h_2 = 0, \\ \text{etc.} \end{aligned} \tag{161}$$

From (159)-(161), one derives

$$\frac{\partial h_1}{\partial \tau} + \frac{3c_0}{2h_0} h_1 \frac{\partial h_1}{\partial \xi} = -\frac{c_0 h_0^2}{6} \frac{\partial^3 h_1}{\partial \xi^3} \tag{162}$$

which is the *Korteweg-de Vries equation.*

Recall the linear dispersion relation for water waves (see (25))

$$\omega^2 = gk \tanh kh_0 \tag{25}$$

from which, for shallow-water waves, one has

$$kh_0 \ll 1 : \omega = kc_0 \left[1 - \frac{h_0^2 k^2}{6} + \cdots \right]. \tag{163}$$

Observe that the coefficient of the dispersive term on the right-hand side of the Korteweg-de Vries equation (162) is the same as the coefficient of k^3 in the linear dispersion relation (163), as to be expected!

(i) Solitary Waves

Write the Korteweg-de Vries equation (162) in the form,

$$\phi_t + \kappa\phi\phi_x + \phi_{xxx} = 0. \tag{164}$$

Looking for steady, progressing, and localized wave solutions of the form

$$\phi(x,t) = \phi(\xi), \quad \xi \equiv x - Ut, \tag{165}$$

equation (164) gives

$$\phi_\xi(\kappa\phi - U) + \phi_{\xi\xi\xi} = 0 \tag{166a}$$

along with the boundary conditions,

$$|\xi| \Rightarrow \infty : \phi, \phi_\xi, \phi_{\xi\xi} \Rightarrow 0. \tag{166b}$$

Integrating equation (166a) using the above boundary conditions yields,

$$\phi_{\xi\xi} = U\phi - \frac{\kappa}{2}\phi^2 \tag{167}$$

and again, equation (167) gives

$$\frac{1}{2}\phi_\xi^2 = \frac{U}{2}\phi^2 - \frac{\kappa}{6}\phi^3 \tag{168}$$

from which we have,

$$\int_{\phi_{max}}^{\phi} \frac{d\phi}{\sqrt{\frac{U}{2}\phi^2 - \frac{\kappa}{6}\phi^3}} = \sqrt{2}\xi. \tag{169}$$

The integral in (169) is of the form

$$I \equiv \int \frac{d\phi}{\phi\sqrt{1-\sigma\phi}}. \tag{170}$$

Introducing

$$\psi \equiv \sqrt{1-\sigma\phi}, \quad \sigma \equiv \frac{\kappa}{3U} \tag{171}$$

we may write (170) as,

$$I = -2\int \frac{d\psi}{1-\psi^2} = \ell n \frac{1-\psi}{1+\psi}. \tag{172}$$

Using (172), (169) leads to

$$\frac{1-\psi}{1+\psi} = e^{\sqrt{U}\xi} \tag{173}$$

from which,

$$\psi = \frac{1 - e^{\sqrt{U}\xi}}{1 + e^{\sqrt{U}\xi}}. \tag{174}$$

Noting from (171) that

$$\phi = \frac{1 - \psi^2}{\sigma} \tag{171}$$

and using (174) we obtain

$$\phi = \frac{1}{\sigma} \frac{4e^{\sqrt{U}\xi}}{\left(1 + e^{\sqrt{U}\xi}\right)^2} = \frac{1}{\sigma}\text{sech}^2 \frac{\sqrt{U}\xi}{2} = \frac{3U}{\kappa}\text{sech}^2\left[\frac{\sqrt{U}}{2}(x - Ut)\right] \tag{175}$$

which represents a unidirectional *solitary wave*. We see from (175) that,

- this solitary wave moves with a velocity larger than c_0, the speed of the gravity waves, by an amount proportional to its amplitude (a consequence of the nonlinearity of the wave);
- the width of the solitary wave is inversely proportional to the square root of its amplitude (a consequence of the spreading of the wave due to dispersion) confirming confirms the condition $\alpha \sim \beta^2$ discussed before (see (146))!

Solitary waves are nonlinear localized dispersive waves propagating without change in shape and velocity (Zabusky and Kruskal, 1965).[7] The nonlinearity tends to steepen the wavefront in consequence of the increase of the wavespeed with amplitude. The dispersive effects on the other hand, tend to spread the wavefront. The essential quality of a solitary wave is the balance between these two seemingly disparate behaviors. Solitary waves are found to both preserve their identity almost like particles and be stable in processes of mutual collisions as illustrated in Figure 10.11. Refer to Section 10.11(iii) for a theoretical development of this aspect. Zabusky and Kruskal (1965) introduced the term *soliton* to describe such structures. Solitary waves are therefore strictly nonlinear phenomena with no counterparts in the linear regime.

The stability of solitary waves has been demonstrated with respect to both one-dimensional perturbations (Jeffrey and Kakutani, 1972, Benjamin, 1972)

---

7 Solitary waves were first observed by J. Scott Russell on the Edinburgh-Glasgow canal as a moving "well-defined heap of water" in 1834. Russell also performed laboratory experiments on solitary waves. He empirically deduced that the speed u of the solitary wave is given by

$$u^2 = g(h_0 + a)$$

a being the amplitude of the wave and h_0 being the undisturbed depth of water. However, the great significance of Russell's discoveries were not properly appreciated until more than 130 years later.

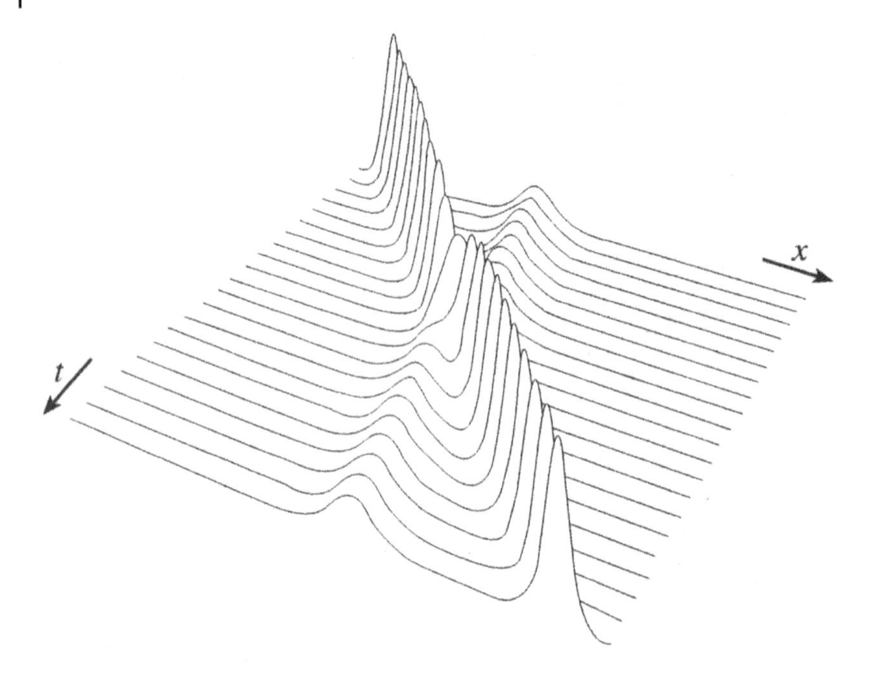

Figure 10.11 Numerical solution of the Korteweg-de Vries equation describing the overtaking collision of solitary waves (from Dodd et al., 1982).

and two-dimensional perturbations (Kadomtsev and Patviashvili, 1970, Oikawa et al., 1974).

(ii) Periodic Cnoidal Waves

Koreweg-de Vries equation also possesses periodic *cnoidal* wave solutions given in terms of *Jacobi elliptic functions*. Consider the Korteweg-de Vries equation in the form,

$$\phi_t + \kappa\phi\phi_x + \phi_{xxx} = 0. \tag{164}$$

Looking for stationary propagating wave solutions of the form,

$$\phi(x,t) = \phi(\xi), \xi \equiv x - Ut \tag{165}$$

equation (164) leads to

$$-U\phi_\xi + \kappa\phi\phi_\xi + \phi_{\xi\xi\xi} = 0. \tag{176}$$

Integrating equation (176) twice, one obtains

$$-\frac{1}{2}U\phi^2 + \frac{\kappa}{6}\phi^3 + \frac{1}{2}\phi_\xi^2 = A\phi + B \tag{177}$$

where A and B are constants of integration.

equation (177) may be written in the form,

$$\frac{3}{\kappa}\phi_\xi^2 = -\phi^3 + \frac{3U}{\kappa}\phi^2 + \frac{6A}{\kappa}\phi + \frac{6B}{\kappa} \equiv f(\phi). \tag{178}$$

Under certain conditions, equation (178) may give a periodic solution between two consecutive real zeros of $f(\phi)$ where $f(\phi) \geq 0$. Note that $f(\phi)$ is a cubic, and therefore, has three zeros. So, one has two separate cases to consider (a) Only one zero is real (Figure 10.12).

For a real solution, one has to restrict ϕ to $\phi \leq \sigma_1$. In this region, one obtains, from equation (176),

$$\frac{d\phi}{d\xi} = \pm\sqrt{\frac{\kappa}{3}f(\phi)}. \tag{179}$$

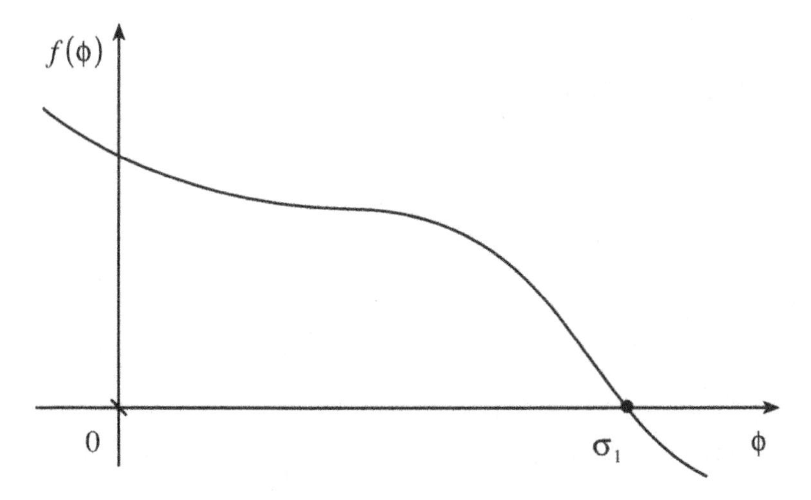

Figure 10.12 The case with one zero.

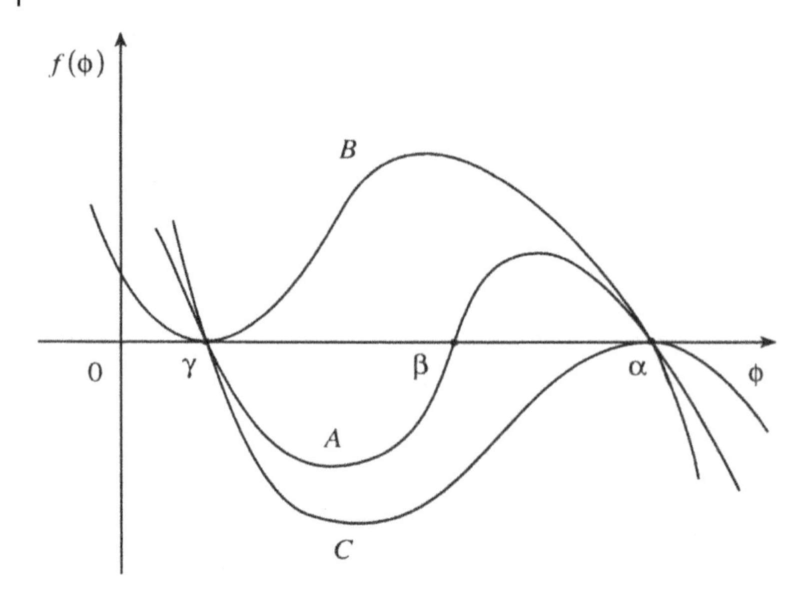

Figure 10.13 The case with three zeros.

However, note that this solution becomes unbounded, since

$$\phi \Rightarrow -\infty : f(\phi) \Rightarrow \infty.$$

(b) All three zeros are real (Figure 10.13).
Here, one has three subcases to consider,

(b1) all three real roots α, β, γ are distinct (curve A);
(b2) two roots are equal $\beta = \gamma$ (curve B);
(b3) two roots are equal $\beta = \alpha$ (curve C).

Writing

$$f(\phi) = (\phi - \gamma)(\phi - \beta)(\alpha - \phi) \tag{180}$$

one has, from (178),

$$\left.\begin{aligned} &\frac{1}{3}(\alpha + \beta + \gamma) = \frac{U}{\kappa} \\ &-\frac{1}{6}(\alpha\beta + \beta\gamma + \gamma\alpha) = \frac{A}{\kappa} \\ &\frac{1}{6}\alpha\beta\gamma = \frac{B}{\kappa} \end{aligned}\right\}. \tag{181}$$

For the solution to be real and bounded, one requires $\beta < \phi < \alpha$ for curve A, $\beta = \gamma < \phi < \alpha$ for curve B. For curve C, $\phi \leq \gamma$ appears acceptable, but the solution becomes unbounded, as in case (a).

Incidentally, one has, from (181),

$$U^2 + 2A\kappa = \frac{\kappa^2}{18}\left[(\alpha - \beta)^2 + (\beta - \gamma)^2 + (\gamma - \alpha)^2\right] \tag{182}$$

from which, in order that the zeros α, β, γ are real, one requires

$$U^2 + 2A\kappa \geq 0. \tag{183}$$

(b1) Here, $(\alpha - \beta)$ can be considered as a measure of the amplitude of the wave. One has, from (179) and (180),

$$-\frac{d\phi}{[(\phi - \gamma)(\phi - \beta)(\alpha - \phi)]^{1/2}} = \frac{d\xi}{\sqrt{3/\kappa}}. \tag{184}$$

Setting

$$p^2 \equiv \alpha - \phi \tag{185}$$

from (184) we obtain

$$\frac{d\xi}{\sqrt{3/\kappa}} = \frac{2dp}{[\{(\alpha - \gamma) - p^2\}\{(\alpha - \beta) - p^2\}]^{1/2}}. \tag{186}$$

Furthermore, setting

$$q = \frac{p}{\sqrt{(\alpha - \beta)}} = \sqrt{\frac{\alpha - \phi}{\alpha - \beta}} \tag{187}$$

equation (186) becomes

$$\sqrt{(\alpha - \gamma)}\frac{d\xi}{\sqrt{3/\kappa}} = \frac{2dq}{[(1 - s^2q^2)(1 - q^2)]^{1/2}} \tag{188}$$

where

$$s^2 \equiv \frac{\alpha - \beta}{\alpha - \gamma}, 0 < s^2 < 1 \tag{189}$$

$s^2 = 0$ corresponds to the *linear* limit, while $s^2 = 1$ corresponds to the *most nonlinear* limit.

Recall that, here, $\beta < \phi < \alpha$, and from (185) and (187), note that

$$\left.\begin{array}{l}\phi = \alpha : q = 0\\ \phi = \beta : q = 1\end{array}\right\}.$$

Taking

$$q = 0 : \xi = 0 \tag{190}$$

equation (188) gives

$$\xi = \sqrt{\frac{12}{\kappa(\alpha-\gamma)}}\int_0^q \frac{dq}{\sqrt{(1-s^2q^2)(1-q^2)}} = \sqrt{\frac{12}{\kappa(\alpha-\gamma)}}sn^{-1}(q,s). \tag{191}$$

From this we obtain

$$\phi(\xi) = \beta + (\alpha-\beta)\,cn^2\left(\xi\sqrt{\frac{\kappa(\alpha-\gamma)}{12}}, s\right) \tag{192}$$

where sn and cn are the *Jacobian elliptic functions* with modulus s.[8]

The period of $\phi(\xi)$ (which is the period of cn^2) is given by

$$P = 2\sqrt{\frac{12}{\kappa(\alpha-\gamma)}}\int_0^1 \frac{dq}{\sqrt{(1-s^2q^2)(1-q^2)}} \equiv 4\sqrt{\frac{3}{\kappa(\alpha-\gamma)}}K(s^2) \tag{193}$$

where $K(s^2)$ is the *complete elliptic integral of the first kind.*[9] Thus, the bounded solution of the Korteweg-de Vries equation represents a periodic wave. Figure 10.14 shows the *cnoidal* wave solutions for $s = 0, .6$, and $.9$. Observe that the crests become narrower and the troughs become wider as s increases. The actual wave profile is a curve called *trochoid.*

Note that the linear case corresponds to the limit $s^2 \Rightarrow 0$ (see 189). Furthermore, noting that

---

8 Compared with sinusoidal wave profile, the wave profile given by (192) has sharpened crests and flattened troughs.

9 $K(s)$ is given by

$$K(s^2) \equiv \int_0^1 \frac{dq}{\sqrt{(1-s^2q^2)(1-q^2)}}.$$

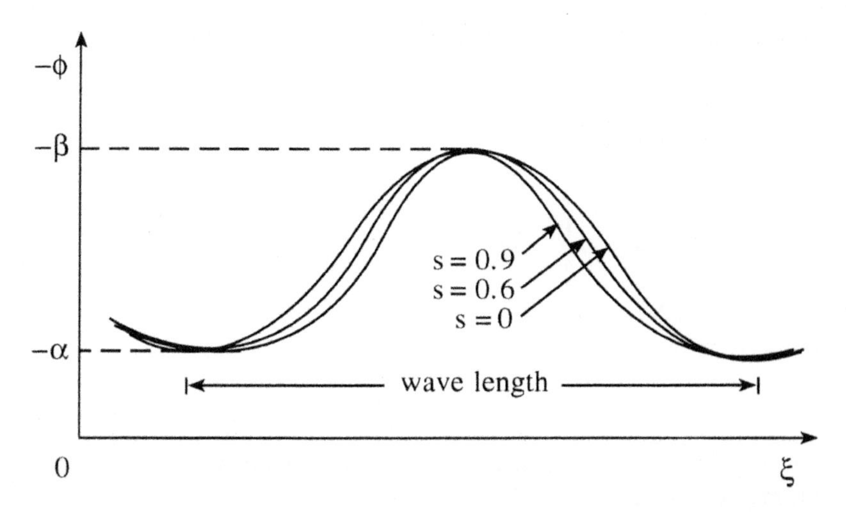

Figure 10.14 One period of the cnoidal wave, for $s = 0.0, 0.6, 0.9$. All three waves have been normalized so that the amplitudes and wavelengths are the same (from Drazin and Johnson, 1989).

$$cn\,(\zeta, 0) \approx \cos\zeta$$

we obtain from (192)

$$\phi(\xi) = \beta + (\alpha - \beta)\cos^2\left(\sqrt{\frac{\kappa(\alpha-\gamma)}{12}}\xi\right) \tag{194}$$

as to be expected!
(b2) Note that the case $\beta = \gamma$ corresponds to $s^2 = 1$ (as per (189)) - the "*most nonlinear*" case. One has now, from (179) and (180),

$$d\xi = \sqrt{\frac{3}{\kappa}}\frac{d\phi}{(\phi-\gamma)\sqrt{(\alpha-\phi)}} \tag{195}$$

which leads to the solitary wave,[10]

---

10 Alternatively, noting that,

$$cn\,(\zeta, 1) = \operatorname{sech}\zeta$$

we obtain from (192)

$$\phi(\xi) = \gamma + (\alpha-\gamma)\operatorname{sech}^2\left(\sqrt{\frac{\kappa(\alpha-\gamma)}{12}}\xi\right)$$

as in (196)!

$$\phi(\xi) = \gamma + (\alpha - \gamma)\ \text{sech}^2\left(\sqrt{\frac{\kappa(\alpha - \gamma)}{12}}\xi\right). \tag{196}$$

The period of this wave, from (193), is given by

$$P = 4\sqrt{\frac{3}{\kappa(\alpha - \gamma)}}\int_0^1 \frac{dq}{1 - q^2} = \infty \tag{197}$$

as to be expected!

Observing from (196) that

$$\xi \Rightarrow \underline{+}\infty : \phi \Rightarrow \gamma \tag{198}$$

one recognizes that $\phi_\infty = \gamma$ denotes the uniform state prevailing at $\xi \Rightarrow \underline{+}\infty$. One may, therefore, write

$$a \equiv \alpha - \gamma \tag{199}$$

and interpret a as the amplitude of the wave. Thus, (196) may be rewritten as

$$\phi(\xi) - \phi_\infty = a\ \text{sech}^2\left[\sqrt{\frac{a\kappa}{12}}\left\{x - \left(\phi_\infty\kappa + \frac{a\kappa}{3}\right)t\right\}\right] \tag{200}$$

which (on identifying a with $3U/\kappa$) is precisely the solitary wave solution (175), given previously.

(iii) Interacting Solitary Waves

We will now consider the behavior of two interacting solitary waves by using a method due to Hirota (1971). This method involves transforming the Korteweg-de Vries equation into homogeneous bilinear forms. The solution is then developed in a formal power series expansion, which *self truncates* and leads to multisoliton solutions in a rather simple form. Consider Korteweg-de Vries equation in the form

$$\phi_t + \kappa\phi\phi_x + \phi_{xxx} = 0. \tag{201}$$

Making the *Hirota transformation*,

$$\phi = \frac{12}{\kappa}(\ell nF)_{xx} \tag{202}$$

equation (201) gives

$$F\left(F_t + F_{xxx}\right)_x - F_x\left(F_t + F_{xxx}\right) + 3\left(F_{xx}^2 - F_x F_{xxx}\right) = 0. \tag{203}$$

Looking for a solution of the form

$$F = 1 + \varepsilon F_1 + \varepsilon^2 F_2 + \cdots, \varepsilon \ll 1 \tag{204}$$

equation (203) leads to the following hierarchy of equations,

$$\left(F_{1t} + F_{1xxx}\right)_x = 0 \tag{205}$$

$$\left(F_{2t} + F_{2xxx}\right)_x = -3\left(F_{1xx}^2 - F_{1x}F_{1xxx}\right) \tag{206}$$

$$\begin{aligned}\left(F_{3t} + F_{3xxx}\right)_x = &-F_1\left(F_{2t} + F_{2xxx}\right)_x + F_{1x}\left(F_{2t} + F_{2xxx}\right) + \\ &- 3\left(2F_{1xx}F_{2xx} - F_{1x}F_{2xxx} - F_{2x}F_{1xxx}\right)\end{aligned} \tag{207}$$

$$\begin{aligned}\left(F_{4t} + F_{4xxx}\right)_x = &-F_2\left(F_{2tx} + F_{2xxxx}\right) + F_{2x}\left(F_{2t} + F_{2xxx}\right) \\ &- 3\left(F_{2xx}^2 - F_{2x}F_{2xxx}\right)\end{aligned} \tag{208}$$

etc.

Equation (205) has the solution

$$F_1 = f_1 + f_2, f_j = e^{-\alpha_j(x-s_j)+\alpha_j^3 t}; j = 1, 2. \tag{209}$$

Using (209), equation (206) becomes

$$\left(F_{2t} + F_{2xxx}\right)_x = 3\alpha_1\alpha_2\left(\alpha_2 - \alpha_1\right)^2 f_1 f_2 \tag{210}$$

from which,

$$F_2 = \frac{\left(\alpha_2 - \alpha_1\right)^2}{\left(\alpha_2 + \alpha_1\right)^2} f_1 f_2. \tag{211}$$

Using (209) and (211), equation (207) becomes

$$\begin{aligned}(F_{3t}+F_{3xxx})_x &= \frac{(\alpha_2-\alpha_1)^2}{(\alpha_2+\alpha_1)^2}\left[-(f_1+f_2)\{(f_1f_2)_{tx}+(f_1f_2)_{xxxx}\}\right.\\
&+(f_{1x}+f_{2x})\{(f_1f_2)_t+(f_1f_2)_{xxx}\}-3\{2(f_{1xx}+f_{2xx})\times\\
&\times(f_1f_2)_{xx}-(f_{1x}+f_{2x})(f_1f_2)_{xxx}\\
&\left.-(f_1f_2)_x(f_{1xxx}+f_{2xxx})\}\right]\\
&=\frac{(\alpha_2-\alpha_1)^2}{(\alpha_2+\alpha_1)^2}[-(f_1+f_2)(6f_{1xx}f_{2xx}+3f_{1x}f_{2xxx}+3f_{1xxx}f_{2x})+\\
&+(f_{1x}+f_{2x})(3f_{1x}f_{2xx}+3f_{1xx}f_{2x})-3\{2(f_{1xx}+f_{2xx})\times\\
&\times(f_{1xx}f_2+2f_{1x}f_{2x}+f_1f_{2xx})-(f_{1x}+f_{2x})\times\\
&\times(f_{1xxx}f_2+3f_{1xx}f_{2x}+3f_{1x}f_{2xx}+f_1f_{2xxx})+\\
&-(f_{1x}f_2+f_1f_{2x})(f_{1xxx}+f_{2xxx})\}]\\
&=\frac{(\alpha_2-\alpha_1)^2}{(\alpha_2+\alpha_1)^2}f_1f_2[-3\alpha_1\alpha_2(\alpha_1+\alpha_2)(f_1+f_2)+\\
&+3\alpha_1\alpha_2(\alpha_1f_1+\alpha_2f_2)-3\{2(\alpha_1+\alpha_2)(\alpha_1^2f_1+\alpha_2^2f_2)+\\
&-(\alpha_1+\alpha_2)^2(\alpha_1f_1+\alpha_2f_2)-(\alpha_1^3f_1+\alpha_2^3f_2)\}\Big]=0,\end{aligned} \tag{212}$$

so that

$$F_3=0. \tag{213}$$

Using (209), (211), and (213), equation (208) becomes

$$(F_{4t}+F_{4xxx})_x=0, \tag{214}$$

so that

$$F_4=0. \tag{215}$$

Thus,

$$F_n=0, n>2. \tag{216}$$

Therefore, (204), (209), (211), and (216) imply that

$$F=1+\varepsilon(f_1+f_2)+\varepsilon^2\frac{(\alpha_2-\alpha_1)^2}{(\alpha_2+\alpha_1)^2}f_1f_2 \tag{217}$$

which is an *exact* solution! Consequently, one may put $\varepsilon=1$.

Thus, one has from (202) and (217),

$$\frac{\kappa}{12}\phi = \frac{\left[\left(\alpha_1^2 f_1 + \alpha_2^2 f_2\right) + 2\left(\alpha_2 - \alpha_1\right)^2 f_1 f_2 + \frac{\left(\alpha_2 - \alpha_1\right)^2}{\left(\alpha_2 + \alpha_1\right)^2}\left(\alpha_2^2 f_1^2 f_2 + \alpha_1^2 f_1 f_2^2\right)\right]}{\left[1 + \left(f_1 + f_2\right) + \frac{\left(\alpha_2 - \alpha_1\right)^2}{\left(\alpha_2 + \alpha_1\right)^2} f_1 f_2\right]^2}. \tag{218}$$

Observe from (218), that a single solitary wave is given by

$$\frac{\kappa}{12}\phi = \frac{\alpha^2 f}{(1+f)^2}.$$

Note that

$$\phi = \phi_{max} = \frac{3\alpha^2}{\kappa}$$

occurs on

$$f = 1, \quad x = s + \alpha^2 t.$$

Next, we have

$$\left.\begin{aligned} f_1 \approx 1, f_2 \ll 1 : \frac{\kappa}{12}\phi &\approx \frac{\alpha_1^2 f_1}{\left(1 + f_1\right)^2} \\ f_1 \approx 1, f_2 \gg 1 : \frac{\kappa}{12}\phi &\approx \frac{\alpha_1^2 \tilde{f}_1}{\left(1 + \tilde{f}_1\right)^2}, \tilde{f}_1 \equiv \frac{\left(\alpha_2 - \alpha_1\right)}{\left(\alpha_2 + \alpha_1\right)^2} f_1 \end{aligned}\right\}.$$

The latter is a solitary wave with s_1 replaced by $\tilde{s}_1$, where

$$\tilde{s}_1 \equiv s_1 - \frac{1}{\alpha_1}\ell n \left(\frac{\alpha_2 + \alpha_1}{\alpha_2 - \alpha_1}\right)^2$$

which signifies a finite displacement of the profile in the x-direction. Similarly, when $f_2 \approx 1$, and f_1 is either large or small, one has the solitary wave α_2 with or without a shift in s_2. Wherever $f_1 \approx 1$ and $f_2 \approx 1$, one has the interaction region. Wherever f_1 and f_2 are both small or large, one has $\phi \approx 0$.

In order to consider the behavior of two interacting solitary waves, let $\alpha_2 > \alpha_1 > 0$. The solitary wave α_2 is bigger and so moves faster than the wave α_1. As $t \Rightarrow -\infty$, one has

$$\left.\begin{aligned} &f_1 \approx 1, f_2 \ll 1 : \text{ wave } \alpha_1 \text{ on } x = s_1 + \alpha_1^2 t \\ &f_2 \approx 1, f_1 \gg 1 : \text{ wave } \alpha_2 \text{ on } x = s_2 - \frac{1}{\alpha_2}\ell n\left(\frac{\alpha_2+\alpha_1}{\alpha_2-\alpha_1}\right)^2 + \alpha_2^2 t \end{aligned}\right\}$$

and elsewhere $\phi \approx 0$.

As $t \Rightarrow \infty$, one has

$$\left.\begin{aligned} &f_1 \approx 1, f_2 \gg 1 : \text{ wave } \alpha_1 \text{ on } x = s_1 - \frac{1}{\alpha_1}\ell n\left(\frac{\alpha_2+\alpha_1}{\alpha_2-\alpha_1}\right)^2 + \alpha_1^2 t \\ &f_2 \approx 1, f_1 \ll 1 : \text{ wave } \alpha_2 \text{ on } x = s_2 + \alpha_2^2 t \end{aligned}\right\}$$

and elsewhere $\phi \approx 0$.

Thus, the solitary waves emerge unchanged in form with the original parameters α_1 and α_2, the faster wave α_2 now being ahead of the slower wave α_1. The only remnant of the collision process is a forward shift

$$\frac{1}{\alpha_2}\ell n\left(\frac{\alpha_2+\alpha_1}{\alpha_2-\alpha_1}\right)^2$$

for the wave α_2, and a backward shift

$$\frac{1}{\alpha_1}\ell n\left(\frac{\alpha_2+\alpha_1}{\alpha_2-\alpha_1}\right)^2$$

for the wave α_1, from where they would have been had there been no interaction.

(iv) Stokes Waves

Irrotational, steady progressive waves were considered first by Stokes (1847) and are called *Stokes waves.* The Korteweg-de Vries equation (164) can be written as

$$\frac{\partial h_1}{\partial t} + c_0\left(1 + \frac{3}{2}\frac{h_1}{h_0}\right)\frac{\partial h_1}{\partial x} + \gamma\frac{\partial^3 h_1}{\partial x^3} = 0 \tag{219}$$

where

$$\gamma \equiv c_0 h_0^2/6.$$

Let us find the next approximation to the linear periodic wave train using the *method of strained parameters* (Shivamoggi, 2003). Thus let

$$\left.\begin{aligned}\frac{h_1}{h_0} &= \varepsilon h_1^{(1)}(\theta) + \varepsilon^2 h_1^{(2)}(\theta) + \varepsilon^3 h_1^{(3)}(\theta) + \cdots \\ \omega &= \omega_0(k) + \varepsilon\omega_1(k) + \varepsilon^2\omega_2(k) + \cdots\end{aligned}\right\} \tag{220}$$

where,

$$\theta \equiv kx - \omega t.$$

Using (220), equation (219) leads to the following hierarchy of equations,

$$O(\varepsilon) : (\omega - c_0 k)\, h_1^{(1)\prime} - \gamma k^3 h_1^{(1)\prime\prime\prime} = 0 \tag{221}$$

$$O\left(\varepsilon^2\right) : (\omega - c_0 k)\, h_1^{(2)\prime} - \gamma k^3 h_1^{(2)\prime\prime\prime} = \frac{3}{2} c_0 k h_1^{(1)} h_1^{(1)\prime} - \omega_1 h_1^{(1)\prime} \tag{222}$$

$$O\left(\varepsilon^3\right) : (\omega - c_0 k)\, h_1^{(3)\prime} - \gamma k^3 h_1^{(3)\prime\prime\prime} = \frac{3}{2} c_0 k \left(h_1^{(1)} h_1^{(2)}\right)' - \omega_2 h_1^{(1)\prime} \tag{223}$$

etc.

Equation (221) gives the linear result,

$$h_1^{(1)} = \cos\theta, \quad \omega_0(k) = c_0 k - \gamma k^3. \tag{224}$$

Using (224), the removal of *secular* terms on the right-hand side in equation (222) requires

$$\omega_1 = 0. \tag{225}$$

Substituting (224) and (225) in equation (222) gives

$$h_1^{(2)} = \frac{c_0}{8\gamma k^2}\cos 2\theta. \tag{226}$$

Using (224)-(226), the removal of *secular* terms on the right-hand side in equation (223) requires

$$\omega_2 = \frac{3c_0^2}{32\gamma k}. \tag{227}$$

Substituting (224)-(227) in (220), we obtain,

$$\left.\begin{aligned} \frac{h_1}{h_0} &= \varepsilon \cos\theta + \frac{3\varepsilon^2}{4k^2h_0^2}\cos 2\theta + \cdots \\ \frac{\omega}{c_0 k} &= 1 - \frac{1}{6}k^2h_0^2 + \frac{9\varepsilon^2}{16k^2h_0^2} + \cdots \end{aligned}\right\}. \tag{228}$$

We see that (228) exhibits two essential effects of nonlinearities,

- periodic solutions of the form $e^{i(kx-\omega t)}$ may exist, but they are no longer sinusoidal;
- the amplitude appears in the dispersion relation.

Stokes (1847) showed that the Stokes waves have a limiting form with a sharp crest enclosing an angle of 120°. A simple and accurate analytical approximation to the Stokes wave profile was given by Longuet-Higgins (1962):

$$\frac{dh}{dx} = \tan x.$$

It should be mentioned that the two-dimensional waves described by the Korteweg-de Vries equation are more likely to be found only under controlled conditions in the laboratory than in a natural setting. Depth variations, nonuniform currents, and other effects lead to three-dimensional wave patterns in the natural setting. The propagation of three-dimensional, finite-amplitude, long waves is described by the *Kadomtsev-Petviashvili* (1970) equation which is a generalization of the Korteweg-de Vries equation to incorporate weak three-dimensional effects. This equation provides a suitable framework to describe oblique interactions of solitary waves (Miles, 1977) and periodic waves (Segur and Finkel, 1985).

(v) Modulational Instability and Envelope Solutions

Let us superimpose a slowly-varying weak modulation on a stationary weakly nonlinear wave, and study the evolution of such a modulation. Let us assume that the basic wave can be taken to be sinusoidal, i.e.,

$$\psi = a_0 \cos\theta_0 \quad \theta_0 = k_0 x - \omega_0 t. \tag{229}$$

Because of the nonlinearities, the dispersion relation is of the form (see (228)),

$$\omega_0 = \omega_0\left(k_0, a_0^2\right). \tag{230}$$

Consequent to the superposition of the modulation, let us assume that the wave is still plane periodic, but with the amplitude and phase varying slowly in x and t as below,

$$\left.\begin{aligned} a &= a(x,t) \\ \theta(x,t) &= k_0 x - \omega_0 t + \varphi(x,t) \end{aligned}\right\}. \tag{231}$$

Thus, one may define a generalized frequency and wavenumber,

$$\left.\begin{aligned} \omega(x,t) &= -\theta_t = \omega_0 - \varphi_t \\ k(x,t) &= \theta_x = k_0 + \varphi_x \end{aligned}\right\}. \tag{232}$$

For weak modulations, one may write

$$\omega = \omega_0 + \frac{\partial\omega}{\partial a_0^2}\left(a^2 - a_0^2\right) + \frac{\partial\omega}{\partial k_0}\cdot(k - k_0) + \frac{\partial^2\omega}{\partial k_0^2}\cdot(k - k_0)^2 + \cdots \tag{233}$$

and using (232), one obtains

$$\varphi_t + u_0\varphi_x + \frac{u_0'}{2}\varphi_x^2 + \left(\frac{\partial\omega}{\partial a_0^2}\right)\cdot\left(a^2 - a_0^2\right) = 0 \tag{234}$$

where u_0 is the group velocity,

$$u_0 \equiv \frac{\partial\omega}{\partial k_0}.$$

Next, note from (232) that

$$u \equiv \frac{\partial\omega}{\partial k} = u_0 + u_0'\cdot\varphi_x. \tag{235}$$

Furthermore, since the given wave has been assumed to be weakly nonlinear, we may make use of the result,

$$\frac{\partial}{\partial t}\left(a^2\right) + \frac{\partial}{\partial x}\left(ua^2\right) = 0. \tag{65}$$

Using (235), equation (65) leads to

$$\frac{\partial}{\partial t}\left(a^2\right) + u_0\frac{\partial}{\partial x}\left(a^2\right) + u_0'\frac{\partial}{\partial x}\left(\varphi_x a^2\right) = 0. \tag{236}$$

Introducing,

$$\xi = x - u_0 t, \quad \tau = u_0' t \tag{237}$$

equations (234) and (236) become

$$\varphi_t + \frac{1}{2}\varphi_\xi^2 + \frac{1}{u_0'}\left(\frac{\partial \omega}{\partial a_0^2}\right)(a^2 - a_0^2) = 0 \tag{238}$$

$$\left(a^2\right)_t + \left(a^2 \varphi_\xi\right)_\xi = 0. \tag{239}$$

Letting

$$\varphi, (a^2 - a_0^2) \sim e^{i(\chi\xi - \Omega\tau)} \tag{240}$$

one then obtains, on linearizing equations (238) and (239),

$$\Omega^2 = \frac{a_0^2}{u_0'}\frac{\partial \omega}{\partial a_0^2}\chi^2. \tag{241}$$

Therefore, the wave becomes unstable if

$$\frac{1}{u_0'}\frac{\partial \omega}{\partial a_0^2} < 0. \tag{242}$$

If in (233), one replaces $(\omega - \omega_0)$ by the operator $i(\partial/\partial t)$ and $(k - k_0)$ by $-i(\partial/\partial x)$, one obtains the *nonlinear Schrödinger equation*,

$$i\left[\frac{\partial a}{\partial t} + \left(\frac{\partial \omega}{\partial k_0}\right)\frac{\partial a}{\partial x}\right] + \frac{1}{2}\left(\frac{\partial^2 \omega}{\partial k_0^2}\right)\frac{\partial^2 a}{\partial x^2} - \frac{\partial \omega}{\partial a_0^2}\left(|a|^2 - |a_0|^2\right)a = 0. \tag{243}$$

In the frame of reference moving with the group velocity, equation (243) becomes

$$i\frac{\partial \tilde{a}}{\partial \tau} + \frac{1}{2}\frac{\partial^2 \tilde{a}}{\partial \xi^2} + \varkappa |a|^2 \tilde{a} = 0 \tag{244}$$

where we have put $a = \tilde{a}e^{i\frac{\partial \omega}{\partial a_0^2}t}$, and

$$\xi \equiv x - \frac{\partial \omega}{\partial k_0}t, \quad \tau \equiv \frac{\partial^2 \omega}{\partial k_0^2}t, \quad \varkappa \equiv \frac{\partial \omega/\partial a_0^2}{\partial^2 \omega/\partial k_0^2}.$$

Plane-wave solutions of equation (244) are modulationally unstable if $\varkappa > 0$, i.e., a ripple on the envelope of the wave will tend to grow (Benjamin, 1967).

In order to see why the sign of $\varkappa$ matters, consider a ripple on the envelope (see Figure 10.15). Suppose $\partial \omega/\partial a_0^2 < 0$. Then, the phase velocity $\omega/k = c$ becomes somewhat smaller in the region of high intensity. This causes the wave crests to pile up on the left in Figure 10.15 and to spread out on the right. If

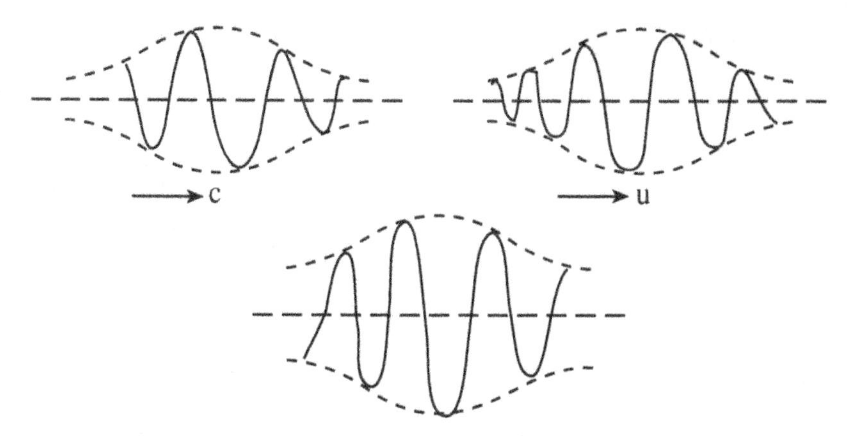

Figure 10.15 Modulational instability.

$\partial^2\omega/\partial k^2 > 0$, the group velocity $u = \partial\omega/\partial k$ will be larger on the left than on the right, so that the wave energy will pile up into a smaller space. Thus, the ripple on the envelope will become narrower and larger and an instability ensues. On the other hand, if $\partial\omega/\partial a_0^2$ and $\partial^2\omega/\partial k^2$ were of the same sign, then this modulational instability will not develop. Alternatively, one may regard equation (244) as the Schrödinger equation for *quasi-particles* whose wave function is given by a. These quasi-particles are trapped by a self-generated localized potential well $\mathcal{V} = \kappa|a|^2$. When $\kappa > 0$, this potential has an attractive sign. Additionally, if the *quasi-particle* density $|a|^2$ increases, then the potential depth increases, and more *quasi-particles* are attracted. This leads to a further increase in the potential depth. In this sense, the instability in question may be regarded as a consequence of the *self-trapping* of the *quasi-particles*.

Let us now construct a localized stationary solution to equation (224). Setting

$$\tilde{a} = v\,(\xi - U\tau)\,e^{i(\gamma\xi - s\tau)} \tag{245}$$

equation (244) gives

$$\frac{1}{2}v'' + \frac{i}{2}(2\gamma - U)\,v' + \left(s - \frac{\gamma^2}{2}\right)v + \kappa|v|^2 v = 0. \tag{246}$$

Letting

$$\gamma = \frac{U}{2}, \quad s = \frac{U^2}{2} - \frac{\alpha}{2}, \tag{247}$$

equation (246) becomes

$$v'' - \alpha v + 2\kappa v^3 = 0. \tag{248}$$

Integrating the above equation once gives us,

$$v'^2 = \alpha v^2 - \kappa v^4 \tag{249}$$

from which we have,

$$\int \frac{dv}{v\sqrt{\alpha - \kappa v^2}} = (\xi - U\tau). \tag{250a}$$

Substituting

$$w \equiv \sqrt{1 - \frac{\kappa v^2}{\alpha}} \tag{251}$$

in (250a) leads to

$$\int \frac{dw}{1 - w^2} = -\sqrt{\alpha}\,(\xi - U\tau) \tag{250b}$$

from which

$$\tanh^{-1} w = -\sqrt{\alpha}\,(\xi - U\tau)$$

or

$$w = -\tanh\sqrt{\alpha}\,(\xi - U\tau). \tag{252}$$

If $\kappa > 0$, $\alpha > 0$, one obtains, from (251) and (252),

$$v = \sqrt{\frac{\alpha}{\kappa}}\,\text{sech}\sqrt{\alpha}\,(\xi - U\tau) \tag{253}$$

which represents an *envelope soliton* (Figure 10.16) that propagates unchanged in shape and with constant velocity. This result arises because, the nonlinearity and dispersion balance each other exactly - a phenomenon unique to one-dimensional solutions. Furthermore, if the wave energy is to move along with the above envelope, then, it must move near the group velocity u of the wave in question.

Note that this solution is possible only if modulational instability occurs, i.e., if $\kappa > 0$. This suggests that the end result of an unstable wavetrain subject to small modulations is a series of envelope solitons.

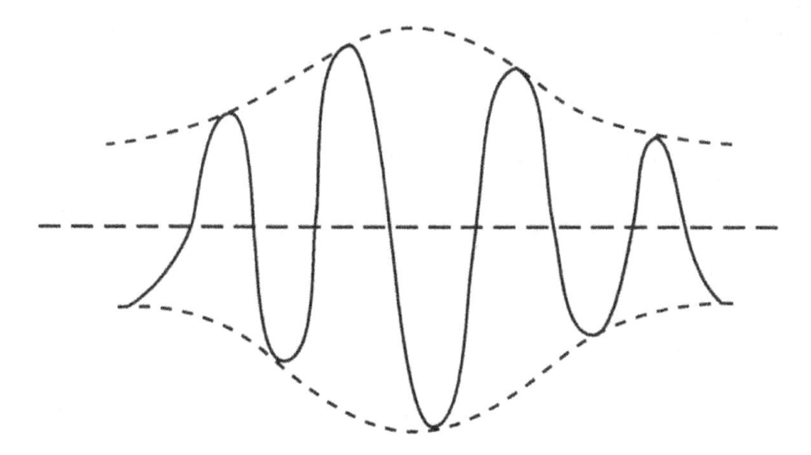

Figure 10.16 An envelope solition.

For nonlinear gravity waves in water, according to (228), one has

$$\frac{\partial \omega}{\partial a_0^2} > 0, \quad \frac{\partial^2 \omega}{\partial k_0^2} < 0 \tag{254}$$

so that according to (242), such waves are unstable to small modulations. Indeed, experiments of Benjamin and Feir (1967) on waves in wavetanks showed that such a modulational instability caused an originally almost uniform wavetrain to degenerate into a series of wavegroups. Figure 10.17 shows the experimental traces of waveheight as a function of distances from the wave-making device at one end of the wavetank. At 200 feet, the amplitude of the wave was almost uniform, while at 400 feet it was not.

It is found that the above modulational instabilities exhibit a non-ergodic behavior in their long-time evolution. The numerical solution of the nonlinear Schrödinger equation (244) with periodic boundary condition and a modulationallly unstable initial condition shows that (Lake et al., 1977) the unstable wave system reaches a state of maximum modulation. Following this, the solution demodulates and eventually returns to an unmodulated state. This process repeats itself in time. Thus, the end state is neither random (no thermalization) nor steady. However, it consists of a time-periodic spreading and regrouping of wave energy initially confined to carrier wavenumber and its linearly unstable harmonics and sidebands (Figure 10.18). This is similar to the Fermi-Pasta-Ulam (1955) recurrence observed in oscillations of an anharmonic lattice (Zabusky and Kruskal, 1965). This is due to the fact that the actively participating modes in this long-time evolution are few and clearly identifiable (with those which are modulationally unstable). These linearly unstable

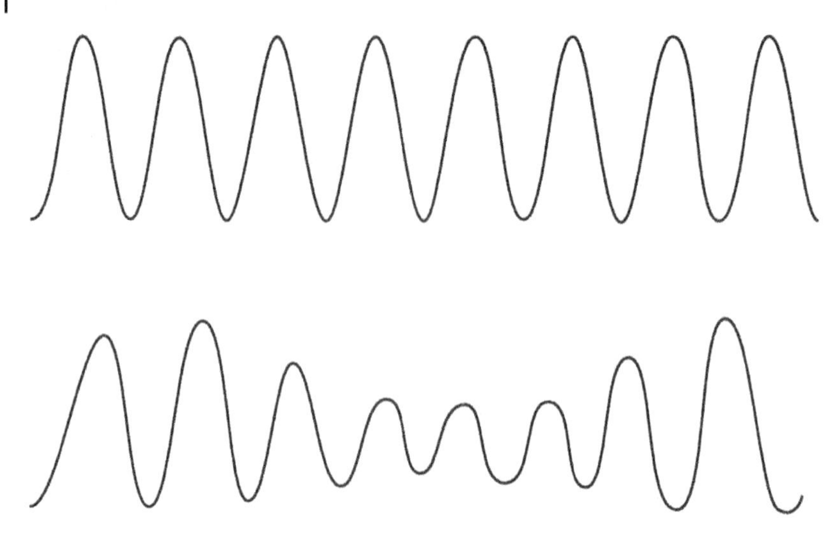

Figure 10.17 Experimental traces of wave height as a function of time : upper trace 200 feet from the wavemaker, lower trace 400 feet (from Brooke Benjamen, 1967).

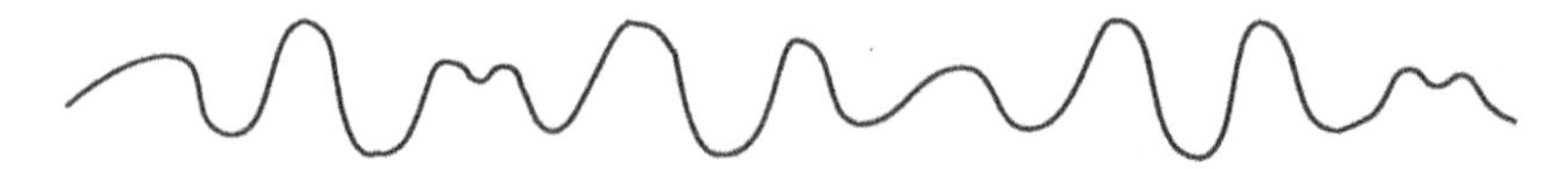

Figure 10.18 Recurring modulation and demodulation of the wave envelope (from Lake et al., 1977).

modes undergo a nonlinear saturation and evolve into a superperiodic state called *supercritical equilibrium* (Shivamoggi, 2014).

According to (242), a finite-amplitude uniform wavetrain is unstable to infinitesimal modulations with sufficiently long wavelengths. However, it is stable to modulations with short wavelengths so that a threshold for instability exists. The long-time behavior of the linearly unstable modulation near this threshold for instability shows that (Janssen, 1981, Shivamoggi, 1990) the nonlinear effects stabilize the linearly unstable modulation and produce a periodic motion. For this purpose, let us set

$$a = e^{-i\varkappa|\phi_0|^2 t} + \phi \tag{255}$$

and first write equation (244) in the form

$$i\frac{\partial\phi}{\partial t} + \frac{1}{2}\frac{\partial^2\phi}{\partial\xi^2} + \varkappa\left(|\phi|^2 - |\phi_0|^2\right)\phi = 0. \tag{256}$$

In order to investigate the modulational instability of the wavetrain whose evolution is governed by equation (256), let us make the *Madelung transformation*,

$$\phi = \rho^{1/2} e^{i\sigma}. \tag{257}$$

Then, equation (256) yields

$$\rho_t + (\rho\sigma_x)_x = 0 \tag{258}$$

$$\sigma_t + \frac{1}{2}\sigma_x^2 + \frac{1}{8\rho^2}\rho_x^2 - \frac{1}{4\rho}\rho_{xx} - \varkappa(\rho - \rho_0) = 0. \tag{259}$$

In order to perform the linear stability analysis, we write,

$$\begin{pmatrix} \rho \\ \sigma \end{pmatrix} = \begin{pmatrix} \rho_0 \\ 0 \end{pmatrix} + \begin{pmatrix} \rho_1 \\ \sigma_1 \end{pmatrix} e^{i(Kx-\Omega t)} \tag{260}$$

where K is the wavenumber and Ω is the frequency of the modulation. Assuming that $|\rho_1| \ll |\rho_0|$, and keeping only the terms linear in ρ_1 and σ_1, we obtain, from equations (258) and (259),

$$\Omega^2 = \frac{1}{4}K^2\left(K^2 - 4\varkappa\rho_0\right). \tag{261}$$

Thus, if $\varkappa > 0$, Ω^2 is negative for $|K| < \sqrt{4\varkappa\rho_0}$.

We will now consider the nonlinear development of the initially linearly unstable modulation. For this purpose, we consider the initial-value problem for modulations with wavenumbers near the threshold for instability given by $K^2 = 4\varkappa\rho_0$.

We look for a solution of the following form:

$$\left.\begin{aligned} \rho(x,t) &= \rho_0 + \varepsilon\rho_1(x,\tau) + \varepsilon^2\rho_2(x,\tau) + \cdots \\ \sigma(x,t) &= \varepsilon\sigma_1(x,\tau) + \varepsilon^2\sigma_2(x,\tau) + \cdots \\ K^2 &= 4\rho_0\varkappa + \varepsilon^2\chi + \cdots \end{aligned}\right\} \tag{262}$$

where ε is a small parameter that characterizes the departure of K^2 from the linear stability threshold value $4\rho_0\varkappa$, and $\tau \equiv \varepsilon t$ is a slow time scale characterizing slow time evolutions near the stability threshold. We have introduced an explicit detuning parameter χ in (262).

Substituting (262) into equations (258) and (259), we obtain the following systems of equations to various orders in ε:

$$O\left(\varepsilon^{n}\right): L\begin{pmatrix}\rho_n\\ \sigma_n\end{pmatrix} = S_n\left(\rho_0,\rho_1,\cdots,\rho_{n-1};\sigma_1,\cdots,\sigma_{n-1}\right),\quad n=1,2,\cdots \tag{263}$$

where

$$L \equiv \begin{bmatrix} 0 & \rho_0\dfrac{\partial^2}{\partial x^2} \\ \dfrac{1}{4}\dfrac{\partial^2}{\partial x^2} + \kappa\rho_0 & 0 \end{bmatrix}$$

and the function S_n depends on the solutions up to $O\left(\varepsilon^{n-1}\right)$.

We obtain from equation (263), to $O\left(\varepsilon\right)$:

$$L\begin{pmatrix}\rho_1\\ \sigma_1\end{pmatrix} = 0 \tag{264}$$

the solution to which corresponds to the neutrally-stable case of the linear problem:

$$\left.\begin{aligned} \rho_1 &= A\left(\tau\right)e^{iKx} + c.c. \\ \sigma_1 &= \alpha\left(\tau\right) \\ K^2 &= 4\rho_0\kappa \end{aligned}\right\} \tag{265}$$

where $c.c.$ means complex conjugate.

Using (261), we obtain from equation (263), to $O\left(\varepsilon^2\right)$:

$$L\begin{pmatrix}\rho_2\\ \sigma_2\end{pmatrix} = \begin{bmatrix} -\dfrac{dA}{d\tau}e^{iKx} + c.c. \\ -\kappa|A|^2 + \rho_0\dfrac{d\alpha}{d\tau} + \left\{-\dfrac{3}{2}\kappa A^2 e^{2iKx} + c.c.\right\} \end{bmatrix} \tag{266}$$

from which,

$$\left.\begin{aligned} \rho_2 &= \frac{1}{\kappa}\frac{d\alpha}{d\tau} - |A|^2 + \frac{A^2}{2\rho_0}e^{2iKx} + c.c. \\ \sigma_2 &= \frac{1}{4\rho_0^2\kappa}\frac{dA}{d\tau}e^{iKx} + c.c. \end{aligned}\right\} \tag{267}$$

Using (265) and (267), we obtain from equation (263), to $O\left(\varepsilon^3\right)$:

$$L\begin{pmatrix}\rho_3\\ \sigma_3\end{pmatrix} = \begin{bmatrix} -\dfrac{1}{\kappa}\dfrac{d^2\alpha}{d\tau^2} + \dfrac{d|A|^2}{d\tau} \\ \left\{\dfrac{1}{4\rho_0\kappa}\dfrac{d^2A}{d\tau^2} + \dfrac{1}{4}\chi A + \dfrac{\kappa}{2\rho_0}|A|^2A + \right. \\ \left. -2\kappa A^2\left(\dfrac{1}{\kappa}\dfrac{d\alpha}{d\tau} - |A|^2\right)\right\}e^{iKx} + c.c. \end{bmatrix} + \text{ nonsecular terms.} \tag{268}$$

Removal of the secular terms in the first member of equation (268) requires

$$\frac{1}{\kappa}\frac{d\alpha}{d\tau} - |A|^2 = const. \tag{269}$$

Let us take the constant above to be zero. Removal of the secular terms in the second member of equation (268) then requires

$$\frac{d^2A}{d\tau^2} + \left(\rho_0\kappa\chi + 2\kappa^2|A|^2\right)A = 0. \tag{270}$$

If we impose the following initial conditions,

$$\tau = 0 : A = \hat{A}, \quad \frac{dA}{d\tau} = 0 \tag{271}$$

and take A to be real, we obtain, from equation (270),

$$\left(\frac{dA}{d\tau}\right)^2 = \kappa^2\left(\hat{A}^2 - A^2\right)\left(A^2 - \beta\right) \tag{272}$$

where,

$$\beta \equiv -\frac{\rho_0\chi}{\kappa} - \hat{A}^2.$$

Equation (272) shows that A is bounded and oscillates between $\hat{A}$ and $\sqrt{\beta}$ if $\beta > 0$, and oscillates between 0 and $\hat{A}$ if $\beta < 0$. This demonstrates the nonlinear saturation of the linearly unstable modulation ($\chi < 0$) near the linear-instability threshold.

10.12 Nonlinear Capillary-Gravity Waves

(i) Resonant Three-Wave Interactions

Consider two waves with frequencies ω_3, ω_2 and wave-vectors k_3, k_2, respectively. Then the nonlinear terms in the equations of fluid flow may contain the product of the wave amplitudes, namely

$$e^{i(k_3-k_2)x-i(\omega_3-\omega_2)t}$$

which is a *beat-frequency* wave. If the frequency $\omega_3-\omega_2 = \omega_1$ and the wave vector $k_3 - k_2 = k_1$ of the beat happen to be a *normal mode* of the fluid, then this mode will be generated by the interaction of the first two waves. An exchange of energy and momentum takes place between these waves, and the process is called *resonant* wave interaction. Thus, the wave interactions are *selective* in that only certain combinations of wave components are capable of significant energy exchange. Furthermore, if the wave amplitudes are small, the wave interactions are *weak*. This is because even for the resonant combinations, the interaction time is large compared with a typical wave period. However, their cumulative dynamical effect is much more pronounced because the existence of resonant wavenumbers and frequencies leads to growth in time of one or more of the wave-amplitudes. The wave-amplitudes show a *slow* modulation in time, the modulation envelopes being given in terms of *elliptic functions*.

Consider an initially quiescent water of infinite depth whose original surface level is given by $z = 0$. The water is assumed to be inviscid and is subjected to gravity $\mathbf{g} = g\,\hat{\mathbf{i}}_z$ acting normal to the surface and directed toward the water. Let us consider the wave propagation to be two-dimensional. If $z = \eta(x,t)$ denotes the disturbed shape of the surface, one has the following boundary value problem,

$$z < \eta : \phi_{xx} + \phi_{zz} = 0 \tag{273a}$$

$$z = \eta : \phi_z = \eta_t + \phi_x \eta_x \tag{274}$$

$$z = \eta : \phi_t + \frac{1}{2}\left(\phi_x^2 + \phi_z^2\right) + g\eta - \frac{T}{\rho}\eta_{xx}\left(1+\eta_x^2\right)^{-3/2} = 0 \tag{275}$$

$$z \Rightarrow -\infty : \phi_z \Rightarrow 0. \tag{276}$$

Let us now transfer the boundary conditions (274) and (275) from the actual location of the surface $z = \eta$ to the original location of the surface $z = 0$. This is accomplished by an expansion of $\phi(x,z,t)$ in Taylor's series about the values at $z = 0$. Let us retain the nonlinearities of only quadratic order since we shall consider only the *second-order interactions*. The system (273a)-(276) now becomes

$$z < 0 : \phi_{xx} + \phi_{zz} = 0 \tag{273b}$$

$$z = 0 : \phi_z - \eta_t = -\phi_{zz}\eta + \phi_x\eta_x \tag{277}$$

$$z = 0 : \phi_t + g\eta - \frac{T}{\rho}\eta_{xx} = \phi_{tz}\eta - \frac{1}{2}\left(\phi_x^2 + \phi_z^2\right) \tag{278}$$

$$z \Rightarrow -\infty : \phi_z \Rightarrow 0 \tag{276}$$

where the left-hand sides represent the linear problem and the right-hand sides contain the nonlinearities that produce the interaction between the waves.

Let us now consider unidirectional wave propagation since this case is well suited for experimental verification. By noting (273a) and (276), we take waves of the form,

$$\left.\begin{array}{l} \phi(x, z, t) \sim e^{kz - ikx} \\ \\ \eta(x, t) \sim e^{-ikx} \end{array}\right\}. \tag{279}$$

Let us now introduce a new dynamical variable,

$$a \equiv \eta - \frac{ik}{\omega}\phi \tag{280}$$

where

$$\omega^2 = gk + \frac{T}{\rho}k^3.$$

Using only the linear terms in (277) and (278), one then obtains

$$\frac{\partial a}{\partial t} - i\omega a = 0 \tag{281}$$

so that a is a *normal mode* of the linearized problem associated with (273a)-(276). The nonlinear analysis of the wave-wave interactions, as seen below, becomes very convenient when formulated in terms of this auxiliary variable a. When the nonlinear terms are included, one obtains, from (277) and (278),

$$\frac{\partial a}{\partial t} - i\omega a = \phi_{zz}\eta - \phi_x\eta_x + \frac{ik}{\omega}\left[\phi_{tz}\eta + \frac{1}{2}\left(\phi_x^2 + \phi_z^2\right)\right]. \tag{282}$$

Let us now consider two capillary-gravity waves of the form $e^{i(\omega_3 t - k_3 x)}$ and $e^{i(\omega_2 t - k_2 x)}$ propagating in the x-direction with

$$\left.\begin{aligned} \omega_3^2 &= gk_3 + \frac{T}{\rho}k_3^3 \\ \omega_2^2 &= gk_2 + \frac{T}{\rho}k_2^3 \end{aligned}\right\}. \tag{283}$$

Due to nonlinear interaction between these two waves, let another capillary-gravity wave of the form $e^{i(\omega_1 t - k_1 x)}$, propagating in the x-direction, be excited such that

$$\omega_1 = \omega_3 - \omega_2,\ k_1 = k_3 - k_2.^{11} \tag{284}$$

One obtains, for the linearized problem,

$$\left.\begin{aligned} \eta &= \frac{a}{2} \\ \phi &= \frac{i\omega}{2k}a \\ a &= \tilde{a}\left(\tilde{t}\right)e^{i(\omega t - kx)} + \tilde{a}^*\left(\tilde{t}\right)e^{i(\omega t - kx)} \end{aligned}\right\} \tag{285}$$

where $\tilde{t}$ is the slow time scale characterizing the amplitude modulations suffered by the waves undergoing resonant interactions.

Using (285), equation (282) gives, on keeping only the resonant terms (according to (284)) on its right-hand side, the following amplitude modulation equations,

$$\frac{d\tilde{a}_3}{d\tilde{t}} = \frac{ik_3\omega_1\omega_2}{2\omega_3}\tilde{a}_1\tilde{a}_2 \tag{286}$$

$$\frac{d\tilde{a}_2}{d\tilde{t}} = -\frac{ik_2\omega_3\omega_1}{2\omega_2}\tilde{a}_3\tilde{a}_1^* \tag{287}$$

$$\frac{d\tilde{a}_1}{d\tilde{t}} = -\frac{ik_1\omega_3\omega_2}{2\omega_1}\tilde{a}_3\tilde{a}_2^* \tag{288}$$

---

11 It may be noted that whether or not solutions of (284) exist depends upon the form of the dispersion relation involved. Thus, it turns out that three-wave resonant interactions are not possible for deep-water gravity waves which instead allow four-wave resonant interactions to occur with the concomitant matching conditions (Phillips, 1960) given below

$$\omega_1 + \omega_2 = \omega_3 + \omega_4$$

$$k_1 + k_2 = k_3 + k_4.$$

and in terms of the ϕ's, as per (285), equations (286)-(288) become

$$\frac{\partial \phi_3}{\partial t} = (k_1 k_2)\, \phi_1 \phi_2 \tag{289}$$

$$\frac{\partial \phi_2}{\partial \tilde{t}} = -\,(k_3 k_1)\, \phi_3 \phi_1^* \tag{290}$$

$$\frac{\partial \phi_1}{\partial \tilde{t}} = -\,(k_3 k_2)\, \phi_3 \phi_2^*. \tag{291}$$

For some prescribed initial values of the amplitudes of the three waves, equations (289)-(291) can be solved by quadratures, and the general solution can be expressed in terms of elliptic functions. As an illustration, let us consider a case wherein only the modes ϕ_1 and ϕ_2 are initially present and mode ϕ_3 is absent, i.e., let

$$t = 0 : \phi_{1,2} = \phi_{1,2}^{(0)}, \phi_3 = 0. \tag{292}$$

Equations (287)-(291) then indicate the generation and growth of the mode ϕ_3 at the expense of the modes ϕ_1 and ϕ_2.

One obtains, from equations (289)-(291), the *Manley-Rowe* (1959) *relations*,

$$\frac{d}{d\tilde{t}}\left(\frac{|\phi_1|^2}{k_2 k_3} + \frac{|\phi_3|^2}{k_1 k_2}\right) = 0$$

or

$$|\phi_1|^2 = |\phi_1^{(0)}|^2 \left(1 - \frac{k_3}{k_1} \frac{|\phi_3|^2}{|\phi_1^{(0)}|^2}\right). \tag{293}$$

Similarly we have,

$$|\phi_2|^2 = |\phi_2^{(0)}|^2 \left(1 - \frac{k_3}{k_2} \frac{|\phi_3|^2}{|\phi_2^{(0)}|^2}\right). \tag{294}$$

Using (293) and (294), equation (289) becomes

$$\left|\frac{\partial \phi_3}{\partial \tilde{t}}\right|^2 = |\phi_1^{(0)}|^2 |\phi_2^{(0)}|^2 k_1^2 k_2^2 \left(1 - \frac{k_3}{k_1} \frac{|\phi_3|^2}{|\phi_1^{(0)}|^2}\right)\left(1 - \frac{k_3}{k_2} \frac{|\phi_3|^2}{|\phi_2^{(0)}|^2}\right) \tag{295}$$

from which,[12]

12 If

$$u(x) = \int_0^x \frac{dt}{\sqrt{(1-t^2)(1-k^2 t^2)}}$$

$$\phi_3 = \phi_2^{(0)} \sqrt{\frac{k_2}{k_3}} sn\,(\tau; k) \tag{296}$$

where

$$\tau \equiv \phi_1^{(0)} k_1 \sqrt{k_2 k_3} t$$

and k is the modulus,

$$k^2 \equiv \frac{k_2 \left|\phi_2^{(0)}\right|^2}{k_1 \left|\phi_1^{(0)}\right|^2}. \tag{297}$$

Let us assume that $k \leq 1$. Noting from (293) and (294) that

$$\frac{\dfrac{\partial}{\partial \tilde{t}} \left| \dfrac{\phi_1}{\phi_1^{(0)}} \right|^2}{\dfrac{\partial}{\partial \tilde{t}} \left| \dfrac{\phi_2}{\phi_2^{(0)}} \right|^2} = \frac{k_2 \left|\phi_2^{(0)}\right|^2}{k_1 \left|\phi_1^{(0)}\right|^2} = k^2, \tag{298}$$

we infer that k may be regarded as measuring the extent to which the resonant partners participate in the interaction. In particular, $k \leq 1$ implies that the mode k_2 is decreasing at a rate faster than the mode k_1.

Using (296), equations (293) and (294) give

$$\phi_2 = \phi_2^{(0)} cn\,(\tau; k) \tag{299}$$

$$\phi_1 = \phi_1^{(0)} dn\,(\tau; k). \tag{300}$$

Here, sn, cn, and dn are *elliptic functions* with real parameter k. The solutions (296), (299), and (300) represent a system for which energy is transfered around periodically among the three waves k_1, k_2, k_3. Initially, the energy in mode k_3 increases at the expense of both modes k_1 and k_2. Eventually, the energy in mode k_2 vanishes since we have assumed the mode k_2 to decay at a rate faster than the mode k_1. Then, the direction of energy transfer is reversed. Modes k_1 and k_2 now increase at the expense of mode k_3 until the initial state is reached

---

then

$$\left.\begin{aligned} x &= snu \\ \sqrt{1-x^2} &= cnu \\ \sqrt{1-k^2x^2} &= dnu. \end{aligned}\right\}$$

again, and then this sequence of energy transfer repeats itself. The period of this energy transfer is

$$T_E = \frac{2K(k)}{k_1 \sqrt{\left|\phi_1^{(0)}\right|^2 k_2 k_3}} \tag{301}$$

where $K(k)$ is the *complete elliptic integral of the first kind* (see footnote 9).

The case $k_1 = k_2 = \sqrt{\rho g/2T} = k_3/2$ and $\omega_1 = \omega_2 = \omega_3/2$, which corresponds to the *second harmonic resonance*[13] is a degenerate case of the triad resonances. Two members of the triad are identical, the *closure* being their second harmonic. Since $k = 1$ for this case, and

$$k \approx 1 : K(k) \sim \ell n \frac{2}{\sqrt{1-k}} \Rightarrow \infty \tag{302}$$

the period of the energy transfer, as per (301) is infinite, and the resonant interaction takes on an asymptotic character. The solutions (296), (299), and (300) then become

$$\phi_3 = \phi_2^{(0)} \sqrt{\frac{k_2}{k_3}} \tanh \tau \tag{303}$$

$$\phi_1 = \phi_1^{(0)} \operatorname{sech} \tau \tag{304}$$

$$\phi_2 = \phi_2^{(0)} \operatorname{sech} \tau. \tag{305}$$

We will discuss the second-harmonic resonance in detail below.

(ii) Second-Harmonic Resonance

Suppose that a typical surface disturbance is characterized by a sinusoidal traveling wave with amplitude a' and wavelength λ'. We nondimensionalize all physical quantities in the following using a reference length $(\lambda'/2\pi)$, a time $(\lambda'/2\pi g')^{1/2}$. Here, g' denotes the acceleration due to gravity and the primes denote the dimensional quantities. The velocity potential of the flow of the liquid is taken to be $g'^{1/2}(\lambda'/2\pi)^{3/2}\phi$. If $y = \eta$ denotes the disturbed shape

13 The second harmonic resonance arises when the fundamental component and its second harmonic have the same linear phase velocity so that the two can interact resonantly with each other.

of the surface (whose original level is given by $y = 0$), one has the following boundary-value problem:

$$y < \eta : \phi_{xx} + \phi_{yy} = 0 \tag{306}$$

$$y = \eta : \phi_y = \eta_t + \eta_x \phi_x \tag{307}$$

$$\phi_t + \frac{1}{2}\left(\phi_x^2 + \phi_y^2\right) + \eta = k^2 \eta_{xx} \left(1 + \eta_x^2\right)^{-3/2} \tag{308}$$

$$y \Rightarrow -\infty : \phi_y \Rightarrow 0 \tag{309}$$

where

$$k^2 \equiv \left(\frac{2\pi}{\lambda'}\right)^2 \frac{T'}{\rho' g'}.$$

Let us now look for traveling waves, and introduce

$$\xi \equiv x - ct \tag{310}$$

so that (306)-(309) become

$$y < \eta : \phi_{\xi\xi} + \phi_{yy} = 0 \tag{311}$$

$$y = \eta : \phi_y = \left(\phi_\xi - c\right)\eta_\xi \tag{312}$$

$$\eta = c\phi_\xi - \frac{1}{2}\left(\phi_\xi^2 + \phi_y^2\right) + k^2 \eta_{\xi\xi}\left(1 + \eta_\xi^2\right)^{-3/2} \tag{313}$$

$$y \Rightarrow -\infty : \phi_y \Rightarrow 0. \tag{309}$$

We seek solutions to (311)-(313) and (309) of the form,

$$\left.\begin{aligned} \phi(\xi, y; \varepsilon) &= \sum_{n=1}^{\infty} \varepsilon^n \phi_n(\xi, y) \\ \eta(\xi; \varepsilon) &= \sum_{n=1}^{\infty} \varepsilon^n \eta_n(\xi) \\ c(k; \varepsilon) &= \sum_{n=0}^{\infty} \varepsilon^n c_n(k) \\ k(\varepsilon) &= \sum_{n=0}^{\infty} \varepsilon^n k_n \end{aligned}\right\} \tag{314}$$

where,

$$\varepsilon \equiv a' \frac{2\pi}{\lambda'} \ll 1.$$

Using (314), one obtains from (309), and (311)-(313), the following hierarchy of boundary-value problems:

$$O(\varepsilon) : y < 0 : \phi_{1\xi\xi} + \phi_{1yy} = 0 \tag{315}$$

$$y = 0 : \phi_{1y} = -c_0 \eta_{1\xi} \tag{316}$$

$$\eta_1 = c_1 \phi_{1\xi} + k_0^2 \eta_{1\xi\xi} \tag{317}$$

$$y \Rightarrow -\infty : \phi_{1y} \Rightarrow 0 \tag{318}$$

$$O(\varepsilon^2) : y < 0 : \phi_{2\xi\xi} + \phi_{2yy} = 0 \tag{319}$$

$$y = 0 : \phi_{2y} + \phi_{1yy}\eta_1 = -c_0 \eta_{2\xi} + (\phi_{1\xi} - c_1)\eta_{1\xi} \tag{320}$$

$$\eta_2 = c_0 (\phi_{2\xi} + \phi_{1y\xi}\eta_1) - \frac{1}{2}\left(\phi_{1\xi}^2 + \phi_{1y}^2\right) + c_1 \phi_{1\xi} + k_0^2 \eta_{2\xi\xi} + 2k_0 k_1 \eta_{1\xi\xi} \tag{321}$$

$$y \Rightarrow -\infty : \phi_{2y} \Rightarrow 0 \tag{322}$$

$$O(\varepsilon^3) : y < 0 : \phi_{3\xi\xi} + \phi_{3yy} = 0 \tag{323}$$

$$y = 0 : \phi_{3y} + \phi_{2yy}\eta_1 + \phi_{1yy}\eta_2 + \tfrac{1}{2}\phi_{1yyy}\eta_1^2 = -c_0\eta_{3\xi} + (\phi_{1\xi} - c_1)\eta_{2\xi} + (\phi_{2\xi} + \phi_{1\xi y}\eta_1 - c_2)\eta_{1\xi} \tag{324}$$

$$\eta_3 = c_0\left(\phi_{3\xi} + \phi_{2\xi y}\eta_1 + \phi_{1\xi y}\eta_2 + \tfrac{1}{2}\phi_{1\xi yy}\eta_1^2\right) + c_1(\phi_{2\xi} + \phi_{1\xi y}\eta_1) + c_2\phi_{1\xi} + -\phi_{1\xi}(\phi_{2\xi} + \phi_{1\xi y}\eta_1) - \phi_{1y}(\phi_{2y} + \phi_{1yy}\eta_1) + k_0^2\eta_{3\xi\xi} + 2k_0k_1\eta_{2\xi\xi} + 2k_0k_2\eta_{1\xi\xi} - \frac{3}{2}k_0^2\eta_{1\xi\xi}\eta_{1\xi}^2 \tag{325}$$

$$y \Rightarrow -\infty : \phi_{3y} \Rightarrow 0. \tag{326}$$

Let

$$\eta_1 = A\cos\xi. \tag{327}$$

Then, from (315)-(318), one obtains

$$\phi_1(\xi, y) = Ac_0 e^y \sin\xi \tag{328}$$

$$c_0^2 = 1 + k_0^2. \tag{329}$$

Next, letting

$$\eta_2 = B \cos 2\xi \tag{330}$$

and using (327)-(330), one then obtains, from (319), (320), and (322),

$$\phi_2 (\xi, y) = c_0 \left(B - \frac{A^2}{2} \right) e^{2y} \sin 2\xi + c_1 A e^y \sin \xi. \tag{331}$$

Using (327)-(331), one finds that the removal of *secular* terms in (321) requires

$$k_1 = c_1 = 0 \tag{332}$$

and

$$B = \frac{c_0^2}{2\left(1 - 2k_0^2\right)} A^2. \tag{333}$$

The case $k_0 = \pm\sqrt{1/2}$, where the above solution breaks down, corresponds to the *second-harmonic resonance*, which we shall treat shortly.

Using (327)-(333), one obtains, from (323), (324), and (326),

$$\phi_3 (\xi, y) = A \left[-c_0 A^2 \left(\frac{c_0^2/4}{1 - 2k_0^2} + \frac{3}{8} \right) + c_2 \right] e^y \sin \xi + \text{higher harmonics}. \tag{334}$$

Using (327)-(334), one finds that the removal of secular terms in (325) requires

$$c_2 = \frac{c_0}{2} \left(\frac{c_0^2/2}{1 - 2k_0^2} + \frac{1}{2} - \frac{3}{8} \frac{k_0^2}{c_0^2} \right) A^2, \quad k_2 = 0 \tag{335}$$

which is again not valid for $k_0 = \pm\sqrt{1/2}$.

In order to treat the case of second-harmonic resonance, wherein $k = \pm\sqrt{1/2}$, first note that the *fundamental* component,

$$\left. \begin{aligned} \eta_1^{(1)} &= A \cos \xi \\ \phi_1^{(1)} &= A c_0 e^y \sin \xi \end{aligned} \right\}$$

and its *second harmonic*

$$\left. \begin{aligned} \eta_1^{(2)} &= \hat{B} \cos 2\xi \\ \phi_1^{(2)} &= \hat{B} c_0 e^{2y} \sin 2\xi \end{aligned} \right\}$$

have the same linear phase velocity c_0 for this case. Consequently, the two can interact *resonantly* with each other. In order to treat this *nonlinear resonant interaction*, set

$$\eta_1 = A\cos\xi + \hat{B}\cos 2\xi \tag{336}$$

$$\phi_1 = c_0\left(Ae^y \sin\xi + \hat{B}e^{2y}\sin 2\xi\right). \tag{337}$$

Using (336) and (337), one obtains, from (319), (320), and (322),

$$\left.\begin{aligned} \phi_2(\xi,y) = {} & \left(-\frac{3A\hat{B}}{2}c_0 + Ac_1\right)e^y \sin\xi + \\ & + \left(-A^2c_0 + 2\hat{B}c_1\right)\frac{1}{2}e^{2y}\sin 2\xi + \text{ higher harmonics.} \end{aligned}\right\} \tag{338}$$

Using (336)-(338), one finds that the removal of *secular* terms in (321) requires

$$-c_0^2\hat{B} + 2c_0c_1 - 2k_0k_1 = 0 \tag{339}$$

$$-\frac{1}{2}c_0^2A + 4c_0\hat{B}c_1 - 8k_0k_1\hat{B} = 0 \tag{340}$$

from which,

$$\hat{B} = \frac{k_0k_1}{c_0^2} \pm \sqrt{\frac{A^2}{4} - c_0^2k_1^2} \tag{341}$$

$$c_1 = \frac{k_1\left(3k_0 + c_0^2/k_0^2\right)}{2c_0} \pm \frac{c_0}{2}\sqrt{\frac{A^2}{4} - c_0^2k_1^2}. \tag{342}$$

These results show that *purely phase-modulated* waves are possible for wavenumbers near the second-harmonic resonant values.

Exercises

1. Show that the wavelength λ of stationary waves on a stream of depth h and flow speed U is given by

$$U^2 = \frac{g\lambda}{2\pi}\tanh\frac{2\pi h}{\lambda}$$

   and hence deduce that such stationary waves exist provided that $U < \sqrt{gh}$.
2. Show that the semivertex angle of the shipwave pattern is 19.5° by merely invoking the fact that the ratio of the group velocity and the phase velocity for deep-water waves is 1/2.

3. The linear dispersion relation $\omega^2 = gk$ for gravity waves in deep water can be generalized to the two-dimensional case by interpreting k as the magnitude of the wavevector $\boldsymbol{k} = (\ell, m)$. This leads to the nonlinear dispersion relation

$$\omega^2 = g\sqrt{\ell^2 + m^2} + \frac{1}{2}\sqrt{g}\left(\ell^2 + m^2\right)^{5/4} a^2.$$

Expanding this about $k_0 = (k_0, 0)$ with perturbation (k_1, k_2), show that

$$\omega = \omega_0 + \frac{\omega_0}{2k_0}k_1 - \frac{\omega_0}{8k_0^2}k_1^2 + \frac{\omega_0}{4k_0^2}k_2^2 + \frac{1}{2}w_0 k_0^2 a^2$$

where $\omega_0 = \sqrt{gk_0}$. Show that this relation, then, leads to the following nonlinear Schrödinger equation describing the evolution of the two-dimensional modulated gravity waves (Zakharov, 1968):

$$i\left(\frac{\partial a}{\partial t} + \frac{\omega_0}{2k_0}\frac{\partial a}{\partial x}\right) - \frac{\omega_0}{8k_0^2}\frac{\partial^2 a}{\partial x^2} + \frac{\omega_0}{4k_0^2}\frac{\partial^2 a}{\partial y^2}$$

$$-\frac{1}{2}\omega_0 k_0^2 \left(|a|^2 - |a_0|^2\right) a = 0.$$

Investigate the stability of the three-dimensional modulations by setting

$$a = \left[\rho\left(\xi, \eta, t\right)\right]^{1/2} e^{i\sigma(\xi,\eta,t)}, \quad \xi = x - \frac{\omega_0}{2k_0}t, \eta = y,$$

$$\rho = \rho_0 + \rho_1\left(\xi, \eta, t\right), \quad \sigma = \sigma_1\left(\xi, \eta, t\right),$$

$$\rho_1\left(\xi, \eta, t\right) \text{ and } \sigma_1\left(\xi, \eta, t\right) \sim e^{i(K_1\xi + K_2\eta - \Omega t)}.$$

By linearizing in ρ_1 and σ_1, show that the growth rate Ω is given by

$$\Omega^2 = \frac{\omega_0}{8k_0^2}\left(K_1^2 - 2K_2^2\right)\left(\frac{K_1^2}{8k_0^2} - \frac{K_2^2}{4k_0^2} - k_0^2|a_0|^2\right),$$

which shows that the instability region for three-dimensional modulations, unlike that for two-dimensional modulations, is unbounded in (K_1, K_2)-space!

11

Applications to Aerodynamics

We shall now deal with the aerodynamic forces which act on a lifting surface in flight. This information is required for performance and structural-strength calculations in general and for purposes of stability and control in the unsteady case in particular.

Though ideal-fluid aerodynamics is predicated on an inviscid model for the fluid, it does incorporate the gross features of the flow which would not be present if the fluid were truly inviscid.[1] The circulation about an airfoil, the *Kutta condition* and the free vortex shedding at the trailing edge of the wing are cases in point. At the trailing edge, while the Kutta condition ensures smooth flow, the free vortex shedding occurs whenever the circulation varies along the wing span. The viscous processes involving boundary layer separation and vortex shedding caused by changes in flow conditions occur so rapidly that the above features can be assumed to hold at each instant. An example of this is a flow over an oscillating airfoil. The vortex sheet in the wake adjusts itself very rapidly to preserve smooth flow condition at the trailing edge. This underscores the success of the two-dimensional airfoil theory.

1 In an attached flow past a wing, the viscous effects are actually confined to a thin boundary layer (see Chapter 21) adjacent to the wing surface and the wake. The presence of the boundary layer changes the effective cross section of the wing, and therefore, influences the pressure distribution over the wing. However, the inviscid theory affords the correct first approximation for most of the flow field, provided the flow remains attached over most of the airfoil.

Introduction to Theoretical and Mathematical Fluid Dynamics, Third Edition.
Bhimsen K. Shivamoggi.

11.1 Airfoil Theory: Method of Complex Variables

Consider a wing of infinite *aspect ratio*, whose generators are parallel to the z-axis. The cross section of the wing is the same in all planes parallel to the x, y-plane, and it will therefore suffice to consider the conditions in that plane alone. Since the third dimension z no longer appears, z may be reassigned to mean something else, as in the following.

(i) Force and Moments on an Arbitrary Body

In the classical approach the force exerted on a fluid by a moving body is found by first calculating the momentum in the fluid, and then, differentiating it with respect to time. This method fails when applied to bodies in uniform steady motion because, the fluid would then have received an infinite momentum from the constant force acting on it over infinite time. A more workable approach is to calculate the forces in steady motion by integrating the pressure on the moving body.

Theorem 11.1 *(Blasius, 1910)* Let C be a simple closed curve in $\mathbb{R}^2$ which constitutes the trace of an arbitrary body in the $x, y-$plane (see Figure 11.1).

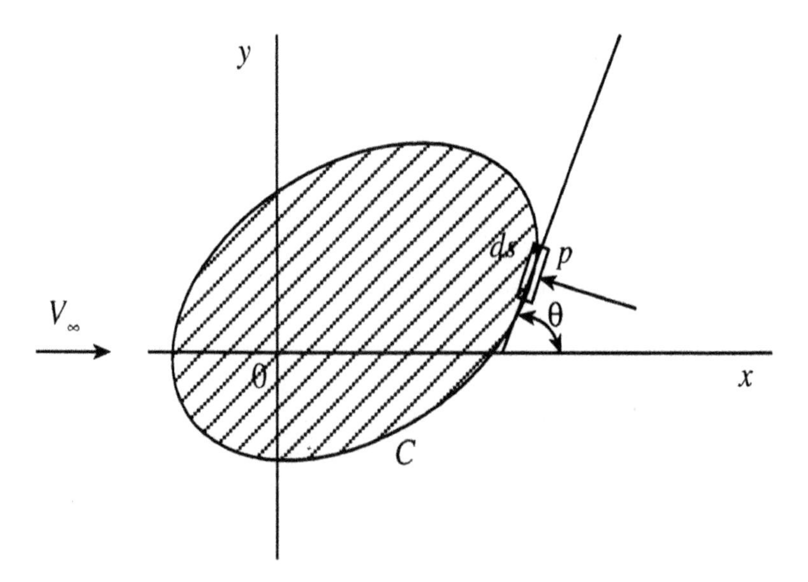

Figure 11.1 Force exerted by a fluid on a body.

Let **v** be a steady two-dimensional velocity field parallel to C at all points on C, defined on the exterior of C. The force exerted by the fluid on the body is given by

$$X - iY = \frac{i\rho}{2} \oint_C [W(z)]^2 \, dz \tag{1}$$

where W is the complex velocity,

$$W(z) = Ve^{-i\theta}, \; z = x + iy.$$

Proof: We have, on using Bernoulli integral,

$$\begin{aligned} X - iY &= \int_C p(-dy - idx) = \frac{\rho}{2} \int_C (u^2 + v^2)(idx + dy) \\ &= \frac{i\rho}{2} \int_C (u - iv)^2 (dx + idy) \\ &= \frac{i\rho}{2} \oint_C [W(z)]^2 \, dz \end{aligned}$$

where we note that **v** is parallel to C at all points on C so that $udy = vdx$. □

The moment about O exerted by the fluid on the body is given by

$$M = \oint_c p(ydy + xdx) = -\frac{\rho}{2} Re \oint_C [W(z)]^2 \, zdz \tag{2}$$

where the moment M is reckoned positive when it is counterclockwise.

Using *Cauchy's Integral Theorem*, the integration may be carried out around an infinitely large circle S enclosing C, provided there are no singularities between S and C.

As an example, consider the flow with velocity V_∞ in the x-direction and a clockwise circulation Γ past a cylinder of arbitrary cross section (Figure 11.2). Then, the complex velocity $W(z)$ is regular except possibly at or inside the contour of the body. It can, therefore, be expanded into a *Laurent series*,

$$W(z) = \sum_{n=0}^{\infty} \frac{A_n}{z^n} \tag{3}$$

where

$$A_0 = V_\infty, \; A_1 = \frac{i\Gamma}{2\pi}, \cdots.$$

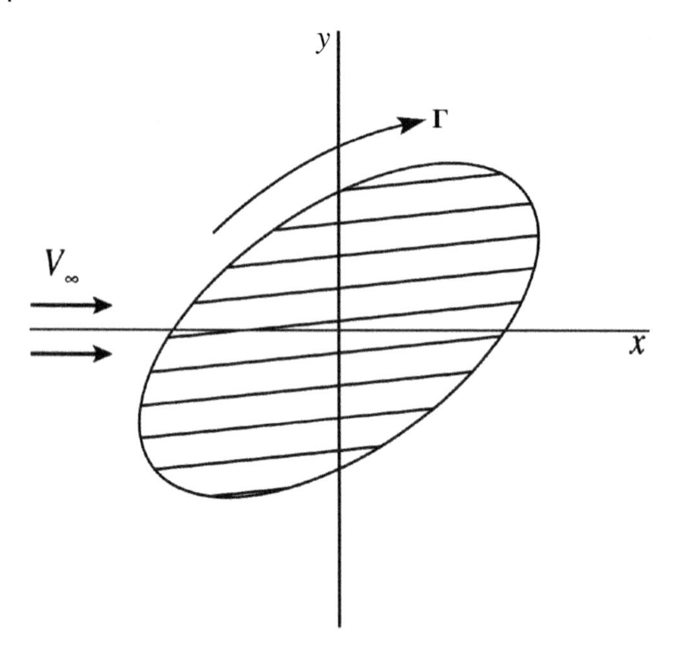

Figure 11.2 Flow with circulation past a cylinder.

Thus

$$[W(z)]^2 = \sum_{n=0}^{\infty} \frac{B_n}{z^n} \tag{4}$$

where

$$B_0 \equiv A_0^2,\; B_1 \equiv 2A_0A_1,\; B_2 \equiv A_1^2 + 2A_0A_2, \cdots .$$

Using (4), (1) gives for the forces exerted by the fluid on the body,

$$X - iY = \frac{i\rho}{2} \cdot 2\pi B_1 = -i\rho V_\infty \Gamma. \tag{5a}$$

So, the lift is given as before by

$$Y = \rho V_\infty \Gamma. \tag{5b}$$

Next, using (4), (2) gives for the moment exerted by the fluid on the body, about O,

$$M = -\frac{\rho}{2} Re\,(2\pi i B_2) = -2\rho V_\infty Re\,(iA_2). \tag{6}$$

(ii) Flow Past an Arbitrary Cylinder

In order to calculate the flow past an arbitrary cylinder, one first writes down the complex potential of the flow past a circular cylinder in an auxilary ζ-plane. One then finds a suitable complex transformation $z = z(\zeta)$ that maps the region outside the circle onto the region outside the cross section of the given cylinder (see Figure 11.3). Such a transformation is possible only when referred to axes fixed in the interior boundaries.

From the relation,

$$W(z) = \frac{\tilde{W}(\zeta)}{dz/d\zeta} \tag{7}$$

and the condition that the flow upstream infinity for the two cases is the same, one requires for the transformation $z = z(\zeta)$,

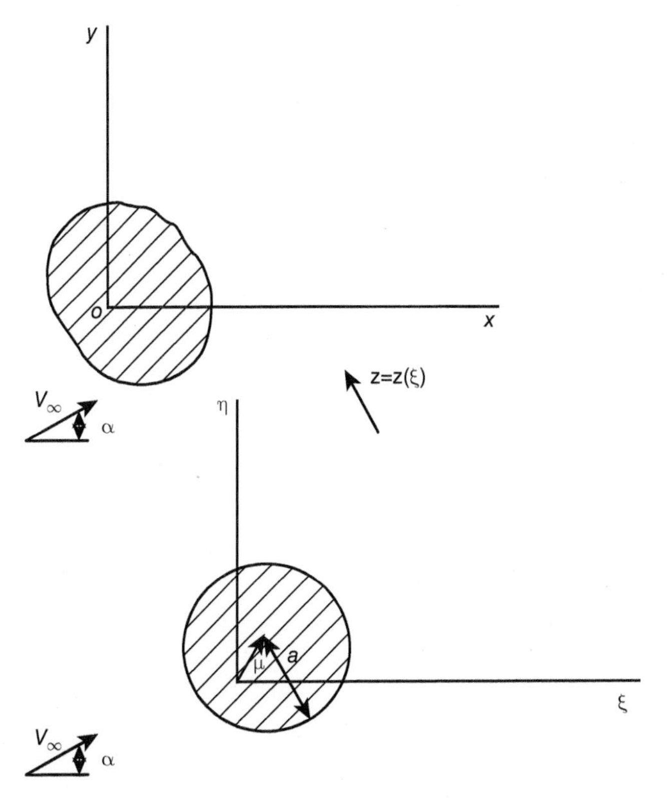

Figure 11.3 Conformal mapping of a flow past an arbitrary cylinder on to a flow past a circular cylinder.

$$\zeta \Rightarrow \infty : \frac{dz}{d\zeta} \Rightarrow 1. \tag{8}$$

Thus, this transformation changes the shape of the interior boundary and the flow in its neighborhood. Note that $z(\zeta)$ is a function analytic everywhere in the region in the ζ-plane outside the circle of radius a. Since it satisfies (8), one may write

$$z = \zeta + \sum_{n=1}^{\infty} \frac{C_n}{\zeta^n}. \tag{9}$$

from which,

$$\frac{dz}{d\zeta} = 1 - \sum_{n=1}^{\infty} \frac{nC_n}{\zeta^{n+1}}$$

or

$$\frac{1}{dz/d\zeta} \approx 1 + \sum_{n=1}^{\infty} \frac{nC_n}{\zeta^{n+1}}. \tag{10}$$

For the flow in the ζ-plane, one has for the complex potential,

$$\hat{F}\left(\hat{\zeta}\right) = V_\infty \left(\hat{\zeta} + \frac{a^2}{\hat{\zeta}}\right) + \frac{i\Gamma}{2\pi} \ell n \frac{\hat{\zeta}}{a} \tag{11}$$

where, referring to Figure 11.3, one has

$$\hat{\zeta} \equiv (\zeta - \mu)\, e^{-i\alpha}$$

α being the angle of attack, and the circular boundary of the cylinder in the ζ-plane is taken to correspond to $\Psi = 0$. Thus,

$$\tilde{F}(\zeta) = V_\infty (\zeta - \mu)\, e^{-i\alpha} + \frac{i\Gamma}{2\pi} \ell n \frac{(\zeta - \mu)}{a e^{i\alpha}} + V_\infty \frac{a^2 e^{i\alpha}}{\zeta - \mu} \tag{12}$$

from which

$$\tilde{W}(\zeta) = \frac{d\tilde{F}}{d\zeta} = V_\infty e^{-i\alpha} + \frac{i\Gamma}{2\pi} \frac{1}{(\zeta - \mu)} - \frac{V_\infty a^2 e^{i\alpha}}{(\zeta - \mu)^2}. \tag{13}$$

Using the expansions

$$\left.\begin{aligned} \frac{1}{\zeta - \mu} &= \frac{1}{\zeta} + \frac{\mu}{\zeta^2} + \frac{\mu^2}{\zeta^3} + \cdots \\ \frac{1}{(\zeta - \mu)^2} &= \frac{1}{\zeta^2} + \frac{2\mu}{\zeta^3} + \cdots \end{aligned}\right\}$$

We may write (13) as

$$\tilde{W}(\zeta) = A_0 + \frac{A_1}{\zeta} + \frac{A_2}{\zeta^2} + \cdots \tag{14}$$

where,

$$A_0 = V_\infty e^{-i\alpha}$$

$$A_1 = \frac{i\Gamma}{2\pi}$$

$$A_2 = \frac{i\Gamma}{2\pi}\mu - V_\infty a^2 e^{i\alpha}$$

$$\vdots$$

Thus,

$$\left[\tilde{W}(\zeta)\right]^2 = B_0 + \frac{B_1}{\zeta} + \frac{B_2}{\zeta^2} + \cdots \tag{15}$$

where

$$B_0 \equiv A_0^2 = V_\infty e^{-2i\alpha}.$$

$$B_1 \equiv 2A_0A_1 = \frac{i\Gamma}{\pi}V_\infty e^{-i\alpha}.$$

$$B_2 \equiv \left(2A_0A_2 + A_1^2\right) = \frac{iV_\infty e^{-i\alpha}\mu\Gamma}{\pi} - 2V_\infty^2 a^2 - \frac{\Gamma^2}{4\pi^2}.$$

$$\vdots$$

Using (7) and (15), (1) gives, for the lift (the *Kutta* (1902)-*Joukowski* (1906) *Theorem*),

$$\mathbb{L} = \rho V_\infty \Gamma \tag{16}$$

and (2) gives, for the moment about O,

$$\begin{aligned} M_0 &= -\frac{\rho}{2} Re\left[2\pi i\left(2B_0C_1 + B_2\right)\right] \\ &= Re\left(\mathbb{L}\mu e^{-i\alpha} - i2\pi\rho V_\infty^2 C_1 e^{-2i\alpha}\right). \end{aligned} \tag{17}$$

Setting

$$\mu = me^{i\delta},\ C_1 = \xi^2 e^{2i\gamma} \tag{18}$$

we obtain,

$$M_0 = \mathbb{L}\, m\cos(\delta - \alpha) + 2\pi\rho V_\infty^2 \xi^2 \sin 2(\gamma - \alpha) \tag{19}$$

(see Figure 11.4).

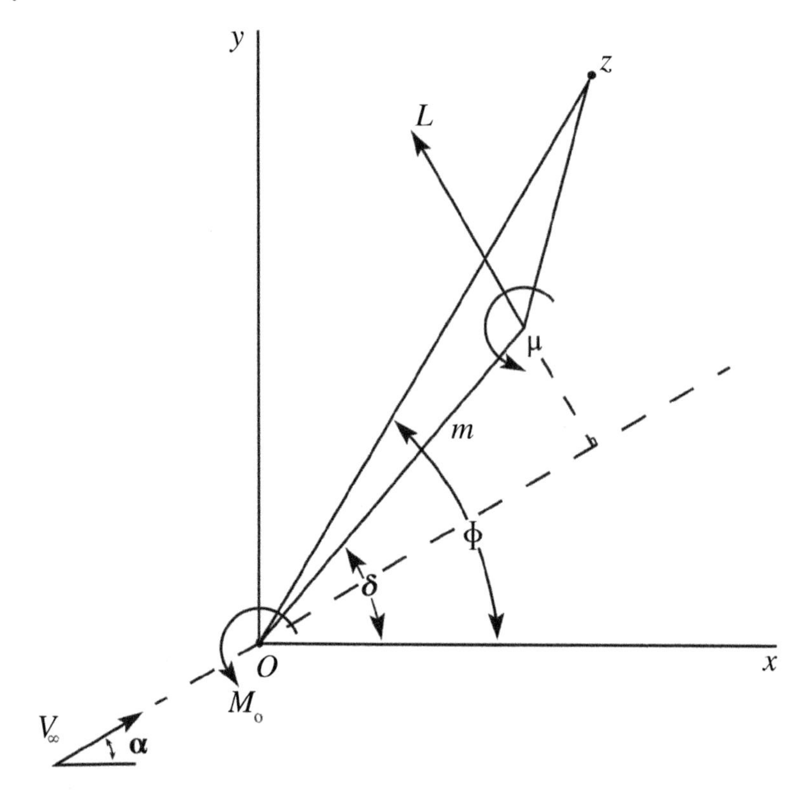

Figure 11.4 Forces and moments on the airfoil.

(iii) Flow Around a Flat Plate

The flow past an airfoil at an angle of attack (see Figure 11.5) near the trailing edge resembles that around a flat plate (see Figure 11.6). One has, for the latter (see Section 6.1),

$$F(z) = Az^{1/2} \tag{20}$$

from which we have for the stream function,

$$\Psi = Ar^{1/2} \sin \frac{\theta}{2}. \tag{21}$$

It turns out that very low pressure near the sharp edge produces a nonzero total force on the boundary. One may see this by calculating the force on a boundary coinciding with a streamline $\Psi = \Psi_0$ (which is a parabola), and then, allowing $\Psi_0 \Rightarrow 0$. The total force $\boldsymbol{F}$ exerted by the fluid on a finite portion of

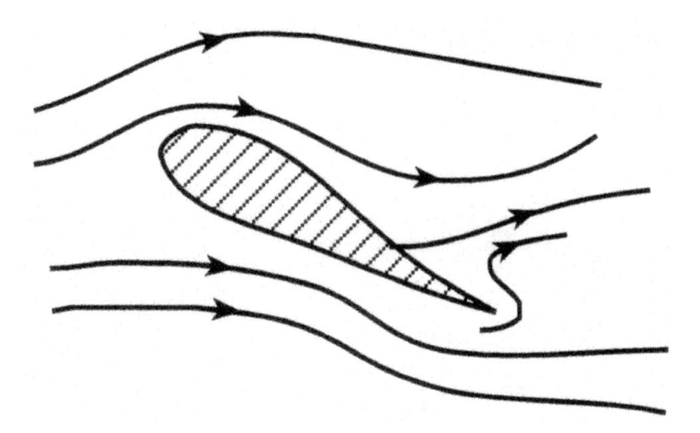

Figure 11.5 Flow near the trailing edge of an airfoil.

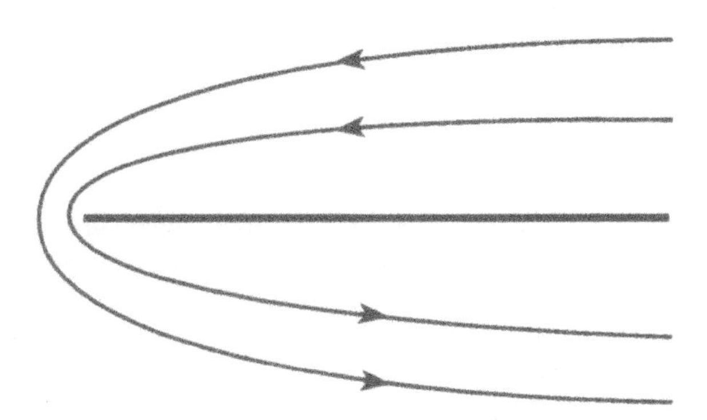

Figure 11.6 Flow aroung a flat plate.

this boundary lying within a small circle is parallel to the x-axis ($\theta = 0$) by symmetry. The x-component of $\boldsymbol{F}$ is given by,

$$F_x = \int p dy. \tag{22}$$

The above integral is taken over the section of the curve defined by

$$Ar^{1/2} \sin \frac{\theta}{2} = \Psi_o \text{ or } y = 2\left(\frac{\Psi_o}{A}\right)^2 \cot \frac{\theta}{2} \tag{23}$$

that lies between $\theta = \varepsilon$ and $\theta = 2\pi - \varepsilon$, where

$$\sin \frac{\varepsilon}{2} = \frac{\Psi_o}{Ar^{1/2}}.$$

Using the Bernoulli integral

$$p = p_o - \rho \frac{\partial \Phi}{\partial t} - \frac{1}{2}\rho |\nabla \Phi|^2 \tag{24}$$

in (22), we obtain

$$F_x = \int_{\varepsilon}^{2\pi-\varepsilon} \left(p_o - \rho \frac{A^4}{8\Psi_o^2} \sin^2 \frac{\theta}{2} \right) \frac{\Psi_o^2}{A^2} \operatorname{cosec}^2 \frac{\theta}{2} d\theta$$

from which

$$F_x \sim -\frac{1}{4}\pi\rho A^2 \text{ as } \varepsilon \Rightarrow 0. \tag{25}$$

This represents a suction force concentrated at the sharp trailing edge which is parallel to the plate and is pointing upstream. The effect of this suction force is to make the drag force on the plate zero. Although the drag force is zero, the plate still experiences a lift because the pressure on either side of the plate acts in a direction normal to it.

(iv) Flow Past an Airfoil

For an airfoil with a sharp trailing edge, the angle ψ_z at the trailing edge is not equal to ψ_ζ at the corresponding point on the circle in the ζ-plane (see Figure 11.7). This implies that the mapping is not conformal at the trailing edge. If the first $(n-1)$ derivatives of the transformation function $z = z(\zeta)$ vanish at $\zeta = \zeta_T$, then, since $\psi_\zeta = \pi$, one has

$$\psi_z = n\pi.$$

Since the airfoil has a sharp trailing edge, we have

$$\tau \equiv 2\pi - n\pi = \pi(2-n) \approx 0$$

or

$$n = 2.$$

Therefore,

$$\zeta = \zeta_T : \frac{dz}{d\zeta} = 0, \frac{d^2z}{d\zeta^2} \neq 0. \tag{26}$$

These imply that the flow velocity at the corresponding point in the z-plane would become infinite unless one imposes from (26) and (7), the following condition,

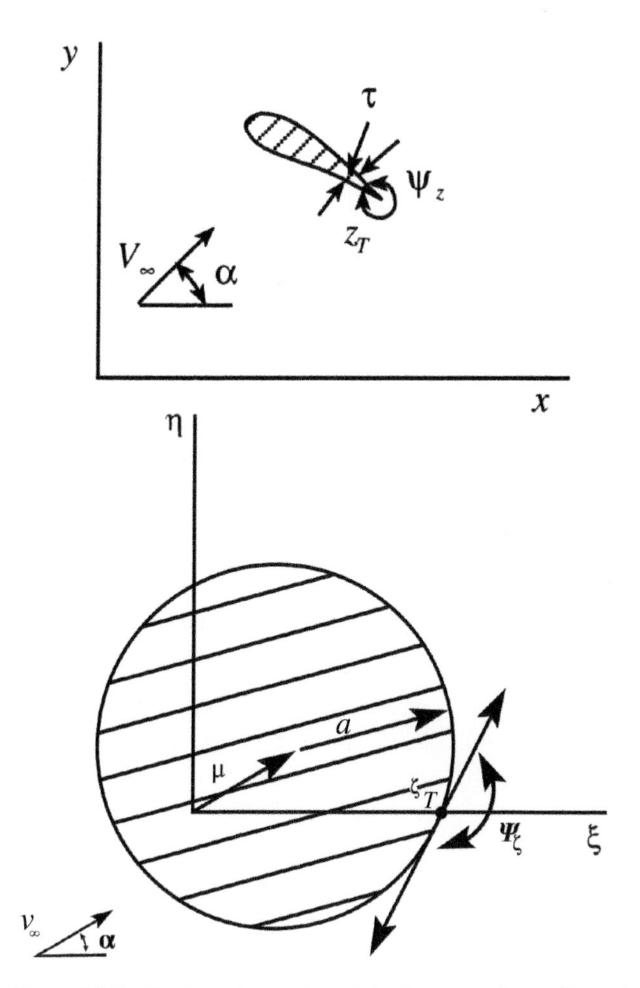

Figure 11.7 Conformal mapping of the flow near the trailing edge of an airfoil (from Karamcheti, 1966).

$$\zeta = \zeta_T \, : \, \tilde{W}(\zeta) = 0. \tag{27}$$

This would lead to a finite flow velocity and hence a smooth flow at the trailing edge of the airfoil (*Kutta condition*).[2] Otherwise, viscous effects would

2 Indeed, the trailing edge of an airfoil needs to be a stagnation point. This is so because, if the stagnation point moves away from the trailing edge, then flow occurs around it with infinite velocity. Consequently, the flow experiences an enormous adverse pressure gradient so that flow separation and vorticity shedding set in at the trailing edge and continue until the stagnation point has moved to the trailing edge. This would reduce both the size of the wake and the drag on the airfoil.

intervene and rapidly restore the Kutta condition. These happen via transient flow adjustments involving flow separation (see Section 21.5) and vorticity shedding.

Using (13), (27) gives

$$\Gamma = 4\pi a V_\infty \sin(\alpha + \beta) \tag{28}$$

where

$$\mu - \zeta_T = a e^{i(\pi - \beta)} \text{ or } \zeta_T - \mu \equiv a e^{-i\beta}$$

(see Figure 11.8). The smooth flow condition (27) implies that the circulation set up around an airfoil have strength just sufficient to make the flow leave the airfoil smoothly at the trailing edge. Thus, for uniform irrotational flow past an airfoil with sharp trailing edge, there is just one value of the circulation Γ, for which the velocity is finite everywhere. The process by which this circulation is generated involves shedding of vorticity into the fluid. In response to this, a

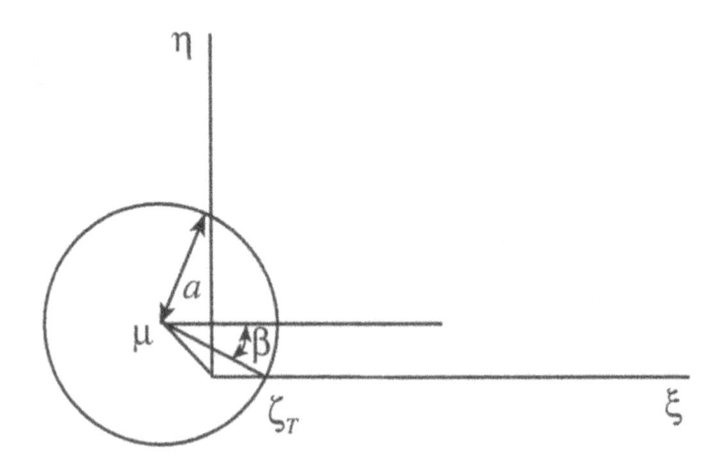

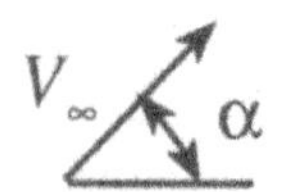

Figure 11.8 Flow past a circular cylinder in the mapping -plane (from Karamcheti, 1966).

circulation around the airfoil in a sense opposite to the starting vortex is generated. This may be seen by applying *Kelvin's circulation theorem* (see Section 5.7) to a closed path enclosing both the initial and present positions of the airfoil, and noting that the circulation around the path must remain zero.

Now, from (19), the moment about any point z is given by

$$M_z = M_\mu - \mathbb{L}h\cos(\varphi - \alpha) \tag{29a}$$

where

$$\left.\begin{aligned} M_\mu &\equiv 2\pi\rho V_\infty^2 \xi^2 \sin 2(\gamma - \alpha) \\ z - \mu &= he^{i\varphi} \end{aligned}\right\}.$$

Using (16) and (28), (29a) may be written as

$$M = 2\pi\rho V_\infty^2 \xi^2 \left[\sin 2(\gamma - \alpha) - \frac{ah}{\xi^2}\{\sin(\beta + \varphi) + \sin(2\alpha + \beta - \varphi)\}\right]. \tag{29b}$$

We see from (29b) that the *aerodynamic center* (the point, about which the moment is independent of the angle of attack), is given by

$$h = \frac{\xi^2}{a}, \varphi = \beta + 2\gamma - \pi \tag{30}$$

and the moment, about the aerodynamic center, is given by

$$M_{AC} = -2\pi\rho V_\infty^2 \xi^2 \sin 2(\gamma + \beta) \tag{31}$$

which can be verified, from (19), to be the same as that corresponding to zero lift, as indeed it must be.

(v) The Joukowski Transformation

It is of interest to note that airfoil-like profiles may be generated from a circle via a transformation,

$$z = \zeta + \sum_{n=1}^{\infty} \frac{C_n}{\zeta^n} \tag{9}$$

which gives,

$$\frac{dz}{d\zeta} = 1 - \frac{C_1}{\zeta^2} - \frac{2C_2}{\zeta^3} + \cdots.$$

If one writes

$$\frac{dz}{d\zeta} = \left(1 - \frac{\zeta_1}{\zeta}\right)\left(1 - \frac{\zeta_2}{\zeta}\right)\cdots\left(1 - \frac{\zeta_k}{\zeta}\right)$$

then, comparison with the above formula yields,

$$\sum_{i=1}^{k} \zeta_i = 0.$$

The *Joukowski transformation* is given by

$$\frac{dz}{d\zeta} = \left(1 - \frac{\zeta_T}{\zeta}\right)\left(1 + \frac{\zeta_T}{\zeta}\right) = 1 - \frac{\zeta_T^2}{\zeta^2} \tag{32}$$

so that

$$z = \zeta + \frac{\zeta_T^2}{\zeta} = \zeta + \frac{C^2}{\zeta}, \text{ for some constant C.} \tag{33}$$

The inverse of equation (33) is

$$\zeta = \frac{1}{2}z \underline{+} \sqrt{\frac{1}{4}z^2 - C^2}$$

which shows that there are *branch points* at $z = \underline{+}2C$. In order to resolve the multi-valued nature of the above expression, we cut the z-plane along the real axis between $z = -2C$ and $z = 2C$. We then interpret $\sqrt{\frac{1}{4}z^2 - C^2}$ as that branch of the function, which behaves like $\frac{1}{2}z$ as $|z| \Rightarrow \infty$ so as to ensure that $\zeta \sim z$ as $|z| \Rightarrow \infty$.

Note that the transformation (33) maps the circle $\zeta = Ce^{i\theta}$ onto the strip $x = 2C\cos\theta, y = 0$. Thus, one generates a *flat-plate* airfoil (see Figure 11.9) by setting $\mu = 0,\ C = a$, and $\beta = 0$. The complex velocity given by (13) then becomes

$$\tilde{W}(\zeta) = V_\infty e^{-i\alpha} + \frac{i\Gamma}{2\pi\zeta} - \frac{V_\infty a^2 e^{i\alpha}}{\zeta^2}, \tag{34}$$

where, we have from (28),

$$\Gamma = 4\pi a V_\infty \sin\alpha.$$

Then (7) gives

$$W(z) = \frac{V_\infty e^{-i\alpha} + \dfrac{i\Gamma}{2\pi\zeta} - \dfrac{V_\infty a^2 e^{i\alpha}}{\zeta^2}}{1 - C^2/\zeta^2} \tag{35}$$

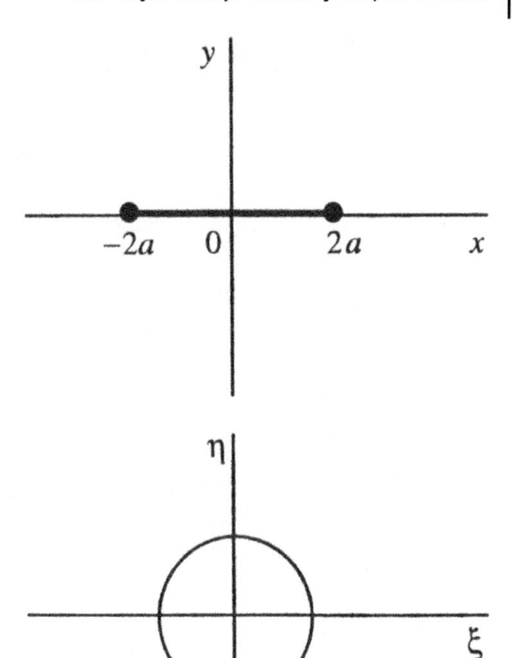

Figure 11.9 Conformal mapping of a strip onto a circle.

and on the plate $\zeta = Ce^{i\theta}$, (35) becomes

$$W(z) = \frac{[\sin\alpha - \sin(\alpha - \theta)]}{\sin\theta} V_\infty = \frac{V_\infty \cos\left(\alpha - \dfrac{\theta}{2}\right)}{\cos\dfrac{\theta}{2}}. \tag{36}$$

Noting that

$$x = 2C\cos\theta \tag{37}$$

from (36), the stagnation points on the plate are given by

$$\frac{x^2}{4C^2} - \frac{x\sin^2\alpha}{C} + \left(\sin^2\alpha - \cos^2\alpha\right) = 0 \tag{38a}$$

which gives

$$x = 2C, -2C\cos 2\alpha. \tag{38b}$$

The velocity at the trailing edge, from (36), is given by

$$W = V_\infty \cos \alpha,$$

so that the flow leaves the plate smoothly and parallel to it.

From (30), the aerodynamic center is at $\frac{1}{4}$ of the chord from the leading edge (chord is the straight segment from leading to trailing edge). Also, at zero incidence, from (29b), the moment acting on the plate (about any point) vanishes since the pressure is constant everywhere. So, the moment about the aerodynamic center vanishes at all incidence. This means that the force on the plate acts through the aerodynamic center.

Setting $\mu = me^{i\pi/2}, a = C \sec \beta, \beta \neq 0$, one generates a *circular-arc* airfoil because, from (33), one derives[3]

$$\frac{(z-2C)}{(z+2C)} = \frac{(\zeta - C)^2}{(\zeta + C)^2}.$$

Thus, if

$$\arg\left(\frac{\zeta - C}{\zeta + C}\right) = \varphi \tag{39}$$

then (see Figure 11.10)

$$\arg\left(\frac{z - 2C}{z + 2C}\right) = 2\varphi. \tag{40}$$

Therefore, the image of the circle $\zeta = im = ae^{i\theta}$ is a circular arc.

---

3 Equation (33) leads to

$$z \pm 2C = \frac{(\zeta \pm C)^2}{\zeta}$$

from which,

$$\frac{(z-2C)}{(z+2C)} = \frac{(\zeta - C)^2}{(\zeta + C)^2}.$$

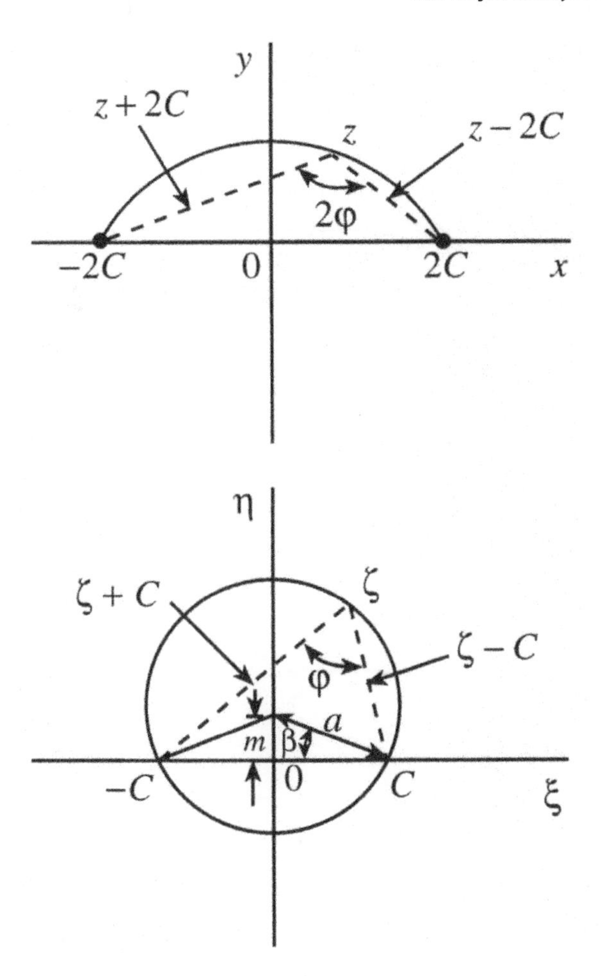

Figure 11.10 Conformal mapping of a circular arc onto a circle (from Karamcheti, 1966).

Next, by setting $\mu = me^{i\pi}, \beta = 0$, one generates a *symmetric* airfoil (see Figure 11.11) given by

$$z = (ae^{i\theta} - m) + \frac{C^2}{(ae^{i\theta} - m)} \tag{41}$$

from which one gets

$$x = 2C\cos\theta, \; y = C\varepsilon\,(2\sin\theta - \sin 2\theta) \tag{42}$$

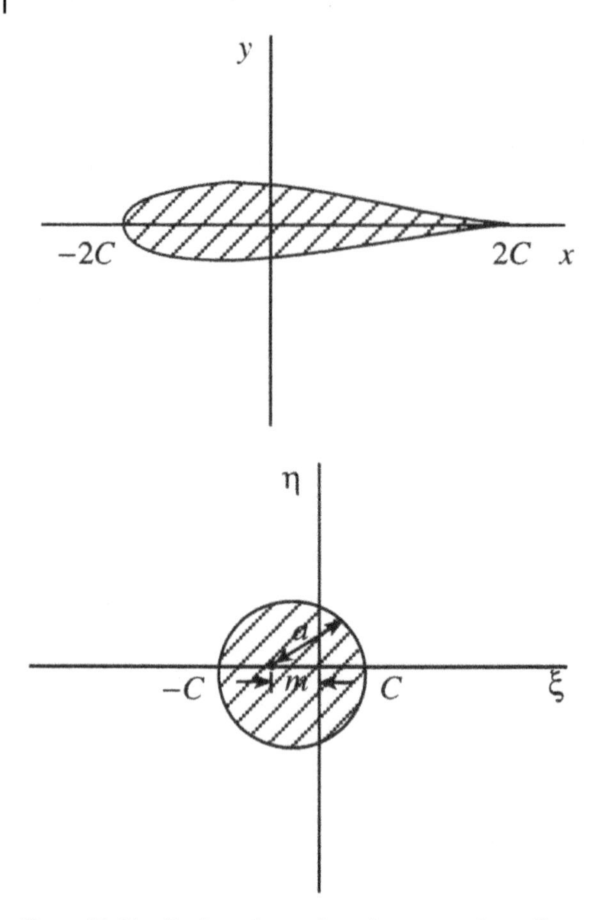

Figure 11.11 Conformal mapping of a symmetric profile onto a circle (from Karamcheti, 1966).

where,

$$a = m + C \equiv C(1 + \varepsilon), \varepsilon \ll 1.$$

Thus, by displacing the center μ of the circle in the ζ-plane along the η-axis from the origin, one produces a *camber* for the airfoil in the z-plane. The *camberline* is the locus of midpoints of segments cut out by the airfoil on the straight lines perpendicular to the chord, and the camber is the maximum distance of the camberline from the chord for the airfoil in the z-plane. By displacing μ along the ξ-axis from the origin, one produces thickness for the airfoil in the z-plane.

11.2 Thin Airfoil Theory

We will now consider the more difficult problem of determining the pressure distribution on a *thin* airfoil of a given profile (Glauert, 1957). The theory in the following gives only an approximate solution to the problem. It involves determination of the singularity distribution representing the airfoil, that satisfies the kinematic boundary conditions in the presence of a wake or a trail of free vortices.

Consider a steady flow past a stationary airfoil. Then, one has the boundary-value problem for the velocity potential Φ:

$$\nabla^2 \Phi = 0 \tag{43}$$

$$\nabla \Phi \cdot \nabla F = 0 \text{ on } F(x, y) = 0 \tag{44}$$

$$|\nabla \Phi| \Rightarrow V_\infty \text{ at infinity} \tag{45}$$

Kutta condition at trailing edge (46)

where the airfoil profile is given by

$$F(x, y) = \varepsilon \eta(x) - y = 0, \ \varepsilon \ll 1. \tag{47}$$

The kinematic condition (44) indicates that the flow should occur tangentially to the surface of the body and implies that

$$\left.\begin{aligned} \frac{\Phi_y}{\Phi_x} &= \varepsilon \frac{d\eta_u}{dx} \text{ on } y = \varepsilon \eta_u(x) \\ \frac{\Phi_y}{\Phi_x} &= \varepsilon \frac{d\eta_l}{dx} \text{ on } y = \varepsilon \eta_l(x) \end{aligned}\right\} \tag{48}$$

where the subscripts u and l refer to the upper and the lower parts of the profile (see Figure 11.12). Note that

$$\left.\begin{aligned} \eta_u(x) &= [\eta_t(x) + \eta_c(x)] - \alpha x \\ \eta_l(x) &= [-\eta_t(x) + \eta_c(x)] - \alpha x \end{aligned}\right\} \tag{49}$$

where the subscripts t and c refer to the thickness part and the camber part, respectively.

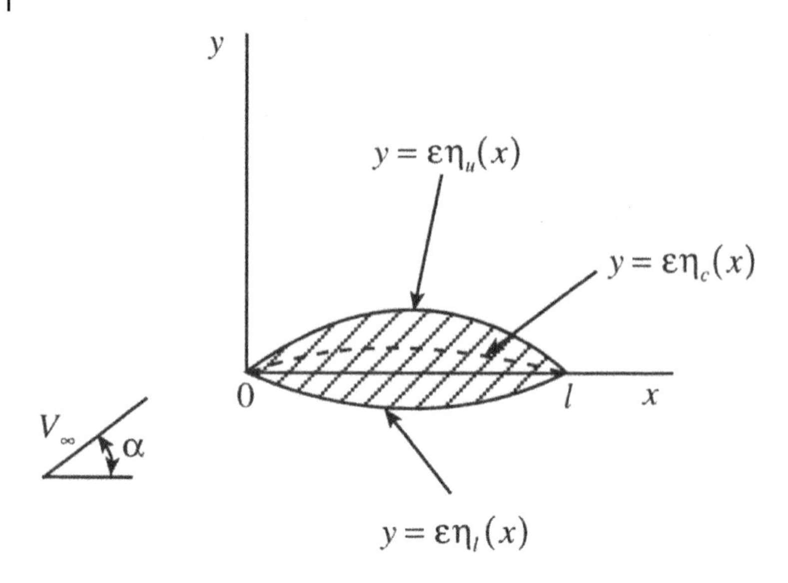

Figure 11.12 Airfoil section and camberline.

Assuming the perturbations produced by the airfoil on the uniform flow to be small, and setting

$$\Phi = V_\infty x + \varepsilon\phi\,(x, y) \tag{50}$$

(43), (45), and (48) give, on linearization in the small parameter ε,

$$\phi_{xx} + \phi_{yy} = 0 \tag{51}$$

$$y = 0^+ : \phi_y = V_\infty \frac{d\eta_u}{dx} \tag{52}$$

$$y = 0^- : \phi_y = V_\infty \frac{d\eta_l}{dx} \tag{53}$$

$$\phi_x, \phi_y \Rightarrow 0 \text{ at infinity.} \tag{54}$$

The assumption of small perturbations requires that both the angle of attack and the airfoil thickness are sufficiently small.

One now splits (51)-(54) into three simpler problems: Let

$$\phi = \phi_1 + \phi_2 + \phi_3$$

where ϕ_1 corresponds to the thickness problem, ϕ_2 to the camber problem, and ϕ_3 to the angle of attack problem.

(i) Thickness problem:

$$\nabla^2 \phi_1 = 0 \tag{55}$$

$$y = 0^{\pm} : \phi_{1y} = \pm V_{\infty} \frac{d\eta_t}{dx}, \; 0 \leq x \leq l \tag{56}$$

$$|\nabla \phi_1| \Rightarrow 0 \text{ at infinity;} \tag{57}$$

(ii) Camber problem:

$$\nabla^2 \phi_2 = 0 \tag{58}$$

$$y = 0^{\pm} : \phi_{2y} = \pm V_{\infty} \frac{d\eta_c}{dx}, \; 0 \leq x \leq l \tag{59}$$

$$|\nabla \phi_2| \Rightarrow 0 \text{ at infinity;} \tag{60}$$

$$\text{Kutta condition at } x = l. \tag{61}$$

(iii) Angle of attack problem:

$$\nabla^2 \phi_3 = 0 \tag{62}$$

$$y = 0^{\pm} : \phi_{3y} = -V_{\infty}\alpha, \; 0 \leq x \leq l \tag{63}$$

$$|\nabla \phi_3| \Rightarrow 0 \tag{64}$$

$$\text{Kutta condition at } x = l. \tag{65}$$

In the kinematic condition (56), the $\pm$ signs correspond to the upper and lower surfaces.

The pressure on the airfoil is given (in terms of the pressure coefficient) by

$$C_p = \frac{p - p_{\infty}}{\frac{1}{2}\rho V_{\infty}^2} = 1 - \frac{V^2}{V_{\infty}^2} = -\frac{2}{V_{\infty}}\phi_x. \tag{66}$$

Let us now consider these three problems separately and then superpose the three solutions to get the complete solution of the thin airfoil problem.

(i) Thickness Problem

This problem corresponds to a flow past a *symmetrical* airfoil at zero angle of attack.

One has

$$\nabla^2 \phi = 0 \tag{55}$$

$$y = 0^{\pm} : \phi_y = \pm V_\infty \frac{d\eta_t}{dx}, \; 0 \le x \le l \tag{56a}$$

$$|\nabla \phi| \Rightarrow 0 \text{ at infinity.} \tag{57}$$

There is no Kutta condition because there is no circulation around the airfoil in the thickness problem.

The thickness problem is equivalent to a superposition of uniform stream and a source distribution (see Figure 11.13) so that one may write

$$\phi(x, y) = \frac{1}{2\pi} \int_0^l q(\xi) \ell n \left[(x - \xi)^2 + y^2 \right]^{1/2} d\xi \tag{67}$$

from which, the velocity components are given by

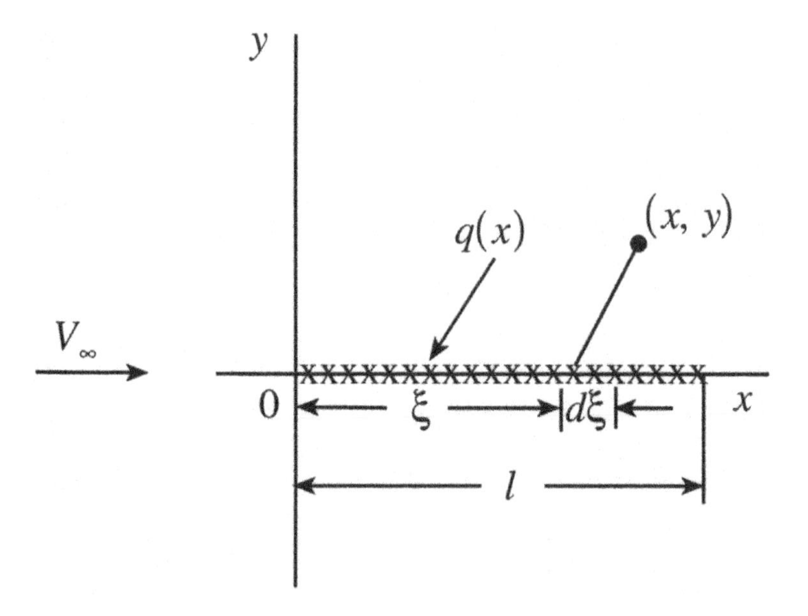

Figure 11.13 Superposition of a uniform stream and a source distribution.

$$\left.\begin{aligned} u(x,y) &= \frac{1}{2\pi}\int_0^l \frac{q(\xi)\cdot(x-\xi)}{(x-\xi)^2+y^2}d\xi \\ v(x,y) &= \frac{1}{2\pi}\int_0^l \frac{q(\xi)\cdot y}{(x-\xi)^2+y^2}d\xi \end{aligned}\right\}. \tag{68}$$

Using (68) and (56a) the kinematic condition gives

$$v\left(x,0\pm\right) = \lim_{y\Rightarrow 0\pm}\left[\frac{1}{2\pi}\int_0^l \frac{q(\xi)\cdot y}{(x-\xi)^2+y^2}d\xi\right] = \underline{+}V_\infty\frac{d\eta_t}{dx}, 0 \le x \le l. \tag{56b}$$

In order to resolve the apparent singularity at $y = 0$, introduce

$$\eta \equiv \frac{\xi - x}{y}$$

so that (56b) becomes

$$v\left(x,0\pm\right) = \lim_{y\Rightarrow 0\pm}\frac{1}{2\pi}\int_{\eta_0}^{\eta_l}\frac{q(\xi)\,d\eta}{1+\eta^2} = \underline{+}V_\infty\frac{d\eta_t}{dx}$$

where

$$\eta_{0,l} = \eta\,(x = 0, l)\,.$$

Thus,

$$v\left(x,0\pm\right) = \mp\frac{1}{2\pi}q(x)\int_{-\infty}^{\infty}\frac{d\eta}{1+\eta^2} = \underline{+}V_\infty\frac{d\eta_t}{dx}$$

from which,

$$\frac{1}{2}q(x) = V_\infty\frac{d\eta_t}{dx}. \tag{69}$$

Note that at any given point on the strip, the jump in v across the strip is equal to the source strength at the point. Thus, (67) becomes

$$\phi(x,y) = \frac{V_\infty}{\pi}\int_0^l \frac{d\eta_t}{dx}(\xi)\,\ell n\left[(x-\xi)^2+y^2\right]^{1/2}d\xi. \tag{70}$$

If one introduces

$$\left.\begin{aligned} x &= \frac{l}{2}(1+\cos\theta), \ -\pi \leq \theta \leq \pi \\ \xi &= \frac{l}{2}(1+\cos\varphi) \end{aligned}\right\} \tag{71}$$

then the x- component of the velocity given by (68) evaluated at $y = 0$ is as follows,

$$u(x,0) = \frac{V_\infty}{\pi}\int_0^l \frac{d\eta_t}{dx}(\xi)\frac{d\xi}{x-\xi}, \ 0 \leq x \leq l.$$

This becomes

$$u(\theta) = -\frac{V_\infty}{\pi}\int_0^\pi \frac{d\eta_t}{dx}(\varphi)\frac{\sin\varphi}{\cos\varphi - \cos\theta}d\varphi. \tag{72}$$

Noting that $d\eta_t/dx\,(\theta)$ is an odd function of θ, one may write

$$\frac{d\eta_t}{dx}(\theta) = \sum_{n=1}^{\infty} A_n \sin n\theta. \tag{73a}$$

This shows that $d\eta_t/dx$ vanishes at $\theta = 0, \pi$ and

$$A_n = \frac{2}{\pi}\int_0^\pi \frac{d\eta_t}{dx}(\theta)\sin n\theta d\theta. \tag{73b}$$

Using (73a, b), (72) becomes

$$\frac{u(\theta)}{V_\infty} = \sum_{n=1}^{\infty} A_n \cos n\theta \tag{74}$$

where we have used the interesting result,

$$\frac{1}{\pi}\int_0^\pi \frac{\sin n\varphi \cdot \sin\varphi}{\cos\theta - \cos\varphi}d\varphi = \cos n\theta. \tag{75}$$

Note that $u(\theta)$ is an even function of θ, as to be expected.

(ii) Camber Problem

This problem corresponds to a flow past a camber profile at zero angle of attack (see Figure 11.14).

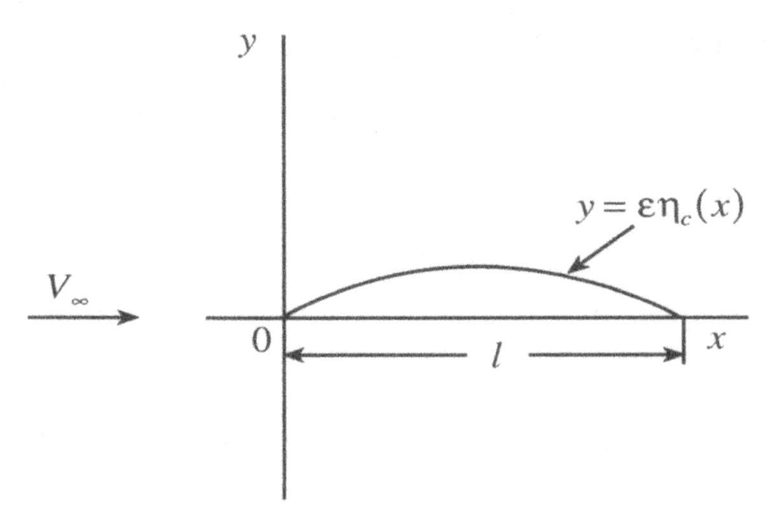

Figure 11.14 Flow past a cambered profile at zero angle of attack.

One has

$$\nabla^2 \phi = 0 \tag{58}$$

$$y = 0^{\pm} : \phi_y = V_\infty \frac{d\eta_c}{dx} \tag{59}$$

$$|\nabla \phi| \Rightarrow 0 \text{ at infinity;} \tag{60}$$

$$\text{Kutta condition at } x = l. \tag{61}$$

The camber problem is equivalent to a superposition of a uniform stream and a vortex distribution so that one may write

$$\phi(x, y) = \frac{1}{2\pi} \int_0^{\pi} \gamma(\xi) \tan^{-1}\left(\frac{y}{x - \xi}\right) d\xi \tag{76}$$

from which, the velocity components are given by

$$\left.\begin{aligned} u(x, y) &= \frac{1}{2\pi} \int_0^{\pi} \frac{\gamma(\xi) \cdot y}{(x - \xi)^2 + y^2} d\xi \\ v(x, y) &= -\frac{1}{2\pi} \int_0^{\pi} \frac{\gamma(\xi) \cdot (x - \xi)}{(x - \xi)^2 + y^2} d\xi \end{aligned}\right\}. \tag{77}$$

Note, from (77), that

$$u\left(x,0^{\pm}\right) = \lim_{y\Rightarrow 0^{\pm}} \frac{1}{2\pi} \int_{0}^{\pi} \frac{\gamma(\xi)\cdot y}{(x-\xi)^{2}+y^{2}} d\xi = \pm \frac{\gamma(x)}{2} \tag{78}$$

so that, at any given point on the strip, the jump in u across the strip is equal to the vortex strength at that point.

Using (77), the kinematic condition (59) gives

$$\lim_{y\Rightarrow 0^{\pm}} \left[-\frac{1}{2\pi} \int_{0}^{l} \frac{\gamma(\xi)\cdot(x-\xi)}{(x-\xi)^{2}+y^{2}} d\xi \right] = V_{\infty} \frac{d\eta_{c}}{dx},\ 0 \le x \le l \tag{79}$$

from which,

$$-\frac{1}{2} \int_{0}^{l} \frac{\gamma(\xi)}{x-\xi} d\xi = V_{\infty} \frac{d\eta_{c}}{dx},\ 0 \le x \le l. \tag{80}$$

The Kutta condition requires that there is no jump in u at $x = l$. This implies from (78), that $\gamma(\ell) = 0$.

Introducing again

$$\left. \begin{aligned} x &= \frac{l}{2}(1+\cos\theta),\ -\pi \le \theta \le \pi \\ \xi &= \frac{l}{2}(1+\cos\varphi) \end{aligned} \right\} \tag{71}$$

equation (80) becomes

$$\frac{1}{\pi} \int_{0}^{\pi} \frac{\gamma(\varphi)}{2V_{\infty}} \frac{\sin\varphi}{\cos\varphi - \cos\theta} d\varphi = \frac{d\eta_{c}}{dx}(\theta). \tag{81}$$

Noting that $d\eta_{c}/dx\,(\theta)$ is an even function of θ, one may write

$$\frac{d\eta_{c}}{d\theta}(\theta) = \frac{B_{0}}{2} + \sum_{n=1}^{\infty} B_{n} \cos n\theta \tag{82}$$

where

$$B_n = \frac{2}{\pi}\int_0^{\pi} \frac{d\eta_c}{dx}(\theta)\cos n\theta \cdot d\theta.$$

Let

$$\gamma(\theta) = \gamma_1(\theta) + \gamma_2(\theta) + \frac{K}{\sin\theta} \tag{83}$$

where, K is an arbitrary constant. Noting the result

$$\int_0^{\pi} \frac{d\varphi}{\cos\varphi - \cos\theta} = 0$$

the last term on the right-hand side in (83) is seen to represent the homogeneous solution to (81).

Using (83) in (81), one may choose γ_1 and γ_2 such that

$$\frac{1}{\pi}\int_0^{\pi} \frac{\gamma_1(\varphi)}{2V_\infty}\frac{\sin\varphi}{\cos\varphi - \cos\theta}d\varphi = \sum_{n=1}^{\infty} B_n \cos n\theta \tag{84}$$

and

$$\frac{1}{\pi}\int_0^{\pi} \frac{\gamma_2(\varphi)}{2V_\infty}\frac{\sin\varphi}{\cos\varphi - \cos\theta}d\varphi = \frac{B_0}{2}. \tag{85}$$

Using (75), (84) leads to

$$\gamma_1(\theta) = -2V_\infty \sum_{n=1}^{\infty} B_n \sin n\theta. \tag{86}$$

Noting another interesting result,

$$\frac{1}{\pi}\int_0^{\pi} \frac{\cos\varphi}{\cos\varphi - \cos\theta}d\varphi = 1, \tag{87}$$

equation (85) gives

$$\gamma_2(\theta) = V_\infty B_0 \frac{\cos\theta}{\sin\theta}. \tag{88}$$

Using (86) and (88), (83) becomes

$$\gamma(\theta) = -2V_\infty \sum_{n=1}^{\infty} B_n \sin n\theta + V_\infty B_0 \frac{\cos\theta}{\sin\theta} + \frac{K}{\sin\theta}. \tag{89}$$

Applying the Kutta condition given by,

$$\gamma\left(x=l\right)=\gamma\left(\theta=0\right)=0 \tag{61}$$

so that there is no jump in u at $x = l$, one obtains from (89),

$$K=-V_{\infty}B_0. \tag{90}$$

Using (90), (89) becomes

$$\gamma\left(\theta\right)=-2V_{\infty}\left(\frac{B_0}{2}\frac{1-\cos\theta}{\sin\theta}+\sum_{n=1}^{\infty}B_n\sin n\theta\right). \tag{91}$$

Next, the pressure on the airfoil, using (78), is given by

$$C_p\left(x,0^{\pm}\right)=-\frac{2u\left(x,0^{\pm}\right)}{V_{\infty}}=\mp\frac{\gamma\left(x\right)}{V_{\infty}}. \tag{92}$$

Using (91), the pressure coefficient becomes

$$C_p\left(\theta\right)=2\left[\frac{B_0}{2}\cdot\frac{1-\cos\theta}{\sin\theta}+\sum_{n=1}^{\infty}B_n\sin n\theta\right]. \tag{93}$$

The lift on the airfoil is given by

$$\mathbb{L}=\frac{\rho V_{\infty}^2}{2}\int_0^l\left[C_p\left(x,0^-\right)-C_p\left(x,0^+\right)\right]dx$$

and using (92), this becomes

$$\mathbb{L}=\rho V_{\infty}\int_0^l\gamma\left(x\right)dx. \tag{94a}$$

The moment (about the leading edge) is given by

$$M=-\rho V_{\infty}\int_0^l\gamma\left(x\right)x dx. \tag{95a}$$

Using (71), (94a) and (95a) become

$$\mathbb{L}=\frac{\rho l}{2V_{\infty}}\int_0^{\pi}\gamma\left(\theta\right)\sin\theta d\theta. \tag{94b}$$

$$M=-\frac{\rho l^2}{4V_{\infty}}\int_0^{\pi}\gamma\left(\theta\right)\cdot\left(1+\cos\theta\right)\sin\theta d\theta. \tag{95b}$$

Furthermore, using (91), (94b) and (95b) become

$$\mathbb{L} = -\left(B_0 + B_1\right)\pi \cdot \frac{\rho l}{2} \tag{96}$$

$$M = \frac{\rho l^2}{4}\left[\frac{\pi}{4}B_0 + \frac{\pi}{2}B_1 + \frac{\pi}{4}B_2\right]. \tag{97a}$$

Writing (97a) as

$$M = \frac{\rho l^2}{4}\left[\frac{\pi}{4}\left(B_0 + B_1\right) + \frac{\pi}{4}\left(B_1 + B_2\right)\right] \tag{97b}$$

and using (96), one has

$$M = -\frac{\ell \mathbb{L}}{4} + \frac{\pi}{4}\frac{\rho l^2}{4}\left(B_1 + B_2\right) \tag{98}$$

which shows the lift acting at the 1/4-chord point - the aerodynamic center.

(iii) Flat Plate at an Angle of Attack

One has

$$\nabla^2 \phi = 0 \tag{62}$$

$$y = 0^{\pm} : \phi_y = -V_\infty \alpha, \ 0 \le x \le l \tag{63}$$

$$|\nabla \phi| \Rightarrow 0 \text{ at infinity;} \tag{64}$$

$$\text{Kutta condition at } x = l. \tag{65}$$

The angle of attack problem, like the camber problem, is equivalent to superposition of a uniform stream and a vortex distribution. So, one may write

$$\phi(x,y) = -\frac{1}{2\pi}\int_0^l \gamma(\xi)\tan^{-1}\left(\frac{y}{(x-\xi)}\right)d\xi \tag{99}$$

from which, the velocity components are given by

$$\left.\begin{aligned} u(x,y) &= \frac{1}{2\pi}\int_0^l \frac{\gamma(\xi)\cdot y}{(x-\xi)^2 + y^2}d\xi \\ v(x,y) &= -\frac{1}{2\pi}\int_0^l \frac{\gamma(\xi)\cdot(x-\xi)}{(x-\xi)^2 + y^2}d\xi \end{aligned}\right\}. \tag{100}$$

Using (100), the kinematic condition (63) gives

$$\lim_{y \Rightarrow 0^\pm}\left[-\frac{1}{2\pi}\int_0^l \frac{\gamma(\xi)\cdot(x-\xi)}{(x-\xi)^2+y^2}d\xi\right] = -V_\infty\alpha,\ 0 \le x \le l$$

from which we obtain,

$$\frac{1}{2\pi}\int_0^l \frac{\gamma(\xi)}{x-\xi}d\xi = V_\infty\alpha,\ 0 \le x \le l. \tag{101}$$

Introducing again

$$\left.\begin{aligned} x &= \frac{l}{2}(1+\cos\theta),\ -\pi \le \theta \le \pi \\ \xi &= \frac{l}{2}(1+\cos\varphi) \end{aligned}\right\} \tag{71}$$

equation (101) becomes

$$\frac{1}{\pi}\int_0^\pi \frac{\gamma(\varphi)\sin\varphi}{\cos\varphi-\cos\theta}d\varphi = -2V_\infty\alpha. \tag{102}$$

equation (102) implies that

$$\gamma(\theta) = \frac{K}{\sin\theta} - 2V_\infty\alpha\frac{\cos\theta}{\sin\theta} \tag{103}$$

where K is an arbitrary constant. The Kutta condition $\gamma(0) = 0$ gives, from (103),

$$K = 2V_\infty\alpha. \tag{104}$$

Substituting this back in (103) we obtain,

$$\gamma(\theta) = \frac{2V_\infty\alpha}{\sin\theta}(1-\cos\theta). \tag{105}$$

The pressure on the airfoil is then given, from (92), by

$$C_p(\theta) = -\frac{\gamma(\theta)}{V_\infty} = -2\alpha\left(\frac{1-\cos\theta}{\sin\theta}\right). \tag{106}$$

Thus, the lift on the airfoil and the moment about the leading edge are given by

$$\mathbb{L} = \frac{\rho l/2}{V_\infty}\int_0^\pi \gamma(\theta)\sin\theta d\theta = \pi\alpha\rho l \tag{107}$$

$$M = -\frac{2}{V_\infty}\frac{\rho l^2}{4}\int_0^\pi \gamma(\theta)(1+\cos\theta)\sin\theta d\theta = -\frac{\rho l^2}{4}\pi\alpha = -\frac{\mathbb{L}}{4}. \tag{108}$$

We see from (108) that the force system on the flat plate is again equivalent to a lift force acting at the quarter chord point. This point is, therefore, the aerodynamic center.

(iv) Combined Aerodynamic Characteristics

Let us now combine the results from the three simpler problems we just considered. We have from (73a) and (82),

$$\frac{d\eta}{dx}(\theta) = \frac{B_0}{2} + \sum_{n=1}^{\infty} B_n \cos n\theta + \sum_{n=1}^{\infty} A_n \sin n\theta \tag{109}$$

where

$$B_n = \frac{2}{\pi}\int_0^\pi \frac{d\eta}{dx}(\theta)\cos n\theta d\theta$$

$$A_n = \frac{2}{\pi}\int_0^\pi \frac{d\eta}{dx}(\theta)\sin n\theta d\theta.$$

Using (74), (93), and (106), one obtains for the pressure on the airfoil,

$$C_p = 2\left[\frac{B_0}{2}\frac{1-\cos\theta}{\sin\theta} + \sum_{n=1}^{\infty} B_n \sin n\theta - \alpha\left(\frac{1-\cos\theta}{\sin\theta}\right) - \sum_{n=1}^{\infty} A_n \cos n\theta\right] \tag{110}$$

from which, the lift on the airfoil and the moment (about the leading edge) are given by

$$\mathbb{L} = \frac{\rho l}{2}\left[2\pi\alpha - \pi\left(B_0 + B_1\right)\right] \tag{111}$$

$$\begin{aligned} M &= \frac{\rho l^2}{4}\left[-\frac{\pi\alpha}{2} + \frac{\pi}{4}\left(B_0 + B_1\right) + \frac{\pi}{4}\left(B_1 + B_2\right)\right] \\ &= -\frac{\mathbb{L}}{4} + \frac{\rho l^2}{4}\left[\frac{\pi}{4}\left(B_1 + B_2\right)\right]. \end{aligned} \tag{112}$$

Note that the quarter chord point is again the aerodynamic center.

(v) The Leading-Edge Problem of a Thin Airfoil

The foregoing theory breaks down at the stagnation points (such as one at the leading edge) because of the violation of the assumption of small disturbances

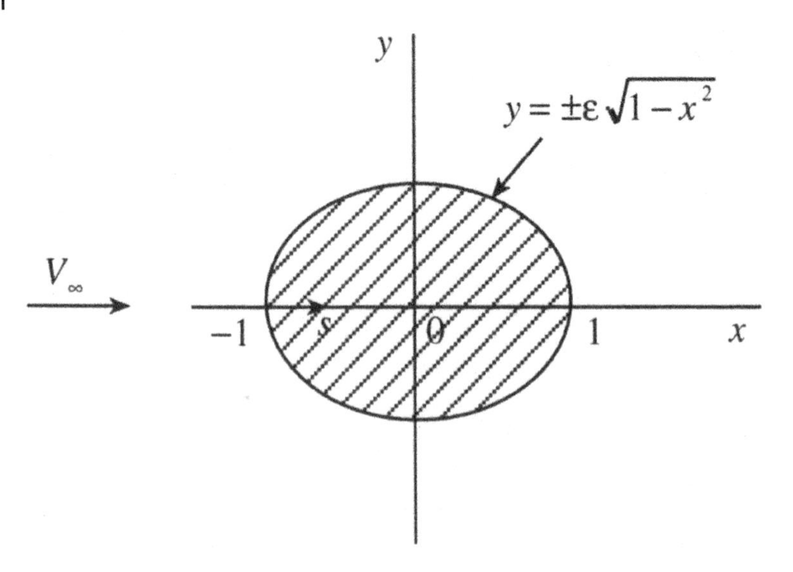

Figure 11.15 Uniform flow past a thin elliptic airfoil.

there. In order to see how one can handle the leading-edge problem, consider a uniform flow past a thin elliptic airfoil (see Figure 11.15). If, the fluid velocity is given by

$$\mathbf{v} = V_\infty \nabla \Phi$$

then, one has the following boundary-value problem,

$$\Phi_{xx} + \Phi_{yy} = 0 \tag{113}$$

$$\frac{\Phi_y}{\Phi_x} = \mp \varepsilon \frac{x}{\sqrt{1-x^2}} \text{ at } y = \pm \varepsilon \sqrt{1-x^2} \equiv \varepsilon T(x) \tag{114}$$

$$\Phi \sim x + O(1) \text{ as } \left(x^2 + y^2\right) \Rightarrow \infty. \tag{115}$$

Seeking a solution of the form,

$$\Phi(x, y; \varepsilon) = x + \varepsilon \phi_1 + \varepsilon_2 \phi_2 + \cdots, \varepsilon \ll 1 \tag{116}$$

the boundary value problem (113)-(115) leads to the following hierarchy of problems:

$$O(\varepsilon) : \phi_{1xx} + \phi_{1yy} = 0 \tag{117}$$

$$y = 0^{\pm} : \phi_{1y} = \mp \frac{x}{\sqrt{1 - x^2}} \tag{118}$$

$$\phi_1 = O(1) \text{ as } \left(x^2 + y^2\right) \Rightarrow \infty \tag{119}$$

$$O\left(\varepsilon^2\right) : \phi_{2xx} + \phi_{2yy} = 0 \tag{120}$$

$$y = 0^{\pm} : \phi_{2y} = \mp \frac{x}{\sqrt{1 - x^2}} \phi_{1x} \mp \sqrt{1 - x^2} \phi_{1yy}$$

or

$$\phi_{2y} = \pm \frac{\partial}{\partial x} \left[\sqrt{1 - x^2} \phi_{1x}\right] \tag{121}$$

$$\phi_2 = O(1) \text{ as } \left(x^2 + y^2\right) \Rightarrow \infty. \tag{122}$$

The flow speed at the airfoil is then given by

$$\frac{V}{V_\infty} = \left[\{1 + \varepsilon\left(\phi_{1x} + \varepsilon T \phi_{1xy}\right) + \varepsilon^2 \phi_{2x} + \cdots\}^2 + \{\varepsilon \phi_{1y} + \varepsilon^2 \phi_{2y} + \cdots\}^2\right]_{y=0}^{1/2} \tag{123}$$

Using (118) and (121), (123) becomes

$$\frac{V}{V_\infty} = 1 + \varepsilon \phi_{1x}(x, 0) + \varepsilon^2 \left[\phi_{2x}(x, 0) - \frac{1}{(1 - x^2)} + \frac{1}{2} \frac{x^2}{(1 - x^2)} + \cdots\right]. \tag{124}$$

Representing the body by a distribution of sources along the x-axis, one has from (68), on nondimensionlizing the lengths using the semichord length l,

$$\left.\begin{aligned} \phi_{nx}(x, y) &= \frac{1}{\pi} \int_{-1}^{1} \frac{(x - \xi)\, \phi_{ny}\left(\xi, 0^{\pm}\right) d\xi}{(x - \xi)^2 + y^2} \\ \phi_{ny}(x, y) &= -\frac{1}{\pi} \int_{-1}^{1} \frac{y \phi_{ny}\left(\xi, 0^{\pm}\right)}{(x - \xi)^2 + y^2},\ n = 1, 2, \cdots \end{aligned}\right\}. \tag{125}$$

On the surface of the airfoil, (125) gives

$$\phi_{nx}(x,0) = \frac{1}{\pi}\int_{-1}^{1}\frac{\phi_{ny}(\xi,0^{\pm})\,d\xi}{x-\xi}. \tag{126}$$

Using the kinematic condition (118), (126) gives

$$\phi_{1x}(x,0) = -\frac{1}{\pi}\int_{-1}^{1}\frac{\xi d\xi}{\sqrt{1-\xi^2}\,(x-\xi)} \tag{127}$$

from which,

$$y=0:\ \phi_{1x} = \begin{cases} 1, & x^2<1 \\ 1-\sqrt{\dfrac{x}{x^2-1}}, & x^2>1. \end{cases} \tag{128}$$

Noting that the problem for ϕ_2 is identical to that of ϕ_1, the flow speed on the airfoil from (124), is given by

$$\frac{V}{V_\infty} = 1+\varepsilon-\varepsilon^2\frac{x^2/2}{1-x^2}+\cdots. \tag{129}$$

Substituting

$$s \equiv 1+x \tag{130}$$

in (129) gives in the neighborhood of the leading edge $x=-1$, where $s \ll 1$,

$$V = V_\infty(1+\varepsilon)\left[1-\frac{\varepsilon^2}{4s(1+\varepsilon)}+\cdots\right]. \tag{131}$$

Note that the assumption of small disturbances has been violated at the stagnation points (such as that at $x=-1$) so that (131) breaks down at those points locally. Note that the region of nonuniformity is of order ε^2.

One way to remove this singularity is to sum the patently divergent series in (131) by pretending temporarily that it is not divergent, i.e., $s \sim O(1)$, though this is not justified. Then, (131) becomes

$$V \approx \frac{V_\infty(1+\varepsilon)}{1+\varepsilon^2/4s} \approx \frac{V_\infty(1+\varepsilon)}{\sqrt{1+\varepsilon^2/2s}} \approx V_\infty(1+\varepsilon)\sqrt{\frac{s}{s+\varepsilon^2/2}} \tag{132}$$

which is an exact result that corresponds to a uniform flow of speed $V_\infty(1+\varepsilon)$ past a parabola of nose radius ε^2! This is how the flow indeed appears near the leading edge for this problem. The result (132) ensues from the leading-edge value of the exact expression for the flow speed on the airfoil

$$\frac{V}{V_\infty} = \frac{1+\varepsilon}{\sqrt{1+\varepsilon^2\left(\dfrac{x^2}{1-x^2}\right)}}. \tag{133}$$

This can be obtained by the method of complex variables (see Van Dyke, 1975).

Another way of removing the singularity in (131) is to introduce a multiplicative correction factor. This converts the formal thin-airfoil solution for the flow speed V on an airfoil of nose radius ε^2 into a *uniformly valid approximation* $\overline{V}$:

$$\overline{V} = \frac{\sqrt{\dfrac{s}{s+\varepsilon^2/2}}}{\left(1-\dfrac{\varepsilon^2}{4s}+\cdots\right)}V. \tag{134}$$

As a result, near the leading edge, the flow speed on the airfoil is nearly that on the *osculating* parabola. Far from the leading edge, the correction factor approaches unity so that the thin-airfoil solution for the airfoil is recovered. One may write (134) as

$$\frac{\overline{V}}{V_\infty} = \sqrt{\frac{s}{s+\varepsilon^2/2}}\left(1+\frac{\varepsilon^2}{4s}\right)\frac{V}{V_\infty} \tag{135a}$$

or on using (133)

$$\frac{\overline{V}}{V_\infty} = \sqrt{\frac{1+x}{1+x+\varepsilon^2/2}}\left(1+\varepsilon+\frac{\varepsilon^2}{4}\frac{1-2x}{1-x}\right). \tag{135b}$$

Though the singularity at the leading edge has been removed, the one at the trailing edge remains. This too can be removed by applying the above correction again by introducing $\tilde{s} \equiv 1-x$. One then obtains the *uniformly valid* result

$$\frac{\overline{V}}{V_\infty} = \sqrt{\frac{1-x^2}{1-x^2+\varepsilon^2+\varepsilon^4/4}}\left(1+\varepsilon+\frac{\varepsilon^2}{2}\right). \tag{136}$$

11.3 Slender-Body Theory

Slender-body theory is concerned with the calculation of flows past bodies whose lateral dimensions change slowly with distance parallel to the flow direction.

One first chooses a singularity distribution along the axis of symmetry of a slender body of a given shape. This is done in such a way that the flow generated by these singularities in combination with a uniform stream satisfies the fluid impenetrability condition at the surface of the body.

Consider a slender body of length l simulated by a doublet distribution of strength $m(x)$, x being the distance along the axis of symmetry. One then has for a uniform stream past this body,

$$\Phi = -V_\infty x + \frac{1}{4\pi}\int_0^l \frac{m(\xi)}{\sqrt{(x-\xi)^2 + r^2}} d\xi. \tag{137}$$

From this, the velocity components are given by

$$\left.\begin{aligned} u &\equiv -\frac{\partial \Phi}{\partial r} = \frac{1}{4\pi}\int_0^l \frac{m(\xi)\cdot r}{\left[(x-\xi)^2 + r^2\right]^{3/2}} d\xi \\ w &\equiv -\frac{\partial \Phi}{\partial x} = V_\infty + \frac{1}{4\pi}\int_0^l \frac{m(\xi)\cdot(x-\xi)}{\left[(x-\xi)^2 + r^2\right]^{3/2}} d\xi \end{aligned}\right\}. \tag{138}$$

One may write (138) as

$$u = \frac{1}{4\pi r}\int_{-x/r}^{(l-x)/r} \frac{m(\xi)}{\left[\left(\frac{\xi - x}{r}\right)^2 + 1\right]^{3/2}} d\left(\frac{\xi - x}{r}\right). \tag{139}$$

For a slender body, r is very small. So, (139) may be approximated by

$$u \approx \frac{1}{4\pi r} m(x) \int_{-\infty}^{\infty} \frac{d\eta}{(\eta^2 + 1)^{3/2}} = \frac{1}{2\pi r} m(x), \eta \equiv \frac{\xi - x}{r}. \tag{140}$$

If the surface of the body is given by $r = F(x)$, then the boundary condition there is,

$$r = F(x) : u = V_\infty \frac{dF}{dx}. \tag{141}$$

Using (140), we obtain from (141),

$$m(x) = 2\pi V_\infty F \frac{dF}{dx}. \tag{142}$$

Using (142), (137) becomes

$$\Phi = -V_\infty x + \frac{V_\infty}{2}\int_0^l \frac{F(\xi)\dfrac{dF(\xi)}{d\xi}}{\sqrt{(x-\xi)^2 + r^2}}d\xi. \tag{143}$$

11.4 Prandtl's Lifting-Line Theory for Wings

For wings of *finite span*, one has to take into account the effects of the three-dimensionality of the flow. This is handled satisfactorily by the *lifting-line theory* of Prandtl (1927) provided that,

- the wing is of sufficiently large *aspect ratio* (which is span/chord);
- the wing section (made by a plane perpendicular to the span) does not vary too rapidly along the span.

The lifting-line model envisages the following simplifying conditions:
(i) The circulation Γ around a wing section varies symmetrically about the center line of the wing. Furthermore, the variation of Γ along the wingspan leads to vortex shedding from the wing. This creates a sheet of trailing line vortices which are parallel to the direction of flow. This sheet extends downstream from the trailing edges of the wing. In reality, it is unstable and "rolls up" at the edges which is ignored here. The circulation is assumed to drop to zero at the wing-tips as given below,

$$\Gamma = \Gamma(y), \quad \Gamma(+y) = \Gamma(-y), \quad \Gamma\left(\pm\frac{b}{2}\right) = 0; \tag{144}$$

(ii) The span b of the wing is sufficiently large compared to the chord l so that the variation of the velocity in the spanwise direction is small compared to the variation in a plane normal to the span;
(iii) For a very large aspect ratio, the wing is replaced by a *lifting line* (see Figure 11.16) having the same distribution of lifting forces along the span as the wing. The disturbances caused by the lifting line are taken to be small.
(iv) The velocity field *induced* at the wing by the trailing vortices consists of a small *downwash* for wings of large aspect ratio. Consequently, the flow at each sectional plane can be considered as a two-dimensional flow around an airfoil. The only additional feature of the flow in the sectional plane is the modification of the angle of attack (see Figure 11.17) as defined by the undisturbed flow, on account of the induced velocity $w(y)$.

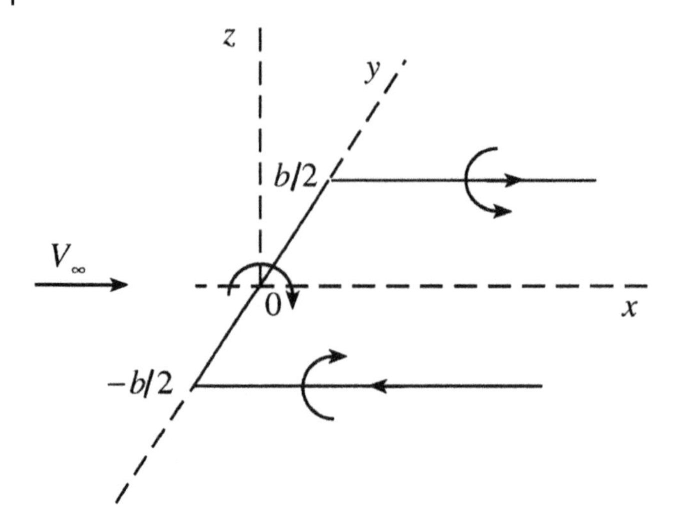

Figure 11.16 Lifting-line configuration (from Karamcheti, 1966).

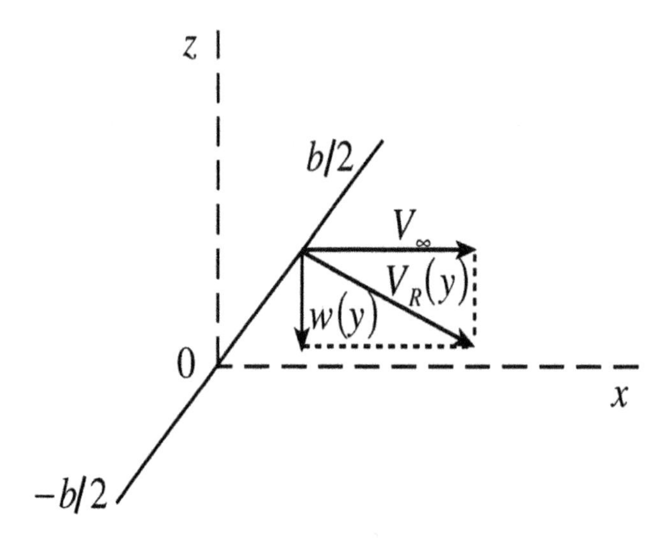

Figure 11.17 Flow at sectional plane perepndicular to thelifting line (from Karamcheti, 1966).

The flow-situation at a wing section is represented in Figure 11.18. Now, from the thin airfoil theory, namely, (28), one has

$$\Gamma(y) = \pi l(y) V_R(y) \alpha_R(y) \tag{145}$$

Figure 11.18 The trailing vortex sheet (from Karamcheti, 1966).

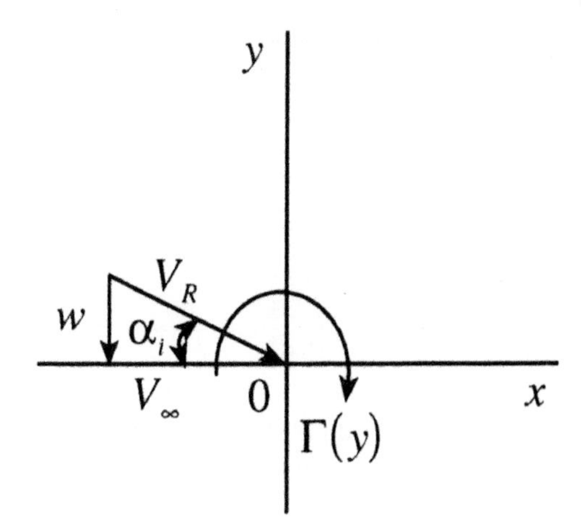

where the effective angle of attack $\alpha_R(y)$ (taken to be small) is given by

$$\alpha_R(y) = \alpha(y) - \alpha_i(y) = \alpha(y) - \tan^{-1}\left(\frac{w(y)}{V_\infty}\right) \tag{146}$$

$\alpha_i(y)$ being the *induced* angle of attack.

The velocity induced by an element of trailing vortex-sheet of strength $\gamma(\eta)$ (see Figure 11.19) is given by (see Section 8.7)

$$\delta w(y) = \frac{1}{4\pi}\left(\frac{\gamma(\eta)}{\eta - y}\right) \tag{147}$$

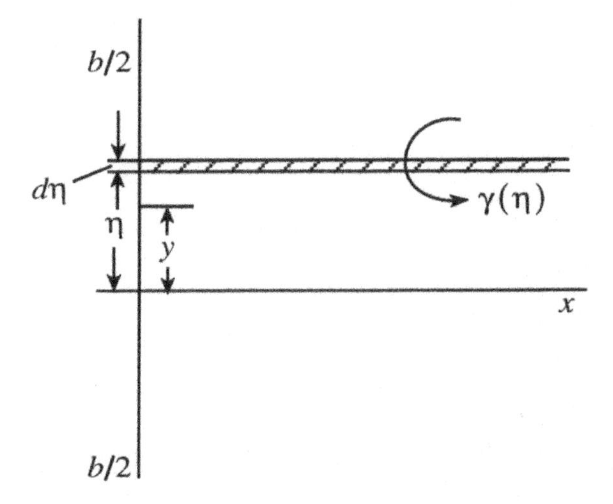

Figure 11.19 Flow situation in the wing section (from Karamcheti, 1966).

where, by the *conservation of vorticity*, the strength $\gamma(\eta)$ of the vortex sheet is given by

$$\gamma(\eta) = -\frac{d\Gamma}{dy}(\eta)\, d\eta. \tag{148}$$

Thus, one obtains for the entire trailing-vortex sheet,

$$w(y) = \frac{1}{4\pi} \int_{-b/2}^{b/2} \frac{d\Gamma}{dy}(\eta) \frac{d\eta}{y - \eta}. \tag{149}$$

Using (146) and (149), and assuming $w/V_\infty \ll 1$, (145) gives an *integral equation* for $\Gamma(y)$ as below,

$$\Gamma(y) \approx \pi l(y) \left[V_\infty \alpha(y) - \frac{1}{4\pi} \int_{-b/2}^{b/2} \frac{d\Gamma}{dy}(\eta) \frac{d\eta}{y - \eta} \right] \tag{150}$$

with the boundary conditions,

$$\Gamma\left(\pm\frac{b}{2}\right) = 0. \tag{144}$$

The force on a section dy is given by

$$\delta \mathbf{F}(y) = \rho \mathbf{V}_R(y) \Gamma(y)\, dy \hat{\mathbf{i}}_y \tag{151}$$

from which, the lift and the *induced drag* on the wing element at y are given by

$$\delta \mathbb{L}(y) \approx \rho V_\infty \Gamma(y)\, dy \tag{152}$$

$$\delta D_i(y) \approx \rho w(y) \Gamma(y)\, dy. \tag{153}$$

Thus, for the entire wing, one obtains

$$\mathbb{L} = \rho V_\infty \int_{-b/2}^{b/2} \Gamma(y)\, dy, \tag{154}$$

$$D_i = \rho \int_{-b/2}^{b/2} w(y) \Gamma(y)\, dy. \tag{155}$$

An essential by-product of lift on a three-dimensional body is the existence of a trailing vortex sheet. The energy spent by the body to constantly maintain these trailing vortices shows up as the induced drag.

The moment on the wing is given by

$$\mathbf{M} = \int_{-b/2}^{b/2} y\hat{\mathbf{i}}_y \times \left(\hat{\mathbf{i}}_x \delta D_i + \hat{\mathbf{i}}_z \delta \mathbb{L}\right) = -\hat{\mathbf{i}}_z \int_{-b/2}^{b/2} y\delta D_i + \mathbf{i}_x \int_{-b/2}^{b/2} y\delta \mathbb{L} \equiv \mathbf{M}_Y + \mathbf{M}_R \tag{156}$$

the first term $\mathbf{M}_Y$ on the right-hand side in (156) corresponds to *yawing* of the wing and the second term $\mathbf{M}_R$ to *rolling*.

Consider a loading on the wing given by

$$\Gamma(\theta) = 2bV_\infty \sum_{n=1}^{\infty} A_n \sin n\theta \tag{157}$$

where

$$y = \frac{b}{2}\cos\theta \tag{158}$$

and the boundary conditions (144) have been incorporated.

Substituting (157) in (149), we get,

$$w(\theta) = V_\infty \sum_{n=1}^{\infty} nA_n \frac{\sin n\theta}{\sin\theta}. \tag{159}$$

Using (157)-(159), (154)-(156) give for the lift induced drag and moments on the wing,

$$\mathbb{L} = \rho V_\infty^2 b^2 \frac{\pi}{2} A_1 \tag{160}$$

$$D_i = \rho V_\infty^2 b^2 \frac{\pi}{2} \sum_{n=1}^{\infty} nA_n^2 \tag{161}$$

$$M_R = -\frac{\pi}{8}\rho V_\infty^2 b^3 A_2 \tag{162}$$

$$M_y = \frac{\pi}{2}\rho V_\infty^2 b^3 \sum_{n=1}^{\infty} (2n+1) A_n A_{n+1}. \tag{163}$$

For a body of *minimum induced drag*, one has, from (161),

$$A_n = 0 \text{ for } n > 1 \tag{164}$$

which, from (157), corresponds to an *elliptic lift distribution*.

Furthermore, equation (150) gives, on using (164),

$$l(\theta)\left[\frac{\pi\alpha}{2b} - \frac{\pi A_1}{2b}\right] = A_1 \sin\theta$$

or

$$l(\theta) \sim \sin\theta$$

or

$$l(y) \sim \sqrt{1-(2y/b)^2} \tag{165}$$

which corresponds to an *elliptic plan-form* for the wing.

Thus, for a wing with *elliptic plan-form* and an elliptic lift distribution, namely,

$$\left.\begin{aligned} l(y) &= l_0\sqrt{1-(2y/b)^2} \\ \Gamma(y) &= \Gamma_0\sqrt{1-(2y/b)^2} \end{aligned}\right\} \tag{166}$$

equation (150) gives

$$\Gamma_0 = \pi l_0 \left[V_\infty \alpha - \frac{\Gamma_0}{2b}\right]$$

or

$$\Gamma_0 = \frac{2bV_\infty \alpha}{1 + \dfrac{2b}{\pi l_0}} \tag{167}$$

which reduces to the two-dimensional airfoil result, namely (28), in the infinite aspect ratio limit $b/l_0 \Rightarrow \infty$.

11.5 Oscillating Thin-Airfoil Problem: Theodorsen's Theory

The *unsteady airfoil theory* has important applications, like in the *flutter* problem and the forces experienced by airplanes flying through gusts.

When an airfoil performs oscillations around a given mean position, the circulation around the airfoil also undergoes periodic variations. This, in conjunction with the total vorticity conservation constraint, leads to the shedding of free vorticity from the trailing edge.This vorticity is equal in strength but

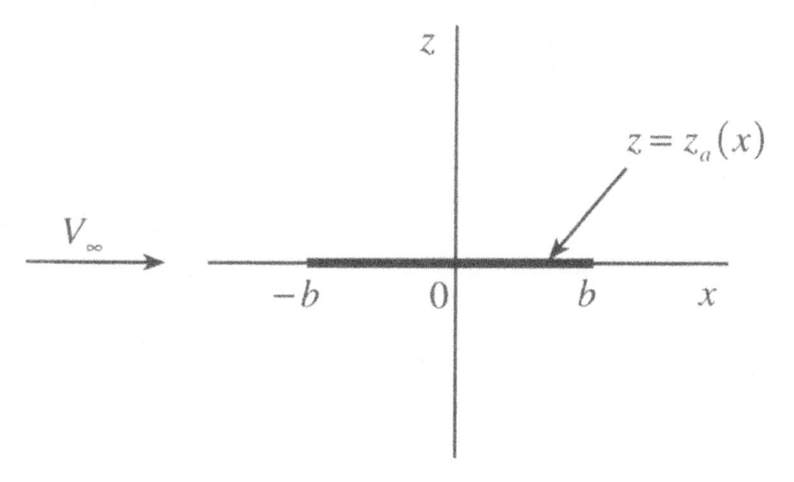

Figure 11.20 Uniform flow past an oscillating airfoil.

opposite in sign to the change in the circulation around the airfoil and is carried downstream by the flow. The velocity field induced around the airfoil by these free vortices causes changes in the instantaneous angles of attack of the airfoil Consequently, the oscillating part of the lift *lags* behind the motion of the airfoil.

Within the framework of a linearized theory, solutions may be superposed to generate another solution. Consider an oscillating airfoil with finite but small thickness and camber at a given mean angle of attack. The solution to this problem can be obtained by a superposition of an unsteady motion for an oscillating airfoil of zero thickness, zero camber at zero mean angle of attack, and a steady-state solution for an airfoil of the given thickness and camber at the given mean angle of attack (Theodorsen, 1932, 1935).

One has, for a uniform flow past an oscillating airfoil (see Figure 11.20), the following boundary-value problem:

$$\nabla^2 \phi = 0 \tag{168}$$

$$z = 0 : w = w_a = \frac{\partial z_a}{\partial t} + V_\infty \frac{\partial z_a}{\partial x}, \ -b \leq x \leq b \tag{169}$$

$$x = b : \ \text{Kutta condition.} \tag{170}$$

Here, $z = z_a(x, t)$ describes the surface of the airfoil, denoted by the subscript a. In (170), we overlook any possibility of lag in the adjustment of flow at the trailing edge.[4] In order to find a solution to (168)-(170) one chooses an appropriate distribution of sources and sinks just above and below the line $z = 0$. We choose a distribution of vortices on this line with countervortices along the wake to infinity. This distribution is such that Kutta's condition is satisfied without disturbing the boundary conditions at the surface of the airfoil. The flow components due to the sources and vortices are most conveniently obtained by using *Joukowski's transformation*, given by,

$$x + iz = (X + iZ) + \frac{b^2/4}{(X + iZ)}. \tag{171}$$

This maps a circle of radius $b/2$ onto the flat-plate airfoil. On the surface of the airfoil and its mapped image, one has from (171),[5]

$$X + iZ = \frac{b}{2}e^{i\theta}: \qquad x = b\cos\theta, \qquad z = 0, \qquad 0 \leq \theta \leq \pi \tag{172}$$

and the complex velocity,

$$u - iw = \frac{q_x - iq_z}{2i\sin\theta e^{-i\theta}} = \frac{-q_\theta - iq_r}{2\sin\theta} \tag{173}$$

from which we have,

$$u = -\frac{q_\theta}{2\sin\theta}, \qquad w = \frac{q_r}{2\sin\theta}, \qquad 0 \leq \theta \leq \pi. \tag{174}$$

One may distribute sources on the upper half of the circle, and sinks of equal strength $H^+(\xi, t)$ on the lower half. Corresponding to the source sheet on the airfoil in the x, z-plane, one has

$$\phi(x, z, t) = \frac{1}{4\pi}\int_{-b}^{b} H^+(\xi, t)\,\ell n\left[(x - \xi)^2 + z^2\right]d\xi. \tag{175}$$

4 The success of this model, as it turns out, indicates that such adjustment is, in fact, very rapid.
5 One inserts a "*cut*" along the slit $-b \leq x \leq b$, with two *Riemannian sheets*, which one is forbidden to cross.

From this, the velocity component along the z-direction, on the upper surface of the airfoil, is given by

$$w\left(x,0^{+},t\right)=\frac{1}{2\pi}\lim_{z\Rightarrow 0^{+}}\int_{-b}^{b}\frac{H^{+}\left(\xi,t\right)z}{\left[\left(x-\xi\right)^{2}+z^{2}\right]}d\xi$$

$$=\frac{1}{2\pi}H^{+}\left(x,t\right)\lim_{\substack{z\rightarrow 0^{+}\\ \varepsilon\rightarrow 0}}\left[\tan^{-1}\left(\frac{\varepsilon}{z}\right)-\tan^{-1}\left(-\frac{\varepsilon}{z}\right)\right]$$

$$=\frac{1}{2}H^{+}\left(x,t\right). \tag{176a}$$

Similarly, corresponding to the sink sheet on the airfoil, the velocity component along the z-direction on the lower surface of the airfoil is given by

$$w\left(x,0^{-},t\right)=-\frac{1}{2}H^{-}\left(x,t\right). \tag{176b}$$

In the X,Z-plane, from (174) and (176a), one has

$$\left.\begin{aligned}H^{+}\left(\frac{b}{2},\theta,t\right)&=2q_{r}=4w_{a}\sin\theta\\ H^{-}\left(\frac{b}{2},\theta,t\right)&=-4w_{a}\sin\theta\end{aligned}\right\}. \tag{177}$$

Corresponding to a source of strength $H^{+}\left(b/2\right)d\varphi$ at $\theta=\varphi$ (see Figure 11.21), one has at P, the flow speed

$$\left|dq^{+}\right|=\frac{H^{+}\frac{b}{2}d\varphi}{2\pi b\sin\left(\frac{\varphi-\theta}{2}\right)}=\frac{w_{a}\sin\varphi d\varphi}{\pi\sin\left(\frac{\varphi-\theta}{2}\right)}. \tag{178}$$

Similarly, for a sink of strength $H^{-}\left(b/2\right)d\varphi$ at $\theta=-\varphi$, one has at P, the flow speed

$$\left|dq^{-}\right|=\frac{w_{a}\sin\varphi d\varphi}{\pi\sin\left(\frac{\varphi+\theta}{2}\right)}. \tag{179}$$

Noting

$$\left.\begin{aligned}dq_{\theta}&=-\left|dq^{+}\right|\cos\left(\frac{\varphi-\theta}{2}\right)-\left|dq^{-}\right|\cos\left(\frac{\varphi+\theta}{2}\right)\\ dq_{r}&=\left|dq^{+}\right|\sin\left(\frac{\varphi-\theta}{2}\right)-\left|dq^{-}\right|\sin\left(\frac{\varphi+\theta}{2}\right)\end{aligned}\right\} \tag{180}$$

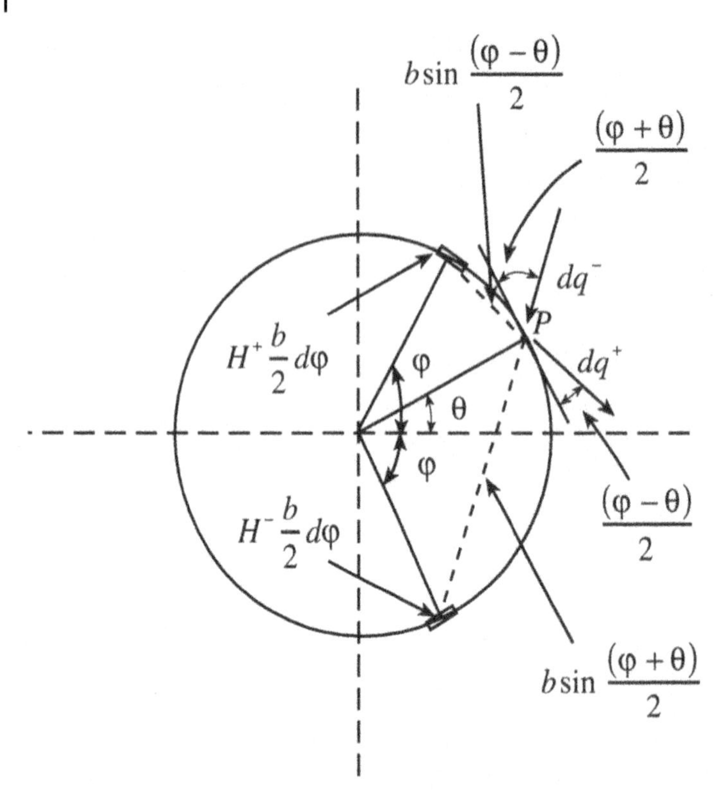

Figure 11.21 Source and sink distribution on the circle generated by Joukowski's mapping of the airfoil (from Bisplinghoff, Ashley and Halfman, 1955).

and using (178) and (179), one has for the flow velocity components at P,

$$dq_\theta = \frac{2w_a \sin^2 \varphi d\varphi}{\pi (\cos \varphi - \cos \theta)} \tag{181}$$

$$dq_r = 0. \tag{182}$$

Equation (182) implies that the circle is a streamline.

One obtains, from (181), for the whole airfoil,

$$q_\theta = \frac{2}{\pi} \int_0^\pi \frac{w_a \sin^2 \varphi d\varphi}{(\cos \varphi - \cos \theta)}. \tag{183}$$

Now, one has

$$\phi(\pi,t)-\phi^{+}(\theta,t)=\int_{0}^{\pi} q_{\theta}\frac{b}{2}d\theta \tag{184}$$

where

$$\phi(\pi,t)=0 \text{ since } \phi^{-}(-\theta,t)=-\phi^{+}(\theta,t). \tag{185}$$

Using (183) and (185), (184) gives for the *noncirculatory* part of the flow,

$$\phi^{+}_{NC}(\theta,t)=-\frac{b}{\pi}\int_{\theta}^{\pi}\int_{0}^{\pi}\frac{w_a\sin^2\varphi d\varphi d\theta}{(\cos\varphi-\cos\theta)}. \tag{186}$$

On the airfoil, one has, from the *Bernoulli integral* upon using (185), for the pressure jump across the airfoil,

$$(p^{+}-p^{-})=-2\rho\left(V_{\infty}\frac{\partial\phi^{+}}{\partial x}+\frac{\partial\phi^{+}}{\partial t}\right).$$

Furthermore, on using (172), we obtain

$$(p^{+}-p^{-})=-2\rho\left(\frac{\partial\phi^{+}}{\partial t}-\frac{V_{\infty}}{b\sin\theta}\frac{\partial\phi^{+}}{\partial\theta}\right). \tag{187}$$

The lift and the moment on the airfoil corresponding to the noncirculatory part of the flow are then given by

$$(\mathbb{L})_{NC}=-\int_{-b}^{b}(p^{+}-p^{-})\,dx=2\rho b\frac{\partial}{\partial t}\int_{0}^{\pi}\phi^{+}\sin\theta d\theta \tag{188}$$

$$\begin{aligned}(M_y)_{NC}&=\int_{-b}^{b}(p^{+}-p^{-})(x-ba)\,dx\\&=2\rho V_{\infty}\int_{-b}^{b}\phi^{+}dx-2\rho\frac{\partial}{\partial t}\int_{-b}^{b}\phi^{+}\cdot(x-ba)\,dx\\&=2\rho V_{\infty}b\int_{0}^{\pi}\phi^{+}\sin\theta d\theta-2\rho b^2\frac{\partial}{\partial t}\int_{0}^{\pi}\phi^{+}\cdot(\cos\theta-a)\sin\theta d\theta\end{aligned} \tag{189}$$

where $0\le a\le 1$.

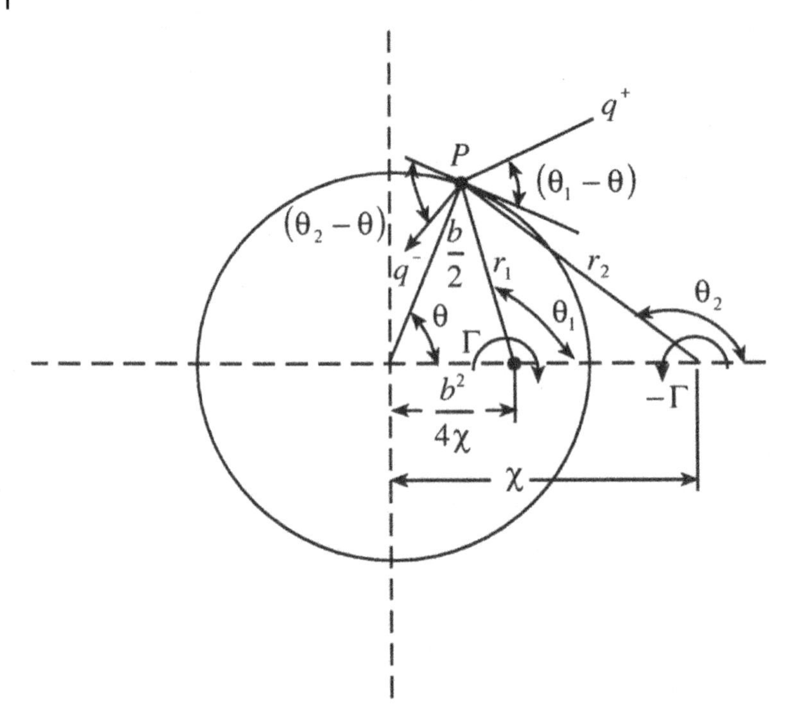

Figure 11.22 Flow generated by a system of wake vortices and image vortices in the circle (from Bisplinghoff, Ashley and Halfman, 1955).

Now, the temporal variation of the total circulation around the airfoil generates a wake of two-dimensional countervortices along the x-axis from the trailing edge to infinity. These are continually moving away from the airfoil at the free-stream velocity. One pairs with each wake vortex, a *bound* one of opposite circulation at the *"image"* position inside the circle. This ensures that the circle and the airfoil slit remain streamlines of the vortex flow so that the boundary conditions remain undisturbed.

Consider the flow due to a single bound vortex of strength Γ and its image $-\Gamma$ (see Figure 11.22). At P one has, by the *Biot-Savart Law* (see Section 8.3),

$$\begin{aligned} q_\theta &= |q^-| \cos(\theta_2 - \theta) - |q^+| \cos(\theta_1 - \theta) \\ &= \frac{\Gamma}{2\pi}\left[\frac{r_2 \cos(\theta_2 - \theta)}{r_2^2} - \frac{r_1 \cos(\theta_1 - \theta)}{r_1^2}\right]. \end{aligned} \tag{190}$$

Noting from Figure 11.22 that,

$$r_1^2 = \left(\frac{b^2}{4\chi}\right)^2 + \left(\frac{b}{2}\right)^2 - 2\frac{b^2}{4\chi}\cdot\frac{b}{2}\cos\theta$$

$$r_2^2 = \chi^2 + \left(\frac{b}{2}\right)^2 - 2\chi\cdot\frac{b}{2}\cos\theta$$

$$\begin{aligned} r_2\cos(\theta_2 - \theta) &= \frac{b}{2} - \chi\cos\theta \\ r_1\cos(\theta_1 - \theta) &= \frac{b}{2} - \frac{b^2}{4\chi}\cos\theta \end{aligned} \tag{191}$$

we write (190) as below,

$$q_\theta = -\frac{\Gamma}{\pi b}\left[\frac{\chi^2 - (b/2)^2}{\chi^2 + (b/2)^2 - \chi b\cos\theta}\right]. \tag{192}$$

Using (192) and (185), one has, from (184), for the circulatory part of the flow,

$$\begin{aligned} \phi_C^+(\theta, t) &= \frac{\Gamma\left[\chi^2 - (b/2)^2\right]}{2\pi}\int_0^\pi \frac{d\theta}{\chi^2 + (b/2)^2 - \chi b\cos\theta} \\ &= \frac{\Gamma}{\pi}\tan^{-1}\left[\frac{\chi - b/2}{\chi + b/2}\sqrt{\frac{1+\cos\theta}{1-\cos\theta}}\right]. \end{aligned} \tag{193}$$

Noting that, if ξ is the distance of a countervortex from the origin in the x, y-plane, one has

$$\xi = \chi + \frac{b^2}{4\chi}, \qquad \frac{d\xi}{dt} = V_\infty \tag{194}$$

from which one obtains,

$$\sqrt{\frac{\xi - b}{\xi + b}} = \frac{\chi - b/2}{\chi + b/2}. \tag{195}$$

Using (195), (193) becomes

$$\phi_C^+(\theta, t) = \frac{\Gamma}{\pi}\tan^{-1}\sqrt{\frac{(\xi - b)(1+\cos\theta)}{(\xi + b)(1-\cos\theta)}}. \tag{196}$$

Using (196) in the Bernoulli integral, namely (187),

$$(p^+ - p^-)_\Gamma = -2\rho\left[\frac{\partial\phi_c^+}{\partial\xi}\frac{d\xi}{dt} - \frac{V_\infty}{b\sin\theta}\frac{\partial\phi_c^+}{\partial\theta}\right]$$

we obtain for the pressure jump across the airfoil

$$(p^+ - p^-)_\Gamma = -\frac{\rho V_\infty \Gamma(\xi + b\cos\theta)}{\pi b\sin\theta\cdot\sqrt{\xi^2 - b^2}}. \tag{197}$$

The corresponding lift and the moment on the airfoil are given by

$$(\mathbb{L})_\Gamma = -\int_0^\pi b\sin\theta\cdot d\theta\cdot(p^+ - p^-)_\Gamma = \frac{\rho V_\infty\Gamma\xi}{\sqrt{\xi^2 - b^2}} \tag{198}$$

$$(M_y)_\Gamma = \int_0^\pi (b\cos\theta - ba)\,b\sin\theta\cdot d\theta\cdot(p^+ - p^-)_\Gamma = \frac{\rho V_\infty\Gamma b^2}{\sqrt{\xi^2 - b^2}}\left(\frac{\xi}{b}a - \frac{1}{2}\right). \tag{199}$$

Note that, as $\xi \Rightarrow \infty$, the flow approaches that of a single-bound vortex Γ. The lift $\rho V_\infty\Gamma$ then acts at midchord.

For the complete wake, one obtains from (197)-(199):

$$(p^+ - p^-)_C = \frac{\rho V_\infty}{\pi b\sin\theta}\int_b^\infty\left[\frac{\xi}{\sqrt{\xi^2 - b^2}}(1 - \cos\theta) + \right.$$
$$\left. + \sqrt{\frac{\xi + b}{\xi - b}}\cos\theta\right]\gamma_w(\xi, t)\,d\xi \tag{200}$$

$$(\mathbb{L})_C = -\rho V_\infty\int_b^\infty \frac{\xi}{\sqrt{\xi^2 - b^2}}\gamma_w(\xi, t)\,d\xi \tag{201}$$

$$(M_y)_C = \rho V_\infty b\int_b^\infty\left[\frac{1}{2}\sqrt{\frac{\xi + b}{\xi - b}} - \left(a + \frac{1}{2}\right)\frac{\xi}{\sqrt{\xi^2 - b^2}}\right]\gamma_w(\xi, t)\,d\xi. \tag{202}$$

Here, we have replaced Γ by $\gamma_w d\xi, \gamma_w$ being the circulation per unit length of the wake.

From (174), one has the Kutta condition at $\theta = 0$ given by,

$$q_\theta = 0. \tag{203}$$

Using (186) and (196), (203) leads to an *integral equation* for γ_w,

$$\frac{2}{\pi}\int_0^\pi \frac{w_a \sin^2\varphi d\varphi}{\cos\varphi - 1} + \frac{1}{\pi b}\int_b^\infty \sqrt{\frac{\xi + b}{\xi - b}}\gamma_w(\xi, t)\, d\xi = 0. \tag{204}$$

Setting

$$\xi^* \equiv \cos\varphi \tag{205}$$

equation (204) becomes

$$\frac{2}{\pi}\int_{-1}^1 \frac{\sqrt{1 - \xi^{*2}}}{(\xi^* - 1)} w_a(\xi^*, t)\, d\xi^* + \frac{1}{\pi b}\int_b^\infty \sqrt{\frac{\xi + b}{\xi - b}}\gamma_w(\xi, t)\, d\xi = 0. \tag{206}$$

Note that, one has, for the bound vortices distributed on the airfoil, the total circulation,

$$\Gamma_0(t) = \int_{-b}^b \gamma_0(x, t)\, dx. \tag{207}$$

The velocity induced by this bound vortex distribution is given by (see Section 8.7)

$$w_0(x) = -\frac{1}{2\pi}\int_{-b}^b \frac{\gamma(x_1)}{x - x_1} dx_1 \tag{208}$$

which is a *Fredholm integral equation of the first kind*. In order to invert

$$w_0(x) = -\frac{1}{2\pi}\int_0^\ell \frac{\gamma(x_1)}{x - x_1} dx_1 \tag{209}$$

one may write

$$\gamma(x) \equiv \int_0^\ell dx_1 w_0(x_1)\psi(x, x_1). \tag{210}$$

Using (210), (209) implies that

$$-\frac{1}{2\pi}\int_0^\ell dx_1 \frac{\psi(\xi, x_1)}{x - x_1} = \delta(x - \xi) \tag{211}$$

from which, one has

$$x\delta(x - \xi) = -\frac{1}{2\pi}\int_0^\ell dx_1 \psi(\xi, x_1) - \frac{1}{2\pi}\int_0^\ell dx_1 \frac{x_1\psi(\xi, x_1)}{x - x_1}$$

and

$$\xi\delta(x - \xi) = -\frac{1}{2\pi}\int_0^\ell dx_1 \frac{\xi\psi(\xi, x_1)}{x - x_1}. \tag{212}$$

Equations (212) in turn give,

$$-\frac{1}{2\pi}\int_0^\ell dx_1 \frac{(x_1 - \xi)\psi(\xi, x_1)}{x - x_1} = \frac{1}{2\pi}\int_0^\ell dx_1 \psi(\xi, x_1). \tag{213}$$

Let us write

$$(x_1 - \xi)\psi(\xi, x_1) \equiv A(\xi)\gamma_1(x_1) \tag{214}$$

where

$$-\frac{1}{2\pi}\int_0^\ell \frac{\gamma_1(x_1)}{x - x_1} dx_1 = 1. \tag{215}$$

Equation (215) implies that

$$\gamma_1(x_1) = 2\sqrt{\frac{x_1}{\ell - x_1}}. \tag{216}$$

Using equations (214) and (215), (213) gives

$$A(\xi) = \frac{1}{2\pi}\int_0^\ell dx_1 \psi(\xi, x_1). \tag{217}$$

Consider

$$\int_0^\ell \gamma_1(\ell - x)\,\delta(x-\xi)\,dx = \int_0^\ell \left(-\frac{1}{2\pi}\right)\int_0^\ell dx_1 \frac{\gamma_1(\ell - x)\,\psi(\xi, x_1)}{x - x_1}dx \quad (218)$$

from which, one has

$$\gamma_1(\ell-\xi) = -\frac{1}{2\pi}\int_0^\ell dx_1 \psi(\xi, x_1)\left[-\int_0^\ell dx \frac{\gamma_1(\ell - x)}{x - x_1}\right]. \quad (219)$$

Using equations (215) and (217), (219) becomes

$$\gamma_1(\ell-\xi) = -2\pi A(\xi). \quad (220)$$

Using equations (210), (214), (216), and (220), one obtains for the bound vortex distribution,

$$\gamma(x) = -\frac{2}{\pi}\sqrt{\frac{\ell - x}{x}}\int_0^\ell dx_1 \frac{w_0(x_1)}{x_1 - x}\sqrt{\frac{x_1}{\ell - x_1}}. \quad (221)$$

Thus, by substituting $x = \dfrac{\ell}{2}(1+x'), -1 \leq x' \leq 1$, and dropping the primes, we obtain

$$\gamma_0(x,t) = \frac{2}{\pi}\sqrt{\frac{1-x}{1+x}}\int_{-1}^{1}\sqrt{\frac{1+\xi^*}{1-\xi^*}}\frac{w_a(\xi^*, t)}{x - \xi^*}d\xi^*. \quad (222)$$

Using (222), (207) gives

$$\Gamma_0(t) = -2b\int_{-1}^{1}\sqrt{\frac{1+\xi^*}{1-\xi^*}}w_a(\xi^*, t)\,d\xi^*. \quad (223)$$

Using (223), equation (206) becomes

$$\Gamma_0(t) + \int_b^\infty \sqrt{\frac{\xi + b}{\xi - b}}\gamma_w(\xi, t)\,d\xi = 0. \quad (224)$$

Example 1: Consider an airfoil in vertical translation $h(t)$, and a rotation about an axis at $x = ba$, through an angle α. One then has

$$w_a(x,t) = -\dot{h} - V_\infty \alpha - \dot{\alpha}(x - ba). \tag{225}$$

Using (225), (186) gives

$$\phi_{NC}^{+}(\theta,t) = \frac{b}{\pi}\left(\dot{h} + V_\infty \alpha\right)\int_\theta^\pi\int_0^\pi \frac{\sin^2\varphi d\varphi d\theta}{(\cos\varphi - \cos\theta)} +$$

$$+\frac{b^2\dot{\alpha}}{\pi}\int_\theta^\pi\int_0^\pi \frac{\sin^2\varphi\cdot(\cos\varphi - a)\,d\varphi d\theta}{(\cos\varphi - \cos\theta)} \tag{226}$$

$$= b\left(\dot{h} + V_\infty\alpha\right)\sin\theta + b^2\dot{\alpha}\sin\theta\cdot\left(\frac{\cos\theta}{2} - a\right).$$

The lift and the moment on the airfoil corresponding to the non-circulatory part of the flow are given below using (226) in (188) and (189),

$$\mathbb{L}_{NC} = \pi\rho b^2\left(\ddot{h} + V_\infty\dot{\alpha} - ba\ddot{\alpha}\right) \tag{227}$$

$$\left(M_y\right)_{NC} = \pi\rho b^2\left[V_\infty\dot{h} + ba\ddot{h} + V_\infty^2\alpha - b^2\left(\frac{1}{8} + a^2\right)\ddot{\alpha}\right]. \tag{228}$$

In order to calculate the corresponding quantities for the circulatory part of the flow, let us take

$$w_a(x,t) = \overline{w}_a(x)\,e^{i\omega t} \tag{229}$$

so that one has for the wake vortices,

$$\gamma_w(\xi,t) = \overline{\gamma}_w e^{i\omega(t-\xi/V_\infty)}. \tag{230}$$

Using (230), note that

$$\frac{\int_b^\infty \frac{\xi}{\sqrt{\xi^2 - b^2}}\gamma_w(\xi,t)\,d\xi}{\int_b^\infty \sqrt{\frac{\xi+b}{\xi-b}}\gamma_w(\xi,t)\,d\xi} = \frac{\int_1^\infty \frac{\xi^*}{\sqrt{\xi^{*2} - 1}}e^{-ik\xi^*}d\xi}{\int_1^\infty \sqrt{\frac{\xi^*+1}{\xi^*-1}}e^{-ik\xi^*}d\xi} \tag{231}$$

$$= \frac{H_1^{(2)}(k)}{H_1^{(2)}(k) + iH_0^{(2)}(k)} \equiv C(k),\ \text{say}$$

where $H_n^{(2)}(x)$ is *Hankel's function of the second kind*, and

$$k \equiv \frac{\omega b}{V_\infty}, \ \xi^* \equiv \frac{\xi}{b}.$$

In order to insure the convergence of the integrals in $C(k)$, one needs to impose an appropriate *radiation condition*, namely, that k is a complex quantity with a negative imaginary part.

The Kutta condition (224), using (225), leads to

$$-\frac{1}{\pi b}\int_b^\infty \sqrt{\frac{\xi+b}{\xi-b}}\gamma_w(\xi,t)\,d\xi$$

$$= \frac{2}{\pi}\int_0^\pi (1+\cos\varphi)\left[\dot{h} + V_\infty\alpha + \dot{\alpha}b(\cos\varphi - a)\right]d\varphi \tag{232}$$

$$= 2\left[\dot{h} + V_\infty\alpha + b\left(\frac{1}{2} - a\right)\dot{\alpha}\right].$$

Thus, using equations (227), (228), (201), (202), (231), and (232), one obtains for the total lift and the moment of the airfoil,

$$\mathbb{L} = (\mathbb{L})_{NC} + (\mathbb{L})_C = \pi\rho b^2\left(\ddot{h} + V_\infty\dot{\alpha} - ba\ddot{\alpha}\right)$$
$$+ 2\pi\rho V_\infty bC(k)\left[\dot{h} + V_\infty\alpha + b\left(\frac{1}{2} - a\right)\dot{\alpha}\right] \tag{233}$$

$$M_y = (M_y)_{NC} + (M_y)_C$$

$$= \pi\rho b^2\left[ba\ddot{h} - V_\infty b\left(\frac{1}{2} - a\right)\dot{\alpha} - b^2\left(\frac{1}{8} + a^2\right)\ddot{\alpha}\right] \tag{234}$$

$$+ 2\pi\rho V_\infty b^2\left(a + \frac{1}{2}\right)C(k)\left[\dot{h} + V_\infty\alpha + b\left(\frac{1}{2} - a\right)\dot{\alpha}\right].$$

In deriving (234), note that certain cancellations have occurred between circulatory and noncirculatory parts of the moment expressions.

Exercises

1. Using the Joukowski transformation, calculate the flow about an elliptic cylinder with its major axis making an angle α with the flow direction, and the moment about the leading edge acting on the cylinder. Show that this moment is oriented so as to set the elliptic cylinder broadside onto the incident flow.
2. Calculate the flow at the trailing edge of a symmetric airfoil (with $\mu = me^{i\pi}, \beta = 0$) at zero incidence in a uniform flow.
3. Calculate the camber-line shape of an airfoil, given the loading $\gamma(\theta) = const = k$, by thin-airfoil theory.
4. Using equation (150), solve the coefficients A_n in the prescription $\Gamma(\theta) = 2bU \sum_{n=1}^{\infty} A_n \sin n\theta$.

Part III

Dynamics of Compressible Fluid Flows

12

Review of Thermodynamics

Thermodynamics is a phenomenalogical theory of *equilibrium* states and *transitions* among them. This is based on the concept of energy and identification of the *Laws* governing the conversion of one form of energy into another, and the use of various media to produce such transformations. The *Laws of Thermodynamics* establish the relation between the *changes* in the thermodynamic state of a system and its interactions with the environment. The *First Law* postulates the *conservation of energy*, while the *Second Law* establishes the *trend toward equilibrium*, as dictated by the existence of *entropy* S. These Laws are concerned with connecting the initial and final states of a system undergoing a process rather than the detailed evolution of the system during the process. *Equilibrium Thermodynamics* is directly applicable to the mechanics of ideal fluids. In contrast, the mechanics of real fluids has to include the various transport phenomena which disturb the state of thermodynamic equilibrium.

12.1 Thermodynamic System and Variables of State

A *thermodynamic system* is a quantity of matter isolated from its surroundings for the purpose of observation. The system we consider is a simple, homogeneous one composed of a single fluid.

An isolated system reaches a state of equilibrium, i.e., the macroscopic state becomes steady. *Variables of state* determine the state of the system. A variable of state is uniquely defined for any equilibrium state of the system and is independent of the process[1] by which the system arrived at that state in the first place.

1 A process is the path of succession of states through which the system passes.

Introduction to Theoretical and Mathematical Fluid Dynamics, Third Edition.
Bhimsen K. Shivamoggi.

For a simple system, one has an equation of state of the form,

$$p = p(V, T) \tag{1}$$

p being the pressure, V the volume, and T the temperature. The *Zeroth Law* states that there exists a variable of state called the temperature T such that, two systems that are in thermal contact are in equilibrium only if T is the same in both.[2]

In the context of work or heat transfer of a system with its surroundings, one introduces another variable of state, the *internal energy* E. This measures the energy stored in the system, with an equation of state given by

$$E = E(V, T). \tag{2}$$

An *extensive* variable of state depends linearly on the mass of the system, ex: E, V, S. An *intensive* variable of state does not depend on the mass of the system, ex: p, T.

12.2 The First Law of Thermodynamics and Reversible and Irreversible Processes

Consider a system that is transformed from one state of equilibrium A to another, B. Suppose this is accomplished by a process, in which a certain amount of work W is done by the surroundings, and a certain quantity of heat Q leaves the surroundings.[3] One may then define the internal energy E by its change during this process given by,

$$E_B - E_A = Q + W. \tag{3}$$

For a small change of state, the process is reversible, and the system passes through a succession of equilibrium states. Then, equation (3) gives,

$$dE = dQ + dW = dQ - pdV \tag{4}$$

or in terms of specific quantities,

$$de = dq - pdv. \tag{5}$$

---

2 This is predicated on the fact that the total energy of the interacting systems is shared between them in such a way that the mean energy per degree of freedom is the same for both systems.

3 Heat and work are identified as different forms of energy only when they cross the boundary of a system, but not when they are contained in it.

Note that unlike E, Q and W are not variables of state because they depend on the process followed in changing the state.[4] So, dq and dw are not *exact differentials* while dE is.

The *First Law* dictates only the magnitude of the change of state of a system as described by equation (4). However, it does not dictate the direction of this change of state.

However, all natural or spontaneous processes are *irreversible*.[5] Note that, during an irreversible process, the system deviates from equilibrium.

A *perfect gas* is taken to have no interparticle forces of repulsion or attraction, so $e \neq e(v)$. Although the particles have mass, they are assumed to occupy no space. A perfect gas is described by the *equation of state*

$$p = \rho RT \tag{6}$$

where, R is the universal gas constant, ρ the density of the gas.

It turns out to be useful to introduce the *enthalpy* h, which is a *Legendre transform* of $e = e(v, T)$, as per

$$h \equiv e + pv. \tag{7}$$

So, we have

$$h = h(p, T). \tag{8}$$

4 This may be seen by considering a perfect gas for which, equation (5) becomes, on using (6)-(10) below,

$$\begin{aligned} dq &= C_v dT + p dv = \frac{C_v}{R} d(pv) + p dv \\ &= \left(1 + \frac{C_v}{R}\right) p dv + \frac{C_v}{R} v dp = \frac{C_p}{R} p dv + \frac{C_v}{R} v dp. \end{aligned}$$

This shows that dq is not an exact differential because

$$\frac{\partial}{\partial p}\left(\frac{C_p}{R} p\right) \neq \frac{\partial}{\partial v}\left(\frac{C_v}{R} v\right)$$

on account of $C_p \neq C_v$. Therefore, Q is not a variable of state.

5 A reversible process is one which, once having taken place, can be reversed and leaves no change in either the system or the surroundings.

One may define the *specific heats at constant volume* and *constant pressure* according to

$$\left.\begin{aligned} C_v &= \left(\frac{\partial e}{\partial T}\right)_v \\ C_p &= \left(\frac{\partial h}{\partial T}\right)_p \end{aligned}\right\}. \tag{9}$$

Using (6)-(9), one has for a perfect gas,

$$C_p = C_v + R. \tag{10}$$

For an *adiabatic* and *reversible* (also called *isentropic*, see Section 12.4) process, one has from (5) and (7),

$$\left.\begin{aligned} de &= -pdv \\ dh &= vdp \end{aligned}\right\} \tag{11}$$

or

$$\left.\begin{aligned} \frac{\partial e}{\partial v}dv + \frac{\partial e}{\partial T}dT &= -pdv \\ \frac{\partial h}{\partial p}dp + \frac{\partial h}{\partial T}dT &= vdp. \end{aligned}\right\} \tag{12}$$

Using (9), (12) leads to

$$\left.\begin{aligned} \frac{\partial T}{dv} &= -\frac{1}{C_v}\left(\frac{\partial e}{\partial v} + p\right) \\ \frac{\partial T}{dp} &= -\frac{1}{C_p}\left(\frac{\partial h}{\partial p} - v\right) \end{aligned}\right\}. \tag{13}$$

For a perfect gas given by equation (6), we have $e = e(T)$ and $h = h(T)$. So, equation (13) yields

$$\left.\begin{aligned} \frac{v}{T}\frac{dT}{dv} &= -\frac{R}{C_v} \\ \frac{p}{T}\frac{dT}{dp} &= \frac{R}{C_p} \end{aligned}\right\}. \tag{14}$$

From (14), we obtain,

$$\frac{v}{p}\frac{dp}{dv} = \frac{C_p}{C_v} \tag{15}$$

which gives,

$$pv^{\gamma} = const \tag{16}$$

where γ is the *ratio of specific heats*,[6]

$$\gamma \equiv \frac{C_p}{C_v}.$$

Let us next consider irreversible processes. For an isolated system, one has for the two end states denoted by A and B,

$$e_B = e_A. \tag{17}$$

For an adiabatic flow, on the other hand, one has

$$h_B = h_A \text{ or } e_B + p_B v_B = e_A + p_A v_A. \tag{18}$$

12.3 The Second Law of Thermodynamics

The *Second Law* expresses the common fact that heat does not travel uphill. More precisely, we have two versions formally summarizing this fact:

Clausius Statement: There is no thermodynamic transformation whose sole effect is to deliver heat from a reservoir at lower temperature to a reservoir at higher temperature.
Kelvin Statement: There is no thermodynamic transformation whose sole effect is to extract heat from a reservoir and convert it entirely into work.

Kelvin's statement immediately implies that we need at least two reservoirs at different temperatures T_1 and T_2 in order to deliver work by extracting heat from a reservoir. This constitutes the *Carnot cycle*. The *Second Law* also implies that only part of the heat that is extracted by the system at $T_2 > T_1$ is transformed into work by the Carnot cycle. The rest of the heat Q_1, instead of being transformed into work, is returned to the reservoir at T_1 (see Figure 12.1). So, the *Carnot engine* cannot be 100% efficient. Note that the Carnot engine may

6 γ is also a measure of the relative internal complexity of the gas molecules. According to the *kinetic theory of gases*, γ is related to the degrees of freedom of the gas molecules n,

$$\gamma = (n + 2)/n.$$

At normal temperatures, there are 6 possible degrees of freedom - 3 translational, 3 rotational.

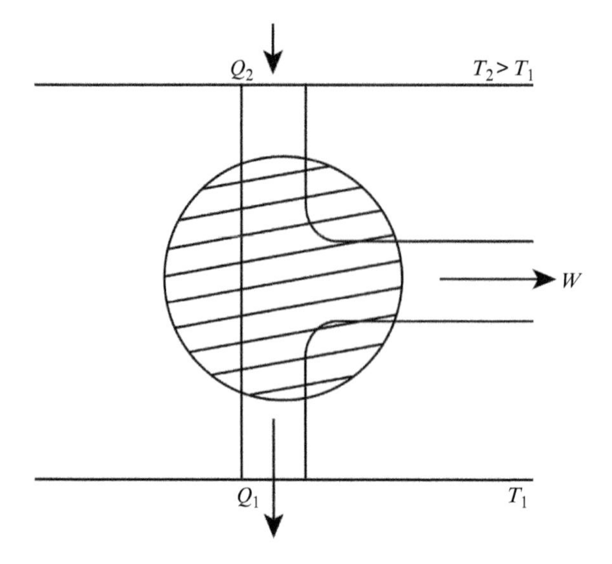

Figure 12.1 The Carnot engine.

be run in reverse as a *refrigerator* by reversing the signs (or transfer directions) of Q_1 and Q_2, and hence W.

12.4 Entropy

The evolution of a system is dictated by certain natural tendencies, and all natural or spontaneous processes are irreversible. There is considerable departure from equilibrium in the natural processes. By contrast, a system remains in equilibrium during a reversible process. In this context, one introduces a new variable of state, called the *entropy*[7] S, which determines whether or not a state is in stable equilibrium.

We first note that the internal energy has the character of a potential energy. For an adiabatic and reversible change of the state of a system, one has from equation (4),

$$p = -\left(\frac{\partial E}{\partial V}\right)_S. \tag{19}$$

7 *Entropy*, from Greek, roughly implies evolution.

The equation of state

$$E = E(V, S) \tag{20}$$

gives

$$dE = \frac{\partial E}{\partial V} dV + \frac{\partial E}{\partial S} dS. \tag{21}$$

On using (19) and defining

$$T \equiv \left(\frac{\partial E}{\partial S} \right) \tag{22}$$

equation (21) becomes

$$dE = -pdV + TdS. \tag{23}$$

On comparing (23) with (4), we have

$$TdS = dQ. \tag{24}$$

Due to the manner of its definition, S, which is a variable of state,[8] is relevant only for a system in equilibrium, or for reversible changes of state. The *Second Law* states that, for an irreversible process,[9] the entropy can only increase, i.e.,

$$S_B - S_A \geq \int_A^B \frac{dQ}{T} \tag{25}$$

where equality prevails for reversible processes.

8 This may be seen by noting (see Footnote 4) that, for a perfect gas

$$dS = \frac{dq}{T} = \frac{C_p}{v} dv + \frac{C_v}{p} dp$$

which shows that dS is a perfect differential because

$$\frac{\partial}{\partial p} \left(\frac{C_p}{v} \right) = \frac{\partial}{\partial v} \left(\frac{C_v}{p} \right) = 0.$$

So, T is an *integrating factor* that turns dQ into an *exact differential* dQ/T.

9 The *Second Law* does not provide a complete formulation to address arbitrary processes.

A system reaches a state of equilibrium if no further spontaneous processes are possible for which we have,

$$dS \geq \frac{dQ}{T}.$$

So, a system is in stable equilibrium if for every process compatible with the constraints of the system, we have,

$$\delta S \leq \frac{\delta Q}{T}.$$

For an isolated system, this gives

$$\delta S \leq 0$$

i.e., the entropy for an isolated system reaches a maximum. Conversely, the state of maximum entropy is the most stable state for an isolated system.

One obtains from equation (23),

$$S = \int \frac{de}{T} + \int p\frac{dv}{T} + const \tag{26}$$

Considering a perfect gas given by (6) and using (9), equation (26) leads to

$$S = \int C_v \frac{dT}{T} + R\, ln\, v + const. \tag{27}$$

or alternatively, using (8) and (9),

$$S = \int C_p \frac{dT}{T} - R\, ln\, p + const. \tag{28}$$

If the specific heats C_p and C_v are constants, then (28) leads to

$$S_2 - S_1 = C_p\, ln \left(\frac{T_2}{T_1}\right)\left(\frac{p_2}{p_1}\right)^{-(\gamma-1)/\gamma}. \tag{29}$$

Next, considering a perfect gas given by (6) and using (9), (23) leads to

$$dS = C_v \frac{dp}{p} + C_p \frac{dv}{v} \tag{30}$$

and on integrating, (30) gives

$$ln\frac{p}{p_0} + \gamma ln\frac{v}{v_0} = \frac{S - S_0}{C_v}. \tag{31}$$

In terms of the mass density $\rho \equiv 1/v$, (31) becomes

$$p = k\rho^{\gamma} \tag{32}$$

where,

$$k \equiv \frac{p_0}{\rho_0^\gamma} e^{(S-S_0)/C_v}.$$

Note that, for an isentropic process, (32) gives

$$p \sim \rho^\gamma \tag{33}$$

in agreement with (16).

12.5 Liquid and Gaseous Phases

A given material may exist in the liquid phase for some values of the two parameters of state (p and v, say) and in the gaseous phase for other values. The occurrence of these two distinct phases is influenced by the manner in which the intermolecular force varies with molecular spacing.

At low temperatures, some compression can bring the molecules in a gas so close together that in many cases actual condensation to the liquid state will occur. Concomitantly, the formation of a free surface separating vapor and liquid can form. However, above a certain temperature, no amount of compression can produce such a condensation.

Figure 12.2 shows the *isothermals* for CO_2. The lower curves all pass through a horizontal step which signifies a change in volume at constant pressure. To the left of this step, one has a liquid state, wherein it takes a large increase in pressure to produce a small change in volume implying that liquids are nearly incompressible. To the right is the vapor or gaseous region, and these curves eventually become hyperbolas $pv = const$. The constant pressure on an isothermal through the polyphase region is the saturated vapor pressure p_v. This is the pressure existing in pure vapor which is in contact with the liquid at the given temperature. The effect of reducing the pressure of a liquid below the saturated vapor pressure is of importance in fluid dynamics, since this leads to the formation of vapor packets distributed throughout the liquid. The occurrence of such vapor packets, called *cavitation*, has important mechanical consequences on objects placed in such flows (see Section 5.5).

At $31.5^\circ C$ for CO_2, the constant pressure step is reduced to a pause in passing. Above this temperature, the isothermal passes steadily from low to high pressure without any condensation to the liquid state. $T_c = 31.5^\circ C$ is called the *critical* temperature for CO_2. The isothermals for $T \gg T_c$ become *hyperbolae*, because the gas behaves more and more like an *ideal gas* ($pv = const$).

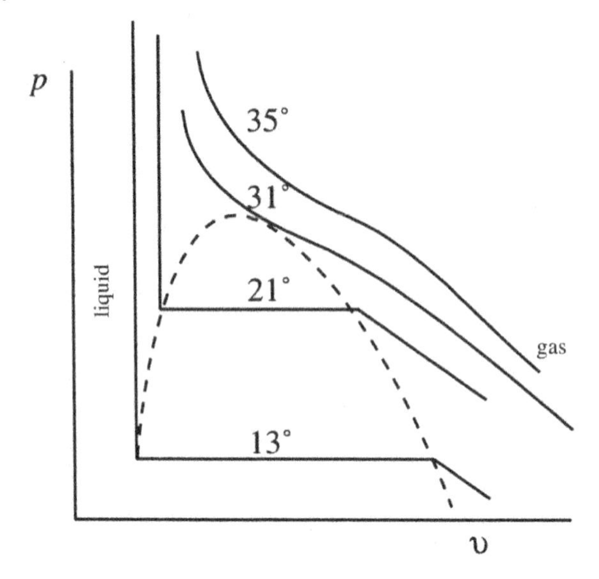

Figure 12.2 Isothermals for CO_2.

Exercise

1. Show that the *Clausius* and the *Kelvin statements* of the *Second Law* are equivalent.

13

Isentropic Fluid Flows

Liquids do not change their density due to changes in pressure and temperature. So, for all practical purposes they can be treated as incompressible fluids. On the contrary, gases experience density changes when pressure and temperature vary, and hence are compressible. This necessitates coupling thermodynamics with hydrodynamics to describe the dynamics of gases. The pressure changes in a gas get propagated as sound waves via a physical mechanism similar to that of elastic waves in a long slender bar. If the variations in the flow properties are small, then this process may be assumed to be adiabatic and reversible (i.e. isentropic). More importantly, fluid compressibility has a profound effect on the streamlines in the fluid. When the flow velocity exceeds the velocity of sound, the entire character of the flow dynamics changes. An ingenious application of this aspect is the *de Laval* nozzle, where the gas is accelerated to supersonic speeds.

13.1 Applications of Thermodynamics to Fluid Flows

Classical thermodynamics deals with a uniform, static system which is specified by certain variables of state, and exchanges energy with its surroundings. While applying classical thermodynamics to fluid flows, a fluid particle, assumed to be small, and hence uniform, is taken to be the system in *local* thermodynamic equilibrium. Strictly speaking, the conditions of equilibrium cannot be attained in a real, nonuniform flow. This is because it takes time for a fluid particle to adjust itself continuously to the new conditions that it encounters. The rate at which the adjustments must be made depends on the gradients in the flow and is a measure of the departure from equilibrium. One may then ascribe state variables such as T, ρ, p, h, s, etc. to the fluid particle and, as before, relate them to each other by equations of state. If the fluid is compressible, then p is the thermodynamic pressure and is a full-fledged

Introduction to Theoretical and Mathematical Fluid Dynamics, Third Edition.
Bhimsen K. Shivamoggi.

dynamical variable. On the contrary, if the fluid is incompressible, then p is simply a passive variable and plays no dynamical role.

13.2 Linear Sound Wave Propagation

In an incompressible fluid, a pressure disturbance is transmitted instantaneously to all points in the fluid (i.e., at almost infinte speed). However, in a compressible fluid a pressure disturbance travels at a finite speed since the inertia of the displaced fluid particles hinders the transmission of the disturbance.

When the pressure disturbance is small, one may use linearized equations to describe the evolution of such a disturbance. These equations are given below,

- mass conservation:

$$\frac{\partial \rho}{\partial t} + \rho_o \frac{\partial V}{\partial x} = 0 \tag{1}$$

- momentum conservation:

$$\rho_o \frac{\partial V}{\partial t} = -\frac{\partial p}{\partial x} = -a^2 \frac{\partial \rho}{\partial x}. \tag{2}$$

Here, a is the speed of sound in the gas given by,

$$a \equiv \sqrt{\left(\frac{dp}{d\rho}\right)_s}$$

assuming that the sound wave propagation is an isentropic process.

The disturbances produced in a fluid by a sound wave are so small that each fluid particle undergoes a nearly isentropic process. Besides, the frequencies of the sound wave are assumed to be small enough to prevent large departures from thermodynamic equilibrium during wave propagation.

Equations (1) and (2) lead to

$$\frac{\partial^2 V}{\partial t^2} = a^2 \frac{\partial^2 V}{\partial x^2} \tag{3}$$

describing the sound wave propagation in a compressible fluid.

13.3 The Energy Equation

The conservation of energy of a fluid particle in an adiabatic flow may be expressed by the energy equation,

$$h + \frac{1}{2}V^2 = const. \tag{4}$$

For a perfect gas, one has

$$h = C_p T = \frac{C_p}{R}\frac{p}{\rho} = \frac{\gamma}{\gamma - 1}\frac{p}{\rho} = \frac{a^2}{\gamma - 1} \tag{5}$$

where the speed of sound a is now given by

$$a \equiv \sqrt{\left(\frac{dp}{d\rho}\right)_S} = \sqrt{\frac{\gamma p}{\rho}} = \sqrt{\gamma RT}. \tag{6}$$

Using (5), equation (4) gives

$$\frac{a^2}{\gamma - 1} + \frac{V^2}{2} = const = \frac{a_0^2}{\gamma - 1} \tag{7}$$

where the subscript 0 refers to the *stagnation* values. Bringing the fluid particle to rest isentropically results in these stagnation values. These are different from the values (called the *static* properties) measured by an observer moving with the particle. This is due to the exchanges occurring between the kinetic energy, the internal energy, and the potential energy (due to the pressure) of the fluid particle with those of the surrounding fluid particles.

One obtains from equation (7), upon using (6),

$$\frac{T_0}{T} = \left(1 + \frac{\gamma - 1}{2}M^2\right) \tag{8}$$

where, M is the Mach number,

$$M \equiv \frac{V}{a}.$$

Using (29) and (32) in Chapter 12, one obtains the following relations,

$$\frac{p_0}{p} = \left(\frac{T_0}{T}\right)^{\gamma/(\gamma-1)} = \left(1 + \frac{\gamma - 1}{2}M^2\right)^{\gamma/(\gamma-1)},$$

$$\frac{\rho_0}{\rho} = \left(\frac{p_0}{p}\right)^{1/\gamma} = \left(1 + \frac{\gamma - 1}{2}M^2\right)^{1/(\gamma-1)}. \tag{9a,b}$$

If $M \ll 1$, (9a,b) leads to the following expansion,

$$\frac{p_0}{p} = 1 + \frac{1}{2}\gamma M^2 + \frac{1}{8}\gamma M^4 + \cdots$$

or

$$p_0 - p = \frac{1}{2}\rho v^2 \left(1 + \frac{1}{4}M^2\right) + \cdots \tag{10}$$

so that, when $M = 0$, one obtains

$$p_0 \approx p + \frac{1}{2}\rho v^2 \tag{11}$$

as for the incompressible flow.

Let $a = a^*$ (called the *critical sound speed*) correspond to the location where $V = a$, then equation (7) gives for a^*,

$$a^{*2} = \frac{2}{\gamma + 1} a_0^2. \tag{12}$$

13.4 Stream-Tube Area and Flow Velocity Relations

The gas flows in propulsive devices such as jet engines undergo either acceleration or deceleration. The variable-area passages such as *de Laval* nozzles convert the thermal energy of the gas into the kinetic energy of the high-speed propulsive jet. While nozzles accelerate the flows, diffusers decelerate them. The initial velocity of the gas (such as subsonic ($M < 1$) or supersonic ($M > 1$)) determines whether the variable-area passage accelerates or decelerates the flow. This is due to the fact that the contour of the flow section is the same for both nozzles and diffusers.

It is of interest to note that a *de Laval*-type nozzle mechanism (Parker, 2001; see also Shivamoggi, 2020) was surmised to be implicit in Parker's hydrodynamic model (Parker, 1958) to describe the solar wind flow. This model provides for a mechanism to enable the solar wind to accelerate continuously from subsonic speeds at the sun to supersonic speeds away from the sun.

The conservation of mass for steady flow in a stream tube requires

$$\rho V A = const \tag{13}$$

where A is the cross-sectional area of the stream tube. Alternatively, one may write equation (13) as

$$\frac{d\rho}{\rho} + \frac{dV}{V} + \frac{dA}{A} = 0. \tag{14}$$

The equation of motion of the fluid in the stream tube, assuming the flow to be one-dimensional, is

$$V \frac{dV}{dx} = -\frac{1}{\rho}\frac{dp}{dx} = -\frac{a^2}{\rho}\frac{d\rho}{dx} \tag{15}$$

where x is the distance along the stream tube.

Using equation (15), equation (14) gives

$$\frac{dA}{A} = \frac{dV}{V}\left(M^2 - 1\right) \tag{16}$$

which shows that, in subsonic flow an increase in speed is produced by a decrease in the stream-tube area, and in a supersonic flow an increase in speed is produced by an increase in area. This is so because in supersonic flows, the decrease in density outweighs the increase in stream-tube area. Furthermore, note that the acceleration of a fluid from a subsonic to a supersonic speed requires that the flow passes through a *throat*, where the flow is sonic ($M = 1$).

Noting, from (7) and (29) in Chapter 12 that for an isentropic flow we have,

$$\frac{p}{p_0} = \left(\frac{T}{T_0}\right)^{\gamma/(\gamma-1)} = \left[1 - \frac{\gamma-1}{2}\left(\frac{V}{a_0}\right)^2\right]^{\gamma/(\gamma-1)}. \tag{17}$$

We obtain from (17),

$$V = \sqrt{\frac{2\gamma RT_0}{\gamma-1}\left[1 - \left(\frac{p}{p_0}\right)^{(\gamma-1)/\gamma}\right]}. \tag{18}$$

The mass flux G is then given by

$$G \equiv \rho V = \rho_0 \sqrt{\frac{2\gamma RT_0}{\gamma-1}\left[\left(\frac{p}{p_0}\right)^{2/\gamma} - \left(\frac{p}{p_0}\right)^{(\gamma+1)/\gamma}\right]} \tag{19}$$

which reaches its maximum, when

$$\left(\frac{p}{p_0}\right)_c : \frac{dG}{d(p/p_0)} = 0. \tag{20}$$

Equation (20) leads to

$$\left(\frac{p}{p_0}\right)_c = \left(\frac{2}{\gamma+1}\right)^{\gamma/(\gamma-1)}. \tag{21}$$

Using (21) and (12), (18) yields

$$V_c = \sqrt{\frac{2\gamma RT_0}{\gamma+1}} = a^* \tag{22}$$

i.e., the flow becomes *sonic*. On the other hand, in order to show that the sonic flow conditions prevail at the throat, we have on using equation (32) in Chapter 12 and equation (7) that

$$a^2 = \frac{dp}{d\rho} = k\gamma\rho^{\gamma-1} = a_0^2 - \frac{1}{2}(\gamma-1)V^2. \tag{23}$$

Using (23), equation (13) gives

$$A^{\gamma-1}\left[a_0^2 - \frac{1}{2}(\gamma-1)V^2\right]V^{\gamma-1} = const. \tag{24}$$

Differentiating equation (24) with respect to V, one obtains

$$(\gamma-1)\,A^{\gamma-2}\frac{dA}{dV}\left[a_0^2-\frac{1}{2}(\gamma-1)\,V^2\right]V^{\gamma-1}+$$
$$+\,A^{\gamma-1}\left[(\gamma-1)\,a_0^2V^{\gamma-2}-\frac{1}{2}\left(\gamma^2-1\right)V^{\gamma}\right]=0 \qquad (25)$$

which shows that A is extremized, i.e.,

$$\frac{dA}{dV}=0 \qquad (26)$$

when

$$a_0^2=\frac{1}{2}(\gamma+1)\,V^2 \text{ or } V=a^*.$$

Furthermore, at the location, $V=a^*$, (25) gives, upon differentiating again,

$$V=a^*:\ \frac{d^2A}{dV^2}\left[a_0^2-\frac{1}{2}(\gamma-1)\,V^2\right]V-A\,(\gamma+1)\,V=0$$

from which,

$$V=a^*:\ \frac{d^2A}{dV^2}=\frac{A\,(\gamma+1)}{a_0^2-\frac{1}{2}(\gamma-1)\,a^{*2}}=\frac{A\,(\gamma+1)^2}{2a_0^2}>0. \qquad (27)$$

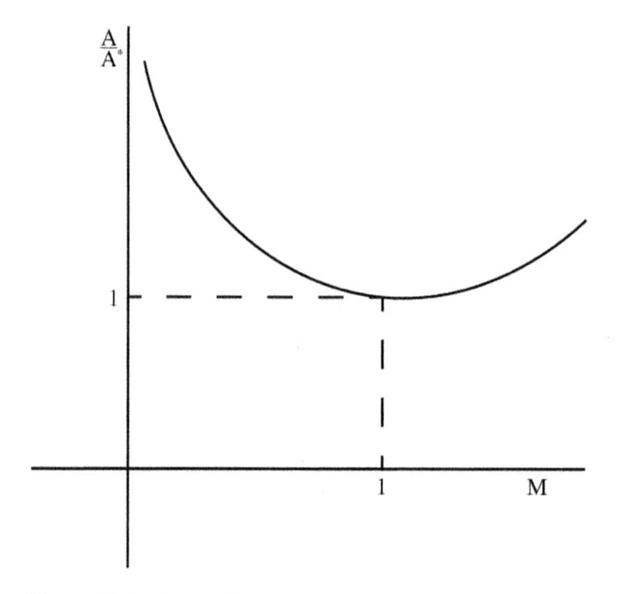

Figure 13.1 Nozzle flow.

Thus, A indeed attains a minimum at $V = a^*$. Next, noting

$$\rho VA = \rho_c a^* A^* = \left(\frac{2}{\gamma+1}\right)^{\frac{1}{2}\left(\frac{\gamma+1}{\gamma-1}\right)} \rho_0 a_0 A^*$$

one obtains

$$\frac{A}{A^*} = \frac{1}{M}\left[\frac{2}{\gamma+1}\left(1+\frac{\gamma-1}{2}M^2\right)\right]^{(\gamma+1)/2(\gamma-1)}. \tag{28}$$

For a given value of (A/A^*), in general, (28) gives two values of M, one subsonic and the other supersonic. However, at $A/A^* = 1$, the two roots for M coalesce at $M = 1$ (see Figure 13.1).

If $A_{min} > A^*$, then V decreases for $A > A_{min}$ and the gas flow remains subsonic everywhere. On the other hand, if $A_{min} < A^*$, then the flow *chokes* and sonic conditions shift to A_{min}.

Exercises

1. Show that for very small Mach-number flows, the changes in the flow-velocity are much larger than those in the speed of sound in the fluid. On the other hand, show that the converse is true for very large Mach-number flows.
2. Show that for an isothermal flow ($a = const$ or $\gamma \to 1$), the area ratio of a nozzle becomes

$$\frac{A}{A_*} = \frac{1}{M}e^{(M^2-1)/2}.$$

(One may obtain this result either from equation (16) or directly from (28)). For $M \approx 1$, deduce from this result that

$$\frac{A}{A_*} \approx 1 + (M-1)^2 + O(M-1)^4,$$

so that A has a minimum at $M = 1$.

14

Potential Flows

In this chapter, the general equations governing the flows of an ideal gas are established. We will discuss some mathematical properties like characteristics in the hyperbolic case. Additional information on the dynamics of gas flows can be obtained by reformulating the governing equations in natural coordinates. We consider conical flows, and as an example, discuss the Prandtl-Meyer flow. For the case of a body placed in the flow producing small disturbances, the method of solution uses small perturbation theory.

14.1 Governing Equations

The equations expressing the conservation of mass, momentum, and energy are

$$\frac{\partial \rho}{\partial t} + \frac{\partial}{\partial x_j}(\rho v_j) = 0 \tag{1}$$

$$\frac{\partial v_i}{\partial t} + v_j \frac{\partial v_i}{\partial x_j} = -\frac{1}{\rho}\frac{\partial p}{\partial x_i} \tag{2}$$

$$\frac{\partial}{\partial t}\left(e + \frac{v^2}{2}\right) + v_j \frac{\partial}{\partial x_j}\left(e + \frac{v^2}{2}\right) = -\frac{1}{\rho}\frac{\partial}{\partial x_i}(p v_i). \tag{3a}$$

Equation (3a) may be expressed alternatively as

$$\frac{\partial e}{\partial t} + v_j \frac{\partial e}{\partial x_j} = -\frac{p}{\rho}\frac{\partial v_i}{\partial x_i} \tag{3b}$$

$$\frac{\partial}{\partial t}\left(h + \frac{v^2}{2}\right) + v_j \frac{\partial}{\partial x_j}\left(h + \frac{v^2}{2}\right) = \frac{1}{\rho}\frac{\partial p}{\partial t} \tag{3c}$$

$$\frac{\partial h}{\partial t} + v_j \frac{\partial h}{\partial x_j} = \frac{1}{\rho}\left(\frac{\partial p}{\partial t} + v_j \frac{\partial p}{\partial x_j}\right). \tag{3d}$$

Introduction to Theoretical and Mathematical Fluid Dynamics, Third Edition.
Bhimsen K. Shivamoggi.

For a *potential* flow, one may write

$$v_i = \frac{\partial \Phi}{\partial x_i} \tag{4}$$

then equation (1) becomes

$$\frac{1}{\rho}\frac{\partial \rho}{\partial t} + \frac{\partial^2 \Phi}{\partial x_i \partial x_i} + \frac{1}{\rho} v_j \frac{\partial \rho}{\partial x_j} = 0. \tag{5}$$

Equation (2) may be rewritten as

$$-\frac{v_j}{a^2}\left(\frac{\partial v_j}{\partial t} + v_k \frac{\partial v_j}{\partial x_k}\right) = \frac{v_j}{a^2 \rho}\frac{\partial p}{\partial x_j} = \frac{v_j}{\rho}\frac{\partial \rho}{\partial x_j} \tag{6}$$

a being the speed of sound,

$$a \equiv \sqrt{(dp/d\rho)_s}.$$

On the other hand, the *Bernoulli integral* of equation (2) (see Section 5.3)

$$\frac{\partial \Phi}{\partial t} + \frac{v^2}{2} + \int \frac{dp}{\rho} = const \tag{7}$$

gives

$$\frac{\partial^2 \Phi}{\partial t^2} + v_j \frac{\partial v_j}{\partial t} + \frac{a^2}{\rho}\frac{\partial \rho}{\partial t} = 0. \tag{8}$$

Using equations (6) and (8), equation (5) gives the potential-flow equation

$$\frac{\partial^2 \Phi}{\partial x_i^2} = \frac{1}{a^2}\frac{\partial^2 \Phi}{\partial t^2} + \frac{2}{a^2} v_i \frac{\partial^2 \Phi}{\partial t \partial x_i} + \frac{1}{a^2} v_i v_j \frac{\partial^2 \Phi}{\partial x_i \partial x_j} \tag{9a}$$

which in cartesian coordinates becomes,

$$\left(1 - \frac{\Phi_x^2}{a^2}\right)\phi_{xx} + \left(1 - \frac{\Phi_y^2}{a^2}\right)\Phi_{yy} + \left(1 - \frac{\Phi_z^2}{a^2}\right)\Phi_{zz} - 2\frac{\Phi_x \Phi_y}{a^2}\Phi_{xy}$$
$$- 2\frac{\Phi_y \Phi_z}{a^2}\Phi_{yz} - 2\frac{\Phi_z \Phi_x}{a^2}\Phi_{zx} = \frac{1}{a^2}\left(\Phi_{tt} + 2\Phi_x \Phi_{xt} + 2\Phi_y \Phi_{yt} + 2\Phi_z \Phi_{zt}\right). \tag{9b}$$

For an isentropic process with the relation,

$$p/\rho^{\gamma} = const \tag{10}$$

Equation (7) leads to

$$a^2 + \frac{\gamma - 1}{2}\left(\Phi_x^2 + \Phi_y^2 + \Phi_z^2\right) + (\gamma - 1)\Phi_t = const = a_0^2. \tag{11}$$

14.2 Streamline Coordinates

Consider a two-dimensional flow expressed in *streamline coordinates* (see Figure 14.1). The equations expressing the conservation of mass and momentum in a steady flow are

$$\frac{1}{\rho}\frac{\partial \rho}{\partial s} + \frac{1}{V}\frac{\partial V}{\partial s} + \frac{1}{\triangle n}\frac{\partial \triangle n}{\partial s} = 0 \tag{12}$$

$$\rho V \frac{\partial V}{\partial s} = -\frac{\partial p}{\partial s} \tag{13}$$

$$-\frac{\rho V^2}{R} = -\frac{\partial p}{\partial n} \tag{14}$$

R being the radius of curvature of the streamline in question.

Noting (see Figure 14.1) from geometry that

$$\triangle s \frac{\partial \triangle n}{\partial s} = \left(\frac{\partial \theta}{\partial n} \triangle n\right) \triangle s \tag{15}$$

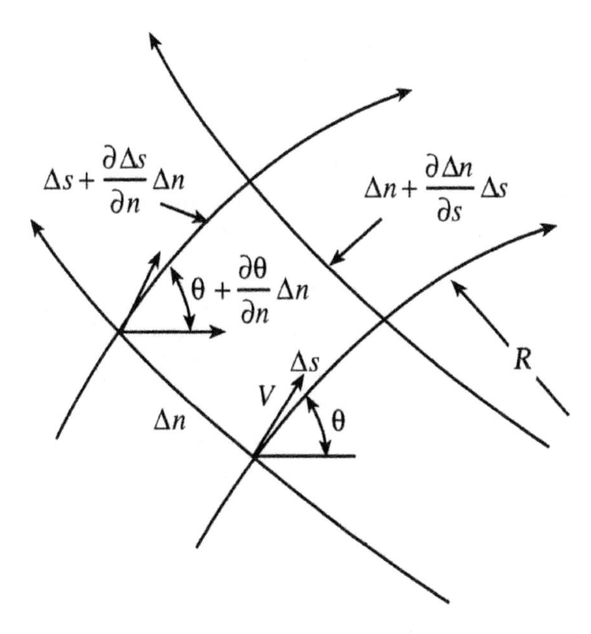

Figure 14.1 The streamline coordinate system.

equations (12) and (13) give

$$\left(\frac{V^2}{a^2} - 1\right)\frac{1}{V}\frac{\partial V}{\partial s} - \frac{\partial \theta}{\partial n} = 0. \tag{16}$$

Next, the energy-conservation equation is given by (see equations (7) and (23) in Chapter 12)

$$\frac{\partial h}{\partial n} = T\frac{\partial S}{\partial n} + \frac{1}{\rho}\frac{\partial p}{\partial n} \tag{17a}$$

and using equation (14) the above equation becomes,

$$\frac{\partial h_0}{\partial n} = T\frac{\partial S}{\partial n} + V\Omega \tag{17b}$$

where h_0 is the stagnation enthalpy, and Ω is the vorticity,

$$\Omega \equiv \frac{\partial V}{\partial n} - \frac{V}{R}. \tag{18}$$

Equation (17b) shows that the variation of total enthalpy and entropy across streamlines is related to the vorticity of the flow. An example of a flow with vorticity is the flow downstream of a *curved shock wave*. The inclination of the shock wave relative to the oncoming flow determines its strength (see Chapter 16). This influences the entropy change experienced by the fluid. The entropy then continually varies across the streamlines on the downstream side. This leads to the production of vorticity in the flow there (*Crocco, 1937*).

14.3 Conical Flows: Prandtl-Meyer Flow

In plane polar coordinates for a steady flow, equation (9a) becomes

$$\left(a^2 - \frac{\Phi_\theta^2}{r^2}\right)\frac{\Phi_{\theta\theta}}{r^2} - 2\Phi_r\Phi_\theta\frac{\Phi_{r\theta}}{r^2} + \left(a^2 - \Phi_r^2\right)\Phi_{rr} + \frac{\Phi_r}{r}\left(a^2 + \frac{\Phi_\theta^2}{r^2}\right) = 0. \tag{19}$$

For a conical flow, the flow properties are constant on the rays from the corner which is taken to be the origin. One then has for such flows,[1]

$$\frac{\partial}{\partial r}V_r = 0, \quad \frac{\partial}{\partial r}V_\theta = 0 \tag{20a}$$

or

$$\Phi_{rr} = 0, \quad \frac{1}{r}\Phi_\theta = \Phi_{r\theta}. \tag{20b}$$

---

1 This type of flow typically prevails when there is no characteristic length scale in the problem.

It follows, from (20b), that

$$\Phi(r,\theta) = r\varphi(\theta) \tag{21}$$

and, equation (19) gives

$$\left(\Phi_r + \frac{\Phi_{\theta\theta}}{r}\right)\left(a^2 - \frac{\Phi_\theta^2}{r^2}\right) = 0 \tag{22}$$

from which,

$$\frac{\Phi_\theta}{r} = a. \tag{23a}$$

Note that (see Figure 14.2)

$$\Phi_r = \sqrt{V^2 - \frac{\Phi_\theta^2}{r^2}} = \sqrt{V^2 - a^2} = V\sqrt{1 - \frac{a^2}{V^2}} = V\cos\mu \tag{23b}$$

where

$$\sin\mu \equiv \frac{a}{V} \tag{24}$$

so that the radius vector intersects the streamline at the Mach angle and so must be a *Mach line*. The *Prandtl-Meyer flow* (Prandtl, 1936) (Figure 14.3) is such an example. This flow consists of two regions of uniform flow separated by a fan-shaped region of expansion where the positive *characteristics* (see Chapter 15) are all straight lines through the corner O.

Note that using (23a), one obtains from energy conservation (see equation (7) in Chapter 13)

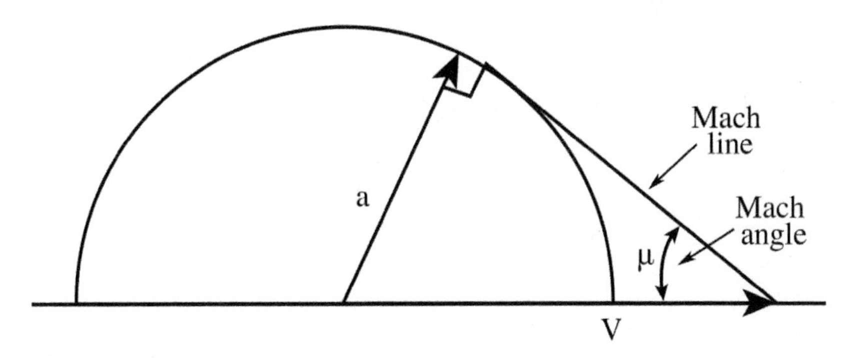

Figure 14.2 The Mach angle.

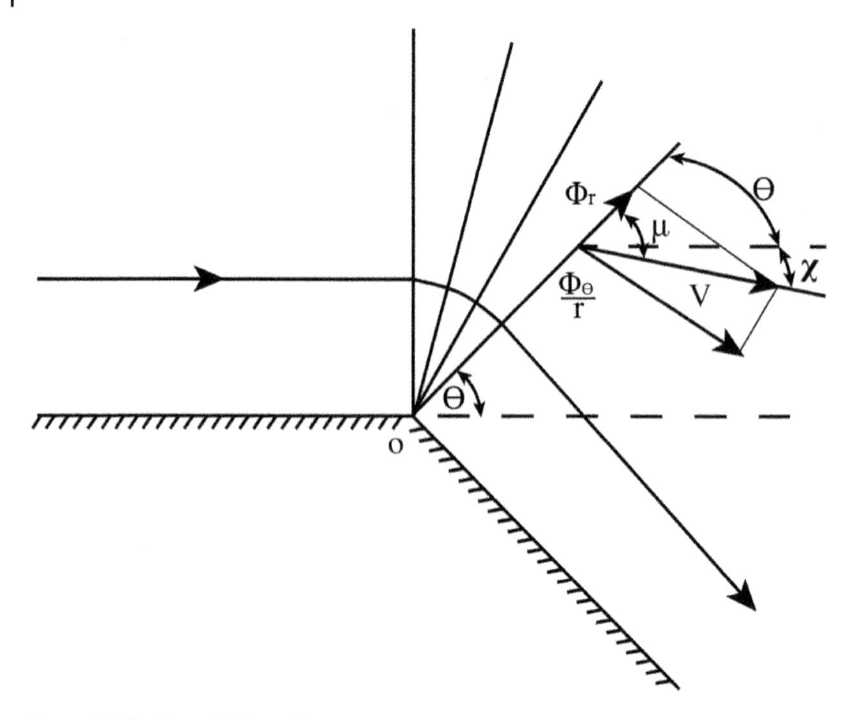

Figure 14.3 Prandtl-Meyer flow.

$$\frac{\Phi_\theta^2}{r^2} = \left(\frac{\gamma - 1}{\gamma + 1}\right)\left(V_{max}^2 - \Phi_r^2\right). \tag{25}$$

Using (20b) and (23a), and the boundary condition,

$$\theta = \frac{\pi}{2} \ : \ \Phi_r = 0 \tag{26}$$

equation (25) leads to

$$\left.\begin{aligned} \Phi_r &= V_{max} \sin \alpha \\ \frac{\Phi_\theta}{r} &= a = V_{max}\sqrt{\frac{\gamma - 1}{\gamma + 1}} \cos \alpha \end{aligned}\right\} \tag{27}$$

where

$$\alpha \equiv \sqrt{\frac{\gamma - 1}{\gamma + 1}}\left(\frac{\pi}{2} - \theta\right).$$

Note that, according to (27), the streamlines become radial when $\Phi_\theta = 0$, i.e.,

$$\theta = \theta_{max} = -\frac{\pi}{2}\left(\sqrt{\frac{\gamma+1}{\gamma-1}} - 1\right). \tag{28}$$

If the angle turned through by the wall is larger than θ_{max}, then a region of zero pressure will form between the fluid and the wall. Referring to Figure 14.3, the angle turned by a streamline from $M = 1$ to $M > 1$ is given by

$$\chi = \mu - \theta = \tan^{-1}\left(\frac{\Phi_\theta / r}{\Phi_r}\right) - \theta. \tag{29}$$

On using (27), (29) becomes

$$\chi = \tan^{-1}\left[\sqrt{\frac{\gamma-1}{\gamma+1}}\cot\alpha\right] - \theta. \tag{30}$$

Using the identity,

$$\tan^{-1}\beta + \tan^{-1}(1/\beta) = \frac{\pi}{2} \tag{31}$$

in (30), we obtain,

$$\chi = \left(\frac{\pi}{2} - \theta\right) - \tan^{-1}\left[\sqrt{\frac{\gamma+1}{\gamma-1}}\tan\alpha\right]. \tag{32}$$

On using (27) again, we have

$$M^2 = \frac{\sin^2\alpha + \left(\frac{\gamma-1}{\gamma+1}\right)\cos^2\alpha}{\left(\frac{\gamma-1}{\gamma+1}\right)\cos^2\alpha} \tag{33a}$$

from which,

$$\tan\alpha = \sqrt{\left(\frac{\gamma-1}{\gamma+1}\right)(M^2-1)}. \tag{33b}$$

Using (33b), (32) becomes

$$\chi = \sqrt{\frac{\gamma+1}{\gamma-1}}\tan^{-1}\sqrt{\frac{\gamma-1}{\gamma+1}(M^2-1)} - \tan^{-1}\sqrt{M^2-1}. \tag{34}$$

The maximum turning angle which corresponds to $M \Rightarrow \infty$ is then given by

$$\chi_{max} = \frac{\pi}{2}\left(\sqrt{\frac{\gamma+1}{\gamma-1}} - 1\right). \tag{35a}$$

So, we have from (30) and (35a),

$$\theta_{max} = -\chi_{max} = -\frac{\pi}{2}\left(\sqrt{\frac{\gamma+1}{\gamma-1}} - 1\right) \tag{35b}$$

in agreement with (28), as expected.

14.4 Small Perturbation Theory

Consider small disturbances introduced into a steady stream by a small body placed in it. Let

$$\Phi = U_\infty x + \phi(x, y, z). \tag{36}$$

Using (36), equation (9b) gives on linearization in ϕ

$$\left(1 - M_\infty^2\right)\phi_{xx} + \phi_{yy} + \phi_{zz} = 0 \tag{37}$$

where

$$M_\infty \equiv \frac{U_\infty}{a_\infty}$$

and the subscript ∞ denotes conditions in the free stream.

The pressure is given, in terms of the pressure coefficient C_p, by

$$C_p \equiv \frac{p - p_\infty}{\rho_\infty U_\infty^2} = \frac{2}{\gamma M_\infty^2}\left(\frac{p}{p_\infty} - 1\right) = \frac{2}{\gamma M_\infty^2}\left[\left(\frac{T}{T_\infty}\right)^{\gamma/(\gamma-1)} - 1\right]. \tag{38}$$

Using (36) in the energy-conservation equation (see equation (4) in Section 13.3) we obtain,

$$\frac{\left(U_\infty + \phi_x\right)^2 + \phi_y^2 + \phi_z^2}{2} + \frac{a^2}{\gamma - 1} = \frac{a_\infty^2}{\gamma - 1} + \frac{U_\infty^2}{2}. \tag{39}$$

Using this, (38) can be expanded in powers of ϕ to give,

$$C_p = -\left[\frac{2\phi_x}{U_\infty} + \left(1 - M_\infty^2\right)\frac{\phi_x^2}{U_\infty^2} + \frac{\phi_y^2 + \phi_z^2}{U_\infty^2} + \cdots\right]. \tag{40}$$

Note that this expansion is not valid either when $M_\infty \approx 1$ or when $M_\infty \gg 1$.

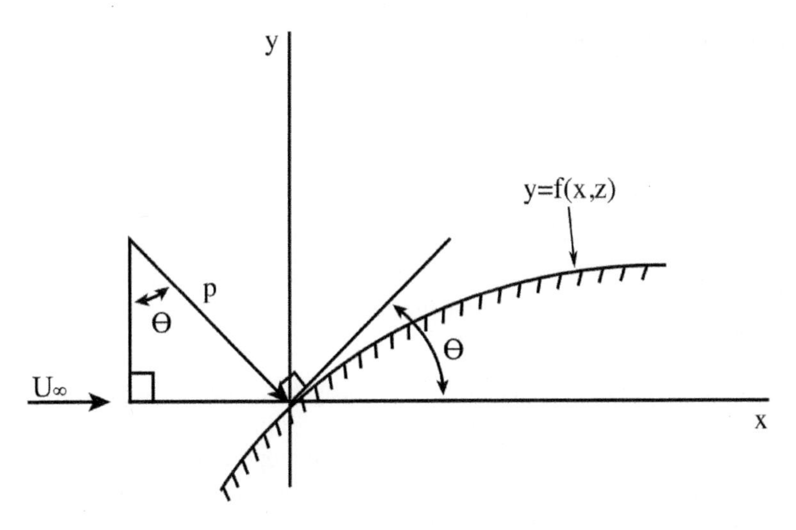

Figure 14.4 Flow past a body.

The requirement that the streamline near the surface of the body be tangential to it leads, upon linearization, to the boundary condition,

$$y = 0 : \frac{\phi_y}{U_\infty} = \frac{\partial f}{\partial x} \tag{41}$$

where $y = f(x, z)$ describes the surface of the body (see Figure 14.4).

Example 1: Consider the flow past a wavy wall given by

$$y = \varepsilon \sin \alpha x = f(x)$$

for which, one has, from (37) and (41), the following boundary-value problem:

$$\begin{aligned} &\left(1 - M_\infty^2\right) \phi_{xx} + \phi_{yy} = 0, -\infty < x < \infty \\ &y = 0 : \phi_y = \varepsilon U_\infty \alpha \cos \alpha x, -\infty < x < \infty \end{aligned}.$$

For the case $M_\infty < 1$, this problem has the solution,

$$\phi(x, y) = -\frac{\varepsilon U_\infty}{\sqrt{1 - M_\infty^2}} \exp\left(-y\alpha\sqrt{1 - M_\infty^2}\right) \cos \alpha x.$$

So, the pressure coefficient at the wall is given by

$$y = 0 : C_p = -\frac{2\varepsilon\alpha}{\sqrt{1 - M_\infty^2}} \sin \alpha x.$$

On the other hand, for the case $M_\infty > 1$, this problem has the solution,

$$\phi(x,y) = -\frac{\varepsilon U_\infty}{\sqrt{M_\infty^2 - 1}} \sin\alpha\left(x - y\sqrt{M_\infty^2 - 1}\right).$$

So, the pressure coefficient at the wall is given by

$$y = 0 : C_p = \frac{2\varepsilon\alpha}{\sqrt{M_\infty^2 - 1}} \cos\alpha x.$$

Note the differences between the subsonic-flow and the supersonic-flow solutions, as indicated by Example 1:

(i) the disturbances in a supersonic flow propagate unattenuated to infinity, whereas those in a subsonic flow attenuate at infinity;
(ii) the pressure changes in a supersonic flow are proportional to the *slope* of the boundary, whereas those in a subsonic flow are proportional to the *curvature* of the boundary;
(iii) as a consequence of (ii), a nonzero drag force, called wave drag exists on the wavy wall in a supersonic flow whereas there is no drag force on the wavy wall in a subsonic flow[2];
(iv) thanks to the limited upstream influence in a supersonic flow, the analysis of the latter becomes simpler than that of a subsonic flow.

14.5 Characteristics

A distinguishing property of *hyperbolic* partial differential equations is the existence of certain *characteristic surfaces* across which there can exist discontinuities in the normal derivatives of the dependent variables. The governing system of partial differential equations imposes certain conditions on the relative magnitudes of these jumps. These conditions form a linear, homogeneous system, and the solvability condition for this prescribes the possible orientation of the characteristic surface. The admissibility of discontinuities in the normal derivatives of the velocity on the characteristic surfaces, implies that the velocity itself must be continuous. Furthermore, the dependent variables satisfy certain *compatibility* relations on the characteristic surfaces which can be made the basis of a *graphical* procedure to compute the flow

2 Referring to Figure 14.4, the drag force, in the supersonic flow case is given by

$$D \sim (p - p_\infty)\sin\theta \sim (p - p_\infty)\tan\theta \sim [f'(x)]^2$$

which may be interpreted as being due to the energy carried by the sound waves - *wave drag*.

(Prandtl and Busemann, 1929). Such a procedure is not possible for an elliptic problem where each point is influenced by all the other points in the region.

For a steady two-dimensional potential flow, equation (9b) becomes,

$$\left(a^2 - \Phi_x^2\right)\Phi_{xx} + \left(a^2 - \Phi_y^2\right)\Phi_{yy} - 2\Phi_x\Phi_y\Phi_{xy} = 0. \tag{42}$$

This may be rewritten as in the first equation below, along with two additional differential relations.

$$\begin{aligned} A\Phi_{xx} + 2B\Phi_{xy} + C\Phi_{yy} &= 0 \\ d\Phi_x &= \Phi_{xx}dx + \Phi_{xy}dy \\ d\Phi_y &= \Phi_{yx}dx + \Phi_{yy}dy \end{aligned} \tag{43}$$

from which, one has by using *Cramer's rule*,

$$\Phi_{xy} = \frac{\begin{vmatrix} A & 0 & C \\ dx & d\Phi_x & 0 \\ 0 & d\Phi_y & dy \end{vmatrix}}{\begin{vmatrix} A & 2B & C \\ dx & dy & 0 \\ 0 & dx & dy \end{vmatrix}} \equiv \frac{\triangle_1}{\triangle}. \tag{44}$$

Using this, the characteristics (which correspond to discontinuities in the flow-velocity gradients) are given by

$$\triangle = 0$$

or

$$\left(\frac{dy}{dx}\right)_{\pm} = \frac{B \pm \sqrt{B^2 - AC}}{A}. \tag{45}$$

Comparing equation (43) with equation (42), equation (45) becomes

$$\left(\frac{dy}{dx}\right)_{\pm} = \frac{-\dfrac{\Phi_x\Phi_y}{a^2} \pm \sqrt{\dfrac{\Phi_x^2 + \Phi_y^2}{a^2} - 1}}{1 - \Phi_x^2/a^2}. \tag{46}$$

equation (46) implies that the characteristics can exist only in a supersonic flow.

Expressing the velocity components in polar coordinates,

$$\Phi_x = V\cos\theta,\quad \Phi_y = V\sin\theta \tag{47}$$

(46) becomes

$$\left(\frac{dy}{dx}\right)_{\pm} = \frac{-M^2\sin\theta\cdot\cos\theta\pm\sqrt{M^2-1}}{1-M^2\cos^2\theta} \tag{48}$$

or

$$\left(\frac{dy}{dx}\right)_{\pm} = \tan(\theta\mp\mu) \tag{49}$$

where

$$\tan\mu \equiv \frac{1}{\sqrt{M^2-1}}. \tag{50}$$

From (49) we observe that the Mach lines are the characteristics of the flow system.

(i) Compatibility Conditions in Streamline Coordinates

It turns out that along the characteristics, the dependent variables satisfy certain compatibility relations. Let us now proceed to determine these relations. Using (50) in the equation of motion in *streamline* coordinates (16) we obtain,

$$\frac{\cot^2\mu}{V}\frac{\partial V}{\partial s} - \frac{\partial\theta}{\partial n} = 0 \tag{51}$$

while one has from the *irrotationality* condition, (see equation (18))

$$\Omega = \frac{\partial V}{\partial n} - \frac{V}{R} = \frac{\partial V}{\partial n} - V\frac{\partial\theta}{\partial s} = 0. \tag{52}$$

Let us set

$$d\nu \equiv \cot\mu\cdot\frac{dV}{V} \tag{53}$$

ν may be interpreted as the angle deflected by the flow upon crossing a Mach line[3] (see Section 16.2). Equations (51) and (52) then become,

$$\frac{\partial \nu}{\partial s} - \tan \mu \cdot \frac{\partial \theta}{\partial n} = 0 \tag{54}$$

$$\tan \mu \cdot \frac{\partial \nu}{\partial n} - \frac{\partial \theta}{\partial s} = 0. \tag{55}$$

We have from equations (54) and (55),

$$\frac{\partial}{\partial s}(\nu \mp \theta) \pm \tan \mu \cdot \frac{\partial}{\partial n}(\nu \mp \theta) = 0. \tag{56}$$

---

3 One may find an explicit expression for ν as follows. Using the relation,

$$a = V \sin \mu$$

in the energy-conservation equation,

$$\frac{V^2}{2} + \frac{a^2}{\gamma - 1} = const,$$

one obtains

$$\frac{dV}{V} + \frac{2 \sin \mu \cos \mu}{(\gamma - 1) + 2 \sin^2 \mu} d\mu = 0.$$

On using (53), this equation becomes

$$d\nu = \frac{2 \cot^2 \mu}{(\gamma - 1) \cot^2 \mu + (\gamma + 1)} d\mu.$$

Setting

$$t \equiv \cot \mu$$

this equation, in turn, becomes

$$d\nu = \left[\frac{1}{\left(\frac{\gamma - 1}{\gamma + 1}\right) t^2 + 1} - \frac{1}{t^2 + 1} \right] dt.$$

Integrating this, one obtains

$$\nu = \sqrt{\frac{\gamma + 1}{\gamma - 1}} \tan^{-1}\left(\sqrt{\frac{\gamma - 1}{\gamma + 1}} \cot \mu\right) - \tan^{-1}(\cot \mu) + const$$

in agreement with (34), as expected.

We now introduce the *characteristic coordinates* η and ξ, which measure distances along the two characteristics $C^{\pm}$, given by

$$C^{\pm} : (\eta, \xi) = \frac{1}{2}\left(s \pm \int \frac{dn}{\tan\mu}\right). \tag{57}$$

Then, noting,

$$\left(\frac{\partial}{\partial\eta}, \frac{\partial}{\partial\xi}\right) = \left(\frac{\partial}{\partial s} \pm \tan\mu \frac{\partial}{\partial n}\right),$$

equation (56) leads to

$$\left.\begin{aligned} &\frac{\partial}{\partial\eta}(\nu - \theta) = 0, \text{ along } C^+ \\ &\frac{\partial}{\partial\xi}(\nu + \theta) = 0, \text{ along } C^- \end{aligned}\right\} \tag{58}$$

from which we obtain

$$C^{\pm} : \nu \mp \theta = const \tag{59}$$

which are called the *Riemann (1860) invariants* (see also Section 15.1, where the Riemann invariants are developed for a one-dimensional gas flow). The power of the streamline coordinate formulation is reflected in the development of these invariants in two-dimensional gas flows.

Each equation in characteristic form involves a particular linear combination of the derivatives of ν and θ. This feature can be used to gain insight into the structure of solutions of the equations. Considering a construction of the solution at successive small time increments gives information about the correct number of boundary conditions and the *domain of dependence*. In order to see this, let us express the equations in the characteristic form,

$$\frac{d\psi_k}{dt} + f_k(x, t, \boldsymbol{\psi}) = 0 \text{ on } \frac{dx}{dt} = c_k(x, t, \boldsymbol{\psi}). \tag{60}$$

Consider the initial value problem in $x > 0, t > 0$, with data prescribed on the x-axis (which is *transverse* to the characteristics, i.e., nowhere tangent to them) at $t = 0$[4]. Let P be the point of interest in the x-t plane. Suppose that Q_k is a neighboring point on the k-th characteristic passing through P. Then, we obtain from equation (60),

4 If data are prescribed on the characteristic, the differential equation does not determine the solution at any point *not* on the characteristic.

$$\left.\begin{aligned} \psi_k(P) - \psi_k(Q_k) + f_k(Q_k)\,[t(P) - t(Q_k)] = 0 \\ x(P) - x(Q_k) = c_k(Q_k)\,[t(P) - t(Q_k)] \end{aligned}\right\}. \tag{61}$$

Furthermore, the values at P will depend only on the data between P_1 and P_2 on the x-axis where PP_1 and PP_2 are the two characteristics passing through P (see Figure 14.5). In other words, P_1P_2 is the *domain of dependence* of P. Thus, for the full initial value problem, with ψ_k given on $t = 0, -\infty < x < \infty$, a unique solution can be constructed for $t > 0$.

Therefore, it is as though the characteristics carry information from the boundaries into the region of interest. Physically, the characteristics correspond to the paths of waves propagating with velocities c_k.

(ii) A Singular-Perturbation Problem for Hyperbolic Systems

Consider (Kevorkian and Cole, 1980),

$$\varepsilon\left(\frac{\partial^2 v}{\partial x^2} - \frac{\partial^2 v}{\partial t^2}\right) = a\frac{\partial v}{\partial x} + b\frac{\partial v}{\partial t} \tag{62}$$

which has real characteristics (see Figure 14.6) given by,

$$r = t - x, \quad s = t + x. \tag{63}$$

The characteristics serve to define the *region of influence* propagating into the future of a disturbance at a point Q (see Figure 14.6). The specification of boundary conditions on an arc for a fully posed boundary-value problem

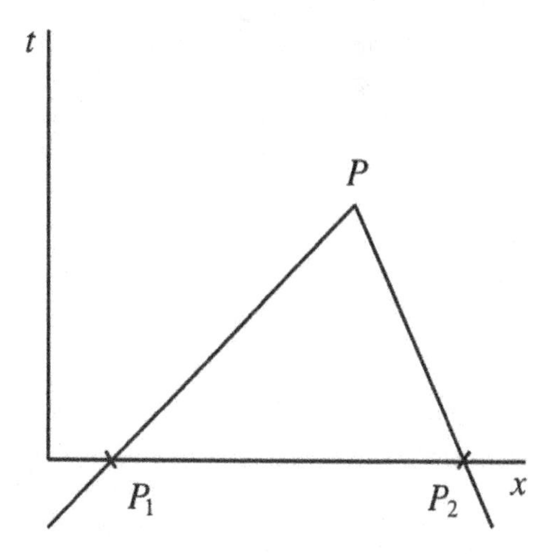

Figure 14.5 Characteristics through a point.

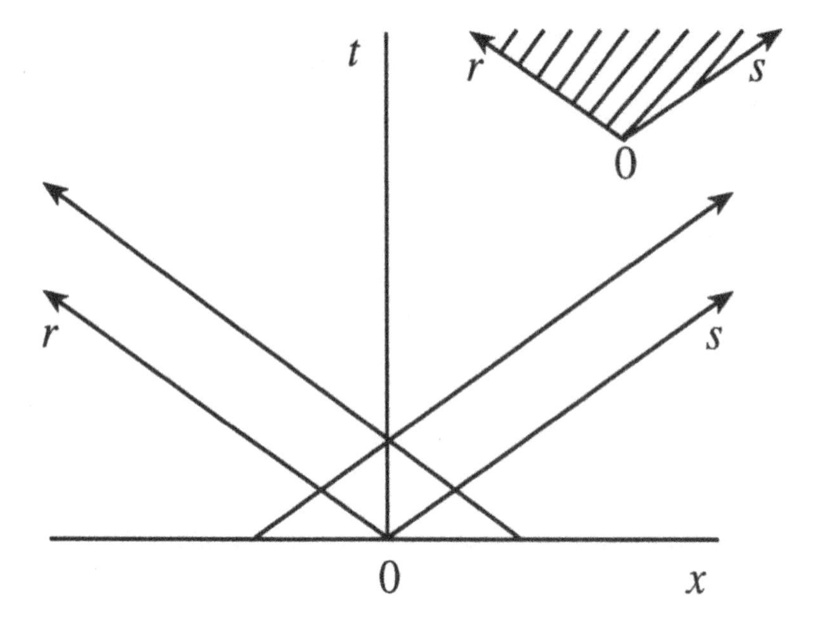

Figure 14.6 Characteristics and region of influence (from Kevorkian and Cole, 1980).

depends on the orientation of the arc with respect to the characteristic directions of propagation. *One* boundary condition is specified on the *time-like* arc (see Figure 14.7) corresponding to *one* characteristic leading into the adjacent region in which the solution is defined. *Two* boundary conditions are given on the *space-like* arc (see Figure 14.7) corresponding to the *two* characteristics leading into the adjacent domain. When the boundary curves are along the characteristic curves, only one condition can be prescribed, and the characteristic relations must hold. The characteristic initial-value problem describes a condition on each of the characteristics AB and AC. These conditions define the solution in the region $ABCD$ (see Figure 14.8).

Consider the initial value problem corresponding to equation (62) for $-\infty < x < \infty$ with initial conditions,

$$t = 0 : v = F(x), \quad v_t = G(x). \tag{64}$$

According to the general theory of characteristics, the solutions at a point $P(x, t)$ (see Figure 14.9) can depend only on that part of the initial data which can send a signal to P. This is the part $(x_1 < x < x_2)$ of the initial line contained between the backward-running characteristics through P.

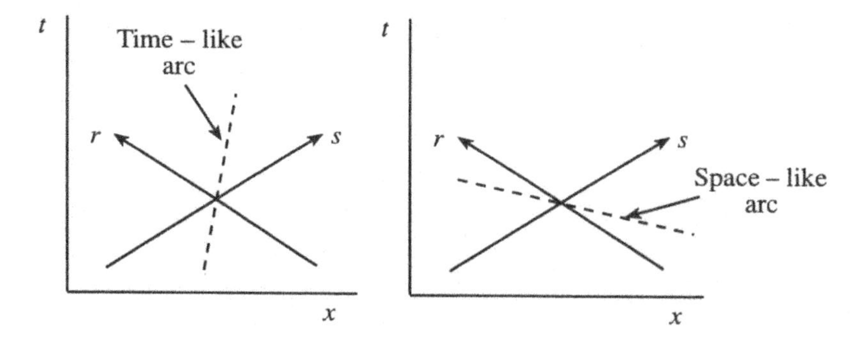

Figure 14.7 Time-like arc and space-like arc (from Kevorkian and Cole, 1980).

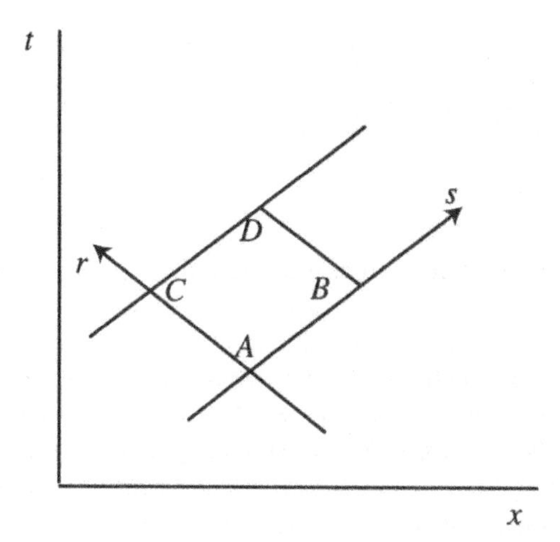

Figure 14.8 Domain of influence in a characteristic initial-value problen (from Kevorkian and Cole, 1980).

Now, the outer problem associated with equation (62) corresponds to taking the limit $\varepsilon \Rightarrow 0$ and it is given by

$$a\frac{\partial v^{(0)}}{\partial x} + b\frac{\partial v^{(0)}}{\partial t} = 0. \tag{65}$$

The solution of equation (65) is

$$v^{(0)}(x,t) = f\left(x - \frac{a}{b}t\right). \tag{66}$$

In the limit $\varepsilon \Rightarrow 0$, the solution $v(x,t)$ depends therefore only on the data connected to P along a *subcharacteristic* of equation (62) given by

$$bx - at = const. \tag{67}$$

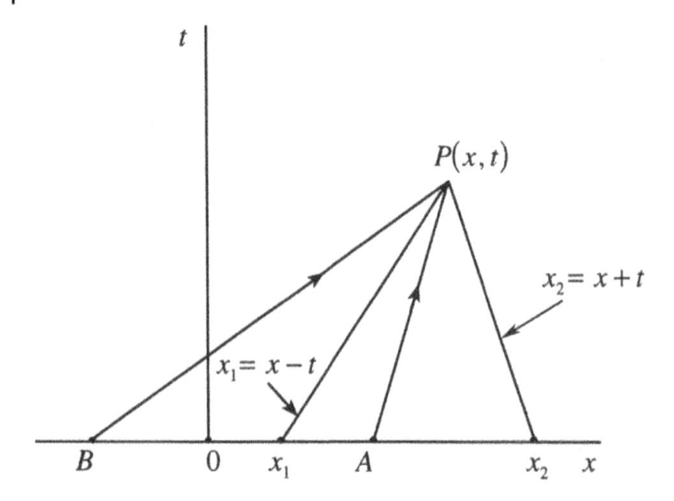

Figure 14.9 Domain of influence and time-like and space-like subcharacteristics (from Kevorkian and Cole, 1980).

The subcharacteristic reaching P originates at point A between x_1, x_2 if $|b/a| > 1$, i.e., if it is time-like. Then, the limit $\varepsilon \Rightarrow 0$ preserves the domain of influence. However, if $|b/a| < 1$ the subcharacteristic reaching P is space-like and lies outside the domain of influence, originating at B (see Figure 14.9). In this case, the limit $\varepsilon \Rightarrow 0$ increases the domain of influence. This implies that the outer solution cannot be obtained by taking the outer limit of the solution to the full problem in equation (62), contrary to what might be expected. For the case $|b/a| = 1$, the subcharacteristics reduce to the real characteristics of the full problem.

It turns out that the issue of stability of the solutions $v\,(x, t)$ is also related to whether the subcharacteristics are time-like or space-like. In order to see this, note that in terms of the characteristic coordinates (63), equation (62) becomes

$$-4\varepsilon \frac{\partial^2 v}{\partial r \partial s} = (b - a)\frac{\partial v}{\partial r} + (b + a)\frac{\partial v}{\partial s}. \tag{68}$$

Consider the propagation of a jump in the normal derivative $\partial v / \partial r$ along $r = r_0 = const$. Let

$$K \equiv \left[\frac{\partial v}{\partial r}\right]_{r=r_0} \equiv \left(\frac{\partial v}{\partial r}\right)_{r_0^+} - \left(\frac{\partial v}{\partial r}\right)_{r_0^-}. \tag{69}$$

Assuming that v itself is continuous across $r = r_0$, one finds from equation (68),

$$-4\varepsilon \frac{\partial K}{\partial s} = (b - a)K$$

from which,

$$K = K_0 \exp\left[-\left(\frac{b-a}{4\varepsilon}\right)(s-s_0)\right]. \tag{70}$$

Recall that, a jump across a characteristic propagates to infinity along that characteristic. Using (69) and (70), this implies

$$\left.\begin{array}{r}(b-a) > 0 \Rightarrow \text{ stability} \\ (b-a) < 0 \Rightarrow \text{ instability}\end{array}\right\}. \tag{71}$$

Similarly, a jump in the normal derivative $\partial v/\partial s$ across a characteristic $s = s_0$ leads to

$$\left.\begin{array}{r}(b+a) > 0 \Rightarrow \text{ stability} \\ (b+a) < 0 \Rightarrow \text{ instability}\end{array}\right\}. \tag{72}$$

Combining (71) and (72), one obtains

$$|b/a| > 1 \qquad \text{for stability} \tag{73}$$

which applies to time-like subcharacteristics. We restrict further discussion to the stable case.

Now, note that the outer solution $v^{(0)}(x,t)$ given in (66) can satisfy only one initial condition. Consequently, one may expect the existence of a boundary layer on the space-like arc $t = 0$.

Assume the following inner expansion, valid near the space-like arc $t = 0$,

$$v^{(i)}(x,\tilde{t};\varepsilon) = v_0^{(i)}(x,\tilde{t}) + \beta_1(\varepsilon)\, v_1^{(i)}(x,\tilde{t}) + \cdots \tag{74}$$

where $\tilde{t}$ is the time scale relevant for $t \approx 0$,

$$\tilde{t} \equiv \frac{t}{\delta(\varepsilon)}$$

and

$$\beta_1, \delta \Rightarrow 0 \text{ as } \varepsilon \Rightarrow 0$$

with x and $\tilde{t}$ held fixed in the associated inner limit process, $\varepsilon \Rightarrow 0$.

Taking

$$\beta_1(\varepsilon) = \delta(\varepsilon) \tag{75}$$

the initial conditions (64) give

$$\left.\begin{array}{c} \tilde{t} = 0 : v_0^{(i)} = F(x), \quad v_n^{(i)} = 0 \text{ for } n > 0 \\ \tilde{t} = 0 : \dfrac{\partial v_0^{(i)}}{\partial \tilde{t}} = 0, \quad \dfrac{\partial \tilde{v}_1^{(i)}}{\partial \tilde{t}} = G(x); \\ \text{etc.} \end{array}\right\} \tag{76}$$

Choosing $\delta(\varepsilon) = \varepsilon$, and substituting (74) and (75) into equation (62) gives

$$O(1) : \frac{\partial^2 v_0^{(i)}}{\partial \tilde{t}^2} + b\frac{\partial v^{(i)}}{\partial \tilde{t}} = 0 \tag{77}$$

$$O(\varepsilon) : \frac{\partial^2 v_1^{(i)}}{\partial \tilde{t}^2} + b\frac{\partial v_1^{(i)}}{\partial \tilde{t}} = -a\frac{\partial v_0^{(i)}}{\partial x} \tag{78}$$

etc.

Notice that the boundary-layer equations (77) and (78) are *ordinary-differential equations*. This is a feature of the boundary layers occurring on a space-like arc which can never be a subcharacteristic.

Using the initial conditions (76), equations (77) and (78) give

$$v_0^{(i)}(x, \tilde{t}) = F(x) \tag{79}$$

$$v_1^{(i)}(x, \tilde{t}) = \left[G(x) + \frac{a}{b}F'(x)\right]\left(1 - e^{-\tilde{t}}\right) - \frac{a}{b}\tilde{t}F'(x) \tag{80}$$

so that we have

$$v^{(i)}(x, \tilde{t}; \varepsilon) \sim F(x) + \varepsilon\left[\left\{G(x) + \frac{a}{b}F'(x)\right\}\left(1 - e^{-\tilde{t}}\right) - \frac{a}{b}\tilde{t}F'(x)\right] + \cdots \tag{81}$$

Note that (81) contains terms that persist in the limit $t \Rightarrow \infty$, as well as terms that decay in time which are typical of a boundary layer (further details on this are found in Chapter 21).

Next, we construct an outer expansion, with x and t held fixed in the associated outer limit process $\varepsilon \Rightarrow 0$ given by

$$v^{(0)}(x,t,\varepsilon) = v_0^{(0)}(x,t) + \varepsilon v_1^{(0)}(x,t) + \cdots \tag{82}$$

The above expansion gives

$$a\frac{\partial v_0^{(0)}}{\partial x} + b\frac{\partial v_0^{(0)}}{\partial t} = 0 \tag{83}$$

$$a\frac{\partial v_1^{(0)}}{\partial x} + b\frac{\partial v_1^{(0)}}{\partial t} = \left(\frac{\partial^2 v_0^{(0)}}{\partial x^2} - \frac{\partial^2 v_0^{(0)}}{\partial t^2}\right). \tag{84}$$

We obtain from equation (83),

$$v_0^{(0)} = f(\xi), \quad \xi \equiv x - \frac{a}{b}t. \tag{85}$$

Using (85), equation (84) becomes

$$a\frac{\partial v_1^{(0)}}{\partial x} + b\frac{\partial v_1^{(0)}}{\partial t} = \left(1 - \frac{a^2}{b^2}\right) f''(\xi) \tag{86}$$

from which,

$$v_1^{(0)} = \frac{a}{b^2}\frac{b^2 - a^2}{b^2 + a^2}\left(x + \frac{b}{a}t\right) f''(\xi) + f_1(\xi). \tag{87}$$

Combining (85) and (87) we obtain,

$$v^{(0)}(x,t;\varepsilon) = f(\xi) + \varepsilon\left[f_1(\xi) + \frac{a}{b^2}\frac{b^2 - a^2}{b^2 + a^2}\left(x + \frac{b}{a}t\right) f''(\xi)\right] + \cdots. \tag{88}$$

The asymptotic matching between $v^{(i)}$ and $v^{(0)}$ requires (see Chapter 21)

$$\begin{aligned} &v_0^{(0)}(x,0) + \varepsilon\left[t v_{0_t}^{(0)}(x,0) + v_1(x,0)\right] + \cdots \\ &= v_0^{(i)}(x,\infty) + \varepsilon v_1^{(i)}(x,\infty) + \cdots \end{aligned}. \tag{89}$$

Substituting (81) and (88) into (89) gives

$$f(x) = F(x)$$

$$f_1(x) + \frac{a}{b^2}\frac{b^2 - a^2}{b^2 + a^2} x f''(x) = G(x) + \frac{a}{b}F'(x). \tag{90}$$

So, the outer solution (88) becomes

$$v^{(0)}(x,t;\varepsilon) = F\left(x - \frac{a}{b}t\right) + \varepsilon\left[\frac{b^2 - a^2}{b^3}tF''\left(x - \frac{a}{b}t\right) + \right.$$
$$\left. + \frac{a}{b}F'\left(x - \frac{a}{b}t\right) + G\left(x - \frac{a}{b}t\right)\right] + \cdots. \tag{91}$$

Consider next, a *radiation problem* in which the boundary conditions are prescribed on a time-like arc and propagate into the quiescent medium in $x > 0$. When the boundary condition is prescribed for instance, on a time-like arc $x = 0$ (Figure 14.10), two cases arise depending on whether the subcharacteristics run into or out of the boundary $x = 0$. Recall that the subcharacteristics are given by

$$\xi = x - \frac{a}{b}t = const. \tag{67}$$

Note, that the subcharacteristics are incoming or outgoing depending on whether $a \lessgtr 0$ (we assume $b > 0$). Let the boundary condition be

$$x = 0 : v = F(t), \quad t > 0. \tag{92}$$

Outgoing Characteristics: Assume an outer solution,

$$v^{(0)}(x,t;\varepsilon) = v_0^{(0)}(x,t) + \varepsilon v_1^{(0)}(x,t) + \cdots \tag{93}$$

where

$$v_0^{(0)} = f(\zeta), \quad \zeta \equiv t - \frac{b}{a}x. \tag{66}$$

Substituting (66) into the boundary condition (92), we obtain

$$v_0^{(0)} = \begin{cases} 0, & t < \frac{b}{a}x \\ F\left(t - \frac{b}{a}x\right), & t > \frac{b}{a}x \end{cases}. \tag{94}$$

This solution has a discontinuity on the particular subcharacteristic through the origin $\zeta = 0$. However, such a discontinuity is not permitted in the solution to equation (62) with $\varepsilon \neq 0$. In order to resolve this, a suitable boundary layer must be introduced on the subcharacteristic $\zeta = 0$ which supports the discontinuity in the outer solution $v^{(0)}$. Assume an inner expansion given by,

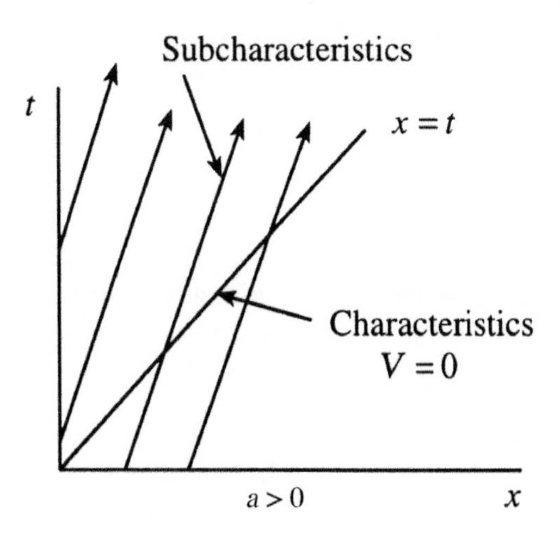

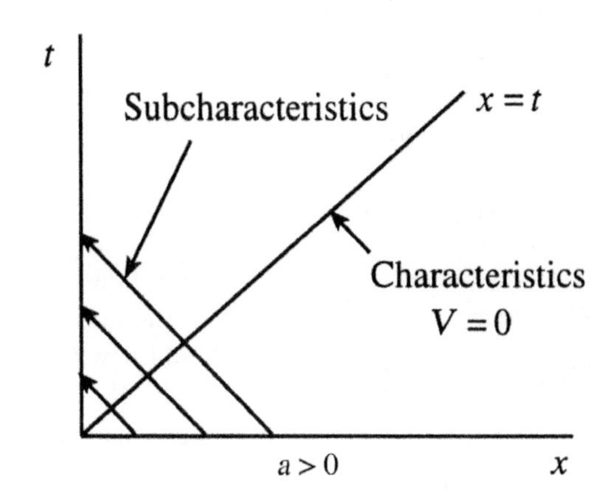

Figure 14.10 Radiation problem with boundary conditions prescribed on a finite portion of the boundary (from Kevorkian and Cole, 1980).

$$v^{(i)}(\tilde{x}, \tilde{t}; \varepsilon) = v_0^{(i)}(\tilde{x}, \tilde{t}) + \mu(\varepsilon) v_1^{(i)}(\tilde{x}, \tilde{t}) + \cdots$$

where

$$\tilde{x} \equiv \frac{x - \frac{a}{b}t}{\delta(\varepsilon)}, \tilde{t} = t \tag{95}$$

and

$$\delta \Rightarrow 0 \text{ as } \varepsilon \Rightarrow 0,$$

with $\tilde{x}$ and $\tilde{t}$ held fixed, in the associated inner limit process, $\varepsilon \Rightarrow 0$. Choosing $\delta = \sqrt{\varepsilon}$, and substituting (95) into equation (62) gives, in the limit $\varepsilon \Rightarrow 0$,

$$K\frac{\partial^2 v_0^{(i)}}{\partial \tilde{x}^2} = \frac{\partial v_0^{(i)}}{\partial \tilde{t}}. \tag{96}$$

Recalling that the subcharacteristics are time-like (i.e., $|a/b| < 1$), we have

$$K \equiv \frac{1 - \frac{a^2}{b^2}}{b} > 0$$

which ensures that $\tilde{t}$ is a positive time-like variable. So, equation (96) is a *diffusion* equation that describes the spreading of the discontinuity in the outer expansion $v^{(0)}$ on the subcharacteristic $\zeta = 0$. Matching $v^{(i)}$ to $v^{(0)}$ asymptotically as before, one obtains

$$v_0^{(i)}(\tilde{x}, \tilde{t}) = \frac{F(0^+)}{2} erfc\left(\frac{\tilde{x}}{2\sqrt{K\tilde{t}}}\right). \tag{97}$$

Incoming Characteristics: Assume an outer expansion given by,

$$v^{(0)}(x, t; \varepsilon) = v_0^{(0)}(x, t) + \varepsilon v_1^{(0)}(x, t) + \cdots. \tag{98}$$

Since the disturbances now propagate along the subcharacteristics from the quiescent region to the boundary, one has

$$v^{(0)} \equiv 0. \tag{99}$$

Then, the discontinuity in $v^{(0)}$ occurs at the boundary $x = 0$. So, one introduces a boundary layer at $x = 0$. Since the line $x = 0$ is not a subcharacteristic, the boundary layer equations should now be ordinary differential equations. Assume an inner expansion,

$$v^{(i)}(\tilde{x}, \tilde{t}; \varepsilon) = v_0^{(i)}(\tilde{x}, \tilde{t}) + \nu_1(\varepsilon) v_1^{(i)}(\tilde{x}, \tilde{t}) + \cdots \tag{100}$$

where

$$\tilde{x} = \frac{x}{\delta(\varepsilon)}, \quad \tilde{t} = t \tag{101}$$

and

$$\nu_1 \Rightarrow 0 \quad \text{as } \varepsilon \Rightarrow 0$$

with $\tilde{x}$ and $\tilde{t}$ held fixed, in the associated inner limit process, $\varepsilon \Rightarrow 0$. If one chooses $\delta = \varepsilon$, substituting (100) into equation (62), we obtain

$$\frac{\partial^2 v_0^{(i)}}{\partial \tilde{x}^2} = a\frac{\partial v_0^{(i)}}{\partial \tilde{x}}. \tag{102}$$

Using boundary condition (92), one obtains from equation (102),

$$v_0^{(i)}(\tilde{x}, \tilde{t}) = F(\tilde{t}) \exp(a\tilde{x}), \quad a < 0. \tag{103}$$

Exercises

1. Show that the rate of change of circulation Γ around a closed curve C made up of the same fluid particles, for non-isentropic cases, is

$$\frac{D\Gamma}{Dt} = \int_C T dS, \textit{ Bjerknes' Theorem.}$$

   (Thus, entropy generation leads to change in circulation in the flow.)
2. Show that the *hodograph curve* described by (34) is an *epicycloid*.
3. Derive the compatibility conditions for flow variables in the non-isentropic cases along the characteristics. Note: This illustrates the utility of characteristics in calculating more general flows.

15

Nonlinear Theory of Plane Sound Waves

While dealing with linear acoustics (Section 13.2), one considers the disturbances on a constant ambient state to be small. This allows the linearization of the governing equations by retaining only the first-order terms in the small disturbances. These disturbances then satisfy the classical wave equation. This process raises two important questions- what are the essential *nonlinear* features that were lost in the linearization of the original equations and how do they materialize? We aim to address these issues in this chapter.

15.1 Riemann Invariants

Consider a one-dimensional propagation of plane sound waves in a compressible fluid. The equations expressing the conservation of mass and momentum are

$$\frac{\partial \rho}{\partial t} + \frac{\partial}{\partial x}(\rho u) = 0. \tag{1}$$

$$\rho\left(\frac{\partial u}{\partial t} + u\frac{\partial u}{\partial x}\right) = -\frac{\partial p}{\partial x}. \tag{2}$$

Introducing a new variable P, which may be called the *potential sound speed*,

$$P \equiv \int_0^\rho a\frac{d\rho}{\rho}, \quad a^2 \equiv \left(\frac{\partial p}{\partial \rho}\right)_s, \tag{3}$$

equations (1) and (2) give

$$\frac{\partial P}{\partial t} + u\frac{\partial P}{\partial x} + a\frac{\partial u}{\partial x} = 0, \tag{4}$$

Introduction to Theoretical and Mathematical Fluid Dynamics, Third Edition.
Bhimsen K. Shivamoggi.

$$\frac{\partial u}{\partial t} + u\frac{\partial u}{\partial x} + a\frac{\partial P}{\partial x} = 0. \tag{5}$$

These equations lead to

$$\frac{\partial}{\partial t}(u \pm P) + (u \pm a)\frac{\partial}{\partial x}(u \pm P) = 0. \tag{6}$$

Equation (6) gives the *Riemann* (1860) *invariants*[1]

$$\mathscr{I}_{\pm} \equiv u \pm P = const \text{ along } C^{\pm} : \frac{dx}{dt} = u \pm a. \tag{7}$$

The lines $dx/dt = u \pm a$ are called the characteristics $C^{\pm}$.

15.2 Simple Wave Solutions

The existence of two families of characteristics greatly facilitates finding solutions to hyperbolic systems. For a linear problem,the characteristics are found

---

1 Equation (7) may also be derived alternatively by noting that equations (1) and (2) may be rewritten as follows

$$\begin{pmatrix} \rho \\ u \end{pmatrix}_t + A \cdot \begin{pmatrix} \rho \\ u \end{pmatrix}_x = 0,$$

where

$$A \equiv \begin{bmatrix} u & \rho \\ a^2/\rho & u \end{bmatrix}.$$

The eigenvalues of A are $(u \pm a)$ and the corresponding eigenvectors are $\begin{bmatrix} \pm a/\rho \\ 1 \end{bmatrix}$. If the characteristics are given by

$$C^{\pm} : t = t(\tau_{\pm}), x = x(\tau_{\pm})$$

with,

$$\frac{dt}{d\tau_{\pm}} = 1 \text{ and } \frac{dx}{d\tau_{\pm}} = u \pm a.$$

Then the Riemann invariants $\mathscr{I}_{\pm}(\rho, u)$ given by

$$\begin{bmatrix} u & a^2/\rho \\ \rho & u \end{bmatrix}\begin{bmatrix} \partial\mathscr{I}_{\pm}/\partial\rho \\ \partial\mathscr{I}_{\pm}/\partial u \end{bmatrix} = (u \pm a)\begin{bmatrix} \partial\mathscr{I}_{\pm}/\partial\rho \\ \partial\mathscr{I}_{\pm}/\partial u \end{bmatrix}$$

---

can be shown to be constant along the characteristics $C^{\pm}: \dfrac{dx}{dt} = u \pm a$.

$$
\begin{aligned}
\frac{\partial \mathcal{I}_\pm}{\partial \tau_\pm} &= \frac{\partial \mathcal{I}_\pm}{\partial \rho}\left[\frac{\partial \rho}{\partial t} + (u \pm a)\frac{\partial \rho}{\partial x}\right] + \frac{\partial \mathcal{I}_\pm}{\partial u}\left[\frac{\partial u}{\partial t} + (u \pm a)\frac{\partial u}{\partial x}\right] \\
&= \begin{pmatrix} \dfrac{\partial \mathcal{I}_\pm}{\partial \rho} \\ \dfrac{\partial \mathcal{I}_\pm}{\partial u} \end{pmatrix}^T \begin{pmatrix} \dfrac{\partial \rho}{\partial t} \\ \dfrac{\partial u}{\partial t} \end{pmatrix} + (u \pm a)\begin{pmatrix} \dfrac{\partial \mathcal{I}_\pm}{\partial \rho} \\ \dfrac{\partial \mathcal{I}_\pm}{\partial u} \end{pmatrix}^T \begin{pmatrix} \dfrac{\partial \rho}{\partial x} \\ \dfrac{\partial u}{\partial x} \end{pmatrix} \\
&= -\begin{pmatrix} \dfrac{\partial \mathcal{I}_\pm}{\partial \rho} \\ \dfrac{\partial \mathcal{I}_\pm}{\partial u} \end{pmatrix}^T \begin{pmatrix} u & \rho \\ a^2/\rho & u \end{pmatrix} \begin{pmatrix} \dfrac{\partial \rho}{\partial x} \\ \dfrac{\partial u}{\partial x} \end{pmatrix} \\
&\quad + (u \pm a)\begin{pmatrix} \dfrac{\partial \mathcal{I}_\pm}{\partial \rho} \\ \dfrac{\partial \mathcal{I}_\pm}{\partial u} \end{pmatrix}^T \begin{pmatrix} \dfrac{\partial \rho}{\partial x} \\ \dfrac{\partial u}{\partial x} \end{pmatrix} \\
&= -\left[\begin{pmatrix} u & \rho \\ a^2/\rho & u \end{pmatrix}^T \begin{pmatrix} \dfrac{\partial \mathcal{I}_\pm}{\partial \rho} \\ \dfrac{\partial \mathcal{I}_\pm}{\partial u} \end{pmatrix}\right]^T \begin{pmatrix} \dfrac{\partial \rho}{\partial x} \\ \dfrac{\partial u}{\partial x} \end{pmatrix} \\
&\quad + (u \pm a)\begin{pmatrix} \dfrac{\partial \mathcal{I}_\pm}{\partial \rho} \\ \dfrac{\partial \mathcal{I}_\pm}{\partial u} \end{pmatrix}^T \begin{pmatrix} \dfrac{\partial \rho}{\partial x} \\ \dfrac{\partial u}{\partial x} \end{pmatrix} \\
&= -(u \pm a)\begin{pmatrix} \dfrac{\partial \mathcal{I}_\pm}{\partial \rho} \\ \dfrac{\partial \mathcal{I}_\pm}{\partial u} \end{pmatrix}^T \begin{pmatrix} \dfrac{\partial \rho}{\partial x} \\ \dfrac{\partial u}{\partial x} \end{pmatrix} \\
&\quad + (u \pm a)\begin{pmatrix} \dfrac{\partial \mathcal{I}_\pm}{\partial \rho} \\ \dfrac{\partial \mathcal{I}_\pm}{\partial u} \end{pmatrix}^T \begin{pmatrix} \dfrac{\partial \rho}{\partial x} \\ \dfrac{\partial u}{\partial x} \end{pmatrix} = 0.
\end{aligned}
$$

by simple integration, and a knowledge of the Riemann invariants along the characteristics then completes the solution. However, this is not possible for equations (4) and (5) because their non-linearity forces the Riemann invariants to depend on unknown variables u and a, making integration impossible. Nonetheless, for certain special geometries having special boundary conditions, one of the Riemann invariants is constant everywhere (i.e., it takes the same value on every characteristic of its family). Consequently, one of these two families of characteristics reduces to a set of straight lines, making integration of equations (4) and (5) possible.

A *simple wave* situation with one family of straight-line characteristics occurs when all the members of the other family come from a region of uniform flow conditions in the (x, t) plane. An example for this is a uniform flow of infinite extent occurring at infinity upstream. A special case is the *Prandtl-Meyer expansion* of a supersonic flow around a convex corner. If the corner is sharp, one family of characteristics comprise of straight lines bunched into a fan centered on the corner itself (Section 14.3). Suppose for instance that the condition,

$$\mathcal{I}_- = u - \int a\frac{d\rho}{\rho} = const \tag{8}$$

holds not just on C^-, but everywhere. Then, since $u + \int a\,(d\rho/\rho)$ is constant on a given characteristic $C^+ : dx/dt = u + a$, it follows that u and $\int a\,(d\rho/\rho)$ must separately be constant on C^+. If the flow is *homentropic*,[2] then $a = a\,(\rho)$, and therefore, a is constant on C^+. Thus, $u + a$ is also constant on C^+. This makes C^+ a straight line. The family of characteristics $C^+ : dx/dt = u + a$ hence consists of straight lines. Since one of these passes through each point of

Noting that

$$\begin{bmatrix} \dfrac{\partial \mathcal{I}_\pm}{\partial \rho} \\[2ex] \dfrac{\partial \mathcal{I}_\pm}{\partial u} \end{bmatrix} = \begin{bmatrix} \pm a/\rho \\ 1 \end{bmatrix}$$

we obtain

$$C^\pm : \mathcal{I}_\pm = u \pm \int a\frac{d\rho}{\rho} = const$$

as given in (7).

2 A flow is called *homentropic* if the entropy of each fluid particle is the same and remains so for all times.

(x, t) space (as does one of the other family), the solutions of equations (1) and (2) for this region, (called the *simple wave* region), take the form

$$u = f\left[x - (u + a)t\right] \tag{9}$$

corresponding to the initial condition,

$$t = 0 : u = f(x). \tag{10}$$

If $f(x) \geq 0$, $u + a$ will be positive everywhere. Thus, the family of characteristics carrying the wave has positive slope in the (x, t) plane, and the wave is said to be *forward-progressing*.

Example 1: Consider plane sound waves propagating isentropically in an ideal gas. Using the equation of state

$$\frac{p}{p_0} = \left(\frac{\rho}{\rho_0}\right)^{\gamma}$$

we have

$$a^2 = \left(\frac{dp}{d\rho}\right)_S = \frac{\gamma p}{\rho}.$$

Hence,

$$\frac{a}{a_0} = \left(\frac{\rho}{\rho_0}\right)^{(\gamma-1)/2}.$$

Using this, (3) gives for the *potential sound speed*,

$$P = \int_{\rho_0}^{\rho} \frac{a}{\rho} d\rho = 2\left(\frac{a - a_0}{\gamma - 1}\right).$$

Then, for a simple wave, the Riemann invariant,

$$\mathscr{I}_{-} = u - P = const = 0$$

gives

$$a = a_0 + \frac{1}{2}(\gamma - 1)u, \text{ everywhere.}$$

Now, the signal is propagated at the velocity

$$\frac{dx}{dt} = u + a = a_0 + \frac{1}{2}(\gamma + 1)u$$

which implies that the solution becomes multivalued unless a *shock wave* (see Chapter 16) is introduced to prevent it. Thus, finite-amplitude compressive waves continually steepen with time (see Figure 15.1) until a discontinuity

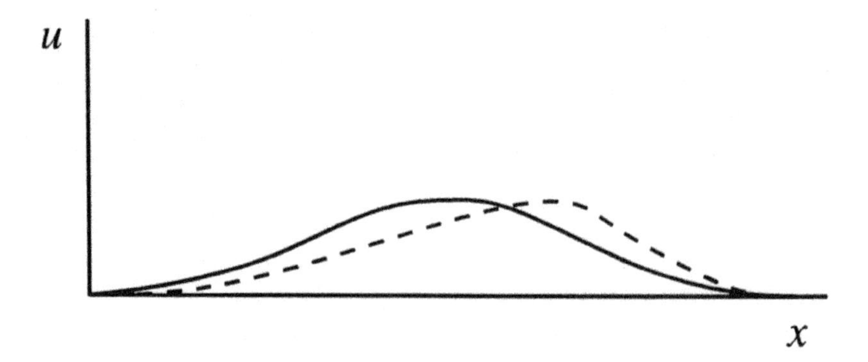

Figure 15.1 Nonlinear steepening of plane waves.

develops. One then needs to include the effects of viscosity and heat conductivity (see Section 19.10) so as to produce a balance between the effects of *nonlinear steepening* and *dissipative spreading*, leading to a steady profile. It turns out that the transition layer (the *shock wave*) over which this balance takes effect has a thickness of the order of a few mean free paths. This shock region may however be idealized into a discontinuity in the inviscid theory. One may then simply add the jump conditions across the disconuity of the flow variables (dictated by the integral form of the governing equations) to the inviscid theory.[3]

It turns out that the isentropic flow considered above has an exact explicit solution (*Fubini-Ghiron*, 1935) type. First, note that using the relation,

$$a = a_0 + \frac{1}{2}(\gamma - 1)\,u \tag{11}$$

(9) becomes

$$u = f\left[x - \left(a_0 + \frac{\gamma + 1}{2}u\right)t\right]. \tag{12}$$

Example 2: Consider the initially sinusoidal motion defined by the condition,

$$t = 0 : u = u_0 \sin kx.$$

3 Such a discontinuous solution is called a *weak* solution and is interpreted in the sense of *distributions* (see Section 16.1).

Then, (12) becomes

$$u = u_0 \sin k \left[x - \left(a_0 + \frac{\gamma + 1}{2} u \right) t \right].$$

In order to solve for u, it turns out to be useful to expand u/u_0 as a *Fourier series*,

$$\frac{u}{u_0} = \sum_{n=1}^{\infty} B_n \sin n (kx - \omega t)$$

where $\omega \equiv k a_o$. The Fourier coefficients B_n are then given by

$$B_n = \frac{1}{\pi} \int_0^{2\pi} \frac{u}{u_0} \sin n (kx - \omega t) \, d (kx - \omega t).$$

Introducing

$$\zeta \equiv k \left[x - \left(a_0 + \frac{\gamma + 1}{2} u \right) t \right]$$

these become

$$\begin{aligned} B_n &= \frac{1}{\pi} \int_0^{2\pi} \sin \zeta \cdot \sin \left(n\zeta + n \frac{\gamma + 1}{2} k u_0 t \sin \zeta \right) \cdot \left(1 + \frac{\gamma + 1}{2} k u_0 t \cos \zeta \right) d\zeta \\ &= 2 \frac{(-1)^{n-1} J_n \left[n \left(\frac{\gamma + 1}{2} \right) k u_0 t \right]}{n \left(\frac{\gamma + 1}{2} \right) k u_0 t}. \end{aligned}$$

Substituting this into the Fourier series for u we obtain for the solution,

$$u = 2u_0 \sum_{n=1}^{\infty} (-1)^{n-1} \frac{J_n \left[n \left(\frac{\gamma + 1}{2} \right) k u_0 t \right]}{n \left(\frac{\gamma + 1}{2} \right) k u_0 t} \sin n (kx - \omega t).$$

In the linear-wave limit (u_0 small),[4] this reduces to the expected result,

$$u \approx u_0 \sin (kx - \omega t).$$

---

4 We use the result,

$$J_1 (x) \approx \frac{1}{2} x \text{ for small } x.$$

Example 3: Consider the waves produced by the prescribed motion of a piston at the end of a long tube. The gas is assumed to be at rest with a uniform state $u = 0, a = a_0$ for $x \geq 0$ at $t = 0$. Since the piston is itself a particle path, all particle paths originate on the x-axis in the uniform region, and end on the piston. The flow is assumed to be isentropic.

The C^- characteristics start on the x-axis in the uniform region. On each of them we have

$$C^- : \mathscr{I}_- = u - \frac{2a}{\gamma - 1} = -\frac{2a_0}{\gamma - 1} = const$$

which holds throughout the flow.

The C^+ characteristics start on the piston, and on each of them, we have

$$C^+ : \mathscr{I}_+ = u + \frac{2a}{\gamma - 1} = \frac{2a_0}{\gamma - 1} = const \text{ along } C^+ : \frac{dx}{dt} = u + a.$$

So

$$u = const \text{ along } C^+ : \frac{dx}{dt} = a_0 + \frac{\gamma + 1}{2}u.$$

Hence, the C^+ characteristics will be a family of straight lines (see Figure 15.2). The leading C^+ characteristic starts on the x-axis and is given by $x = a_0 t$.

Note that since the C^+ characteristics are straight lines with slope dx/dt increasing with u, these characteristics will intersect if u ever increases on

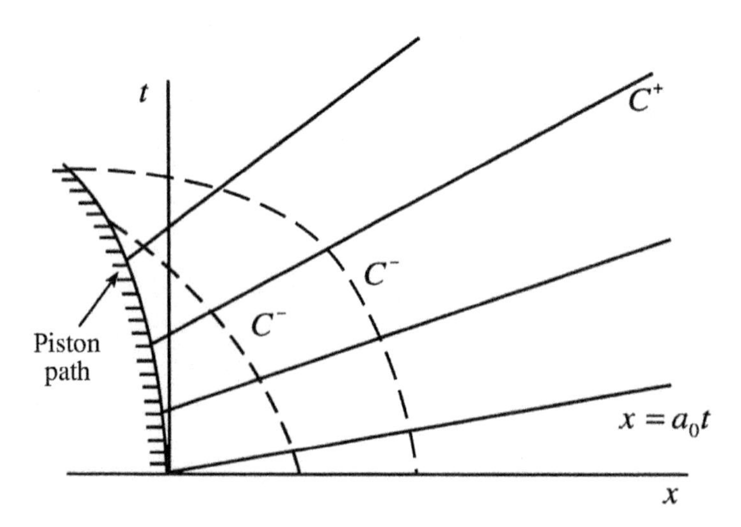

Figure 15.2 Loci of wave fronts produced by the motion of a piston.

the piston. This is the typical *nonlinear breaking* that is resolved by the introduction of shocks (Chapter 16). This happens in the compressive parts of the disturbance (see Example 4).

Example 4: Consider the movement of a piston in a tube. One family of characteristics namely C^+, then converge in the compressive case. An envelope of this family may be determined as follows. If the path of the piston is given by $x_p = f(t)$, then the speed of the piston is $u_p = f'(t)$.

We have along the characteristic C^-,

$$C^- : \mathscr{I}_- = u - \frac{2a}{\gamma - 1} = -\frac{2a_0}{\gamma - 1} = const.$$

Thus, at the piston, we have

$$a_p = a_0 + \frac{\gamma - 1}{2} u_p = a_0 + \frac{\gamma - 1}{2} f'(t).$$

The slope of the characteristic C^+ is then given by

$$\left(\frac{dx}{dt}\right)_{C+} = u_p + a_p = a_0 + \frac{\gamma + 1}{2} f'(t).$$

Therefore, the C^+ family of characteristics through the point $(t_1, f(t_1))$ is given by

$$g(x, t; t_1) \equiv x - f(t_1) - \left[a_0 + \frac{\gamma + 1}{2} f'(t_1)\right](t - t_1) = 0$$

where t_1 is now a parameter. The envelope of this family is then found by eliminating t_1 between $g(x, t; t_1) = 0$ and $\frac{\partial}{\partial t_1} g(x, t; t_1) = 0$.

For further illustration, suppose the path of the piston is given by

$$x_p = f(t) = \alpha t^2.$$

Then, we have

$$g(x, t; t_1) = x - \alpha t_1^2 - [a_0 + (\gamma + 1)\alpha t_1](t - t_1) = 0$$

$$\frac{\partial}{\partial t_1} g(x, t; t_1) = -2\alpha t_1 + a_0 + (\gamma + 1)\alpha t_1 - \alpha(\gamma + 1)(t - t_1) = 0.$$

The envelope is obtained by eliminating t_1 from the above equations

$$x - \frac{a_0^2}{\alpha(\gamma + 1)} = \frac{\alpha(\gamma + 1)^2}{4\gamma}\left[t + \frac{a_0(\gamma - 1)}{\alpha(\gamma + 1)}\right]^2.$$

On the envelope (Figure 15.3), the adjoining C^+ characteristics at infinitesimal distance from each other intersect, and therefore, at such points the

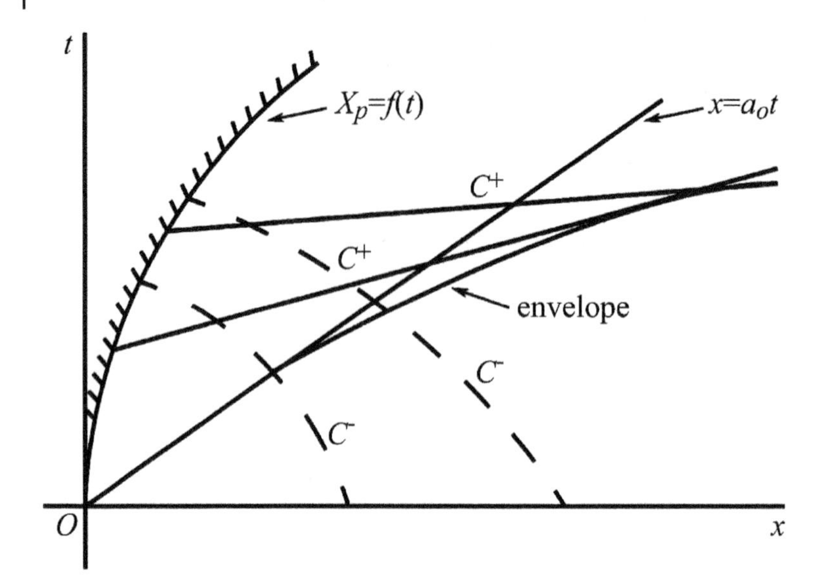

Figure 15.3 Loci of wave fronts produced by the motion of a piston.

flow variables u and a become *multivalued.* This physically impossible situation is then resolved by inserting a shock (Chapter 16) ahead of the envelope.

15.3 Nonlinear Propagation of a Sound Wave

Consider the unidirectional propagation of sound waves in an ideal gas. The governing equations are

$$\frac{\partial \rho}{\partial t} + \frac{\partial}{\partial x}(\rho u) = 0 \tag{13}$$

$$\rho\left(\frac{\partial u}{\partial t} + u\frac{\partial u}{\partial x}\right) = -\frac{\partial p}{\partial x} \tag{14}$$

$$\frac{\partial}{\partial t}\left(\frac{p}{\rho^\gamma}\right) + u\left(\frac{p}{\rho^\gamma}\right) = 0. \tag{15}$$

Noting, from equation (15), that

$$\left(\frac{a}{a_0}\right) = \left(\frac{\rho}{\rho_0}\right)^{\frac{\gamma-1}{2}} \tag{16}$$

and setting

$$\rho = \rho_0 (1 + \hat{\rho}), u \equiv a_0 \hat{u}$$

equations (13) and (14) become

$$\frac{1}{a_0}\frac{\partial \hat{\rho}}{\partial t} + \frac{\partial \hat{u}}{\partial x} = -\hat{u}\frac{\partial \hat{\rho}}{\partial x} - \hat{\rho}\frac{\partial \hat{u}}{\partial x} \tag{17}$$

$$\frac{1}{a_0}\frac{\partial \hat{u}}{\partial t} + \frac{\partial \hat{\rho}}{\partial x} = -\hat{u}\frac{\partial \hat{u}}{\partial x} + (2-\gamma)\frac{\hat{\rho}}{1+\hat{\rho}}\frac{\partial \hat{\rho}}{\partial x}. \tag{18}$$

Let us now treat equations (17) and (18) by using the method of *perturbed characteristics* (Lin, 1954) and (Fox, 1955). We look for a solution of the form (Shivamoggi, 1978),

$$\left.\begin{aligned} \hat{u}(x,t;\varepsilon) &= \varepsilon \hat{u}_1 (s_1, s_2, \tilde{t}) + \varepsilon^2 \hat{u}_2 (s_1, s_2, \tilde{t}) + \cdots \\ \hat{\rho}(x,t;\varepsilon) &= \varepsilon \hat{\rho}_1 (s_1, s_2, \tilde{t}) + \varepsilon^2 \hat{\rho}_2 (s_1, s_2, \tilde{t}) + \cdots \end{aligned}\right\} \tag{19}$$

where $s_{1,2}$ are the characteristic coordinates,

$$s_{1,2} \equiv a_0 t \mp x. \tag{20}$$

Here, $\varepsilon \ll 1$ characterizes the amplitude of the sound wave and $\tilde{t} \equiv \varepsilon t$ represents the *slow time* scale characterizing the *modulation* of the given sound wave.

Substituting (19) and (20) in equations (17) and (18), one obtains on equating coefficients of equal powers of ε to zero,

$$O(\varepsilon): \left(\frac{\partial}{\partial s_2} + \frac{\partial}{\partial s_1}\right)\hat{\rho}_1 + \left(\frac{\partial}{\partial s_2} - \frac{\partial}{\partial s_1}\right)\hat{u}_1 = 0 \tag{21}$$

$$\left(\frac{\partial}{\partial s_2} + \frac{\partial}{\partial s_1}\right)\hat{u}_1 + \left(\frac{\partial}{\partial s_2} - \frac{\partial}{\partial s_1}\right)\hat{\rho}_1 = 0 \tag{22}$$

$$\begin{aligned} O(\varepsilon^2): &\left(\frac{\partial}{\partial s_2} + \frac{\partial}{\partial s_1}\right)\hat{\rho}_2 + \left(\frac{\partial}{\partial s_2} - \frac{\partial}{\partial s_1}\right)\hat{u}_2 \\ &= -\frac{\partial \hat{\rho}_1}{\partial \tilde{t}} - \hat{u}_1\left(\frac{\partial}{\partial s_2} - \frac{\partial}{\partial s_1}\right)\hat{\rho}_1 - \hat{\rho}_1\left(\frac{\partial}{\partial s_2} - \frac{\partial}{\partial s_1}\right)\hat{u}_1 \end{aligned} \tag{23}$$

$$\begin{aligned} &\left(\frac{\partial}{\partial s_2} + \frac{\partial}{\partial s_1}\right)\hat{u}_2 + \left(\frac{\partial}{\partial s_2} - \frac{\partial}{\partial s_1}\right)\hat{\rho}_2 \\ &= -\frac{\partial \hat{u}_1}{\partial \tilde{t}} - \hat{u}_1\left(\frac{\partial}{\partial s_2} - \frac{\partial}{\partial s_1}\right)\hat{u}_1 + (2-\gamma)\hat{\rho}_1\left(\frac{\partial}{\partial s_2} - \frac{\partial}{\partial s_1}\right)\hat{\rho}_1 \end{aligned} \tag{24}$$

$$\vdots$$

One has from equations (21) and (22),

$$4\frac{\partial^2 u_1}{\partial s_1 \partial s_2} = 0. \tag{25}$$

Assuming the given wave to be *right-running*, one has from equations (25) and (21),

$$\hat{u}_1 = f(s_1, \tilde{t}) = \hat{\rho}_1. \tag{26}$$

Using (26), one has from equations (23) and (24),

$$4\frac{\partial^2 \hat{u}_2}{\partial s_1 \partial s_2} = (\gamma + 1)\left[f\frac{\partial^2 f}{\partial s_1^2} + \left(\frac{\partial f}{\partial s_1}\right)^2\right] - 2\frac{\partial^2 f}{\partial s_1 \partial \tilde{t}}. \tag{27}$$

Removal of the *secular* terms on the right-hand side in equation (27) requires

$$(\gamma + 1)\left[f\frac{\partial^2 f}{\partial s_1^2} + \left(\frac{\partial f}{\partial s_1}\right)^2\right] - 2\frac{\partial^2 f}{\partial s_1 \partial \tilde{t}} = 0$$

or

$$\frac{\partial f}{\partial \tilde{t}} - \left(\frac{\gamma + 1}{2}\right) f \frac{\partial f}{\partial s_1} = 0 \tag{28}$$

which gives

$$f = F\left(s_1 + \frac{\gamma + 1}{2} f \tilde{t}\right). \tag{29}$$

Using (29), we have from (26)

$$u \approx \varepsilon F\left[\left(a_0 + \frac{\gamma + 1}{2} u\right) t - x\right]. \tag{30}$$

One observes from (30) that the uniformly valid first-order solution to the nonlinear problem is simply the solution to the corresponding linear problem. However, the linear characteristic is replaced by the characteristic calculated by including the first-order nonlinearities in the problem. In other words, the solution to the linear problem, when the first-order nonlinearities are included, may therefore still have the right form, but not quite at the right place (Whitham, 1952).

Incidentally, as we saw earlier (equation (12)), (30) becomes the exact result if one considers "*simple*" waves for which

$$u + \int a \frac{d\rho}{\rho} \text{ or } u - \int a \frac{d\rho}{\rho} \tag{31}$$

is constant for all x.[5]

15.4 Nonlinear Resonant Three-Wave Interactions of Sound Waves

In an ideal gas, sound waves are nondispersive in the linear regime. So, the frequency and wave vector matching conditions (see (39) below) imply that *resonant wave interaction* in a *triad* of sound waves occurs when the wavevectors are *collinear* (Shivamoggi, 1994).

Consider again unidirectional propagation of sound waves in an ideal fluid. The equations governing this situation are (13)-(15). Setting

$$\rho = \rho_0 (1 + \hat{\rho})$$

and using the relation (16), equations (13) and (14) may be rewritten as

$$\frac{\partial \hat{\rho}}{\partial t} + \frac{\partial u}{\partial x} = -u\frac{\partial \hat{\rho}}{\partial x} - \hat{\rho}\frac{\partial u}{\partial x} \tag{32}$$

5 Suppose

$$\mathcal{I}_- = u - \int a \frac{d\rho}{\rho} = const = 0, \forall x$$

from which,

$$a = a_0 + \frac{\gamma - 1}{2} u, \forall x.$$

Therefore,

$$\mathcal{I}_+ = u + 2\left(\frac{a - a_0}{\gamma - 1}\right) = 2u = const \text{ along } \frac{dx}{dt} = u + a = a_0 + \frac{\gamma + 1}{2} u$$

from which,

$$u = \varepsilon F\left[\left(a_0 + \frac{\gamma + 1}{2} u\right) t - x\right].$$

$$\frac{\partial u}{\partial t} + a_0^2 \frac{\partial \hat{\rho}}{\partial x} = -u\frac{\partial u}{\partial x} + a_0^2 \frac{(2-\gamma)\hat{\rho}}{1+\hat{\rho}}\frac{\partial \hat{\rho}}{\partial x}. \tag{33}$$

Observe that, in equations (32) and (33), the left-hand sides represent the linear problem and the right-hand sides contain the nonlinearities that give rise to the interactions among the sound waves.

Consider the waves of the form,

$$u(x,t) \text{ and } \hat{\rho}(x,t) \sim e^{i(kx-\omega t)} \tag{34}$$

where

$$\frac{\omega}{k} = a_0.$$

Let us use the method introduced in Chapter 10, to treat the nonlinear *three-wave resonant interactions* of sound waves. In this method, one introduces a new dynamical variable,

$$b \equiv u + \frac{\omega}{k}\hat{\rho} \tag{35}$$

which turns out to be a normal mode of the linearized problem associated with equations (17) and (18). This may be verified by noting that the linear parts of equations (32) and (33) lead to

$$\frac{\partial b}{\partial t} + i\omega b = 0. \tag{36}$$

So, b is a *normal mode* of the linearized problem associated with equations (32) and (33).

The nonlinear analysis of the wave-wave interactions (as will be seen below) becomes very convenient when formulated in terms of this auxiliary variable b. When the nonlinear terms are included, one obtains from equations (32) and (33),

$$\frac{\partial b}{\partial t} + i\omega b = -u\frac{\partial u}{\partial x} + \frac{\omega^2}{k^2}(2-\gamma)\hat{\rho}(1-\hat{\rho})\frac{\partial \hat{\rho}}{\partial x} + \frac{\omega}{k}\left(-u\frac{\partial \hat{\rho}}{\partial x} - \hat{\rho}\frac{\partial u}{\partial x}\right). \tag{37}$$

Let us now consider two sound waves of the form $e^{i(k_3 x-\omega_3 t)}$ and $e^{i(k_2 x-\omega_2 t)}$ propagating in the x-direction with equal phase velocities,

$$\frac{\omega_3}{k_3} = \frac{\omega_2}{k_2} \equiv a_0. \tag{38}$$

Due to nonlinear interaction between these two waves, let another sound wave of the form $e^{i(k_1 x - \omega_1 t)}$, propagating with the same phase velocity in the x-direction, be excited such that

$$\omega_3 - \omega_2 = \omega_1, \quad k_3 - k_2 = k_1 \tag{39}$$

where $\omega_1/k_1 = a_0$.

The nonlinear terms on the right hand side in equation (37) represent coupling of these sound waves. If one sets

$$b_j(x,t) = \tilde{b}_j(t) e^{i(k_j x - \omega_j t)} + \tilde{b}_j^*(t) e^{-i(k_j x - \omega_j t)}, j = 1, 2, 3, \tag{40}$$

then these nonlinear terms lead to slow variations with time in the amplitudes $\tilde{b}_j(t)$. Now, note that we have from the linearized problem associated with equations (32) and (33),

$$u = \frac{b}{2}, \hat{\rho} = \frac{k}{2\omega} b. \tag{41}$$

Using (41) on the right-hand side in equation (37), and keeping only the resonant terms (according to (39)), we obtain the following amplitude modulation equations,

$$\begin{aligned}\frac{\partial \tilde{b}_3}{\partial t} = \frac{i/4}{\omega_1 \omega_2 k_3} &[\omega_3 k_2 (\omega_3 k_1 - \omega_1 k_3) - \omega_2 k_3 (\omega_3 k_1 + \omega_1 k_3) \\ &- (\gamma - 1) \omega_3 (\omega_1 k_2^2 + \omega_2 k_1^2)] \tilde{b}_1 \tilde{b}_2\end{aligned} \tag{42}$$

$$\begin{aligned}\frac{\partial \tilde{b}_2}{\partial t} = \frac{-i/4}{\omega_3 \omega_1 k_2} &[\omega_2 k_1 (\omega_2 k_3 - \omega_3 k_2) - \omega_1 k_2 (\omega_2 k_3 + \omega_3 k_2) \\ &- (\gamma - 1) \omega_2 (\omega_3 k_1^2 + \omega_1 k_3^2)] \tilde{b}_3 \tilde{b}_1^*\end{aligned} \tag{43}$$

$$\begin{aligned}\frac{\partial \tilde{b}_1}{\partial t} = \frac{-i/4}{\omega_2 \omega_3 k_1} &[\omega_1 k_3 (\omega_1 k_2 - \omega_2 k_1) - \omega_3 k_1 (\omega_1 k_2 + \omega_2 k_1) \\ &- (\gamma - 1) \omega_1 (\omega_2 k_3^2 + \omega_3 k_2^2)] \tilde{b}_3 \tilde{b}_2^*.\end{aligned} \tag{44}$$

Expressing $\tilde{b}_j$'s in terms of $\tilde{\rho}_j$'s, recalling (38), and setting

$$\tilde{\rho}_j(t) = \left(i \frac{2}{\gamma + 1} \sqrt{\omega_j}\right) \tilde{\phi}_j(\hat{t}), \hat{t} \equiv \sqrt{\omega_1 \omega_2 \omega_3} t \tag{45}$$

equations (42)-(44) become

$$\frac{d\tilde{\phi}_3}{d\hat{t}} = \tilde{\phi}_1 \tilde{\phi}_2 \tag{46}$$

$$\frac{d\tilde{\phi}_2}{d\hat{t}} = -\tilde{\phi}_3 \tilde{\phi}_1^* \tag{47}$$

$$\frac{d\tilde{\phi}_1}{d\hat{t}} = -\tilde{\phi}_3\tilde{\phi}_2^*. \tag{48}$$

Equations (46)-(48) show that the three sound waves have *positive* energy which means that the wave interactions under consideration are *nonexplosive*.[6]

One obtains from equations (46)-(48), the following *Manley-Rowe* (1959) *relations*:

$$\frac{d}{d\hat{t}}\left(\left|\tilde{\phi}_1\right|^2 + \left|\tilde{\phi}_3\right|^2\right) = 0 \tag{49}$$

$$\frac{d}{d\hat{t}}\left(\left|\tilde{\phi}_2\right|^2 + \left|\tilde{\phi}_3\right|^2\right) = 0 \tag{50}$$

$$\frac{d}{d\hat{t}}\left(\left|\tilde{\phi}_1\right|^2 - \left|\tilde{\phi}_2\right|^2\right) = 0. \tag{51}$$

Equations (49)-(51) imply a periodic exchange of energy among the three sound waves $\tilde{\phi}_1, \tilde{\phi}_2$, and $\tilde{\phi}_3$. Consider, for instance, a case wherein only the low-frequency modes $\tilde{\phi}_1$ and $\tilde{\phi}_2$ are present initially, and the high-frequency mode $\tilde{\phi}_3$ is absent, i.e.,

$$\hat{t} = \hat{t}_0 \ : \ \tilde{\phi}_{1,2} = \tilde{\phi}_{1,2}^{(0)}, \tilde{\phi}_3 = 0. \tag{52}$$

Suppose that $\left|\tilde{\phi}_2^{(0)}\right|^2 < \left|\tilde{\phi}_1^{(0)}\right|^2$. Equations (49) and (50) then show that initially the energy in mode $\tilde{\phi}_3$ increases at the expense of both modes $\tilde{\phi}_1$ and $\tilde{\phi}_2$. Eventually, the energy in mode $\tilde{\phi}_2$ vanishes $\left(\text{since } \left|\tilde{\phi}_2^{(0)}\right|^2 < \left|\tilde{\phi}_1^{(0)}\right|^2\right)$. Then, the direction of energy transfer is reversed: modes $\tilde{\phi}_1$ and $\tilde{\phi}_2$ now increase at the expense of the mode $\tilde{\phi}_3$ until the initial state is attained again, and this sequence of energy transfer repeats itself.

Using (49) and (50), (46) leads to

$$\left|\frac{\partial\tilde{\phi}_3}{\partial\hat{t}}\right|^2 = \left|\tilde{\phi}_1^{(0)}\right|^2\left|\tilde{\phi}_2^{(0)}\right|^2\left[1 - \frac{\left|\tilde{\phi}_3\right|^2}{\left|\tilde{\phi}_1^{(0)}\right|^2}\right]\left[1 - \frac{\left|\tilde{\phi}_3\right|^2}{\left|\tilde{\phi}_2^{(0)}\right|^2}\right] \tag{53}$$

from which we have,

$$\tilde{\phi}_3 = \tilde{\phi}_2^{(0)} sn\,(\xi; m)\,. \tag{54a}$$

---

6 In an *explosive* interaction, all interacting waves attain infinite amplitude in a *finite* time.

Using (54a), equations (49) and (50) give

$$\left.\begin{aligned}\tilde{\Phi}_2 &= \tilde{\Phi}_2^{(0)} cn\,(\xi; m)\\ \tilde{\Phi}_1 &= \tilde{\Phi}_1^{(0)} dn\,(\xi; m)\end{aligned}\right\}. \tag{54b}$$

Here sn, cn, dn are elliptic functions with real parameter, and

$$\xi \equiv \left|\tilde{\Phi}_1^{(0)}\right| (\hat{t} - \hat{t}_0) \tag{55a}$$

and m is the *modulus*,

$$m^2 \equiv \left|\frac{\tilde{\Phi}_2^{(0)}}{\tilde{\Phi}_1^{(0)}}\right|^2. \tag{55b}$$

Let us assume that $m < 1$. Noting that

$$\frac{\dfrac{\partial}{\partial \hat{t}}\left|\dfrac{\tilde{\Phi}_1}{\tilde{\Phi}_1^{(0)}}\right|^2}{\dfrac{\partial}{\partial \hat{t}}\left|\dfrac{\tilde{\Phi}_2}{\tilde{\Phi}_2^{(0)}}\right|^2} = \frac{\left|\tilde{\Phi}_2^{(0)}\right|^2}{\left|\tilde{\Phi}_1^{(0)}\right|^2} = m^2 \tag{56}$$

m may be regarded as a measure of the extent to which resonant partners participate in the interaction. In particular, $m < 1$ implies that the mode $\tilde{\Phi}_2$ is decreasing at a rate faster than the mode $\tilde{\Phi}_1$. Using the solutions (54a) and (55a), the period of the resonant energy exchange among the three modes $\tilde{\Phi}_1, \tilde{\Phi}_2$, and $\tilde{\Phi}_3$ is given by

$$\hat{T} = \frac{2}{\left|\tilde{\Phi}_1^{(0)}\right|} K\,(m) \tag{57}$$

where $K\,(m)$ is the *complete elliptic integral of the first kind.*

The case $\omega_1 = \omega_2 = \dfrac{\omega_3}{2}$ and $\tilde{\Phi}_1^{(0)} = \tilde{\Phi}_2^{(0)}$ constitutes the degenerate case of the triad resonances for which two members of the triad are identical, the closure (the third member) being their second harmonic.[7] Since $m = 1$ for this case, and

$$m \approx 1 : K\,(m) \sim ln \frac{2}{\sqrt{1-m}} \Rightarrow \infty \tag{58}$$

7 For this case, the fundamental component and its second harmonic have the same linear phase velocity so that the two can interact *resonantly* with each other.

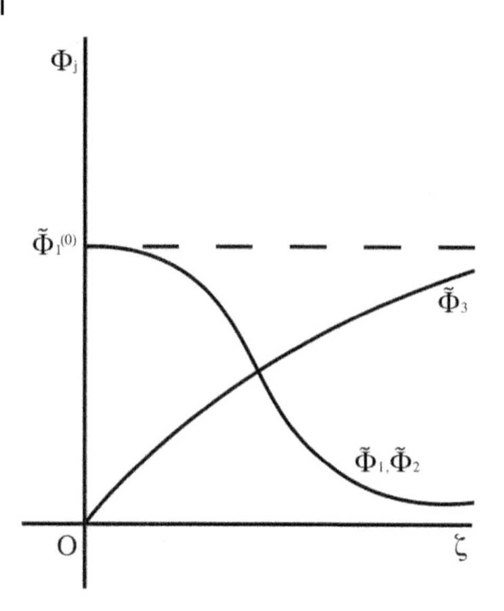

Figure 15.4 The second harmonic reconance.

the period $\hat{T}$ of the energy transfer from (57) is infinite. The resonant interactions now take on an asymptotic character, and (54a) then becomes

$$\left.\begin{aligned}\tilde{\phi}_3 &= \tilde{\phi}_1^{(0)} \tanh \xi \\ \tilde{\phi}_2 = \tilde{\phi}_1 &= \tilde{\phi}_1^{(0)} \operatorname{sech} \xi\end{aligned}\right\}. \tag{59}$$

Equations (46)-(48) also indicate the development of a *nonlinear saturation* state under certain conditions (Figure 15.4). In order to see this, we first derive, from equations (46)-(48), the following equations:

$$\frac{d^2\tilde{\phi}_3}{d\hat{t}^2} + \left(\left|\tilde{\phi}_2\right|^2 + \left|\tilde{\phi}_1\right|^2\right)\tilde{\phi}_3 = 0 \tag{60}$$

$$\frac{d^2\tilde{\phi}_2}{d\hat{t}^2} + \left(\left|\tilde{\phi}_1\right|^2 - \left|\tilde{\phi}_3\right|^2\right)\tilde{\phi}_2 = 0 \tag{61}$$

$$\frac{d^2\tilde{\phi}_1}{d\hat{t}^2} + \left(\left|\tilde{\phi}_2\right|^2 - \left|\tilde{\phi}_3\right|^2\right)\tilde{\phi}_1 = 0. \tag{62}$$

Consider next, the case wherein the high-frequency mode $\tilde{\phi}_3$ is initially much stronger than the low-frequency modes $\tilde{\phi}_1$ and $\tilde{\phi}_2$, i.e.,

$$\hat{t} = \hat{t}_0 : \left|\tilde{\phi}_3\right| >> \left|\tilde{\phi}_1\right| \text{ and } \left|\tilde{\phi}_2\right|. \tag{63}$$

One may then take $\tilde{\phi}_3$ to be constant for small times subsequent to $\hat{t} = \hat{t}_0$ in the evolution of $\tilde{\phi}_1$ and $\tilde{\phi}_2$. If we assume further that,

$$\tilde{\phi}_1 \text{ and } \tilde{\phi}_2 \sim e^{\gamma\hat{t}} \tag{64}$$

we obtain from equations (61) and (62):

$$\gamma \approx \left|\tilde{\phi}_3\right|. \tag{65}$$

We observe from (64) and (65) that, for small times subsequent to $\hat{t} = \hat{t}_0$, the high-frequency mode $\tilde{\phi}_3$ undergoes a *decay instability* into two low-frequency modes $\tilde{\phi}_1$ and $\tilde{\phi}_2$. These grow exponentially until the nonlinear regime is reached where a *saturation* state develops. In this *nonlinear saturation* state, the amplitudes of the three modes are

$$\left|\tilde{\phi}_1\right| \approx \left|\tilde{\phi}_2\right| \approx \left|\tilde{\phi}_3\right|. \tag{66}$$

Thus, depending on the particular set of initial conditions, nonlinear resonant three-wave interactions of sound waves exhibit two different evolution scenarios as given below,

- a *periodic* exchange of energy among three sound waves (an exception occurs for the degenerate case with two members of the triad identical, for which the period of the energy exchange become infinite);
- a *decay instability* of the high-frequency mode into two low-frequency modes followed by a nonlinear saturation state.

15.5 Burgers Equation

A real fluid cannot sustain an actual discontinuity like a shock wave (Chapter 16). So, a shock wave is only an idealization of the sharp gradients in the flow variables that occur in reality. Consequent to these flow gradients, various transport processes such as those due to viscosity and heat conductivity show up inside the shock. The evolution of the flow structure within the shock when viscosity and heat conductivity are included can in fact be satisfactorily described by an approximate model represented by *Burgers* (1948) *equation*. We will now divert from the inviscid fluid model considered in this chapter to discuss Burgers equation. This is because this discussion is done within the framework of nonlinear acoustics rather than shock waves or viscous fluids.

Let us consider here, weak shock waves so that the flow of interest is a perturbation on a uniform sonic flow of velocity a^*. The equations governing the flow are those describing the conservation of mass, momentum, and energy,[8]

$$\frac{D\rho}{Dt} + \rho\frac{\partial u}{\partial x} = 0 \tag{67}$$

---

8 Equation (69) is one-dimensional version of equation (22) in Chapter 2 along with the assumption of perfect gas model for the fluid.

$$\rho \frac{Du}{Dt} = -\frac{\partial p}{\partial x} + \frac{4}{3}\mu \frac{\partial^2 u}{\partial x^2} \tag{68}$$

$$\frac{1}{\gamma - 1}\frac{Dp}{Dt} - \frac{\gamma p/\rho}{\gamma - 1}\frac{D\rho}{Dt} = \frac{4}{3}\mu \left(\frac{\partial u}{\partial x}\right)^2 + K\frac{\partial^2 T}{\partial x^2} \tag{69}$$

where μ, the coefficient of viscosity, and K, the coefficient of heat conductivity are taken to be constants, and

$$\frac{D}{Dt} \equiv \frac{\partial}{\partial t} + u\frac{\partial}{\partial x}.$$

Using equation (67), equation (69) becomes

$$\frac{Dp}{Dt} + a^2 \rho \frac{\partial u}{\partial x} - (\gamma - 1) K \frac{\partial^2 T}{\partial x^2} - \frac{4}{3}\mu (\gamma - 1)\left(\frac{\partial u}{\partial x}\right)^2 = 0. \tag{70}$$

Suppose that the temporal variations and dissipative effects are both of first order. Then, to the zeroth order, we have

$$\left.\begin{aligned} dp &= -\rho u du \\ C_p dT &= -u du \end{aligned}\right\} \tag{71}$$

and dropping $(\partial u/\partial x)^2$ in comparison with $u\,(\partial^2 u/\partial x^2)$, equation (70) gives

$$-\rho u \frac{\partial u}{\partial t} + u\frac{\partial p}{\partial x} + \rho a^2 \frac{\partial u}{\partial x} + \frac{K}{C_p}(\gamma - 1)\, u \frac{\partial^2 u}{\partial x^2} = 0. \tag{72}$$

Now, on multiplying equation (68) through by u gives,

$$\rho u \frac{\partial u}{\partial t} + \rho u^2 \frac{\partial u}{\partial x} + u\frac{\partial p}{\partial x} - \frac{4}{3}\mu u \frac{\partial^2 u}{\partial x^2} = 0. \tag{73}$$

One obtains from equations (72) and (73),

$$2\rho u \frac{\partial u}{\partial t} + \rho\left(u^2 - a^2\right)\frac{\partial u}{\partial x} - \frac{4}{3}\mu u \left(1 + \frac{\gamma - 1}{Pr}\right)\frac{\partial^2 u}{\partial x^2} = 0 \tag{74}$$

where

$$Pr \equiv \frac{\mu C_p}{K}.$$

For weak shocks, one may write[9]

$$u^2 - a^2 \approx (\gamma + 1) u (u - a^*) \tag{75}$$

so that equation (74) becomes

$$\frac{\partial u}{\partial t} + \left(\frac{\gamma + 1}{2}\right)(u - a^*)\frac{\partial u}{\partial x} = \nu \frac{\partial^2 u}{\partial x^2} \tag{76}$$

where

$$\nu \equiv \frac{2}{3}\frac{\mu}{\rho}\left(1 + \frac{\gamma - 1}{Pr}\right).$$

Setting

$$u(x,t) = \left(\frac{2}{\gamma + 1}\right) W(X,t), X \equiv x - \left(\frac{\gamma + 1}{2}\right) a^* t \tag{77}$$

equation (76) becomes

$$\frac{\partial W}{\partial t} + W\frac{\partial W}{\partial X} = \nu \frac{\partial^2 W}{\partial X^2} \tag{78}$$

which is *Burgers equation*. Burgers equation is the simplest equation combining the *nonlinear convection* effects and the *dissipative diffusion* effects.

9 On using equation (7) in Chapter 13, we have

$$\begin{aligned} u^2 - a^2 &= u^2 - \left(a_0^2 - \frac{\gamma - 1}{2} u^2\right) \\ &= \frac{\gamma + 1}{2}\left(u^2 - a^{*2}\right) \\ &\approx (\gamma + 1) u (u - a^*) \end{aligned}$$

An explicit solution to the initial-value problem associated with Burgers equation can in fact be found, thanks to the *Hopf* (1950)-*Cole* (1951) *tranformation*. Setting

$$W \equiv -2\nu \frac{\psi_X}{\psi} \tag{79}$$

one finds that equation (78) gives the heat equation,

$$\psi_t = \nu \psi_{XX}. \tag{80}$$

Let us prescribe an initial condition,

$$t = 0 : W = F(X), -\infty < X < \infty \tag{81}$$

and from (79), this gives

$$t = 0 : \psi = \Psi(X) \equiv \exp\left(-\frac{1}{2\nu}\int_{0^+}^{X} F(\eta)\, d\eta\right). \tag{82}$$

The lower-limit of the integral in (82) is arbitrary, and we take it to be 0^+ to suit the following development.

One may now solve equation (80), using the initial condition (82), to obtain

$$\psi(X,t) = \frac{1}{\sqrt{4\pi\nu t}} \int_{-\infty}^{\infty} \Psi(\eta) \exp\left[-\frac{(X-\eta)^2}{4\nu t}\right] d\eta. \tag{83}$$

Using (82) and (83), (79) gives

$$W(X,t) = \frac{\int_{-\infty}^{\infty} \left(\frac{X-\eta}{t}\right) \exp\left(-\frac{G}{2\nu}\right) d\eta}{\int_{-\infty}^{\infty} \exp\left(-\frac{G}{2\nu}\right) d\eta} \tag{84}$$

where

$$G(\eta) \equiv \int_{0^+}^{\eta} F(\eta')\, d\eta' + \frac{(X-\eta)^2}{2t}. \tag{85}$$

As an example, let us consider a single-hump solution,

$$F(X) = A\delta(X). \tag{86}$$

Then, (85) gives

$$G(\eta) = \frac{(X-\eta)^2}{2t} - AH(-\eta). \tag{87}$$

$H(x)$ being the *Heaviside unit step function*. Using (87), (84) gives

$$W = \sqrt{\frac{\nu}{t}} \frac{(e^R - 1)\exp\left(-\frac{X^2}{4\nu t}\right)}{\sqrt{\pi} + (e^R - 1) \int\limits_{X/\sqrt{4\nu t}}^{\infty} e^{-\xi^2} d\xi} \tag{88}$$

where R is the *Reynolds number*,

$$R \equiv \frac{A}{2\nu}.$$

Note, from (88), that

$$R \ll 1 : W \approx \sqrt{\frac{\nu}{\pi t}} R \exp\left(-\frac{X^2}{4\nu t}\right) \tag{89}$$

which is the same as the result one obtains on neglecting the nonlinear convective term in equation (78)! In this limit, the initial discontinuity spreads without propagation as t increases.

In order to investigate the inviscid limit $R \gg 1$, first introduce

$$Z \equiv \frac{X}{\sqrt{2At}} \tag{90}$$

so that (88) becomes

$$W = \sqrt{\frac{2A}{t}} \frac{(e^{R-1})}{2\sqrt{R}} \frac{e^{-Z^2 R}}{\sqrt{\pi} + (e^R - 1) \int\limits_{Z\sqrt{R}}^{\infty} e^{-\xi^2} d\xi}. \tag{91}$$

Now, note from (91) that

$$\left.\begin{aligned} R \gg 1 : Z < 0 : W &\sim \sqrt{\frac{2A}{t}} \frac{1}{2\sqrt{\pi R}} e^{-Z^2 R} \\ Z > 0 : W &\sim \sqrt{\frac{2A}{t}} \frac{Z}{1 + 2Z\sqrt{\pi R} e^{R(Z^2 - 1)}} \end{aligned}\right\} \tag{92}$$

where we have used the results,

$$\int_{-\infty}^{\infty} e^{-\xi^2} d\xi = \sqrt{\pi}, \quad \int_{\eta}^{\infty} e^{-\xi^2} d\xi \approx \frac{e^{-\eta^2}}{2\eta} \text{ as } \eta \Rightarrow \infty.$$

Thus, we have

$$R \gg 1 : \left. \begin{array}{r} Z < 0 : W \approx 0 \\ 0 < Z < 1 : W \sim \sqrt{\dfrac{2A}{t}} Z \\ Z > 1 : W \Rightarrow 0 \end{array} \right\} \tag{92a}$$

or

$$W \sim \begin{cases} X/t, & 0 < X < \sqrt{2At} \\ 0, & \text{outside} \end{cases} \tag{92b}$$

which represents a shock at $X = \sqrt{2At}$ with velocity $U = \sqrt{A/2t}$ (see Figure 15.5). The *shock condition* (see Section 16.1),

$$U = \frac{W_+ + W_-}{2} \tag{93}$$

is satisfied because W jumps from 0 to $\sqrt{2A/t}$ across the shock.

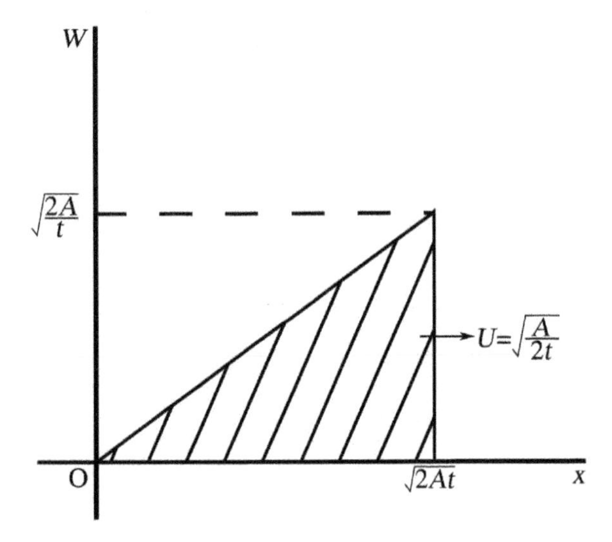

Figure 15.5 Shock structure in the R ≫ 1 limit.

Note that the area under the shock profile remains constant even when diffusion is included because we have, from equation (78),

$$\frac{d}{dt}\int_{-\infty}^{\infty} W dx = \left(\nu W_x - \frac{1}{2}W^2\right)\Big|_{-\infty}^{\infty} = 0. \tag{94}$$

Indeed, the area under the shock profile at any t is

$$\frac{1}{2}\sqrt{2At}\cdot\sqrt{\frac{2A}{t}} = A$$

which is equal to the value of this area at $t = 0$:

$$\int_{-\infty}^{\infty} A\delta(x)\,dx = A.$$

Let us next find a stationary solution of equation (78) with,

$$W(X,t) = W(\xi) \tag{95}$$

where

$$\xi \equiv X - Ut.$$

Equation (78) now becomes

$$-UW_\xi + WW_\xi = \nu W_{\xi\xi}. \tag{96}$$

On integrating once, equation (96) gives

$$\frac{1}{2}W^2 - UW + C = \nu W_\xi \tag{97}$$

where C is an arbitrary constant.

Let us impose infinity conditions appropriate for a shock wave,

$$\xi \Rightarrow \pm\infty : W \Rightarrow W_{1,2}. \tag{98}$$

Using (98), equation (97) gives

$$U = \frac{1}{2}(W_1 + W_2),\ \ C = \frac{1}{2}W_1W_2. \tag{99a,b}$$

We observe that (99a,b) is in agreement with (93), as it should be.

Using (99a,b), equation (97) becomes

$$(W - W_1)(W_2 - W) = -2\nu W_\xi \tag{100}$$

from which we obtain,

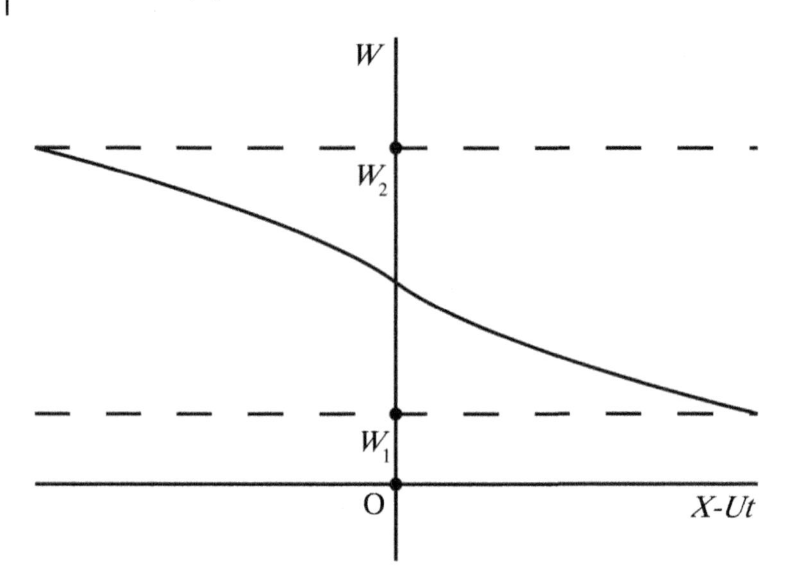

Figure 15.6 The dispersed shock.

$$\frac{\xi}{\nu} = \frac{2}{W_2 - W_1} ln \left(\frac{W_2 - W}{W - W_1} \right) \tag{101a}$$

or

$$W = W_1 + \frac{W_2 - W_1}{1 + \exp\left(\frac{W_2 - W_1}{2\nu}\right)(X - Ut)} \tag{101b}$$

which represents a shock wave connecting two different uniform states given by (98) (see Figure 15.6).

Thus, the diffusive term on the right-hand side in equation (78) prevents the development of steep wave profiles and tends to spread the sharp discontinuities into smooth profiles.

Exercises

1. A source of finite duration produces a wavemotion only in the plane case. Show that in the spherical case, the time-integral of the wave-amplitude at any fixed location vanishes.
2. Consider a semi-infinite tube containing a gas which is at rest at $t = 0$. Suppose that waves are produced by a piston which moves out of this tube for time $t > 0$, with constant velocity u_0. Show that the ensuing flow consists of two uniform regions in the (x, t)-plane, joined by an expansion fan for

$-a_0 t < x < \left(\frac{\gamma+1}{2}u_0 - a_0\right)t$. Furthermore, deduce that this expansion fan region has a similarity solution (depending only on the variable x/t) given by

$$u = \frac{2}{\gamma+1}\left(a_0 + \frac{x}{t}\right), \quad a = \frac{2a_0}{\gamma+1} - \left(\frac{\gamma-1}{\gamma+1}\right)\frac{x}{t}.$$

3. Consider the piston problem, in which a piston at rest begins to move smoothly at time $t = 0$ with a given displacement $X(t)$, so that one has the boundary condition,

$$x = X(t) : u = X'(t)H(t).$$

Show that the solution to this problem is given by

$$u = X'(\phi)H\left(t \mp \frac{x}{a_0}\right)$$

where

$$\phi = t - \frac{x - X(\phi)}{\beta X'(\phi) \pm a_0}, \quad \beta \equiv \frac{\gamma+1}{2}.$$

Note that the parameter ϕ represents the time at which a given signal left the piston.

16

Shock Waves

As we saw in Chapter 15, in the absence of dissipative effects such as viscosity and heat conductivity, compressional disturbances propagate with speeds that increase with compression. This leads to a continual steepening of wave forms that eventually cannot be expressed by single-valued functions of position. This difficulty is resolved by introducing a discontinuity like a shock wave, into the flow. A *shock wave* is a surface in a flow field across which the flow variables change discontinuously. Additionally, the existence of shock waves allows one to admit certain types of boundary conditions that cannot be satisfied in a continuous flow.

However, an actual discontinuity is impossible in a real fluid. So, the discontinuity is merely an idealization of the sharp gradients in the flow variables that occur in reality in a shock wave. As a consequence of these flow gradients, various transport processes such as those due to viscosity and heat conductivity materialize inside the shock. When such processes are included, an exact solution for the flow structure within the shock can be found for the particular case of Prandtl number $P_r \equiv \mu C_p/k = 3/4$. (see Section 19.10).

16.1 The Normal Shock Wave

(i) The Jump Conditions

The jumps in flow properties which occur across a shock wave are prescribed by the equations governing the ideal fluid flow *without* reference to the specific dissipation processes occurring within the shock wave. The dissipation processes influence only the structure within the shock wave.[1] Consider a

1 It may be noted that *real gas effects* can provide additional *thermodynamic relaxation processes* to support the flow gradients in a shock wave. In case of the very weakest shocks, the

Introduction to Theoretical and Mathematical Fluid Dynamics, Third Edition.
Bhimsen K. Shivamoggi.

stationary shock wave with its plane normal to the flow. One has for a one-dimensional flow of an ideal gas,

$$\frac{\partial \rho}{\partial t} + u\frac{\partial \rho}{\partial x} + \rho\frac{\partial u}{\partial x} = 0 \tag{1}$$

$$\rho\left(\frac{\partial u}{\partial t} + u\frac{\partial u}{\partial x}\right) + \frac{\partial p}{\partial x} = 0 \tag{2}$$

$$\frac{\partial e}{\partial t} + u\frac{\partial e}{\partial x} - \frac{p}{\rho^2}\left(\frac{\partial \rho}{\partial t} + u\frac{\partial \rho}{\partial x}\right) = 0. \tag{3}$$

Equations (1)–(3) may be rewritten in *conservation* form as follows,

$$\frac{\partial \rho}{\partial t} + \frac{\partial}{\partial x}(\rho u) = 0 \tag{4}$$

$$\frac{\partial}{\partial t}(\rho u) + \frac{\partial}{\partial x}(\rho u^2 + p) = 0 \tag{5}$$

$$\frac{\partial}{\partial t}\left(\frac{1}{2}\rho u^2 + \rho e\right) + \frac{\partial}{\partial x}\left[\left(\frac{1}{2}\rho u^2 + \rho e\right)u + pu\right] = 0. \tag{6}$$

Equations (4)–(6) may be represented by a nonlinear hyperbolic equation of the form,

$$\frac{\partial}{\partial x}[A(u)] + \frac{\partial u}{\partial t} = 0 \tag{7}$$

which may be rewritten as

$$\nabla \cdot \mathbf{F} = 0 \tag{8}$$

where

$$\mathbf{F} \equiv [A(u), u]$$
$$\nabla \equiv \hat{\mathbf{i}}_x\frac{\partial}{\partial x} + \hat{\mathbf{i}}_t\frac{\partial}{\partial t}. \tag{9}$$

---

process with the longest relaxation time is adequate to support the concomitant flow gradient. However, when the shock becomes stronger, the next longest relaxation-time process is called into play to sustain the main shock transition. This is followed by a slower relaxation involving only the longer relaxation-time process. Eventually, the shock becomes so strong that all these thermodynamic relaxations are insufficient to support the steep flow gradients occurring inside the shock. Strongest dissipative processes like viscosity and thermal conductivity are then called into play to sustain the main shock transition, followed by slower thermodynamic relaxations.

If φ is a C^∞ test function with compact support in the (x, t) plane, then equation (8) is equivalent to

$$\iint \varphi \nabla \cdot \mathbf{F}\, dxdt = 0, \quad \forall \varphi. \tag{10}$$

On integrating by parts, equation (10) leads to

$$\iint \nabla \varphi \cdot \mathbf{F}\, dxdt = 0, \quad \forall \varphi. \tag{11}$$

If u is smooth, equations (8)–(11) are, of course, equivalent. However, if u is not smooth, equation (11) may remain valid even when equation (8) does not. Equation (8) is then said to be valid in the sense of *distributions*. In fact, u is defined to be a *weak* solution of equation (8), if it satisfies equation (11) for all smooth functions φ with compact support. Note that a *weak* solution is capable of exhibiting a discontinuous behavior.

Let us now consider properties of a *weak* solution u of equation (8) near a jump discontinuity given by a smooth curve Σ in the (x, t) plane. If φ is a smooth function vanishing outside a region S that is divided into S_1 and S_2 by the curve Σ (see Figure 16.1), so that $S = S_1 U S_2$, one has, from equation (11),

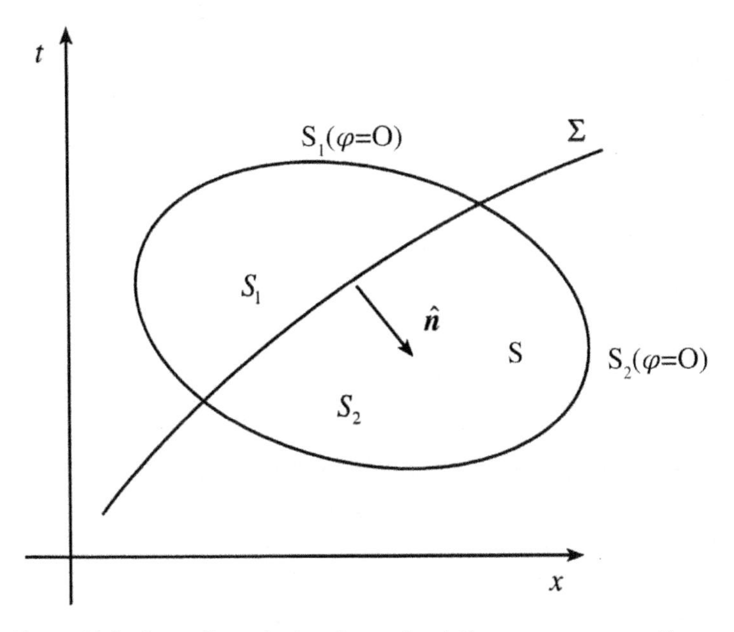

Figure 16.1 Jump discontinuity of a weak solution q across a curve Σ.

$$\iint_S \nabla\varphi \cdot \mathbf{F}\,dxdt = \iint_{S_1} \nabla\varphi \cdot \mathbf{F}\,dxdt + \iint_{S_2} \nabla\varphi \cdot \mathbf{F}\,dxdt = 0. \tag{12}$$

Note that

$$\iint_{S_1} \nabla\varphi \cdot \mathbf{F}\,dxdt = \iint_{S_1} \nabla \cdot (\varphi\mathbf{F})\,dxdt - \iint_{S_1} \varphi\nabla \cdot \mathbf{F}\,dxdt.$$

If u is assumed to be smooth in S_1, then equation (8) prevails in S_1, and one obtains

$$\iint_{S_1} \nabla\varphi \cdot \mathbf{F}\,dxdt = \int_\Sigma \varphi\mathbf{F}_1 \cdot \hat{\mathbf{n}}ds + \int_{\partial S_1} \varphi\mathbf{F} \cdot \hat{\mathbf{n}}ds = \int_\Sigma \varphi\mathbf{F}_1 \cdot \hat{\mathbf{n}}ds$$

where $\mathbf{F}_1$ denotes the value taken by $\mathbf{F}$ on Σ as the limit is taken from the region S_1. Similarly, one obtains

$$\iint_{S_2} \nabla\varphi \cdot \mathbf{F}\,dxdt = -\int_\Sigma \varphi\mathbf{F}_2 \cdot \hat{\mathbf{n}}ds$$

where the negative sign on the right-hand side signifies the fact that the outward normal $\hat{\mathbf{n}}$ for S_1 is the inward normal for S_2.

Substituting these results, equation (12) leads to

$$\int_\Sigma \varphi\,(\mathbf{F}_1 - \mathbf{F}_2) \cdot \hat{\mathbf{n}}ds = 0, \quad \forall\varphi$$

from which, one obtains the *jump condition*,

$$[\mathbf{F} \cdot \hat{\mathbf{n}}] = 0 \quad \text{on} \quad \Sigma \tag{13}$$

where the rectangular bracket denotes the jump of the contents across Σ.

If the curve Σ is parametrized by t and is given by $x = x\,(t)$, one has for the normal vector

$$\hat{\mathbf{n}} = \frac{1}{\sqrt{1+U^2}} < 1, -U >$$

where $U \equiv dx/dt$ is the speed of the discontinuity. The jump condition (13) then becomes

$$[A(u)] - U[u] = 0 \quad \text{on } \Sigma. \tag{14}$$

Thus, a *weak* solution u satisfies equation (7) where possible, and (14) across a *jump discontinuity*[2] Σ moving with speed U.

Example 1: Consider a nonlinear hyperbolic equation (of the convective nonlinearity type),

$$u_t + uu_x = 0.$$

Here, we have

$$A(u) = \frac{1}{2}u^2.$$

So, the speed of the discontinuity in this case is given by

$$U = \frac{u_+ + u_-}{2}$$

which is the average of the values of u ahead and behind the shock.

We now apply the jump conditions (14) that are valid across a discontinuity $\sum$ to equations (4)–(6) to obtain

$$-U[\rho] + [\rho u] = 0 \tag{15}$$

$$-U[\rho u] + \left[\rho u^2 + p\right] = 0 \tag{16}$$

$$-U\left[\frac{1}{2}\rho u^2 + \rho e\right] + \left[\left(\frac{1}{2}\rho u^2 + \rho e\right)u + pu\right] = 0. \tag{17}$$

These equations describe the jump conditions across a shock wave moving with speed $dx/dt = U$.

Transforming to a frame of reference moving with the shock wave, i.e., setting

$$V \equiv U - u \tag{18}$$

equations (15)–(17) become

$$[\rho V] = 0 \tag{19}$$

$$\left[p + \rho V^2 - \rho V U\right] = 0 \tag{20}$$

---

2 A discontinuity in a nonlinear hyperbolic system is an idealization of a thin region in reality where the flow variables vary rapidly. These variations are caused by dissipative effects (see Section 15.5) no matter how small.

$$\left[\rho V\left(h+\frac{1}{2}V^2\right)-(p+\rho V^2)U+\frac{1}{2}\rho VU^2\right]=0 \tag{21}$$

or

$$[\rho V]=0 \tag{22}$$

$$[p+\rho V^2]=0 \tag{23}$$

$$\left[h+\frac{1}{2}V^2\right]=0 \tag{24}$$

which merely highlight (compare with equations (15)–(17)) the fact that equations (1)–(3) are *Galilean invariant*. Thus we have,

$$\rho_1 V_1=\rho_2 V_2 \tag{25}$$

$$p_1+\rho_1 V_1^2=p_2+\rho_2 V_2^2 \tag{26}$$

$$h_1+\frac{1}{2}V_1^2=h_2+\frac{1}{2}V_2^2 \tag{27}$$

where the subscripts 1 and 2 refer to the flow conditions in front of the shock and behind it. Equations (25)–(27) are called the *Rankine* (1870) – *Hugoniot* (1889) *relations.*

The Rankine-Hugoniot relations are not sufficient to determine a unique, physically correct weak solution to equations (1)–(3). It is necessary to impose further conditions, like the *causality condition*, for this purpose. The causality condition stipulates that, when a shock separates characteristics of a family[3] (see Example 2), the characteristics on each side can be traced back to the initial data. The shock is then determined by the given initial data and not by future events. For a perfect gas, the causality condition is equivalent to the *thermodynamic entropy condition* (See Equation (40)). This stipulates that the entropy increases across a shock making the flow transition across a shock an irreversible process.

From equations (25) and (26), one obtains

$$V_2^2-V_1^2=(p_1-p_2)\left(\frac{1}{\rho_1}+\frac{1}{\rho_2}\right). \tag{28}$$

Using (28), equation (27) gives

---

3 A discontinuity separates a family of characteristics if, through each point of the graph of the discontinuity in the (x,t) plane, there exist two characteristics, both of which either point forward in time or can be traced backward in time.

$$\frac{(p_1 - p_2)}{2}\left(\frac{1}{\rho_1} + \frac{1}{\rho_2}\right) = \frac{\gamma}{\gamma - 1}\left(\frac{p_1}{\rho_1} - \frac{p_2}{\rho_2}\right)$$

from which, one has

$$\frac{p_2}{p_1} = \frac{\left(\frac{\gamma + 1}{\gamma - 1}\right)\frac{\rho_2}{\rho_1} - 1}{\left(\frac{\gamma + 1}{\gamma - 1}\right) - \frac{\rho_2}{\rho_1}} \tag{29a}$$

or

$$\frac{\rho_2}{\rho_1} = \frac{\left(\frac{\gamma + 1}{\gamma - 1}\right)\frac{p_2}{p_1} + 1}{\frac{p_2}{p_1} + \left(\frac{\gamma + 1}{\gamma - 1}\right)} \tag{29b}$$

or

$$\frac{p_2 - p_1}{\rho_2 - \rho_1} = \gamma\frac{p_2 + p_1}{\rho_2 + \rho_1}.$$

Equation (29a) is represented (called the *Hugoniot curve*)[4] along with an isentrope (for which $p \sim \rho^\gamma$) in Figure 16.2. Note that

$$\frac{p_2}{p_1} \gg 1 : \frac{\rho_2}{\rho_1} \Rightarrow \frac{\gamma + 1}{\gamma - 1}. \tag{30}$$

So, the density ratio has an upper bound as p_2/p_1 increases indefinitely.

Next, one has, from equation (27),

$$\frac{T_2}{T_1} = \frac{1 + \frac{\gamma - 1}{2}M_1^2}{1 + \frac{\gamma - 1}{2}M_2^2}, \tag{31}$$

4 The Hugoniot curve is the focus of all states in the (p, ρ)-plane that can be reached by a shock transition from the state (p_1, ρ_1).

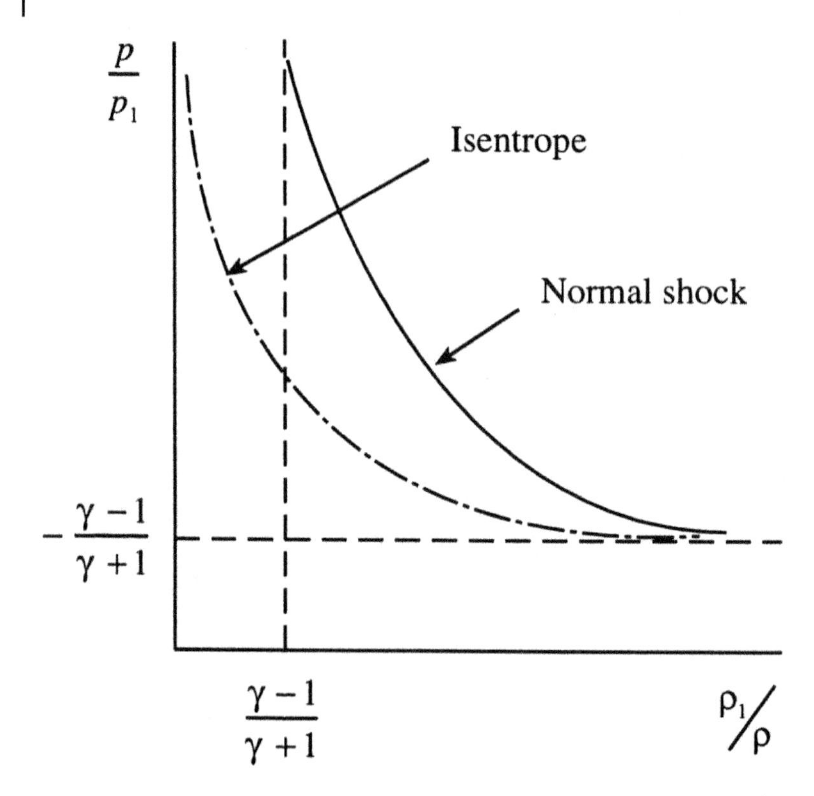

Figure 16.2 The Hugoniot curve.

and from the perfect gas equation of state and equation (25), one has

$$\frac{T_2}{T_1} = \frac{p_2}{p_1} \cdot \frac{\rho_1}{\rho_2} = \frac{p_2}{p_1} \cdot \frac{V_2}{V_1} = \frac{p_2}{p_1} \cdot \frac{M_2}{M_1}\sqrt{\frac{T_2}{T_1}}. \tag{32}$$

Using (31) and (32), one obtains

$$\frac{p_2}{p_1} = \frac{M_1\sqrt{1+\left(\frac{\gamma-1}{2}\right)M_1^2}}{M_2\sqrt{1+\left(\frac{\gamma-1}{2}\right)M_2^2}}. \tag{33}$$

In addition, from equation (26), one obtains

$$\frac{p_2}{p_1} = \frac{1+\gamma M_1^2}{1+\gamma M_2^2}. \tag{34}$$

From (33) and (34), one obtains

$$M_2^2 = \frac{1 + \frac{\gamma - 1}{2} M_1^2}{\gamma M_1^2 - \left(\frac{\gamma - 1}{2}\right)} \tag{35a}$$

or

$$M_2^2 = 1 + \frac{\left(\frac{\gamma + 1}{2}\right)(1 - M_1^2)}{\gamma M_1^2 - \left(\frac{\gamma - 1}{2}\right)} \tag{35b}$$

equation (35b) shows that $M_2 < 1$, if $M_1 > 1$.[5]

On using (35a), (34) gives

$$\frac{p_2}{p_1} = 1 + \frac{2\gamma}{\gamma + 1}\left(M_1^2 - 1\right). \tag{36}$$

Using (36), we obtain the following equation from (29b) and (25):

$$\frac{\rho_2}{\rho_1} = \frac{\left(\frac{\gamma + 1}{2}\right) M_1^2}{1 + \left(\frac{\gamma - 1}{2}\right) M_1^2} = 1 + \frac{(M_1^2 - 1)}{1 + \left(\frac{\gamma - 1}{2}\right) M_1^2} = \frac{V_1}{V_2}. \tag{37}$$

Noting, from (8) and (12) in Section 13.3, we have

$$M_1^{*2} \equiv \frac{V_1^2}{a^{*2}} = \frac{\left(\frac{\gamma + 1}{2}\right) M_1^2}{1 + \left(\frac{\gamma - 1}{2}\right) M_1^2}. \tag{38}$$

Using (37) and (38), one obtains the *Prandtl relation*,

$$V_1 V_2 = a^{*2} \quad \text{or} \quad M_1^* M_2^* = 1, \tag{39}$$

so that if the flow ahead of the normal shock is supersonic, the flow behind it is subsonic.

5 This result implies that when a shock wave is passed through a gas which is initially still, the gas is accelerated in the direction of propagation of the shock wave.

Example 2: Suppose there is a gas at rest with speed of sound a_0 for $x > 0$ in a tube $-\infty < x < \infty$ containing a piston at $x = 0$. At $t = 0$, the piston begins to move into the gas in the positive x-direction with velocity $u \equiv \dot{x}_p(t)$. A compressive shock is then generated and travels into the quiescent gas with constant speed U so that, behind this shock, the gas moves with the piston at constant speed u (Figure 16.3). Then, the Rankine-Hugoniot relations across the shock (using subscripts 0 and 1 to denote states ahead of the shock and behind it respectively) are given by,

$$\left.\begin{aligned} \rho_1(U-u) &= \rho_0 U \\ p_1 + \rho_1(U-u)^2 &= p_0 + \rho_0 U^2 \\ \frac{\gamma}{\gamma-1}\frac{p_1}{\rho_1} + \frac{1}{2}(U-u)^2 &= \frac{\gamma}{\gamma-1}\frac{p_0}{\rho_0} + \frac{1}{2}U^2 \end{aligned}\right\}$$

from which, on eliminating p_1 and ρ_1, one deduces

$$U^2 - \frac{\gamma+1}{2}uU - a_0^2 = 0.$$

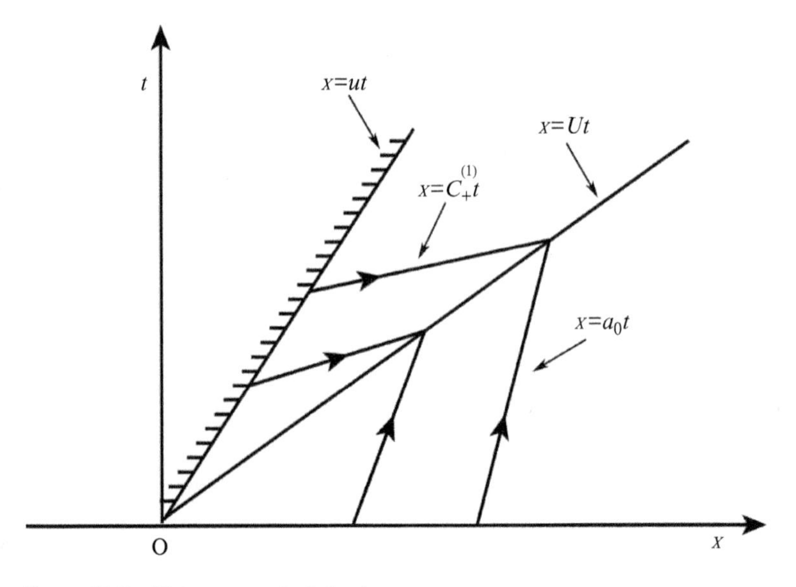

Figure 16.3 Piston generated shock wave.

Thus

$$U = \frac{\gamma+1}{4}u + \sqrt{\frac{(\gamma+1)^2}{16}u^2 + a_0^2}.$$

Observe that $c_+^{(1)} \geq U \geq c_+^{(0)} = a_0$[6] in accordance with the *causality principle*. In the limit $u \Rightarrow 0$, the shock becomes very weak and lies along the characteristic $x = a_0 t$!

Next, the entropy change across the shock wave is given by (see equation (31) in Section 12.4)

$$(S_2 - S_1) = C_v \ln\left(\frac{p_2}{p_1}\right)\left(\frac{\rho_1}{\rho_2}\right)^\gamma. \tag{40a}$$

Using (29a), (40a) becomes

$$S_2 - S_1 = C_v \ln\left[\frac{(\gamma+1)\left(\frac{\rho_2}{\rho_1}\right) - (\gamma-1)}{(\gamma+1)\left(\frac{\rho_2}{\rho_1}\right)^\gamma - (\gamma-1)\left(\frac{\rho_2}{\rho_1}\right)^{\gamma+1}}\right]. \tag{40b}$$

Now, the dissipative effects associated with the transport processes inside a shock wave will raise the entropy of the fluid, so that from (40b) we require

$$g\left(\frac{\rho_2}{\rho_1}\right) \equiv (\gamma-1)\left(\frac{\rho_2}{\rho_1}\right)^{\gamma+1} - (\gamma+1)\left(\frac{\rho_2}{\rho_1}\right)^\gamma + (\gamma+1)\left(\frac{\rho_2}{\rho_1}\right) - (\gamma-1) > 0. \tag{41}$$

The function $g(\rho_2/\rho_1)$ is sketched in Figure 16.4. It is seen that

$$g(\rho_2/\rho_1) > 0 \quad \text{if} \quad \rho_2/\rho_1 > 1 \quad \text{or, if} \quad M_1 > 1, \quad \text{from (37)}, \tag{42}$$

so that only the compression shocks are admissible and the expansion shocks are not.

6 We have from

$$U = \frac{c_+^{(1)} + a_0}{2}$$

the following result,

$$c_+^{(1)} = U + (U - a_0) \geq U.$$

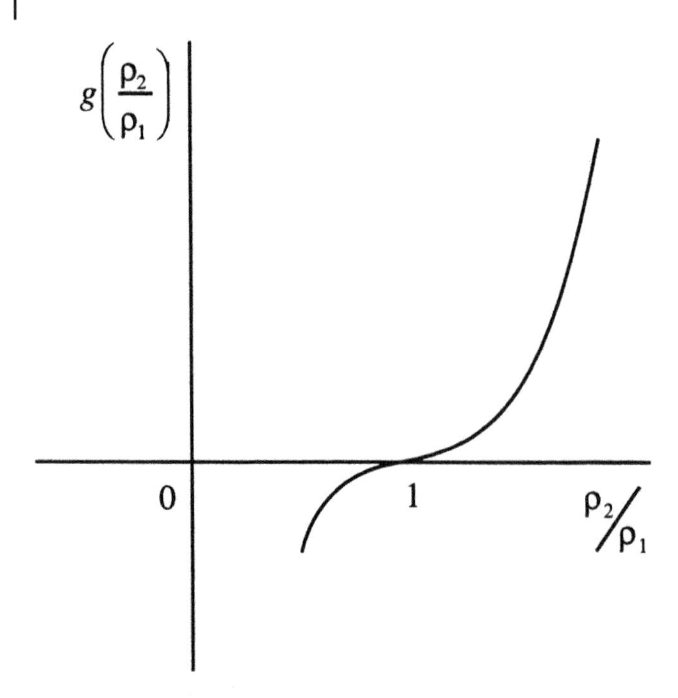

Figure 16.4 Entropy variation across a shock.

(ii) Weak Shock Waves

Let us next consider weak shock waves. One has from equations (7) and (23) in Chapter 12,

$$T \triangle S = h - h_1 - \int_{p_1}^{p} \frac{dp}{\rho}. \tag{43}$$

Using equations (27) and (28), one has

$$h - h_1 = \frac{1}{2}(p - p_1)\left(\frac{1}{\rho} + \frac{1}{\rho_1}\right). \tag{44}$$

In order to evaluate the integral in (43), we use the *trapezoidal rule* given by,

$$\begin{aligned}\int_{x_1}^{x_2} f(x)\,dx =& \frac{1}{2}(x_2 - x_1)\left[f(x_1) + f(x_2)\right] + \\ & - \frac{1}{12}(x_2 - x_1)^3 f''(x_1) + O\left[(x_2 - x_1)^4\right] \text{ as } x_2 \Rightarrow x_1\end{aligned} \tag{45}$$

so that we obtain

$$\int_{p_1}^{p} \frac{dp}{\rho} = \frac{1}{2}\left(\frac{1}{\rho}+\frac{1}{\rho_1}\right)(p-p_1) - \frac{1}{12}\left[\frac{\partial^2}{\partial p^2}\left(\frac{1}{\rho}\right)\right]_1 (p-p_1)^3 + O\left[(p-p_1)^4\right]. \tag{46}$$

Using equations (44) and (46), (43) becomes

$$T \triangle S = \frac{1}{12}\left[\frac{\partial^2}{\partial p^2}\left(\frac{1}{\rho}\right)\right]_1 (p-p_1)^3 + O\left[(p-p_1)^4\right]. \tag{47}$$

This means that the Hugoniot curve for the normal shock and the isentrope in Figure 16.2 have the same tangent and curvature at the point (p_1, ρ_1).

Furthermore, for a weak shock, equations (25)–(27) reduce to

$$\rho dV + V d\rho = 0 \tag{48}$$

$$dp + \rho V dV = 0 \tag{49}$$

$$C_p dT + V dV = 0. \tag{50}$$

We then obtain,[7]

$$\begin{aligned} \frac{\rho-\rho_1}{\rho_1} &\approx \frac{1}{\gamma}\frac{p-p_1}{p_1} \\ \frac{V-V_1}{V_1} &\approx -\frac{1}{\gamma}\frac{p-p_1}{p_1} \\ \frac{T-T_1}{T_1} &\approx \frac{\gamma-1}{\gamma}\frac{p-p_1}{p_1}. \end{aligned} \tag{51}$$

7 Alternatively, by letting

$$\triangle \equiv M_1^2 - 1, M_1 \gtrsim 1$$

we obtain

$$\frac{p_2}{p_1} \approx 1 + \frac{2\gamma}{\gamma+1}\triangle$$

$$\frac{\rho_2}{\rho_1} \approx 1 + \frac{2}{\gamma+1}\triangle$$

$$M_2^2 \approx 1 - \triangle.$$

Comparison of equation (51) with equation (47) shows that to a good approximation, one may treat the weak shock waves as being isentropic.

16.2 The Oblique Shock Wave

Consider a stationary shock wave with its plane now oblique to the flow direction. The equations expressing the conservation of mass, momenta (both tangential and normal to the shock), and energy are

$$\rho_1 V_{n1} = \rho_2 V_{n2} \tag{52}$$

$$(\rho_1 V_{n1}) V_{t1} = (\rho_2 V_{n2}) V_{t2} \tag{53}$$

$$p_1 + \rho_1 V_{n1}^2 = p_2 + \rho_2 V_{n2}^2 \tag{54}$$

$$h_1 + \frac{V_1^2}{2} = h_2 + \frac{V_2^2}{2} \tag{55}$$

where the subscripts n and t denote the values perpendicular and parallel to the shock wave.

We have from equations (52) and (53),

$$V_{t1} = V_{t2} = V_t. \tag{56}$$

From equation (55), it follows that

$$\frac{p_1}{\rho_1} = \left[\frac{\gamma+1}{2\gamma}a^{*2} - \frac{\gamma-1}{2\gamma}\left(V_{n1}^2 + V_{t1}^2\right)\right] \tag{57a}$$

$$\frac{p_2}{\rho_2} = \left[\frac{\gamma+1}{2\gamma}a^{*2} - \frac{\gamma-1}{2\gamma}\left(V_{n2}^2 + V_{t2}^2\right)\right]. \tag{57b}$$

Using (57a), equation (54) gives the *generalized Prandtl relation*,

$$V_{n1} V_{n2} = a^{*2} - \frac{\gamma-1}{\gamma+1} V_t^2. \tag{58}$$

Since the superposition of a uniform velocity V_t does not affect the static properties of the flow, the normal-shock jump relations can be carried over by merely replacing M_1 by $M_1 \sin\sigma$ (see Figure 16.5).

Thus, we have from (35a)–(37),

$$\frac{\rho_2}{\rho_1} = \frac{\left(\frac{\gamma+1}{2}\right) M_1^2 \sin^2 \sigma}{\left(\frac{\gamma-1}{2}\right) M_1^2 \sin^2 \sigma + 1} \tag{59}$$

$$\frac{p_2}{p_1} = 1 + \frac{2\gamma}{\gamma+1}\left(M_1^2 \sin^2 \sigma - 1\right) \tag{60}$$

$$M_2^2 \sin^2(\sigma - \delta) = \frac{1 + \left(\frac{\gamma-1}{2}\right) M_1^2 \sin^2 \sigma}{\gamma M_1^2 \sin^2 \sigma - \left(\frac{\gamma-1}{2}\right)}. \tag{61}$$

Noting, from equation (52) and Figure 16.5, that

$$\frac{\rho_2}{\rho_1} = \frac{V_{n1}}{V_{n2}} = \frac{\tan \sigma}{\tan(\sigma - \delta)} \tag{62}$$

and using (59), we obtain for the flow deflection across the shock,

$$\tan \delta = \frac{\left(M_1^2 \sin^2 \sigma - 1\right) \cot \sigma}{1 + M_1^2 \left(\frac{\gamma+1}{2} - \sin^2 \sigma\right)}. \tag{63}$$

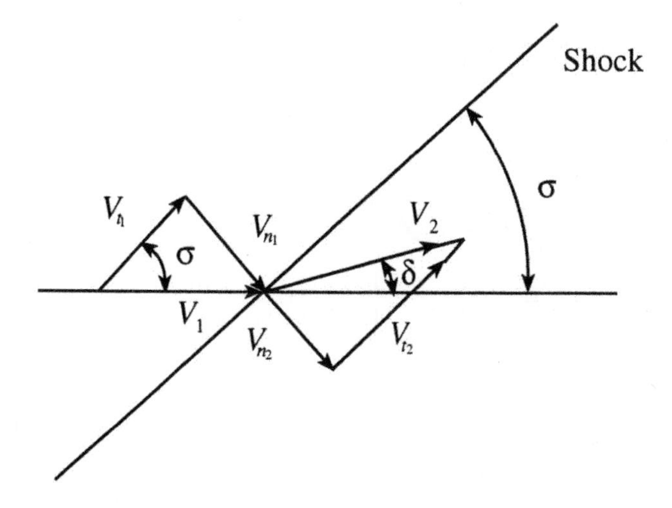

Figure 16.5 The oblique shock.

Now, since

$$M_1 \sin\sigma > 1$$

we require

$$\sin^{-1}\left(\frac{1}{M_1}\right) \leq \sigma \leq \frac{\pi}{2}. \tag{64}$$

Equation (63) shows that

$$\sigma = \frac{\pi}{2}, \sin^{-1}\left(\frac{1}{M_1}\right) : \delta = 0. \tag{65}$$

For $\sin^{-1}(1/M_1) < \sigma < \pi/2, \delta(\sigma)$ is positive and reaches a maximum (see Figure 16.6). δ_{max} is the maximum angle of flow deflection for which an attached shock can exist. Furthermore, as indicated by (63), δ_{max} increases with M_1. Note that, when $\delta < \delta_{max}$, for each value of δ and M_1, there are two possible shock waves corresponding to the two different values of σ - one corresponds to a strong shock ($M_2 < 1$), and the other to a weak shock ($M_2 > 1$).

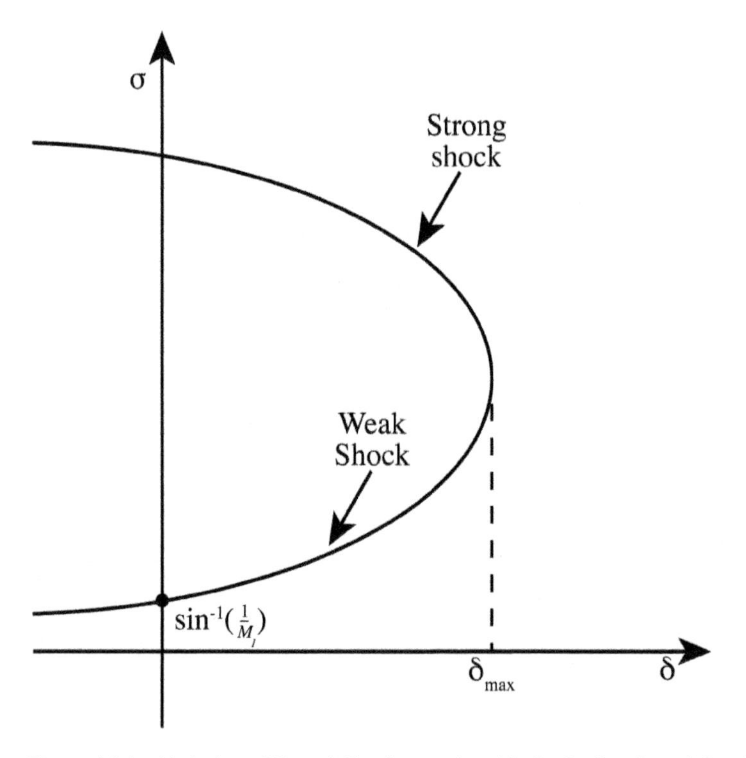

Figure 16.6 Variation of flow-deflection angle with the inclination of the shock to the incoming flow.

Let us now consider weak oblique shock waves. Then, (63) gives

$$M_1^2 \sin^2 \sigma - 1 \approx \frac{\gamma + 1}{2} \frac{M_1^2}{\sqrt{M_1^2 - 1}} \delta. \tag{66}$$

Using (66), (60) gives

$$\frac{p - p_1}{p_1} \approx \frac{\gamma M_1^2}{\sqrt{M_1^2 - 1}} \delta. \tag{67}$$

Noting, from (51), that

$$(p - p_1) \approx -\rho_1 V_1 (V - V_1)$$

one obtains from (67),

$$\frac{V - V_1}{V_1} \approx -\frac{\delta}{\sqrt{M_1^2 - 1}}. \tag{68}$$

A centered-fan of such weak oblique waves may be used to simulate the flow past a gentle convex corner – the *Prandtl-Meyer flow* (see Section 14.3).

16.3 Blast Waves: Taylor's Self-similarity and Sedov's Exact Solution

Strong explosions involve instantaneous release of a large quantity of energy E in a small volume.[8] Consider the propagation of a very strong spherical shock wave *(blast wave)* produced by such a strong explosion. Let the wave be located such that its distance from the source is both small enough to keep its amplitude large, and large enough compared with the dimensions of the source, so that the latter is treated as a point. Since the shock wave is strong, one may neglect the pressure p_1 of the undisturbed stagnant gas in front of it in comparison with the pressure p_2 immediately behind it. The density ratio ρ_2/ρ_1 then approaches its limiting value $(\gamma + 1) / (\gamma - 1)$. Thus, as Taylor (1950) argued, the gas flow pattern is essentially determined by two parameters – E and ρ_1.

8 A nuclear explosion releases a very large amount of energy in an unconfined space without generating any gas in the process. By contrast, a chemical explosion generates a large amount of gas at a high temperature in a confined space.

This enables one to find some *self-similar* solutions to the flow. First, let us form a dimensionless parameter,

$$\xi \equiv R\left(\frac{\rho_1}{Et^2}\right)^{1/5} \tag{69}$$

where R is the radial distance.

We assume that p, ρ and V are all functions only of ξ. Then, the choice of ξ in (69) results in the conservation of total energy of the gas within the sphere bounded by the shock. This is given by,

$$E = \int_0^{R_0} \left(\frac{p}{\gamma - 1} + \frac{1}{2}\rho V^2\right) 4\pi R^2 dR \tag{70}$$

At any time t_0, R_0 is the shock radius and ξ_0 is the corresponding value of ξ. Then, we have

$$R_0 = \xi_0 \left(\frac{Et_0^2}{\rho_1}\right)^{1/5} \tag{71a}$$

from which, the rate of propagation of the shock wave is given by

$$u_1 \equiv \frac{dR_0}{dt} = \frac{2R_0}{5t}. \tag{72}$$

As Taylor (1950) originally deduced, equation (72) implies

$$R_0 \sim t^{2/5}. \tag{71b}$$

This relationship was very impressively confirmed by the observations of the mushroom cloud generated by the explosion of the first atomic bomb (see Figure 16.7).

The equations governing the flow are

$$\frac{\partial \rho}{\partial t} + \frac{\partial}{\partial R}(\rho V) + \frac{2\rho V}{R} = 0 \tag{73}$$

$$\frac{\partial V}{\partial t} + V\frac{\partial V}{\partial R} = -\frac{1}{\rho}\frac{\partial p}{\partial R} \tag{74}$$

$$\left(\frac{\partial}{\partial t} + V\frac{\partial}{\partial R}\right)\ln\left(\frac{p}{\rho^\gamma}\right) = 0. \tag{75}$$

The state of the gas immediately behind the shock is given (using the results from Exercise (1)) by

$$\left.\begin{aligned} &V_2 \equiv u_1 - u_2 \approx \frac{2u_1}{\gamma + 1}, \quad p_2 \approx 2\rho_1\frac{u_1^2}{\gamma + 1} \\ &\rho_2 \approx \rho_1\frac{\gamma + 1}{\gamma - 1}. \end{aligned}\right\} \tag{76}$$

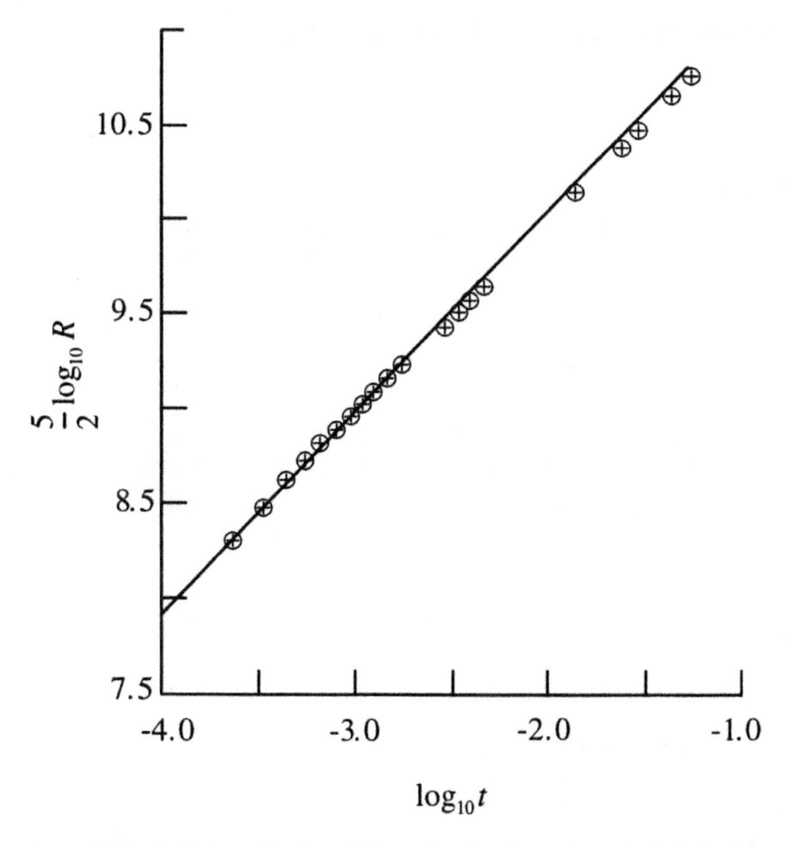

Figure 16.7 Variation with time of the radius R_0 of a strong spherical shock wave (from Faber, 1995). The straight line corresponds with (71b) while the crosses correspond to the observations.

In order to determine the gas flow throughout the region behind the shock, Sedov (1946) made a *similarity transformation*,

$$V = \frac{4}{5(\gamma+1)}\frac{R}{t}V'(\xi), \quad \rho = \left(\frac{\gamma+1}{\gamma-1}\right)\rho_1\rho'(\xi)$$

$$p = \frac{8\rho_1}{25(\gamma+1)}\frac{R^2}{t^2}p'(\xi). \tag{77}$$

On the shock wave, as per (76), the boundary conditions are

$$\xi = \xi_0 : V', p', \rho' = 1. \tag{78}$$

Using (77), equations (73)–(75) become ordinary differential equations

$$\xi\left[\frac{dV'}{d\xi}+\left(V'-\frac{2}{5}\right)\frac{1}{\rho'}\frac{d\rho'}{d\xi}\right]=-3V' \tag{79}$$

$$\xi\left[\left(V'-\frac{2}{5}\right)\frac{dV'}{d\xi}+\frac{1}{\rho'}\frac{dp'}{d\xi}\right]=-V'\left(V'-1\right)-2\frac{p'}{\rho'} \tag{80}$$

$$\xi\left(V'-\frac{2}{5}\right)\left(\frac{1}{p'}\frac{dp'}{d\xi}-\frac{\gamma}{\rho'}\frac{d\rho'}{d\xi}\right)=-2\left(V'-1\right). \tag{81}$$

The integration of equations (79)–(81) is facilitated by noting the existence of another integral. This integral is obtained from the conservation of the total energy between any two similarity lines $R/t^{2/5} = const$, i.e., from[9]

$$\left(\frac{p}{\gamma-1}+\frac{1}{2}\rho V^2\right)\left(V-\frac{2R}{5t}\right)+pV=0 \tag{82a}$$

In terms of similarity variables, (82a) can be rewritten as

$$\frac{p'}{\rho'}=\frac{\gamma+1-2V'}{2\gamma V'-(\gamma+1)}V'^2. \tag{82b}$$

An ingenius solution to equations (79)–(82a) was given by Sedov (1946),

9 The amount of energy that leaves a sphere of radius R in time dt is

$$4\pi R^2 dt\left[\left(\frac{p}{\gamma-1}+\frac{1}{2}\rho V^2\right)V+pV\right].$$

The volume of this sphere increases in time dt by $4\pi R^2\,(2R/5t)\,dt$, and the energy in this volume increment is

$$4\pi R^2 dt\left[\frac{p}{\gamma-1}+\frac{1}{2}\rho V^2\right](2R/5t).$$

Equating the two gives,

$$\left(\frac{p}{\gamma-1}+\frac{1}{2}\rho V^2\right)\left(V-\frac{2R}{5t}\right)+pV=0.$$

$$\rho' = \left[\frac{2\gamma V' - (\gamma+1)}{\gamma-1}\right]^{\upsilon_3} \left[\frac{5(\gamma+1) - 2(3\gamma-1)V'}{7-\gamma}\right]^{\upsilon_4} \left[\frac{\gamma+1-2V'}{\gamma-1}\right]^{\upsilon_5} \tag{83a}$$

$$\left(\frac{\xi_0}{\xi}\right)^5 = V'^2 \left[\frac{5(\gamma+1) - 2(3\gamma-1)V'}{7-\gamma}\right]^{\upsilon_1} \left[\frac{2\gamma V' - (\gamma+1)}{\gamma-1}\right]^{\upsilon_2} \tag{83b}$$

where

$$\left.\begin{aligned} &\upsilon_1 \equiv \frac{13\gamma^2 - 7\gamma + 12}{(3\gamma-1)(2\gamma+1)}, \quad \upsilon_2 \equiv -\frac{5(\gamma-1)}{2\gamma+1}, \quad \upsilon_3 \equiv \frac{3}{2\gamma+1} \\ &\upsilon_4 \equiv \frac{13\gamma^2 - 7\gamma + 12}{(2-\gamma)(3\gamma-1)(2\gamma+1)}, \quad \upsilon_5 \equiv \frac{2}{\gamma-2}. \end{aligned}\right\}$$

The constant ξ_0 is determined by condition (70) which is rewritten as below,

$$\frac{32\pi\xi_0^5}{25(\gamma^2-1)} \int_0^1 (\xi^4 \rho' V'^2 + \xi^4 p')\, d\xi = 1. \tag{84}$$

One has from (83a)

$$\frac{R}{R_0} \Rightarrow 0 : \frac{V}{V_2} \sim \frac{R}{R_0}, \quad \frac{\rho}{\rho_2} \sim \left(\frac{R}{R_0}\right)^{3/(\gamma-1)}. \tag{85}$$

Figure 16.8 shows $V/V_2, p/p_2, \rho/\rho_2$ vs. R/R_0, as given by (82a) and (83a), for air $(\gamma = 1.4)$. Observe the sharp fall in the density away from the shock front. The gas is almost completely concentrated in a thin layer behind the shock. It is as though the shock has swept the gas behind it into a thin layer moving along with it.[10] One may estimate the thickness δ of this thin dense layer by considering the mass balance,

10 This appears to suggest that shock waves erupting from *supernova explosions* could trigger the formation of stars when they crash into dense interstellar clouds of gas and dust, causing them to collapse into *protostars*.

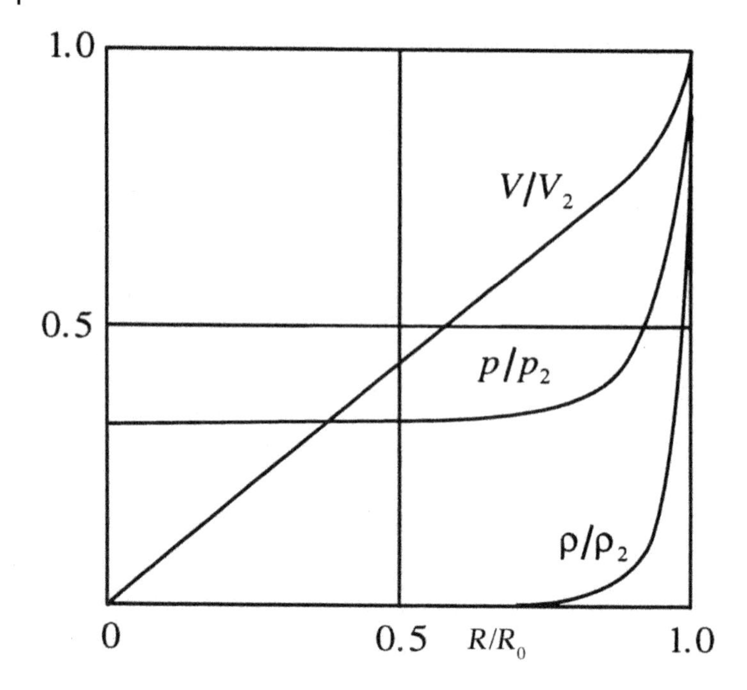

Figure 16.8 Variation of flow properties behind a blast wave (from Sedov, 1959).

$$\frac{4}{3}\pi R^3 \rho_1 \approx 4\pi R^2 \rho_2 \delta \tag{86}$$

from which one has,

$$\frac{\delta}{R} \approx \frac{1}{3}\frac{\rho_1}{\rho_2} \approx \frac{1}{3}\left(\frac{\gamma - 1}{\gamma + 1}\right). \tag{87}$$

For $\gamma = 1.4$, (87) gives $\delta/R \approx 0.05$. So, the thickness of the thin dense layer is about 5% of the shock radius!

Exercise

1. Study the nature of flow transitions across a very strong and very weak shock wave (normal and oblique).

17

The Hodograph Method

It turns out that the problem of plane steady potential flow of a gas becomes linear when the velocity components are used as independent variables (the *hodograph transformation*). However, the advantages of linearity are somewhat offset by the following,

- A lack of global invertibility of the hodograph transformation.
- The practical difficulty in satisfying the boundary conditions prescribed in the physical plane. Indeed, the shape of the body cannot be prescribed in advance, unless the body has a simple representation in the hodograph, like a wedge.
- The fact that the shape of the body changes if either the free stream Mach number or the thickness ratio is altered.

17.1 The Hodograph Transformation of Potential Flow Equations

Recall that the equations of motion in the streamline coordinates for a two-dimensional steady potential flow are (equations (51) and (52) in Section 17.5)

$$\left(1 - M^2\right) \frac{1}{V} \frac{\partial V}{\partial s} + \frac{\partial \theta}{\partial n} = 0 \tag{1}$$

$$\frac{1}{V} \frac{\partial V}{\partial n} - \frac{\partial \theta}{\partial s} = 0. \tag{2}$$

Molenbrock (1890) discovered that equations (1) and (2) become *linear*, if one uses (V, θ), instead of (s, n), as the independent variables. Setting,

$$V = \frac{\partial \Phi}{\partial s}, \quad \frac{\rho V}{\rho_0} = \frac{\partial \Psi}{\partial n} \tag{3}$$

Introduction to Theoretical and Mathematical Fluid Dynamics, Third Edition.
Bhimsen K. Shivamoggi.

equations (1) and (2) become

$$\frac{\rho V}{\rho_0}\frac{\partial \theta}{\partial \Psi}+\left(1-M^2\right)\frac{\partial V}{\partial \Phi}=0 \tag{4}$$

$$V\frac{\partial \theta}{\partial \Phi}-\frac{\rho}{\rho_0}\frac{\partial V}{\partial \Psi}=0. \tag{5}$$

Now, from

$$\left.\begin{aligned} d\Phi &= \frac{\partial \Phi}{\partial V}dV+\frac{\partial \Phi}{\partial \theta}d\theta \\ d\Psi &= \frac{\partial \Psi}{\partial V}dV+\frac{\partial \Psi}{\partial \theta}d\theta \end{aligned}\right\} \tag{6}$$

one obtains

$$\left.\begin{aligned} dV &= \frac{1}{\triangle}\left(\frac{\partial \Psi}{\partial \theta}d\Phi-\frac{\partial \Phi}{\partial \theta}d\Psi\right) \\ d\theta &= \frac{1}{\triangle}\left(-\frac{\partial \Psi}{\partial V}d\Phi+\frac{\partial \Phi}{\partial V}d\Psi\right) \end{aligned}\right\} \tag{7}$$

where $\triangle$ is the *Jacobian* of the transformation from (Φ, Ψ) to (V, θ),

$$\triangle \equiv \frac{\partial(\Phi,\Psi)}{\partial(V,\theta)}=\begin{vmatrix} \dfrac{\partial \Phi}{\partial V} & \dfrac{\partial \Phi}{\partial \theta} \\ \dfrac{\partial \Psi}{\partial V} & \dfrac{\partial \Psi}{\partial \theta} \end{vmatrix}.$$

Using (7), equations (4) and (5) give

$$\frac{\rho V}{\rho_0}\frac{\partial \Phi}{\partial V}+\left(1-M^2\right)\frac{\partial \Psi}{\partial \theta}=0 \tag{8}$$

$$V\frac{\partial \Psi}{\partial V}-\frac{\rho}{\rho_0}\frac{\partial \Phi}{\partial \theta}=0. \tag{9}$$

Equations (8) and (9) are linear because ρ and M are functions of only V.

17.2 The Chaplygin Equation

From equations (8) and (9), we have

$$V\Psi_{VV}+\Psi_V-\left(\frac{\rho}{\rho_0}\right)_V\Phi_\theta-\left(\frac{\rho}{\rho_0}\right)\Phi_{\theta V}=0 \tag{10a}$$

or

$$V\Psi_{VV} + \Psi_V - \left(\frac{\rho}{\rho_0}\right)_V \frac{\rho_0}{\rho} V\Psi_V + \left(\frac{\rho}{\rho_0}\right)\left[\left(1 - M^2\right)\frac{\rho_0}{\rho V}\Psi_\theta\right]_\theta = 0. \qquad (10b)$$

From the momentum conservation,

$$\frac{1}{\rho}\,dp + V dV = 0 \qquad (11)$$

we obtain

$$\frac{d\rho}{dV} = \frac{d\rho}{dp}\frac{dp}{dV} = -\frac{\rho M^2}{V}. \qquad (12)$$

Using (12), equation (10b) becomes

$$V\Psi_{VV} + \left(1 + M^2\right)\Psi_V + \frac{\left(1 - M^2\right)}{V}\Psi_{\theta\theta} = 0 \qquad (13)$$

which is *Chaplygin's* (1904) *equation.*

Introducing

$$\tau \equiv \frac{V^2}{\left(\frac{2}{\gamma - 1}\right) a_0^2} \qquad (14)$$

equation (13) becomes

$$4\tau^2(1 - \tau)\Psi_{\tau\tau} + 4\left[1 + \left(\frac{2 - \gamma}{\gamma - 1}\right)\tau\right]\tau\Psi_\tau + \left[1 - \left(\frac{\gamma + 1}{\gamma - 1}\right)\tau\right]\Psi_{\theta\theta} = 0. \qquad (15)$$

Looking for solutions of the form,

$$\Psi(\tau, \theta) = \tau^{\frac{m}{2}} f(\tau) \sin m\theta \qquad (16)$$

we then obtain

$$\tau(1 - \tau) f'' + \left[(m + 1) - \left(m + \frac{\gamma - 2}{\gamma - 1}\right)\tau\right] f' + \frac{m(m + 1)}{2(\gamma - 1)} f = 0. \qquad (17)$$

Now, the *hypergeometric equation*

$$x(1 - x) y'' + [t - (r + s + 1) x] y' - rsy = 0 \qquad (18)$$

has a solution,

$$y = F(r, s, t; x) = 1 + \frac{r \cdot s}{t \cdot 1!} x + \frac{r(r + 1) \cdot s(s + 1)}{t(t + 1) \cdot 2!} x^2 + \cdots. \qquad (19)$$

Comparing equation (17) with equation (18), we obtain

$$f = F(r, s, t; \tau) \tag{20}$$

where

$$\left.\begin{array}{c} r + s = m - \dfrac{1}{\gamma - 1}, \quad rs = -\dfrac{m(m+1)}{2(\gamma - 1)} \\ \text{and } t = m + 1. \end{array}\right\} \tag{21}$$

Two simple types of separable solutions to the hodograph equations are

- $\Psi \sim \theta, \Phi \sim f(\tau)$ - source flow
- $\Psi \sim f(\tau), \Phi \sim \theta$ - vortex flow.

17.3 The Tangent-Gas Approximation

In the foregoing, the nonlinear terms in the potential flow equation were transformed away via the hodograph transformation. It turns out that the compressibility terms can also be transformed away likewise. In order to see this, set

$$\frac{dW}{W} \equiv \sqrt{1 - M^2}\frac{dV}{V}, M < 1. \tag{22}$$

Equations (8) and (9) then become

$$W\frac{\partial \Phi}{\partial W} = -\frac{\rho_0}{\rho}\sqrt{1 - M^2}\frac{\partial \Psi}{\partial \theta} \tag{23}$$

$$\frac{\partial \Phi}{\partial \theta} = \frac{\rho_0}{\rho}W\sqrt{1 - M^2}\frac{\partial \Psi}{\partial W}. \tag{24}$$

Let us now make the *Chaplygin* (1904) *tangent-gas* approximation, which is tenable for subsonic flows,

$$\left(\frac{\rho_0}{\rho}\right)^2 (1 - M^2) = 1, \quad M < 1. \tag{24a}$$

This may be rewritten as

$$\left[1 - \left(\frac{\rho_0}{\rho}\right)^2\right]\frac{dp}{d\rho} = V^2. \tag{24b}$$

On differentiating with respect to ρ, (24b) gives

$$\left[1 - \left(\frac{\rho_0}{\rho}\right)^2\right]\frac{d^2p}{d\rho^2} - 2\left(\frac{\rho}{\rho_0}\right)\frac{1}{\rho_0}\frac{dp}{d\rho} = 2V\frac{dV}{d\rho}. \tag{25}$$

Using the momentum-conservation relation, written in the form,

$$V\frac{dV}{d\rho} + \frac{1}{\rho}\frac{dp}{d\rho} = 0 \tag{11}$$

equation (25) becomes

$$\left[1 - \left(\frac{\rho}{\rho_0}\right)^2\right]\left[\frac{d^2p}{d\rho^2} + \frac{2}{\rho}\frac{dp}{d\rho}\right] = 0 \tag{26}$$

from which we obtain,

$$\frac{d^2p}{d\rho^2} + \frac{2}{\rho}\frac{dp}{d\rho} = 0. \tag{27}$$

The solution of equation (27) is

$$p = A - \frac{B}{\rho} \tag{28}$$

where A and B are arbitrary constants. This solution amounts to replacing the isentrope by a tangent to it at a certain point (hence, the name, the *tangent-gas approximation*).[1] Let us choose the free stream for the latter (see Figure 17.1). This is the *von Kármán-Tsien* (1939) approximation. The following relations are obtained from (24) and (28),

$$\begin{aligned}
&\bullet\ p - p_\infty = \rho_\infty^2 a_\infty^2\left(\frac{1}{\rho_\infty} - \frac{1}{\rho}\right),\\
&\bullet\ \rho^2 a^2 = \rho_\infty^2 a_\infty^2\\
&\bullet\ V^2 - a^2 = V_\infty^2 - a_\infty^2,\\
&\bullet\ \frac{\rho_0}{\rho} = \sqrt{1 + \left(\frac{V}{a_0}\right)^2}
\end{aligned} \tag{29}$$

1 This idea is similar to the *β-plane approximation* used in oceanic and atmospheric flow dynamics (see Section 9.7). Here, one replaces the curved surface of the earth locally by a tangent plane but allows the local vertical component of the earth's angular velocity (called the *Coriolis parmeter*) to vary *linearly* with latitude.

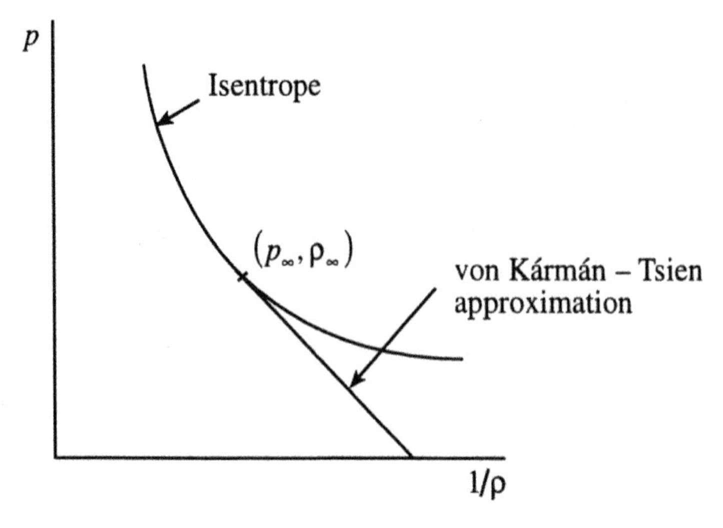

Figure 17.1 The tangent-gas approximation.

where the subscript ∞ denotes the conditions in the free stream. The relations in (29) imply that[2] $V < a$. So, the tangent-gas approximation is valid only for subsonic flows.

Using (29), (21) becomes

$$\frac{dW}{W} = \frac{\sqrt{a^2 - V^2}}{a}\frac{dV}{V} = \frac{a_0}{\rho_0 a_0/\rho}\frac{dV}{V}$$

or

$$\frac{dW}{W} = \frac{1}{\sqrt{1 + V^2/a_0^2}}\frac{dV}{V}$$

2 From the energy-conservation relation,

$$\frac{V^2}{2} + \frac{a^2}{\gamma - 1} = \frac{a_0^2}{\gamma - 1}$$

we obtain, in the tangent-gas approximation,

$$\gamma = -1 : \quad V^2 - a^2 = -a_0^2 = const < 0$$

which implies $V < a$.

which may be rewritten as

$$\frac{dW}{W} = \left[1 - \frac{V^2/a_0^2}{(1 + V^2/a_0^2) + \sqrt{1 + V^2/a_0^2}}\right]\frac{dV}{V} \tag{30}$$

from which we obtain,

$$\ln W = \ln\left(\frac{V}{1 + \sqrt{1 + V^2/a_0^2}}\right) + \ln C. \tag{31}$$

Using the condition implied by (21), namely,

$$a_0 \Rightarrow \infty : \quad W = V, \tag{32}$$

(31) then leads to

$$C = 2. \tag{33}$$

Using (33), (31) leads to

$$W = \frac{2V}{1 + \sqrt{1 + V^2/a_0^2}}, \tag{34a}$$

or alternatively,

$$V = \frac{4a_0^2 W}{4a_0^2 - W^2}. \tag{34b}$$

On the other hand, using the tangent-gas approximation (24a), equations (23) and (24) become

$$W\frac{\partial \Phi}{\partial W} + \frac{\partial \Psi}{\partial \theta} = 0 \tag{35}$$

$$W\frac{\partial \Psi}{\partial W} - \frac{\partial \Phi}{\partial \theta} = 0 \tag{36}$$

which correspond to an equivalent incompressible flow. Thus, the tangent-gas approximation effectively transforms compressibility away.

Next, in order to relate the geometries in the given compressible flow and the equivalent incompressible flow, note

$$\left.\begin{aligned} d\Phi &= \frac{\partial \Phi}{\partial x}dx + \frac{\partial \Phi}{\partial y}dy \\ d\Psi &= \frac{\partial \Psi}{\partial x}dx + \frac{\partial \Psi}{\partial y}dy \end{aligned}\right\} \tag{37}$$

where

$$\left.\begin{aligned} \frac{\partial \Phi}{\partial x} = V\cos\theta, \quad \frac{\partial \Phi}{\partial y} = V\sin\theta \\ \frac{\partial \Psi}{\partial x} = -\frac{\rho V}{\rho_0}\sin\theta, \quad \frac{\partial \Psi}{\partial y} = \frac{\rho V}{\rho_0}\cos\theta \end{aligned}\right\} \tag{38}$$

so that

$$\left.\begin{aligned} dx &= \frac{1}{D}\left(\frac{\rho V}{\rho_0}\cos\theta \cdot d\Phi - V\sin\theta \cdot d\Psi\right) \\ dy &= \frac{1}{D}\left(\frac{\rho V}{\rho_0}\sin\theta \cdot d\Phi + V\cos\theta \cdot d\Psi\right) \end{aligned}\right\} \tag{39a}$$

where D is the Jacobian of the transformation $(\Phi, \Psi) \Rightarrow (x, y)$,

$$D \equiv \frac{\partial(\Phi, \Psi)}{\partial(x, y)} = \begin{vmatrix} V\cos\theta & V\sin\theta \\ -\frac{\rho V}{\rho_0}\sin\theta & \frac{\rho V}{\rho_0}\cos\theta \end{vmatrix} = \frac{\rho}{\rho_0}V^2.$$

Using (34), (39a) may be written alternatively as

$$\left.\begin{aligned} dx &= \frac{U_1}{W^2}\left(1 - \frac{W^2}{4a_0^2}\right)d\Phi - \frac{U_2}{W^2}\left(1 + \frac{W^2}{4a_0^2}\right)d\Psi \\ dy &= \frac{U_2}{W^2}\left(1 - \frac{W^2}{4a_0^2}\right)d\Phi + \frac{U_1}{W^2}\left(1 + \frac{W^2}{4a_0^2}\right)d\Psi \end{aligned}\right\} \tag{39b}$$

where

$$U_1 \equiv W\cos\theta, \quad U_2 \equiv W\sin\theta.$$

If the equivalent incompressible flow is taking place in the (ξ, η)-plane, setting

$$z \equiv x + iy, \quad \zeta \equiv \xi + i\eta, \quad q \equiv U_1 + iU_2, \quad F \equiv \Phi + i\Psi, \quad \overline{q} \equiv \frac{dF}{d\zeta},$$

(39b) may be rewritten as

$$dz = \frac{dF}{\overline{q}} - \frac{q}{4a_0^2} d\overline{F} = d\zeta - \frac{q^2}{4a_0^2} d\overline{\zeta} \tag{39c}$$

which gives a relation between the geometries in the given compressible flow and the equivalent incompressible flow.

17.4 The Lost Solution

The hodograph transformation becomes degenerate when $V = V(\theta)$, i.e., when $\triangle$, the Jacobian of the transformation $(\Phi, \Psi) \Rightarrow (V, \theta)$, vanishes,

$$\begin{aligned}\triangle \equiv \frac{\partial(\Phi,\Psi)}{\partial(V,\theta)} &= \begin{vmatrix} \dfrac{\partial\Phi}{\partial V} & \dfrac{\partial\Phi}{\partial\theta} \\ \dfrac{\partial\Psi}{\partial V} & \dfrac{\partial\Psi}{\partial\theta} \end{vmatrix} = \begin{vmatrix} \dfrac{\partial\Phi}{\partial V} & \left(\dfrac{\partial\Phi}{\partial V}\dfrac{dV}{d\theta}\right) \\ \dfrac{\partial\Psi}{\partial V} & \left(\dfrac{\partial\Psi}{\partial V}\dfrac{dV}{d\theta}\right) \end{vmatrix} \\ &= \frac{dV}{d\theta} \begin{vmatrix} \dfrac{\partial\Phi}{\partial V} & \dfrac{\partial\Phi}{\partial V} \\ \dfrac{\partial\Psi}{\partial V} & \dfrac{\partial\Psi}{\partial V} \end{vmatrix} = 0.\end{aligned} \tag{40}$$

Such a situation prevails for degenerate cases, such as,

- a uniform flow where the flow region maps to a single point on the hodograph plane,
- a *simple-wave* flow where the flow region maps into a single curve in the hodograph plane.

Equations (1) and (2) then become

$$\frac{\partial\theta}{\partial n} + (1 - M^2)\frac{1}{V}\frac{dV}{d\theta}\frac{\partial\theta}{\partial s} = 0 \tag{41}$$

$$\frac{\partial\theta}{\partial s} - \frac{1}{V}\frac{dV}{d\theta}\frac{\partial\theta}{\partial n} = 0 \tag{42}$$

from which we have,

$$1 + (1 - M^2)\frac{1}{V^2}\left(\frac{dV}{d\theta}\right)^2 = 0 \tag{43}$$

or

$$d\nu \equiv \sqrt{M^2-1}\frac{dV}{V} = \pm d\theta. \tag{44a}$$

This implies that the hodograph transformation becomes degenerate along the local characteristics $C^{\pm}$ (see Section 15.1), with the Riemann invariants,

$$C^{\pm} : \nu \mp \theta = const. \tag{44b}$$

17.5 The Limit Line

The limit line corresponds to the case when the hodograph transformation fails to be one-to-one. Then equations (39a) may be rewritten as,

$$\left.\begin{aligned} dx &= \frac{\cos\theta}{V}d\Phi - \frac{\rho_0}{\rho}\frac{\sin\theta}{V}d\Psi \\ dy &= \frac{\sin\theta}{V}d\Phi + \frac{\rho_0}{\rho}\frac{\cos\theta}{V}d\Psi \end{aligned}\right\}. \tag{45a}$$

Using (6), (45a) becomes

$$\begin{aligned} dx &= \left[\frac{\cos\theta}{V}\left(\frac{\partial\Phi}{\partial V}\right) - \frac{\rho_0}{\rho}\frac{\sin\theta}{V}\left(\frac{\partial\Psi}{\partial V}\right)\right]dV + \\ &+ \left[\frac{\cos\theta}{V}\left(\frac{\partial\Phi}{\partial\theta}\right) - \frac{\rho_0}{\rho}\frac{\sin\theta}{V}\left(\frac{\partial\Psi}{\partial\theta}\right)\right]d\theta \\ dy &= \left[\frac{\sin\theta}{V}\left(\frac{\partial\Phi}{\partial V}\right) + \frac{\rho_0}{\rho}\frac{\cos\theta}{V}\left(\frac{\partial\Psi}{\partial V}\right)\right]dV + \\ &+ \left[\frac{\sin\theta}{V}\left(\frac{\partial\Phi}{\partial\theta}\right) + \frac{\rho_0}{\rho}\frac{\cos\theta}{V}\left(\frac{\partial\Psi}{\partial\theta}\right)\right]d\theta. \end{aligned} \tag{45b}$$

Using equations (8) and (9), (45b) becomes

$$dx = \frac{\rho_0}{\rho V}\left[-\left\{\left(1-M^2\right)\frac{\cos\theta}{V}\frac{\partial\Psi}{\partial\theta}+\sin\theta\frac{\partial\Psi}{\partial V}\right\}dV\right.$$

$$\left.+\left\{V\cos\theta\frac{\partial\Psi}{\partial V}-\sin\theta\frac{\partial\Psi}{\partial\theta}\right\}d\theta\right]$$

$$dy = \frac{\rho_0}{\rho V}\left[-\left\{\left(1-M^2\right)\frac{\sin\theta}{V}\frac{\partial\Psi}{\partial\theta}-\cos\theta\frac{\partial\Psi}{\partial V}\right\}dV\right. \tag{46}$$

$$\left.+\left\{V\sin\theta\frac{\partial\Psi}{\partial V}+\cos\theta\frac{\partial\Psi}{\partial\theta}\right\}d\theta\right].$$

On the streamlines defined by,

$$d\Psi = \Psi_V dV + \Psi_\theta d\theta = 0 \tag{47}$$

equation (46) becomes

$$\left.\begin{aligned} dx &= -\frac{\rho_0\cos\theta}{\rho V^2\Psi_\theta}\left[V^2\Psi_V^2-\left(M^2-1\right)\Psi_\theta^2\right]dV \\ dy &= -\frac{\rho_0\sin\theta}{\rho V^2\Psi_\theta}\left[V^2\Psi_V^2-\left(M^2-1\right)\Psi_\theta^2\right]dV \end{aligned}\right\}. \tag{48}$$

The Jacobian of the transformation $(x,y)\Rightarrow(V,\theta)$ is

$$\mathscr{D}\equiv\frac{\partial(x,y)}{\partial(V,\theta)}=\frac{\partial(\Phi,\Psi)/\partial(V,\theta)}{\partial(\Phi,\Psi)/\partial(x,y)}=\frac{\triangle}{D}. \tag{49}$$

Using equations (8) and (9), we have

$$\triangle\equiv\frac{\partial(\Phi,\Psi)}{\partial(V,\theta)}=-\frac{\rho_0}{\rho V}\left[V^2\Psi_V^2-\left(M^2-1\right)\Psi_\theta^2\right]. \tag{50}$$

Using (39a) and (50), (49) becomes

$$\mathscr{D}\equiv\frac{\partial(x,y)}{\partial(V,\theta)}=-\left(\frac{\rho_0}{\rho}\right)^2\frac{1}{V^3}\left[V^2\Psi_V^2-\left(M^2-1\right)\Psi_\theta^2\right]. \tag{51}$$

Solutions of hodograph equations correspond to real flows only when $\mathscr{D}\neq 0$ so that the transformation $(x,y)\Rightarrow(V,\theta)$ is *one-to-one*. The case $\mathscr{D}=0$ corresponds to the *limit line* and can occur only for $M\geq 1$. The occurence of a limit line implies that a continuous flow development throughout the region is impossible and shock waves will appear.

Note that dx and dy along the streamlines change their signs across the limit line, whereas the slope of the streamlines does not change there. Therefore, the streamline must have a *cusp* on the limit line.

Furthermore, the fluid acceleration f on a streamline is given by

$$f \equiv \left(V\frac{\partial V}{\partial s}\right)_\Psi = \left(V^2\frac{\partial V}{\partial \Phi}\right)_\Psi \tag{52a}$$

and using (7), (52a) becomes

$$f = \frac{1}{\triangle}V^2\frac{\partial \Psi}{\partial \theta}. \tag{52b}$$

So, f becomes infinite on a limit line, and the solution there becomes physically impossible. The existence of a limit line implies that the assumption of isentropic processes has broken down (thanks to the onset of shock waves).

Example 1: Consider the case,

$$\Psi = \Psi(V), \quad \frac{\rho_0}{\rho}\frac{d\Psi}{dV} = \frac{K_1}{V}, K_1 = const.$$

We then have from equation (9)

$$\Phi = \Phi(\theta) = K_1\theta.$$

One obtains for this case, from equations (8), (9), and (45b),

$$\left.\begin{aligned} \frac{\partial x}{\partial V} &= -K_1\frac{\sin\theta}{V^2}, & \frac{\partial x}{\partial \theta} &= K_1\frac{\cos\theta}{V} \\ \frac{\partial y}{\partial V} &= -K_1\frac{\cos\theta}{V^2}, & \frac{\partial y}{\partial \theta} &= K_1\frac{\sin\theta}{V}. \end{aligned}\right\} \tag{52}$$

Integrating, one obtains

$$x = K_1\frac{\sin\theta}{V}, \quad y = -K_1\frac{\cos\theta}{V}$$

so that the streamlines ($V = const$) are circles. Choosing $K_1 = r^*a^*$, these circles are given by

$$\frac{r}{r^*} = \frac{a^*}{V}.$$

This solution represents a vortex flow and is confined to the region $r > r^*$, because the curve $r = r^*$ represents the limit line.[3]

Example 2: Consider the case,

$$\Psi = \Psi(\theta).$$

Equation (9) then gives

$$\Phi_\theta = 0.$$

Using this, equation (8) gives

$$\Psi_{\theta\theta} = 0 \quad \text{or} \quad \Psi = K_1\theta + K_2$$

where K_1 and K_2 are arbitrary constants. Substituting this in equation (8) then leads to,

$$\Phi_V = -\frac{\rho_0}{\rho}\frac{K_1}{V}\left(1 - M^2\right).$$

Using the tangent-gas approximation (24a), this becomes

$$\Phi_V = -\frac{\rho}{\rho_0}\frac{K_1}{V}.$$

Using the above results, (45b) leads to

$$\left.\begin{aligned} \frac{\partial x}{\partial V} &= -\frac{\rho}{\rho_0}\frac{K_1}{V^2}\cos\theta, \quad \frac{\partial x}{\partial \theta} = -\frac{\rho_0}{\rho}\frac{K_1}{V}\sin\theta \\ \frac{\partial y}{\partial V} &= -\frac{\rho}{\rho_0}\frac{K_1}{V^2}\sin\theta, \quad \frac{\partial y}{\partial \theta} = \frac{\rho_0}{\rho}\frac{K_1}{V}\cos\theta. \end{aligned}\right\} \tag{53}$$

3 The stream tubes have their minimum cross section where the flow is sonic. Therefore, the radial flow in question must be confined to the region $r > r^*$.

Integrating, one obtains

$$\left.\begin{aligned} x &= K_1 \frac{\rho_0}{\rho} \frac{\cos\theta}{V} + K_3 \\ y &= K_1 \frac{\rho_0}{\rho} \frac{\sin\theta}{V} + K_4 \end{aligned}\right\} \tag{54}$$

where K_3 and K_4 are arbitrary constants. Using these and choosing $K_3, K_4 = 0, K_1 = (\rho^*/\rho_0)\, a^* r^*$, we obtain

$$\frac{r}{r^*} = \frac{\rho^*}{\rho} \frac{a^*}{V}.$$

Note that $r = r^*$ again represents the limit line. Furthermore, thanks to the conservation of mass

$$\rho r V = const$$

exhibited by this solution, one surmises that it represents a source flow and is confined to the region $r > r^*$ (where the flow is either only subsonic or only supersonic).

Next, consider the acceleration of a fluid particle,

$$V \frac{dV}{dr} = \frac{V}{dr/dV}.$$

Using the above solution, we have

$$\frac{dr}{dV} = \frac{\rho^* r^* a^*}{\rho} \left(-\frac{1}{\rho V} \frac{d\rho}{dV} - \frac{1}{V^2} \right).$$

The *tangent-gas* approximation (29), on the other hand, gives

$$\frac{1}{\rho V} \frac{d\rho}{dV} = -\frac{1}{a_0^2 + V^2}.$$

Using this in the above relation and (29) again, we obtain

$$\frac{dr}{dV} = \frac{r^* \rho^* a^*}{\rho} \left[\frac{-a_0^2}{V^2 \left(a_0^2 + V^2 \right)} \right] = \frac{r^* \rho^* a^*}{\rho} \left(\frac{M^2 - 1}{V^2} \right).$$

So, the acceleration of a fluid particle is given by

$$V\frac{dV}{dr} = \frac{\rho V^3}{r^* \rho^* a^* (M^2 - 1)}$$

which becomes infinite at the limit line ($M = 1$). This relation may be rewritten in the form

$$\frac{r}{V}\frac{dV}{dr} = \frac{1}{M^2 - 1}$$

which agrees with the stream-tube area and flow velocity relation (16) given in Section 13.4.

Example 3: Note that[4]

$$\Phi = \frac{\rho_0}{\rho V}\cos\theta, \quad \Psi = \frac{\sin\theta}{V}, \quad \text{with } V \neq a$$

is a particular integral of equations (8) and (9), which is called *Ringleb's* (1940) *solution.*

Using (45a) one obtains along the streamlines ($\Psi = const$),

$$dx = \frac{\cos\theta}{V}d\Phi, \quad dy = \frac{\sin\theta}{V}d\Phi.$$

Now, noting

4 Looking for a solution of the form,

$$\Psi(V, \theta) = V^{\alpha} f(\theta),$$

equation (13) then leads to

$$(1 - V^2/a^2) f'' + \alpha(\alpha + V^2/a^2) f = 0.$$

Setting $\alpha = -1$, this becomes

$$(1 - V^2/a^2)\left(f'' + f\right) = 0$$

from which,

$$f'' + f = 0, \quad \text{if } V \neq a.$$

This leads to

$$f(\theta) = \begin{Bmatrix} \sin\theta \\ \cos\theta \end{Bmatrix}.$$

$$d\Phi = \frac{\partial \Phi}{\partial V} dV + \frac{\partial \Phi}{\partial \theta} d\theta$$

and using equations (8), (9), and (29), one obtains

$$d\Phi = -\frac{\rho_0}{\rho V^2} \left(1 - M^2\right) \cos\theta \cdot dV - \frac{\rho_0}{\rho V} \sin\theta \cdot d\theta.$$

Along the streamlines, one has

$$d\Psi = -\frac{\sin\theta}{V^2} dV + \frac{\cos\theta}{V} d\theta = 0$$

so that

$$d\Phi = -\frac{\rho_0}{\rho V} \left[\cot^2\theta \left(1 - M^2\right) + 1\right] \sin\theta d\theta.$$

Thus, the streamlines ($\Psi = const = k$) are given by

$$\left.\begin{aligned} dx &= -\frac{\rho_0}{\rho} \left[\cot^2\theta \cdot \left(1 - M^2\right) + 1\right] k^2 \cot\theta \cdot d\theta \\ dy &= -\frac{\rho_0}{\rho} \left[\cot^2\theta \cdot \left(1 - M^2\right) + 1\right] k^2 d\theta \end{aligned}\right\}. \tag{55}$$

In the incompressibility ($M \Rightarrow 0$) limit, these equations become

$$\left.\begin{aligned} dx &= -k^2 \frac{\cot\theta}{\sin^2\theta} d\theta \\ dy &= -\frac{k^2}{\sin^2\theta} d\theta \end{aligned}\right\} \tag{56}$$

from which,

$$x = \frac{k^2}{2\sin^2\theta}, \quad y = k^2 \cot\theta.$$

This leads to

$$y^2 = k^4 \left(\frac{2x}{k^2} - 1\right)$$

which represents a family of confocal parabolas (Figure 17.2) about the x-axis with focus at the origin. Thus, in the $M \Rightarrow 0$ limit, Ringleb's solution represents the flow past a semi-infinite wall.

The limit line is given by $\mathscr{D} = 0$, and from (51), it is

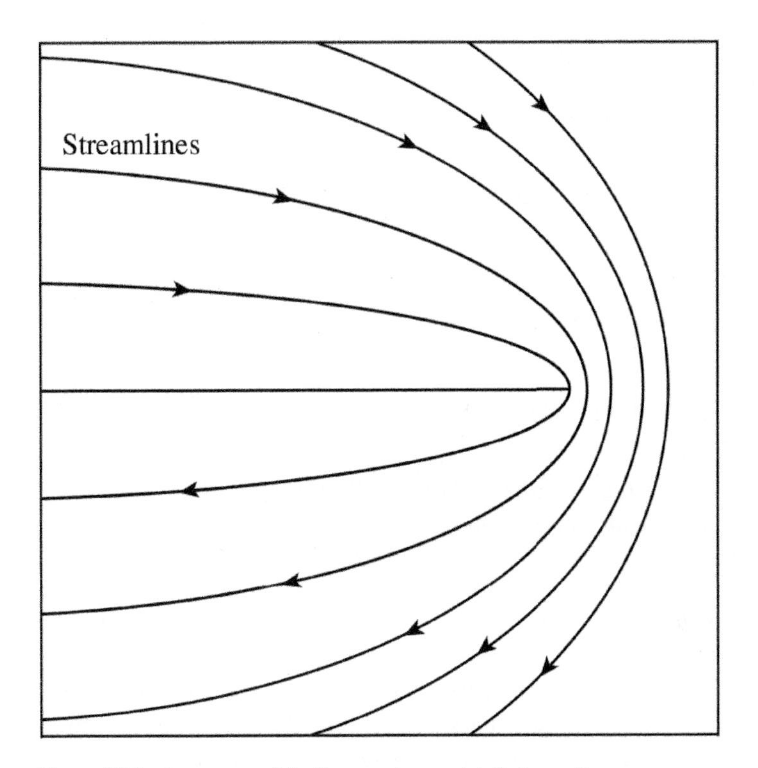

Figure 17.2 Incompressible flow past a semi-infinite wall.

$$(1 - M^2)\left(\frac{\partial \Psi}{\partial \theta}\right)^2 + V^2\left(\frac{\partial \Psi}{\partial V}\right)^2 = 0$$

which, for the present case, becomes

$$M^2 \cos^2 \theta = 1.$$

We have along the streamlines,

$$M^2 = \frac{V^2}{a^2} = \frac{V^2}{a_0^2 - \left(\frac{\gamma - 1}{2}\right) V^2} = \frac{\left(\frac{\sin^2 \theta}{k^2}\right)}{a_0^2 - \left(\frac{\gamma - 1}{2}\right)\left(\frac{\sin^2 \theta}{k^2}\right)}.$$

Thus, the limit line encounter with a streamline is given by

$$\sin^4 \theta - \frac{\gamma + 1}{2} \sin^2 \theta + k^2 a_0^2 = 0,$$

from which,

$$\sin^2\theta = \left(\frac{\gamma+1}{4}\right) \pm \sqrt{\left(\frac{\gamma+1}{4}\right)^2 - k^2 a_0^2}.$$

In order that a streamline does not cross a limit line, one therefore requires[5]

$$k^2 > \left(\frac{\gamma+1}{4a_0}\right)^2.$$

So, a smooth transition from $M < 1$ to $M > 1$, followed by another back to $M < 1$ is possible if $M < M_{max}$, where

$$M_{max} \equiv \left[\left(\frac{\gamma+1}{4}\right)^2 - \left(\frac{\gamma-1}{2}\right)\right]^{-1}.$$

Such a smooth transition is similar to that which can occur on the upper surface of an airfoil in transonic flow (see Section 18.1).

The mapping between the physical plane and the hodograph plane has a *fold* at the limit line so that the fluid returns here with a different velocity than the one with which it left. This solution is therefore physically impossible. However, by choosing the channel walls appropriately it is possible to produce an acceleration of the flow from subsonic to supersonic speeds, followed by a smooth deceleration back to subsonic speeds without the production of a shock.

Exercise

1. In Example 3, derive the range of values for k for which a streamline can have an encounter with a limit line.

---

5 The maximum velocity reached on a streamline must be less than or equal to the maximum velocity attainable via an isentropic flow starting from rest, i.e.,

$$V_{max}^2 = \frac{1}{k^2} \leq \frac{2a_0^2}{\gamma-1}.$$

So, a streamline can have an encounter with a limit line if

$$\frac{\gamma-1}{2a_0^2} \leq k^2 \leq \left(\frac{r+1}{4a_0}\right)^2.$$

18

Applications to Aerodynamics

We now consider the aerodynamic forces acting on a lifting surface in flight. It is interesting to note that the treatment of thin airfoils in supersonic flows becomes actually simpler than that in incompressible (also subsonic) flows because, in the supersonic case,

- The flows above and below the airfoil are independent of each other.
- Sharp corners do not pose any difficulty in determining the flow. For incompressible (also subsonic) flows on the other hand, infinitely large negative pressures develop at the sharp corners.

Furthermore, thanks to the absence of upstream influence, the flow over the airfoil is independent of the conditions in the wake. Thus, one does not impose the *Kutta condition* in supersonic flow unlike in the case of incompressible (also subsonic) flows.

18.1 Thin Airfoil Theory

(i) Thin Airfoil in Linearized Supersonic Flows

If the airfoil is thin and is at a small angle of attack, then one may make use of the small-perturbation theory (Ackeret, 1928). With the assumption of small disturbances, the vorticity generated by any shock wave standing on the airfoil will be small. If, in addition, the flow around the airfoil is assumed to be attached, the vorticity generated by viscous effects in the boundary layer will be confined to the layer. One may then assume the flow to be irrotational. From

Introduction to Theoretical and Mathematical Fluid Dynamics, Third Edition.
Bhimsen K. Shivamoggi.

equation (37), Section 14.4, the general solution to the linearized potential-flow equation in two-dimensions is of the form

$$\phi(x, y) = f(x - \lambda y) + g(x + \lambda y) \tag{1}$$

where $\lambda \equiv \sqrt{M^2 - 1}$. Since the disturbances are carried along only downstream-running Mach lines (see Figure 18.1), one has for the airfoil

$$\phi = \begin{cases} f(x - \lambda y) & y > 0 \\ g(x + \lambda y) & y < 0. \end{cases} \tag{2}$$

The boundary conditions at the airfoil surface $y = F(x)$ are

$$\begin{aligned} y = 0^+ &: \phi_y = -\lambda f'(x) = U_\infty \frac{dF_u}{dx} \\ y = 0^- &: \phi_y = -\lambda g'(x) = U_\infty \frac{dF_l}{dx} \end{aligned} \tag{3}$$

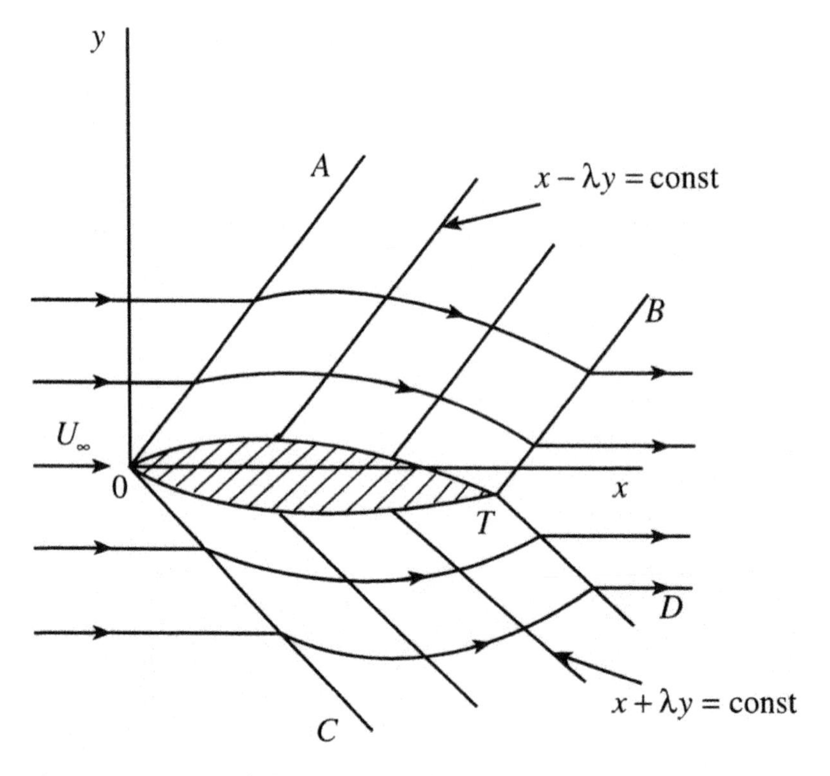

Figure 18.1 Supersonic flow past an airfoil.

where the subscripts u and l denote, respectively, the upper surface and the lower surface of the airfoil, and U_∞ is the free-stream velocity.

Thus, the pressure on the airfoil is then given by

$$C_p = \frac{p - p_\infty}{\frac{1}{2}\rho_\infty U_\infty^2} = -\frac{2\phi_x}{U_\infty} = \begin{cases} -\dfrac{2f'(x)}{U_\infty} & y > 0 \\ \dfrac{2g'(x)}{U_\infty} & y < 0 \end{cases}. \tag{4}$$

Now, note that the potential ϕ is constant outside the wavezone $0 < x - \lambda y < c$ (c being the chordlength of the airfoil). From (4), there is a jump of pressure, for example, across the line $x = \lambda y$, if $F'(0) \neq 0$. This jump is a linearized version of a shock wave.

Furthermore, equation (3) implies that

$$\phi_y = U_\infty F_u'(x - \lambda y) \quad \text{for} \quad y > 0, \quad 0 < x - \lambda y < c.$$

So, in the wavezone $0 < x - \lambda y < c$, the slope of the streamlines (ϕ_y / U_∞ in the linearized theory) is $F_u'(x - \lambda y)$, which is constant along lines $x - \lambda y = const.$ Thus, inside the wavezone $0 < x - \lambda y < c$, each streamline has the same shape as that of the airfoil profile.

Using (3), (4) becomes

$$C_p = \begin{cases} \dfrac{2\left(\dfrac{dF_u}{dx}\right)}{\sqrt{M_\infty^2 - 1}}, y > 0 \\[2ex] -\dfrac{2\left(\dfrac{dF_l}{dx}\right)}{\sqrt{M_\infty^2 - 1}}, y < 0 \end{cases}. \tag{5}$$

If the airfoil is at an angle of attack α, then one obtains

$$C_p = \begin{cases} \dfrac{2\left(\dfrac{dF_u}{dx} - \alpha\right)}{\sqrt{M_\infty^2 - 1}}, y > 0 \\[2ex] \dfrac{2\left(-\dfrac{dF_l}{dx} + \alpha\right)}{\sqrt{M_\infty^2 - 1}}, y < 0 \end{cases}. \tag{6}$$

The lift of the airfoil is given by

$$C_L = \frac{\int_0^c (C_{p_u} - C_{p_l})\, dx}{c} = \frac{4\alpha}{\sqrt{M_\infty^2 - 1}} \tag{7}$$

and the drag of the airfoil is given by

$$\begin{aligned} C_D &= \frac{1/c}{\sqrt{M_\infty^2 - 1}} \left[\int_0^c (C_{p_u}) \left(\frac{dF_u}{dx} - \alpha \right) dx + \int_0^c (C_{p_l}) \left(-\frac{dF_l}{dx} + \alpha \right) dx \right] \\ &= \frac{2}{\sqrt{M_\infty^2 - 1}} \left[\overline{\sigma}_u^2 + \overline{\sigma}_l^2 + 2\alpha^2 \right] \end{aligned} \tag{8}$$

where

$$\overline{\sigma}_{u,l}^2 \equiv \frac{1}{c} \int_0^c \left(\frac{dF_{u,l}}{dx} \right)^2 dx.$$

(ii) Far-Field Behavior of Supersonic Flow Past a Thin Airfoil

Consider a supersonic flow past a thin airfoil. The solution given in (2) is based on the linear equation

$$\phi_{yy} - \lambda^2 \phi_{xx} = 0.$$

Although the solution of this equation is a valid first approximation at or near the airfoil, it fails at large distances from the airfoil. This is because it predicts the disturbances propagating undiminished along the free-stream Mach lines to infinity. However, in reality, the Mach lines are neither straight nor parallel, while shock waves form and decay. The latter is brought about by the cumulative effect of the nonlinear terms in the full potential flow equation.

One has for the corresponding nonlinear problem (from equations (9b) and (36) in Chapter 14),

$$\begin{aligned} \phi_{yy} - \lambda^2 \phi_{xx} = M_\infty^2 \bigg[& \frac{\gamma - 1}{2} (2\phi_x + \phi_x^2 + \phi_y^2)(\phi_{xx} + \phi_{yy}) + (2\phi_x + \phi_x^2)\phi_{xx} + \\ & + 2(1 + \phi_x)\phi_y \phi_{xy} + \phi_y^2 \phi_{yy} \bigg] \end{aligned} \tag{9}$$

with the boundary condition at the airfoil being,

$$y = \varepsilon F(x): \frac{\phi_y}{1+\phi_x} = \varepsilon F'(x), \quad 0 \leq x \leq c, \quad \varepsilon \ll 1 \tag{10}$$

and at upstream infinity,

$$\phi(x,y) \Rightarrow 0. \tag{11}$$

It is convenient to transfer the boundary condition (10) from $y = \varepsilon F(x)$ to $y = 0$ by using the following *Taylor's expansion*,

$$\phi(x, \varepsilon F) = \phi(x,0) + \varepsilon F \phi_y(x,0) + \frac{1}{2}\varepsilon^2 F^2 \phi_{yy}(x,0)\cdots \tag{12}$$

so that (10) gives

$$\frac{\phi_y(x,0) + \varepsilon F \phi_{yy}(x,0) + \cdots}{1 + \phi_x(x,0) + \cdots} = \varepsilon F(x). \tag{13}$$

Seeking solutions to (9), (11), and (13) of the form

$$\phi = \varepsilon\phi_1 + \varepsilon^2\phi_2 + \cdots, \tag{14}$$

one obtains upon equating coefficients of equal powers of ε to zero,

$$O(\varepsilon): \phi_{1yy} - \lambda^2\phi_{1xx} = 0 \tag{15}$$

$$y = 0: \phi_{1y} = F'(x), \quad 0 \leq x \leq c \tag{16}$$

$$\text{upstream}: \quad \phi_1 \Rightarrow 0. \tag{17}$$

$$O(\varepsilon^2): \phi_{2yy} - \lambda^2\phi_{2xx}$$

$$= M_\infty^2 \left[(\gamma+1)\phi_{1x}\phi_{1xx} + (\gamma-1)\phi_{1x}\phi_{1yy} + 2\phi_{1y}\phi_{1xy}\right] \tag{18}$$

$$y = 0: \phi_{2y} = \phi_{1x}F'(x) - \phi_{1yy}F(x), \quad 0 \leq x \leq c \tag{19}$$

$$\text{upstream}: \quad \phi_2 \Rightarrow 0. \tag{20}$$

The boundary value problem (15)-(17) gives

$$\phi_1(x,y) = -\frac{1}{\lambda}F(x - \lambda y). \tag{21}$$

Using (21), equation (19) becomes

$$\phi_{2yy} - \lambda^2\phi_{2xx} = \frac{M_\infty^4(\gamma+1)}{\lambda^2}F'(x-\lambda y)F''(x-\lambda y). \tag{22}$$

Introducing the characteristic coordinates,

$$\xi \equiv x - \lambda y, \quad \eta \equiv x + \lambda y, \tag{23}$$

equation (23) becomes

$$\frac{\partial^2 \phi_2}{\partial \xi \partial \eta} = -\frac{M_\infty^4 (\gamma + 1)}{4\lambda^2} F'(\xi) F''(\xi) \tag{24}$$

from which we have,

$$\phi_2 = -\frac{M_\infty^4 (\gamma + 1)}{8\lambda^2} \left[F'(\xi)\right]^2 \eta + G(\xi). \tag{25}$$

Using (25), (20) gives

$$G'(\xi) = \frac{M_\infty^4 (\gamma + 1)}{4\lambda^4} \xi F' F'' + \frac{1}{\lambda^2}\left[1 - \frac{M_\infty^4 (\gamma + 1)}{8\lambda^2}\right] F'^2 - F' F''. \tag{26}$$

Using (21), (25) and (26) in (14), we obtain

$$\begin{aligned} \frac{u}{U_\infty} = \frac{\phi_x}{U_\infty} = 1 - \varepsilon \frac{F'(\xi)}{\lambda} + \varepsilon^2 \Bigg[& \frac{1}{\lambda^2}\left(1 - \frac{\gamma + 1}{4}\frac{M_\infty^2}{\lambda^2}\right)\left\{F'(\xi)\right\}^2 + \\ & -\frac{\gamma + 1}{2}\frac{M_\infty^4}{\lambda^3} y F'(\xi) F''(\xi) - F(\xi) F''(\xi) \Bigg] + O\left(\varepsilon^3\right). \end{aligned} \tag{27}$$

Observe that (27) is not uniformly valid and becomes inaccurate in the far field.

In order to remove this nonuniformity, let us use *Lighthill's* (1949) *method of strained coordinates* in a modified form due to Pritulo (1962). Thus, we set

$$\xi = \xi_0 + \varepsilon \xi_1 + O\left(\varepsilon^2\right) \tag{28}$$

so that (27) can be written as

$$\begin{aligned} \frac{u}{U_\infty} = 1 - \varepsilon \frac{F'(\xi_0)}{\lambda} + \varepsilon^2 \Bigg[& \frac{1}{\lambda^2}\left(1 - \frac{\gamma + 1}{4}\frac{M_\infty^2}{\lambda^2}\right)\left\{F'(\xi_0)\right\}^2 + \\ & -\left\{\frac{\gamma + 1}{2}\frac{M_\infty^4}{\lambda^2} y F'(\xi_0) + \xi_1\right\} \frac{F''(\xi_0)}{\lambda} - F(\xi_0) F''(\xi_0) \Bigg] + O\left(\varepsilon^3\right). \end{aligned} \tag{29}$$

The nonuniform term can be eliminated by simply choosing

$$\xi_1 = -\frac{\gamma + 1}{2}\frac{M_\infty^4}{\lambda^2} y F'(\xi_0). \tag{30}$$

Then, to first approximation, (29) gives

$$\frac{u}{U} = 1 - \varepsilon \frac{F'(\xi_0)}{\lambda} + \cdots \tag{31}$$

where, using (30) in (28) leads to

$$\xi_0 = \xi + \varepsilon \frac{\gamma + 1}{2} \frac{M_\infty^4}{\lambda^2} y F'(\xi) + \cdots . \tag{32}$$

Thus, as we saw in Section 15.3, the uniformly-valid first-order solution to the nonlinear problem is simply the solution to the corresponding linear problem. However, the linear characteristic is replaced by the characteristic calculated by including the first-order nonlinearities in the problem (Whitham, 1952).

(iii) Thin Airfoil in Transonic Flows

The term *transonic* refers to flows in which the free-stream Mach number M_∞ is not too far removed from 1. There is a need for a special theory for transonic flows because the foregoing linearized treatment (that was valid for supersonic flows) breaks down when applied to transonic flows. The difficulties in solving the transonic flow-field arise from the fact that the governing partial differential equation turns out to be nonlinear and of mixed elliptic-hyperbolic type.

In order to appreciate the latter situation physically, consider the flow pattern past a symmetrical airfoil as the free-stream Mach number M is increased from a subsonic value. When the free-stream Mach number M reaches a critical Mach number M^*, the maximum local Mach number in the flow near the airfoil becomes unity. For $M > M^*$, a supersonic region appears on the airfoil which is then terminated by a shock wave across which the flow is retarded back to a subsonic one. As M increases further, the shock wave moves aft, and the size of the supersonic region as well as the strength of the shock wave increase (see Figure 18.2).

Shock waves in transonic flows are, however, necessarily weak. Along any streamline, recall from Section 16.1 that the change in entropy through a normal shock is proportional to $\left(M_\infty^2 - 1\right)^3$. This means that outside the shock waves and boundary layers, the flow can be considered irrotational.

For simplicity, let us consider only airfoils that are symmetric about the x-axis. One has for a two-dimensional potential flow (see Section 14.1),

$$\left(a^2 - \Phi_x^2\right) \Phi_{xx} + \left(a^2 - \Phi_y^2\right) \Phi_{yy} - 2\Phi_x \Phi_y \Phi_{xy} = 0 \tag{33}$$

$$\frac{a^2}{\gamma - 1} + \frac{1}{2}\left(\Phi_x^2 + \Phi_y^2\right) = \frac{a_\infty^2}{\gamma - 1} + \frac{U_\infty^2}{2} \tag{34}$$

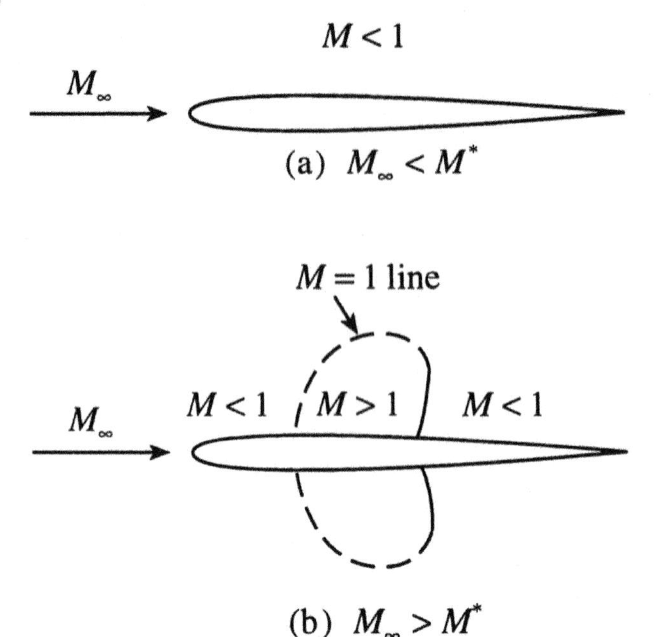

Figure 18.2 Influence of the free-stream Mach number on the flow past a symmetrical airfoil.

$$y = \delta F(x) : \frac{\Phi_y}{\Phi_x} = \delta F'(x) \tag{35}$$

$$\Phi \Rightarrow Ux \quad \text{at upstream infinity} \tag{36}$$

where δ is a small parameter characterizing the thin nature of the airfoil.

In order to treat cases with $M_\infty \approx 1$, set

$$M_\infty^2 \equiv 1 + \alpha\nu(\delta) \tag{37}$$

where

$$\lim_{\delta \Rightarrow 0} \nu(\delta) \Rightarrow 0$$

and α is chosen to be O(1) in a distinguished limit. We seek a solution of the form

$$\Phi(x, y; M_\infty, \delta) = U_\infty [x + \varepsilon(\delta)\phi(x, \tilde{y}, \alpha) + \cdots] \tag{38}$$

where

$$\tilde{y} \equiv \lambda(\delta) y, \quad \lambda \equiv \sqrt{M_\infty^2 - 1}$$

$$\lim_{\delta \Rightarrow 0} \varepsilon(\delta), \lambda(\delta) \Rightarrow 0.$$

In the limit $M_\infty \Rightarrow 1$, the linearized supersonic-flow theory (see Section 14.1) indicates that the disturbances are propagated practically undiminished to infinity in the y-direction but are restricted to a small width in the x-direction. This is reflected in the spatial scalings introduced in (38).

Using (37) and (38), equation (33) becomes

$$-\nu\varepsilon\alpha\phi_{xx} + \varepsilon\lambda^2\phi_{\tilde{y}\tilde{y}} = (\gamma+1)\varepsilon^2\phi_x\phi_{xx} + \text{ higher order terms.} \tag{39}$$

A *distinguished limit* occurs if

$$\varepsilon\lambda^2 \sim \varepsilon\nu \sim \varepsilon^2$$

or

$$\lambda = \sqrt{\varepsilon}, \quad \nu = \varepsilon. \tag{40}$$

Then, equation (39) gives

$$[\alpha + (\gamma+1)\phi_x]\phi_{xx} - \phi_{\tilde{y}\tilde{y}} = 0. \tag{41}$$

This equation is expected to contain the essential features of mixed subsonic-supersonic flow with embedded shock waves.

Next, using (38), (35) becomes

$$\tilde{y} = \lambda\delta F(x) : \varepsilon^{3/2}\phi_{\tilde{y}} = \delta F'(x)$$

from which we have.

$$\varepsilon = \delta^{2/3}. \tag{42}$$

Using (40) and (42), (37) gives

$$\alpha = \frac{M_\infty^2 - 1}{\delta^{2/3}} \tag{43}$$

which is the so-called *transonic similarity parameter* (Murmon and Cole, 1971).

The *parabolic* nature of the transonic-flow equation (41) allows no upstream influence, making $0 \leq x \leq c$ the region of interest. Furthermore, it means that the flow in the region $y \geq 0$ is independent of that in the region $y < 0$.

The nonlinear nature of equation (41), on the other hand, makes it very difficult to solve it exactly. As a first approximation, one may propose a linearized model for transonic flows. However, the calculated flow then does not exhibit the mixed-flow character that is essential for a transonic flow. This is because the differential equation so obtained by assuming ϕ_{xx} to be constant in the non-linear term is everywhere parabolic when $M_\infty = 1$, rather than being so only on the sonic line. Some steps have been taken in the literature to improve upon this method, but they are not a lot more rigorous, although more successful (see Tidjeman and Seebass, 1980).

18.2 Slender Bodies of Revolution

Let us adopt cylindrical coordinates (x, r, θ), with the x-axis aligned with the axis of a body of revolution (Figure 18.3), which is taken to be *slender*. Assuming the flow to be axisymmetric and irrotational and taking the flow velocity to be $\mathbf{v} = (u, v, 0)$, we have for the linearized flow in question,

$$\left(1 - M_\infty^2\right) \frac{\partial^2 \phi}{\partial x^2} + \frac{\partial^2 \phi}{\partial r^2} + \frac{1}{r} \frac{\partial \phi}{\partial r} = 0. \tag{44}$$

We saw in the previous section that in plane flows, the normal velocity at the chord differs little from that at the airfoil surface. However, in axisymmetric

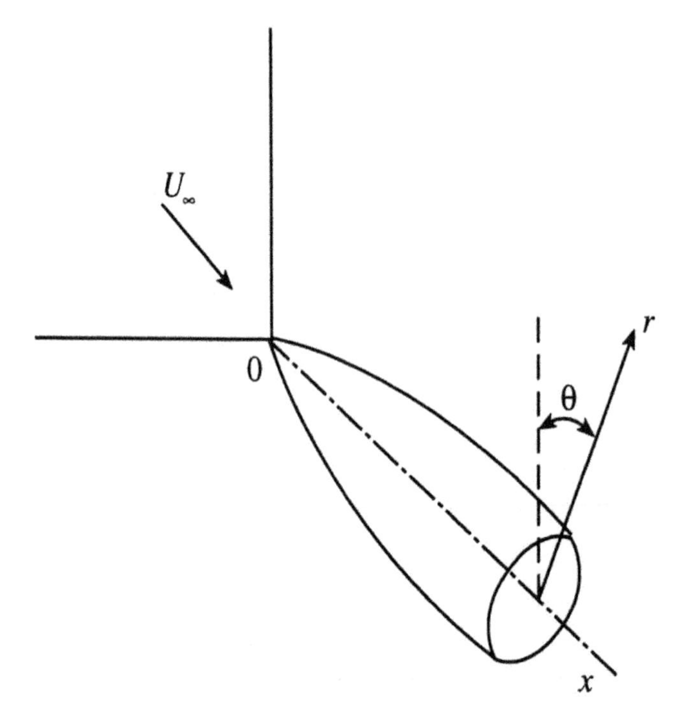

Figure 18.3 Uniform flow past a slender body of revolution.

flows the radial velocity becomes infinite at the axis in order that it becomes finite at the body surface. This result follows from the continuity equation,

$$\frac{\partial u}{\partial x} + \frac{1}{r}\frac{\partial}{\partial r}(vr) = 0 \tag{45}$$

which shows that the product (vr) stays finite. Thus, one writes the boundary condition at the surface of the body given by $r = R(x)$ in the form,

$$r = R(x) : \; R\frac{dR}{dx} = \frac{(vr)_{r=0}}{(U_\infty + u)}. \tag{46}$$

The pressure is then given by (see Section 14.4)

$$C_p = -\frac{2u}{U_\infty} - \frac{v^2}{U_\infty^2} \tag{47}$$

where we have retained the term quadratic in v since the latter is of a *lower* order than u near the axis.

Let us construct the solution to the boundary value problem described by (44) and (46) as a superposition of the singular source solutions to equation (44). Thus, for a subsonic flow we have,

$$\phi(x,r) = -\int_0^L \frac{f(\xi)\,d\xi}{\sqrt{(x-\xi)^2 + m^2 r^2}}, \quad m^2 \equiv 1 - M_\infty^2 > 0. \tag{48}$$

In a supersonic flow, note that the flow conditions at a point (x, r) are affected by the sources on the body only up to the point $\xi = x - \lambda r$. This is because, for the sources downstream of $\xi = x - \lambda r$, the point (x, r) lies ahead of their *Mach cones* (see Figure 18.4). Such sources have no effect on the flow conditions at (x, r). Thus,

$$\phi(x,r) = -\int_0^{x-\lambda r} \frac{f(\xi)\,d\xi}{\sqrt{(x-\xi)^2 - \lambda^2 r^2}}, \quad \lambda^2 \equiv M_\infty^2 - 1 > 0. \tag{49}$$

The region of influence of the body in the three-dimensional case extends over the entire fluid downstream of the Mach cone $x - \lambda r = 0$. By contrast, in the two-dimensional case, the region of influence of the body is bounded by the Mach lines at the leading and trailing edges of the body because the effect of the body is felt only on these Mach lines.

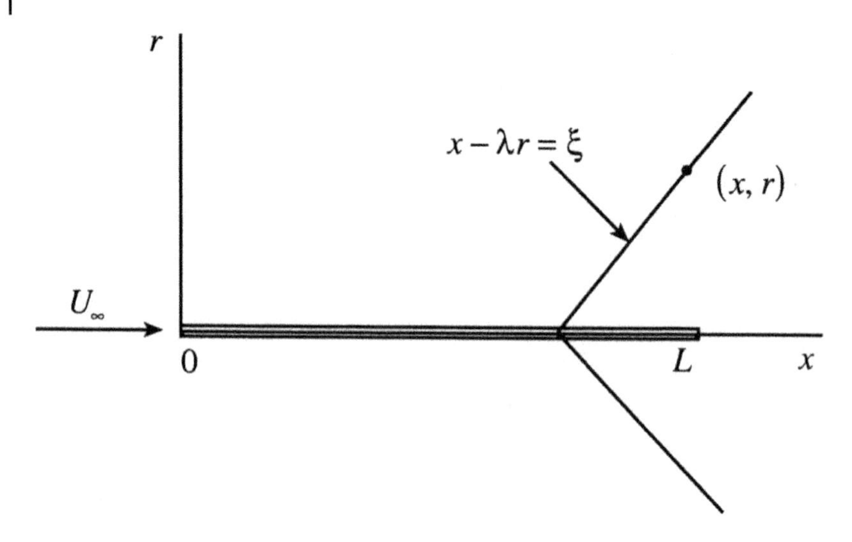

Figure 18.4 Domain of influence in supersonic flows.

In order to obtain the velocity components from (49), note that the integrand in (49) diverges at the upper limit, so that one introduces a smoothing variable,

$$\xi \equiv x - \lambda r \cosh \sigma. \tag{50}$$

Then, (49) becomes

$$\phi(x, r) = -\int_0^{\cosh^{-1}(x/\lambda r)} f(x - \lambda r \cosh \sigma)\, d\sigma \tag{51}$$

from which, if $f(0) = 0$, one obtains

$$\phi_x = -\int_0^{\cosh^{-1}(x/\lambda r)} f'(x - \lambda r \cosh \sigma)\, d\sigma$$

$$\phi_r = -\int_0^{\cosh^{-1}(x/\lambda r)} f'(x - \lambda r \cosh \sigma) \cdot (-\lambda \cosh \sigma)\, d\sigma \tag{52}$$

so that

$$\phi_x = -\int_0^{x-\lambda r} \frac{f'(\xi)\,d\xi}{\sqrt{(x-\xi)^2 - \lambda^2 r^2}}$$

$$\phi_r = \frac{1}{r}\int_0^{x-\lambda r} \frac{(x-\xi)\,f'(\xi)\,d\xi}{\sqrt{(x-\xi)^2 - \lambda^2 r^2}}. \tag{53}$$

Example 1: Consider a supersonic flow past a cone, with $f(\xi) = a\xi$. Then, (51) gives

$$\phi(x,r) = -\int_0^{\cosh^{-1}(x/\lambda r)} (ax - a\lambda r\cosh\sigma)\,d\sigma$$

or

$$\phi(x,r) = -ax\left[\cosh^{-1}\frac{x}{\lambda r} - \sqrt{1-\left(\frac{\lambda r}{x}\right)^2}\,\right]$$

from which, on the surface of the cone, one has

$$\phi_x = -a\cosh^{-1}\frac{x}{\lambda r}, \quad \phi_r = a\lambda\sqrt{\left(\frac{x}{\lambda r}\right)^2 - 1}.$$

Note that both ϕ_x and ϕ_r are functions of x/r, i.e., constant along each ray from the origin - such a flow is called *conical flow* (see Section 14.3). The solution for the flow past a non slender cone may therefore be constructed by fitting a conical flow to a conical shock.

Using these expressions for ϕ_x and ϕ_r, (46) gives

$$\frac{a\sqrt{\cot^2\delta - \lambda^2}}{U_\infty - a\cosh^{-1}\left(\frac{\cot\delta}{\lambda}\right)} = \tan\delta$$

from which, for $\delta \ll 1$ (δ being the semi vertex angle of the cone), one obtains

$$a \approx U_\infty \delta^2$$

so that

$$\phi = -U_\infty \delta^2 x \ell n \frac{2x}{\lambda r} - 1$$

from which, the pressure on the cone is given, using (4), by

$$C_p \approx 2\delta^2 \left(\ell n \frac{2}{\lambda \delta} - \frac{1}{2} \right).$$

Note that for a supersonic flow past a thin wedge of nose angle 2δ, the pressure rise occurs completely at the nose shock, and one has from (4),

$$C_p = \frac{2\delta}{\lambda}.$$

Observe that the pressure rise on the cone is much less than that on the wedge (a result traceable to the three-dimensional flow spreading effect for the cone-see Section 7.1-(ii)), and occurs continuously downstream of the Mach cone at the nose.

Let us now evaluate the integral in (49) for $\lambda r/x \ll 1$. Since the integrand is singular near the upper limit of integration, it proves convenient to write

$$\phi = I_1 + I_2 \tag{54}$$

where

$$I_1 = -\int_0^{x-\lambda r-\varepsilon} \frac{f(\xi)\, d\xi}{\sqrt{(x-\xi)^2 - \lambda^2 r^2}}$$

$$I_2 = -\int_{x-\lambda r-\varepsilon}^{x-\lambda r} \frac{f(\xi)\, d\xi}{\sqrt{(x-\xi)^2 - \lambda^2 r^2}} \qquad \varepsilon \Rightarrow 0^+.$$

In I_1, one may expand the integrand in powers of $\lambda^2 r^2$,

$$\frac{f(\xi)}{\sqrt{(x-\xi)^2 - \lambda^2 r^2}} = f(\xi) \left[\frac{1}{(x-\xi)} + \frac{\lambda^2 r^2}{2} \frac{1}{(x-\xi)^3} + \cdots \right].$$

Then, term-by-term integration gives as $\lambda r \Rightarrow 0^+$,

$$I_1 = f(x)\, \ell n \varepsilon - \int_0^{x-\varepsilon} f'(\xi)\, \ell n (x-\xi)\, d\xi + \cdots \tag{55}$$

where we have noted that

$$f(0) = 0.$$

In order to evaluate I_2, introducing again the smoothing variable,

$$\xi \equiv x - \lambda r \cosh \sigma \tag{50}$$

one obtains

$$I_2 = -\int_0^{\cosh^{-1}\left(\frac{\lambda r + \varepsilon}{\lambda r}\right)} f(x - \lambda r \cosh \sigma)\, d\sigma.$$

Expanding the integrand in powers of λr, one obtains

$$I_2 = -f(x)\int_0^{\cosh^{-1}\left(\frac{\lambda r + \varepsilon}{\lambda r}\right)} d\sigma + \lambda r \int_0^{\cosh^{-1}\left(\frac{\lambda r + \varepsilon}{\lambda r}\right)} f'(x) \cosh \sigma d\sigma + \cdots$$

which, as $\lambda r \Rightarrow 0^+$, leads to

$$I_2 = -f(x)\,\ell n \frac{2}{\lambda r} - f(x)\,\ell n \varepsilon + \cdots. \tag{56}$$

Using (55) and (56) in (54) gives,

$$\phi \approx -f(x)\,\ell n \frac{2}{\lambda r} - \int_0^x f'(\xi)\,\ell n (x - \xi)\, d\xi. \tag{57}$$

The boundary condition at the surface of the body given by $r = R(x)$ then becomes

$$r = R: \frac{\partial \phi}{\partial r} = \frac{f(x)}{R} = U \frac{dR}{dx}$$

or

$$f(x) = \frac{U}{2\pi} \frac{dA}{dx} \tag{58}$$

where

$$A \equiv \pi R^2.$$

The relation (58) implies that the source strength is proportional only to the rate of change of area of the body.

Using (58), (57) becomes

$$\phi \approx -\frac{U}{2\pi} A'(x)\,\ell n \frac{2}{\lambda r} - \frac{U}{2\pi} \int_0^x A''(\xi)\,\ell n\,(x-\xi)\,d\xi. \tag{59}$$

Consider next, the flow past a body of revolution at some angle of attack (see Figure 18.5). Although the flow is no longer axisymmetric, the linearity of the potential-flow equation permits one to write

$$\phi(x,r,\theta) = \phi_a(x,r) + \phi_c(x,r,\theta) \tag{60}$$

where the subscripts a and c denote the axial flow and the cross flow, respectively. One then has

$$\frac{\partial^2 \phi_a}{\partial r^2} + \frac{\partial \phi_a}{r\partial r} - \lambda^2 \frac{\partial^2 \phi_a}{\partial x^2} = 0 \tag{61}$$

$$\frac{\partial^2 \phi_c}{\partial r^2} + \frac{\partial \phi_c}{r\partial r} + \frac{1}{r^2}\frac{\partial^2 \phi_c}{\partial \theta^2} - \lambda^2 \frac{\partial^2 \phi_c}{\partial x^2} = 0. \tag{62}$$

From equations (61) and (62), it is easy to verify that

$$\phi_c = \cos\theta \cdot \frac{\partial \phi_a}{\partial r} \tag{63}$$

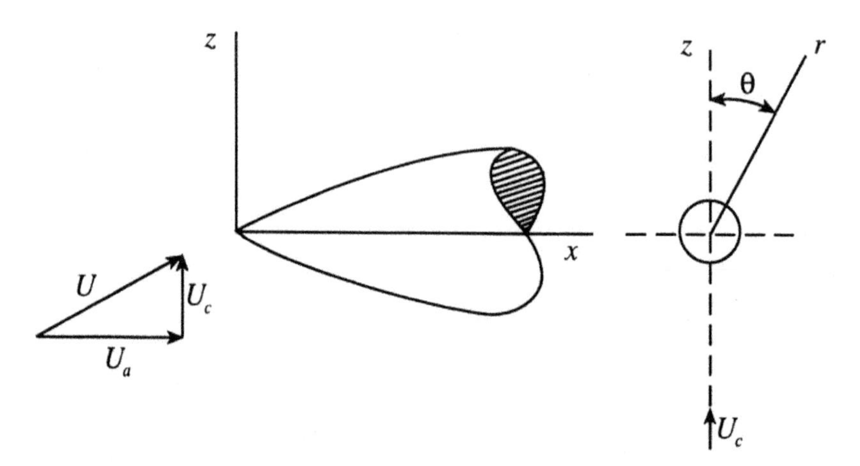

Figure 18.5 Uniform flow past a slender body of revolution at an angle of attack.

so that, using (53), one obtains

$$\phi_c(x,r,\theta) = \frac{\cos\theta}{r} \int_0^{x-\lambda r} \frac{m(\xi)\cdot(x-\xi)\,d\xi}{\sqrt{(x-\xi)^2 - \lambda^2 r^2}}. \tag{64}$$

The boundary conditions are

$$r = R(x) : \left(r\frac{\partial \phi_a}{\partial r}\right) = R\frac{dR}{dx}U_a$$

and

$$\frac{\partial \phi_c}{\partial r} = -U_c \cos\theta. \tag{65}$$

Alternatively, one has, from (64), as $r \Rightarrow 0^+$,

$$\phi_c \approx \frac{\sigma(x)}{r}\cos\theta \tag{66}$$

where $\sigma(x)$, using (65), is given by

$$\sigma(x) = U_c\,[R(x)]^2. \tag{67}$$

So, the strength of the doublet modeling the cross flow is proportional to the local section area. This implies that the cross flow at any section is identical to the flow past the section as if it were part of an infinite cylinder normal to a uniform flow. Thus,

$$\phi_c \approx U_\infty \sin\alpha \frac{[R(x)]^2}{r}\cos\theta, \quad \text{as } r \Rightarrow 0^+, \tag{68}$$

which simply represents the incompressible flow normal to an infinitely long cylinder. This means that in a linearized approximation, the cross flow prevails under incompressible conditions, at least near the body. This may be expected if the cross flow at a cross section is assumed to be independent of the cross flow at other sections.

18.3 Oscillating Thin Airfoil in Subsonic Flows: Possio's Theory

In the oscillating airfoil problem compressibility effects lead to further complications. These are (Possio, 1938),

(i) Additional phase lags appear;
(ii) Flow patterns do not adjust themselves instantaneously to changing boundary conditions. So, the flow properties do not just depend on the

instantaneous accelerations and velocities of the body, but are also affected by their time history.

For a linearized unsteady subsonic flow past a thin airfoil, one has, (see Section 14.1),

$$\frac{1}{a^2}\frac{\partial^2\phi}{\partial t^2}+\frac{2M_\infty}{a}\frac{\partial^2\phi}{\partial x\partial t}-\beta^2\frac{\partial^2\phi}{\partial x^2}=\frac{\partial^2\phi}{\partial y^2} \tag{69}$$

where

$$\beta\equiv\sqrt{1-M_\infty^2}.$$

Introducing

$$\begin{aligned}\tau &= t+\frac{M_\infty}{\beta^2 a}x\equiv t+\alpha x\\ \xi &\equiv \frac{x}{\beta^2 a},\quad \eta\equiv\frac{y}{\beta a}\end{aligned} \tag{70}$$

equation (69) becomes

$$\phi_{\tau\tau}=\phi_{\xi\xi}+\phi_{\eta\eta}. \tag{71}$$

Setting again

$$\xi=r'\cos\theta,\quad \eta=r'\sin\theta \tag{72}$$

equation (71) becomes

$$\phi_{\tau\tau}=\phi_{r'r'}+\frac{1}{r'}\phi_{r'}+\frac{1}{r'^2}\phi_{\theta\theta}. \tag{73}$$

Letting

$$\phi=e^{i\omega\tau}\left\{\begin{array}{c}\sin n\theta\\ \cos n\theta\end{array}\right\}R(r') \tag{74}$$

equation (73) gives

$$\frac{d^2R}{dr'^2}+\frac{1}{r'}\frac{dR}{dr'}+\left(\omega^2-\frac{n^2}{r'^2}\right)R=0 \tag{75}$$

from which,

$$R=A_n\left\{\begin{array}{c}H_n^{(1)}(\omega r')\\ H_n^{(2)}(\omega r')\end{array}\right\}. \tag{76}$$

$H_n^{(1)}(x)$ and $H_n^{(2)}(x)$ are *Hankel functions of the first kind and second kind*, respectively. Note that (76) gives a source at the origin for $n=0$ and a doublet for $n=1$.

Let us introduce the *acceleration potential*, which makes the formulation in the following more simple. The equation of motion is

$$\frac{Dv_i}{Dt} = -\frac{1}{\rho}\frac{\partial p}{\partial x_i}. \tag{77}$$

Letting

$$\frac{Dv_i}{Dt} \equiv \frac{\partial \varphi}{\partial x_i} \tag{78}$$

where φ is the acceleration potential, equation (77) leads to

$$\varphi + \int_{p_0}^{p} \frac{dp}{\rho} = const. \tag{79}$$

For a *barotropic* fluid and thin airfoils, φ , which is the specific enthalpy, will be continuous everywhere. When the wake is idealized into a thin surface, the velocity potential is discontinuous across that surface, although φ is continuous. This means that in the equation for φ, one would have no source terms arising from the wake due to the vorticity shed from the trailing edges.

Consider small disturbances in an otherwise uniform, rectilinear, steady flow. Let

$$\rho = \rho_\infty + \rho', \quad p = p_\infty + p' \quad \rho' \ll \rho_\infty, \quad p' \ll p_\infty. \tag{80}$$

One then obtains from (79),

$$\varphi = -\frac{p'}{\rho_\infty}. \tag{81}$$

The relation (81) indicates that the pressure may be considered to be a potential for the acceleration field, hence the name acceleration potential for φ.

Now, the discontinuity of the pressure field across an oscillating airfoil can be represented by a layer of doublets, the strength of which of course varies with time. For a linearized flow, φ satisfies the same equation as that for ϕ, viz., equation (69). Thus, for a line distribution of doublets (with vertical axes), one has from (74),

$$\varphi(x, y, t) = e^{i\omega t} \int_0^c B(x') \sin\theta H_1^{(2)}(\omega r') e^{i\omega\alpha(x-x')} dx' \tag{82}$$

where we have chosen $H_1^{(2)}(\omega r')$ rather than $H_1^{(1)}(\omega r')$ in accordance with the radiation condition (see Section 10.7), and

$$r' = \frac{1}{\beta^2 a}\sqrt{(x - x')^2 + \beta^2 y^2}$$

$$\theta = \tan^{-1}\left(\frac{\beta y}{x - x'}\right).$$

Consider a small element of length 2ε, at $x = \xi$ (see Figure 18.6). If L is the lift per unit length at $(x, 0)$, then applying the force balance, one finds

$$2\varepsilon L(\xi, t) = \int_C p' dx = -\rho_\infty \int_C \varphi(x, y, t)\, dx$$
$$= 2\rho_\infty \lim_{\delta \Rightarrow 0} \int_{\xi-\varepsilon}^{\xi+\varepsilon} \varphi(x, \delta, t)\, dx. \tag{83}$$

From (82), note that

$$\lim_{\delta \Rightarrow 0} \int_{x-\varepsilon}^{x+\varepsilon} \varphi(x, \delta, t)\, dx = e^{i\omega t} B(x) \int_{x-\varepsilon}^{x+\varepsilon} \lim_{\delta \Rightarrow 0} \sin\theta H_1^{(2)}(\omega r')\, dx' \tag{84}$$

where

$$H_1^{(2)}(z) \sim \frac{2i}{\pi z} \quad \text{for} \quad |z| \ll 1$$

$$\sin\theta = \frac{\beta y}{\sqrt{(x - x')^2 + \beta^2 y^2}}.$$

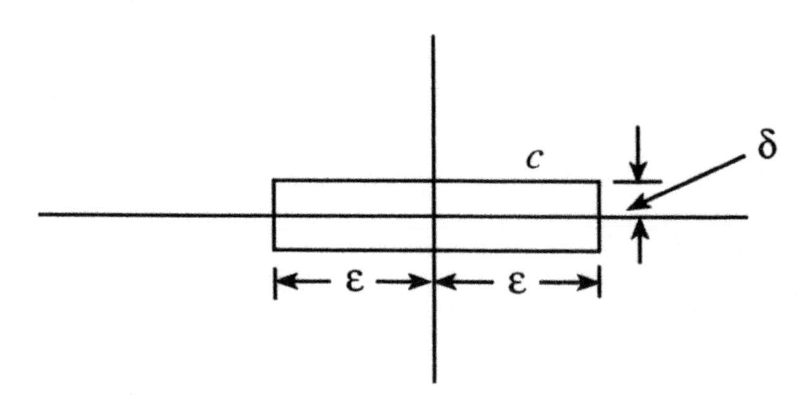

Figure 18.6 An element of the airfoil.

Thus,

$$\begin{aligned}\lim_{\delta\Rightarrow 0}\varphi(x,\delta,t) &= e^{i\omega t}B(x)\lim_{\delta\Rightarrow 0}\int_{x-\varepsilon}^{x+\varepsilon}\frac{2i}{\pi}\frac{\beta^3 a}{\omega}\frac{\delta}{[(x-x')+\beta^2\delta^2]}dx' \\ &= \frac{4i}{\pi}\frac{\beta^2 a}{\omega}e^{i\omega t}B(x)\lim_{\delta\Rightarrow 0}\tan^{-1}\left(\frac{\varepsilon}{\beta\delta}\right) \\ &= \frac{2i\beta^2 a}{\omega}e^{i\omega t}B(x).\end{aligned} \tag{85}$$

Using (85), (83) gives

$$L(\xi,t) = \frac{4i\rho_\infty\beta^2 a}{\omega}e^{i\omega t}B(\xi). \tag{86}$$

Consider next, the velocity field for this flow. From the linearized equation of motion,

$$\frac{\partial\varphi}{\partial y} = \frac{\partial v}{\partial t} + U_\infty\frac{\partial v}{\partial x} \tag{87a}$$

and assuming a harmonic time dependence given by,

$$v,\varphi \sim e^{i\omega t} \tag{87b}$$

one obtains

$$v(x,y) = \frac{1}{U_\infty}\exp\left(-\frac{i\omega x}{U_\infty}\right)\int_{-\infty}^{x}\frac{\partial\varphi}{\partial y'}\exp\left(\frac{i\omega x'}{U_\infty}\right)dx'. \tag{88}$$

Without loss of generality, one may place the doublet at the origin. Then, (88) leads to

$$v(x,y,t) = \frac{1}{U_\infty}\exp\left(-\frac{i\omega x}{U_\infty}\right)\int_{-\infty}^{x}Be^{i\omega\left(t+Kx'\right)}\frac{\partial}{\partial y'}\left[\sin\theta\cdot H_1^{(2)}(\omega r')\right]dx' \tag{89}$$

where

$$K = \alpha + \frac{1}{U_\infty} = \frac{1}{U_\infty \beta^2}.$$

Using the results,

$$\begin{aligned} &\frac{\partial \theta}{\partial y} = \frac{\cos\theta}{\beta a r'}, \ \frac{\partial \theta}{\partial x} = -\frac{\sin\theta}{\beta^2 a r'}, \frac{\partial r'}{\partial y} = \frac{\sin\theta}{\beta a}, \frac{\partial r'}{\partial x} = \frac{\cos\theta}{\beta^2 a} \\ &\frac{n}{z} H_n - H_n'(z) = H_{n+1}(z), \end{aligned} \tag{90}$$

(89) becomes

$$v(x,y,t) = -\frac{B\beta a}{U_\infty \omega} \exp\left[i\omega\left(t - \frac{x}{U_\infty}\right)\right] \int_{-\infty}^{\infty} e^{i\omega K x'} \frac{\partial^2}{\partial y'^2}\left[H_0^{(2)}(\omega r')\right] dx' \tag{91}$$

which is divergent, when $y = 0, x \geq 0$.

Noting that

$$\frac{\partial^2}{\partial y'^2}\left[H_0^{(2)}(\omega r')\right] = -\frac{\omega^2}{\beta^2 a^2} H_0^{(2)}(\omega r') - \beta^2 \frac{\partial^2}{\partial x'^2}\left[H_0^{(2)}(\omega r')\right] \tag{92}$$

(91) becomes

$$\begin{aligned} v(x,y,t) &= A \int_{-\infty}^{\infty} e^{i\omega K x'} \left[\frac{\omega^2}{\beta^2 a^2} H_0^{(2)}(\omega r') + \beta^2 \frac{\partial^2}{\partial x'^2} H_0^{(2)}(\omega r')\right] dx' \\ &\equiv I_1 + I_2 \end{aligned} \tag{93}$$

where

$$A = \frac{B\beta a}{U_\infty \omega} \exp\left[i\omega\left(t - \frac{x}{U_\infty}\right)\right] = \frac{L}{4i\rho_\infty \beta U_\infty} \exp\left(-\frac{i\omega x}{U_\infty}\right)$$

which is obtained using (86).

Noting that

$$\begin{aligned} I_2 = A\beta^2 \Bigg\{ &\omega\left[-H_1^{(2)}(\omega r') \frac{\cos\theta}{\beta^2 a} - iKH_0^{(2)}(\omega r')\right] e^{i\omega K x} \\ &-\omega^2 K^2 \int_{-\infty}^{x} e^{i\omega K x} H_0^{(2)}(\omega r')\, dx' \Bigg\} \end{aligned} \tag{94}$$

(93) becomes

$$v(x,y,t) = A\beta^2\omega e^{i\omega Kx}\left\{-H_1^{(2)}(\omega r')\frac{\cos\theta}{\beta^2 a} + iKH_0^{(2)}(\omega r')\right\} + \\ + A\left(\frac{1}{\beta^2 a^2} - \beta^2 K^2\right)\omega^2 \int_{-\infty}^{x} e^{i\omega Kx'} H_0^{(2)}(\omega r')\, dx' \tag{95}$$

from which,

$$v(x,0,t) = -\frac{iA\omega e^{i\omega Kx}}{U_\infty}\left\{-iM_\infty H_1^{(2)}(|W|)\frac{W}{|W|} + H_0^{(2)}(|W|)\right\} \\ - \frac{A\omega^2}{U_\infty^2}\int_{-\infty}^{x} e^{i\omega Kx'} H_0^{(2)}\left(\omega\frac{|x'|}{\beta^2 a}\right) dx' \tag{96}$$

where

$$W \equiv \frac{\omega}{\beta^2 a}x = \omega M_\infty Kx.$$

Consider

$$I \equiv \int_0^\infty e^{-i\omega Kx'} H_0^{(2)}\left(\frac{\omega}{\beta^2 a}x'\right) dx' = \frac{U_\infty \beta^2}{\omega}\int_0^\infty e^{-iu} H_0^{(2)}(M_\infty u)\, du \tag{97}$$

where

$$u \equiv \omega k x'.$$

Noting that

$$\int_1^\infty e^{-ix\xi}\frac{d\xi}{\sqrt{\xi^2-1}} = -\frac{i\pi}{2}H_0^{(2)}(x), \quad x > 0 \tag{98}$$

(97) becomes

$$I = \frac{2U_\infty\beta^2}{\pi\omega}\int_1^\infty \frac{d\xi}{\sqrt{\xi^2-1}\,(1+M_\infty\xi)}. \tag{99}$$

Setting

$$\xi \equiv \cosh\eta, \tag{100}$$

(99) becomes

$$I = \frac{2U_\infty \beta^2}{\pi\omega} \int_1^\infty \frac{d\eta}{1 + M_\infty \cosh\eta}. \tag{101}$$

Using the result,

$$\int \frac{dx}{1 + \cos\alpha \cosh x} = 2\, cosec\, \alpha \tanh^{-1}\left(\tanh\frac{x}{2}\tan\frac{\alpha}{2}\right)$$

(101) becomes

$$I = \frac{2U_\infty \beta^2}{\pi\omega} \frac{2}{\sqrt{1 - M_\infty^2}} \tanh^{-1}\left(\sqrt{\frac{1 - M_\infty}{1 + M_\infty}} \tanh\frac{\eta}{2}\right)\Bigg|_0^\infty$$

or

$$I = \frac{4U_\infty \beta}{\pi\omega} \tanh^{-1}\sqrt{\frac{1 - M_\infty}{1 + M_\infty}} = \frac{2U_\infty \beta}{\pi\omega} \ell n \frac{1 + \sqrt{1 - M_\infty^2}}{M_\infty}. \tag{102}$$

Using (97) and (102), (96) becomes

$$v(x, 0, t) = \frac{\omega L}{\rho_\infty U_\infty} \kappa(M_\infty, x) \tag{103}$$

where

$$\kappa(M_\infty, x) \equiv \frac{1}{4\beta} e^{iM_\infty W} \left\{ iM_\infty \frac{W}{|W|} H_1^{(2)}(|W|) - H_0^{(2)}(|W|) \right\}$$
$$+ \frac{i\beta}{4} \exp\left(-\frac{i\omega x}{U_\infty}\right) \left\{ \frac{2}{\pi\beta} \ell n \frac{1 + \beta}{M_\infty} + \int_0^{W/M_\infty} e^{iu} H_0^{(2)}(M_\infty |u|)\, du \right\}.$$

Thus, for the entire airfoil, one obtains an integral equation

$$v(x, 0, t) = \frac{\omega}{\rho_\infty U_\infty^2} \oint_0^C L(\xi, t)\, \kappa(M_\infty, x - \xi)\, d\xi \tag{104}$$

with the Kutta condition given in the form of continuity of the pressure at the trailing edge,

$$\xi = c : L = 0. \tag{105}$$

Note that $\kappa(M_\infty, x - \xi)$ has a singularity at $x = \xi$.

18.4 Oscillating Thin Airfoils in Supersonic Flows: Stewartson's Theory

The treatment of oscillating thin airfoils in supersonic flows becomes simpler than that in subsonic flows. This is because, in the supersonic case, the flows above and below the airfoil are independent of each other. Additionally, the flow over the airfoil is independent of the conditions in the wake (Stewartson, 1950).

One has for the linearized unsteady potential flows,[1]

$$\begin{aligned} &\left(1 - M_\infty^2\right)\phi_{xx} + \phi_{yy} - 2M_\infty^2\phi_{xt} - M_\infty^2\phi_{tt} = 0 \\ &y = 0 : \phi_y = v_0(x,t), \quad x \geq 0. \end{aligned} \tag{106}$$

Assuming a harmonic time dependence,

$$v_0(x,t) = v_0(x)e^{i\omega t}, \quad \phi(x,y,t) = \phi(x,y)e^{i\omega t} \tag{107}$$

(106) becomes

$$\begin{aligned} &\left(M_\infty^2 - 1\right)\phi_{xx} - \phi_{yy} + 2M_\infty^2 i\omega\phi_x - M_\infty^2\omega^2\phi = 0 \\ &y = 0 : \phi_y = v_0(x), \quad x \geq 0. \end{aligned} \tag{108}$$

Applying the Laplace transform with respect to x given by,

$$\tilde{\phi}(s,y) \equiv \int_0^\infty e^{-sx}\phi(x,y)\,dx \tag{109}$$

the boundary value problem (108) becomes

$$\begin{aligned} &\phi_{yy} - \mu^2\tilde{\phi} = 0 \\ &y = 0 : \tilde{\phi} = \tilde{v}_0(s) \end{aligned} \tag{110}$$

where

1 Even though transonic flows are patently nonlinear, the nonlinear characteristics in an unsteady situation manifest themselves only in low-frequency airfoil motions. For sufficiently high frequencies, even the unsteady transonic flow problem becomes a linear one like the subsonic- and supersonic-flow cases!

$$\mu^2 \equiv s^2 \left(M_\infty^2 - 1\right) + 2M_\infty^2 i\omega s - \omega^2 M_\infty^2.$$

One obtains from the boundary value problem (110)

$$\tilde{\phi}(s, y) = -\frac{\tilde{v}_0(s)}{\mu} e^{-\mu y}, \quad y > 0. \tag{111}$$

Upon inverting the Laplace transform,[2] (111) gives

$$\phi(x, y) = -\frac{1}{\sqrt{M_\infty^2 - 1}} \int_0^{x-\sqrt{M_\infty^2-1}y} v_0(x_1) \exp\left[-\frac{iM_\infty^2\omega}{M_\infty^2 - 1}(x - x_1)\right] \times$$
$$\times J_0\left[\frac{M_\infty\omega}{M_\infty^2 - 1}\sqrt{(x - x_1)^2 - \left(M_\infty^2 - 1\right) y^2}\right] dx_1. \tag{112}$$

As $M_\infty \Rightarrow 1$, note that (112) becomes

$$\phi(x, y) \approx -\int_0^x v_0(x_1) \frac{\exp\left[-\frac{i\omega}{2}\left\{(x - x_1) + \frac{y^2}{(x - x_1)}\right\}\right]}{\sqrt{2\pi i\omega (x - x_1)}} dx_1 \tag{113}$$

indicating that the transonic case poses no difficulty for the linearized formulation in the unsteady case!

Using (112), the Bernoulli integral gives the pressure on the airfoil,

$$p(x, 0) = \frac{1}{\sqrt{M_\infty^2 - 1}}\left[v_0(x) - \int_0^x v_0(x_1)\frac{\partial G}{\partial x_1} dx_1 + i\omega \int_0^x v_0(x_1) G(x_1) dx_1\right] \tag{114}$$

where

$$G(x, x_1) = \exp\left[-i\frac{M_\infty^2\omega}{M_\infty^2 - 1}(x - x_1)\right] J_0\left[\frac{M_\infty\omega}{M_\infty^2 - 1}(x - x_1)\right].$$

---

2 We note the Laplace inversion result,

$$\mathscr{L}^{-1}\left\{\frac{e^{-b\sqrt{s^2+a^2}}}{\sqrt{s^2 + a^2}}\right\} = \begin{cases} J_0\left(a\sqrt{x^2 - b^2}\right), x > b \\ 0, x < b \end{cases}.$$

Example 2: Consider a sharp-edged gust, given by

$$v_g = \begin{cases} 0, & x > t \\ V, & x < t \end{cases} \tag{115}$$

so that we have,

$$v(\omega) = \int_0^{\infty} v_g(t)\, e^{i\omega t} dt = \frac{iV}{\omega} e^{i\omega x} \tag{116}$$

from which,

$$2\pi v_g(t) = iV \int_{-\infty}^{\infty} \frac{e^{i\omega(x-t)}}{\omega} d\omega. \tag{117}$$

Using (117), (114) becomes

$$\begin{aligned} p(\omega x) = \frac{e^{-i\omega t}}{\sqrt{M_\infty^2 - 1}} \Bigg[& \frac{iV}{\omega} e^{i\omega x} - \int_0^x \frac{iV}{\omega} e^{i\omega x_1} \frac{\partial G}{\partial x_1} dx_1 \\ & + i\omega \int_0 \frac{iV}{\omega} e^{i\omega x_1} G(x_1)\, dx_1 \Bigg] \end{aligned} \tag{118}$$

from which we obtain for the pressure on the airfoil for $t > 0$,

$$p(x,t) = \frac{1}{2\pi} \int_{-\infty}^{\infty} p(\omega, x)\, d\omega. \tag{119}$$

Exercises

1. Estimate the error incurred in writing the boundary condition (46).
2. Obtain the solution for a supersonic flow past a cone by starting with the assumption that the flow is conical (and not using the slender axisymmetric body theory).
3. Linearize the transonic-flow equation using the hodograph transformation and investigate the prospects of its solvability.
4. Set up the boundary-value problem for a nonlinear axisymmetric flow past a cone and investigate the prospects of its solvability.

Part IV

Dynamics of Viscous Fluid Flows

19

Exact Solutions to Equations of Viscous Fluid Flows

The mathematical theory of ideal fluids discussed so far provides a powerful approach to the solution of several problems involving fluid flows. It gives satisfactory descriptions of many characteristics of flows of the real fluid. The primary aspects of wave motion, and the pressure field on streamlined bodies placed in flows are examples of such characteristics. However, this theory is unable to indicate how nearly the flow field (the whole or part of it) will be irrotational. Therefore, application of the results from this theory requires the clarification of the circumstances in which the ideal-fluid assumption is valid. Basically, the ideal-fluid assumption is useful in so far as it may describe the behavior of a real fluid in the limit of vanishing viscosity. However, because of the contamination by vorticity within a *boundary layer* near a solid surface, the ideal-fluid theory does not give an accurate description of the flows near solid boundaries. Consequently, it cannot describe quantities such as *skin friction* and *form drag* of a body placed in a flow. In order to resolve these discrepancies, an understanding and inclusion of the effects of viscosity of a real fluid is essential.

Some of the effects produced by the fluid viscosity are

- generation of shearing stresses in the fluid;
- maintenance of a zero slip-velocity of the fluid at a solid boundary.

In the following, we shall consider only the Newtonian fluids for which the coefficient of viscosity μ is independent of the rate of deformation of the fluid. Furthermore, we shall restrict ourselves to laminar flows wherein the viscous processes have their origin in the molecular transport processes.

Introduction to Theoretical and Mathematical Fluid Dynamics, Third Edition.
Bhimsen K. Shivamoggi.

19.1 Channel Flows

For a steady, unidirectional flow in a channel bounded by a rigid wall, the equation of motion is (Section 2.7),

$$-\frac{dp}{dx} + \mu \frac{d^2u}{dy^2} = 0. \tag{1}$$

In this case, the nonlinear convective term in the equation of motion vanishes identically, making the equation linear.

For a *Couette flow* (see Figure 19.1), with boundary conditions,

$$\left.\begin{array}{l} y = 0 : u = 0 \\ y = h : u = U \end{array}\right\} \tag{2}$$

equation (1) gives

$$u = \frac{y}{h}U - \frac{h^2}{2\mu}\frac{dp}{dx}\frac{y}{h}\left(1 - \frac{y}{h}\right) \tag{3}$$

which gives the linear profile shown in Figure 19.1, when $dp/dx = 0$.

For a *Poiseuille flow* (see Figure 19.2), fluid enters a channel with a uniform velocity over the entire cross section. Thanks to the no-slip condition, the fluid next to the wall will be slowed down and boundary layers form on the channel walls. These boundary layers grow and ultimately meet, producing the Poiseuille flow in the channel with a parabolic velocity profile.

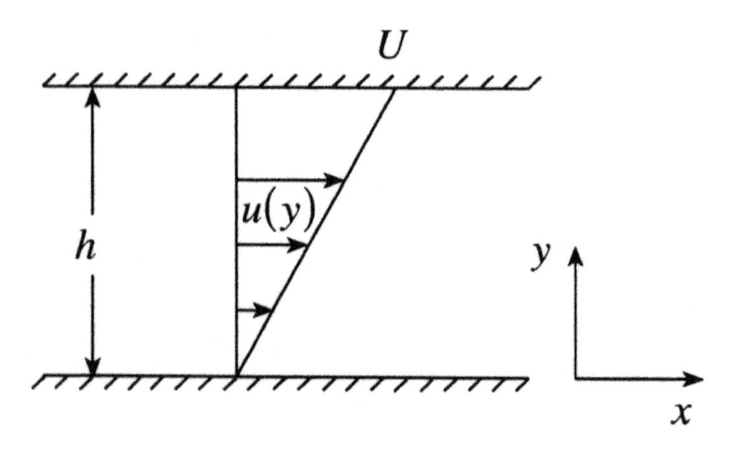

Figure 19.1 Couette flow.

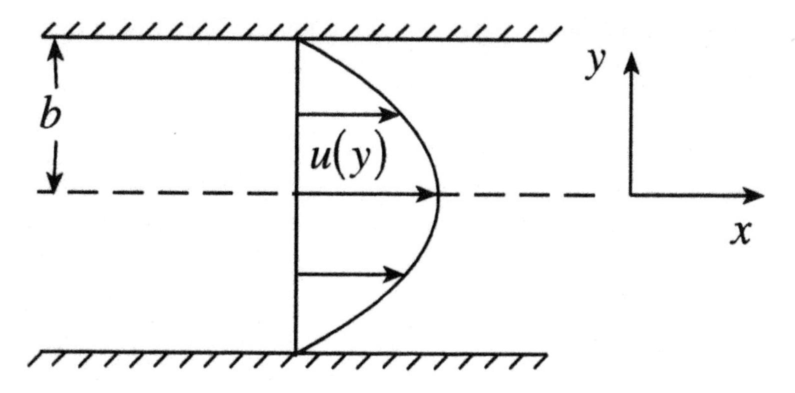

Figure 19.2 Poiseuille flow.

The boundary conditions for a Poiseuille flow are

$$y = \pm b : u = 0. \tag{4}$$

Equation (1) then gives

$$u = -\frac{1}{2\mu}\frac{dp}{dx}\left(b^2 - y^2\right). \tag{5}$$

19.2 Decay of a Line Vortex: The Lamb-Oseen Vortex

Consider the decay of a vortex filament with an initial circulation Γ_0 in an incompressible fluid. We use the cylindrical coordinates and take the z-axis along the axis of the filament. The vorticity is governed by the following equation

$$\frac{\partial \zeta}{\partial t} = \nu \frac{1}{r}\frac{\partial}{\partial r}\left(r\frac{\partial \zeta}{\partial r}\right) \tag{6}$$

where ν is the kinematic viscosity $\nu \equiv \mu/\rho$, and

$$\zeta \equiv \frac{1}{r}\frac{\partial}{\partial r}\left(r u_\theta\right). \tag{7a}$$

The boundary conditions are

$$\left.\begin{aligned} r = 0 &: u_\theta = 0 \\ r \Rightarrow \infty &: u_\theta \Rightarrow 0 \end{aligned}\right\} \text{for } t > 0. \tag{8}$$

In order to find a solution to equation (6), it proves convenient to recast equation (6) in terms of the circulation Γ,

$$\Gamma \equiv 2\pi r u_\theta. \tag{9}$$

Relation (7a) then leads to

$$\zeta = \frac{1}{r}\frac{\partial \Gamma}{\partial r} \tag{7b}$$

and equation (6) becomes

$$\frac{\partial \Gamma}{\partial t} = \nu r \frac{\partial}{\partial r}\left(\frac{1}{r}\frac{\partial \Gamma}{\partial r}\right). \tag{10}$$

Let us look for a *self-similar* solution of the form,

$$\Gamma(r,t) = \Gamma(\eta) \tag{11a}$$

where

$$\eta \equiv \frac{r^2}{4\nu t}. \tag{11b}$$

Equation (10) then leads to the ordinary differential equation

$$\Gamma'' + \Gamma' = 0. \tag{12}$$

The boundary conditions on Γ, reflecting (8), are

$$\left.\begin{array}{l} \eta = 0 : \Gamma = \Gamma_0 \eta \\ \eta \Rightarrow \infty : \Gamma \Rightarrow \Gamma_0 \end{array}\right\}. \tag{13}$$

The solution of equation (12) satisfying (13) is the *Lamb-Oseen* vortex,

$$\Gamma(\eta) = \Gamma_0\left(1 - e^{-\eta}\right) \tag{14a}$$

which describes the build-up of the circulation Γ as one goes outward from the vortex core (Figure 19.3a). Using (9), (14a) leads to

$$u_\theta(r,t) = \frac{\Gamma_0}{2\pi r}\left(1 - e^{-(r^2/4\nu t)}\right). \tag{14b}$$

In the narrow core near the axis of the vortex, i.e., for $r \ll \sqrt{4\nu t}$, (14b) gives

$$u_\theta \approx \frac{\Gamma_0}{8\pi\nu t} r$$

which corresponds to an almost rigid rotation with angular velocity $\Gamma_0/8\pi\nu t$ (Figure 19.3b). The intensity of the vortex thus decreases with time as the core spreads radially outward. The flow field outside the core is irrotational. The vortex core becomes narrower as the viscosity becomes smaller.

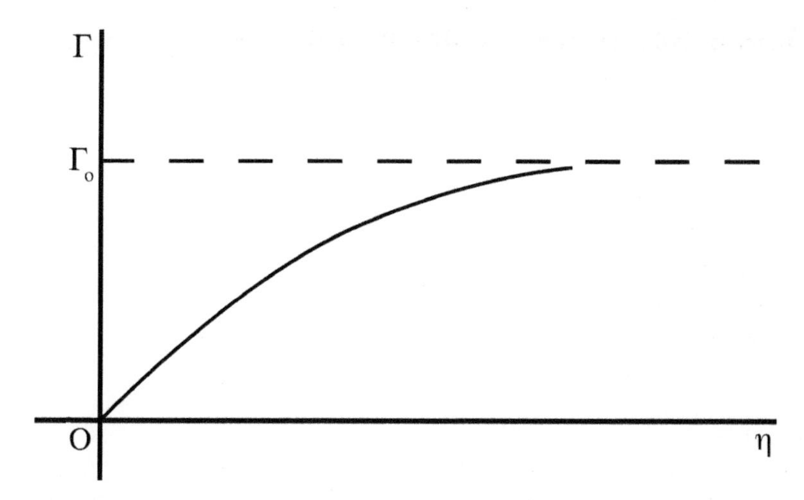

Figure 19.3a Circulation distribution for Lamb-Oseen vortex.

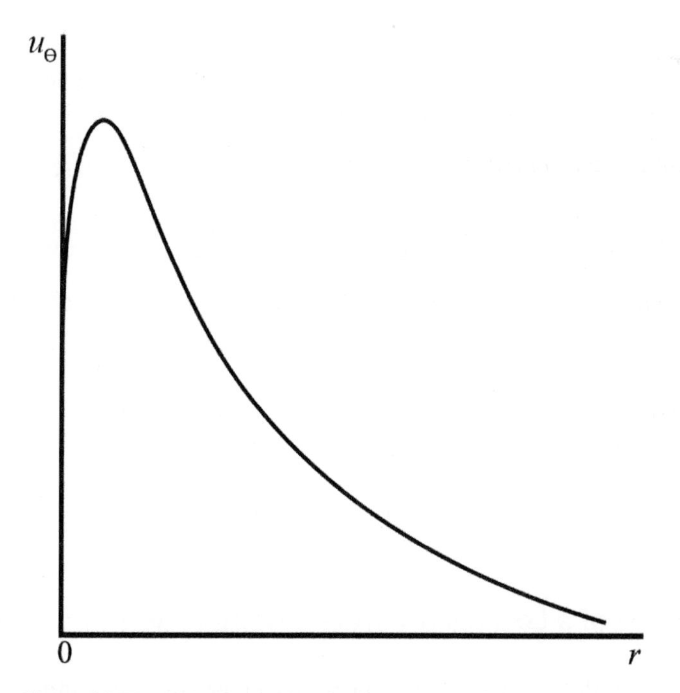

Figure 19.3b Velocity distribution for Lamb-Oseen vortex.

19.3 Line Vortex in a Uniform Stream

Consider a uniform flow with velocity U_∞ parallel to the x-axis. Suppose that this flow is slightly perturbed by a unit line source of vorticity placed at the origin. The equation governing this flow is given by,

$$\frac{\partial \zeta}{\partial x} = \frac{\nu}{U_\infty} \nabla^2 \zeta. \tag{15}$$

Looking for a solution of the form,

$$\zeta = e^{kx} f(x, y), \quad k \equiv \frac{U_\infty}{2\nu} \tag{16}$$

equation (15) gives

$$\nabla^2 f = k^2 f \tag{17}$$

from which, we have

$$f(r) = K_0(kr) \tag{18}$$

where $K_0(z)$ is the modified Bessel function of the second kind of zeroth order. Thus

$$\zeta = \exp(kr\cos\theta) K_0(kr). \tag{19}$$

For $kr \gg 1$, (19) gives

$$\zeta \approx \{\exp[-kr(1 - \cos\theta)]\} \sqrt{\frac{\pi}{2kr}} \tag{20}$$

and for $\theta \approx 0$, (20) gives

$$\zeta \sim \sqrt{\frac{\pi}{2kx}} \exp\left(-kx\frac{\theta^2}{2}\right). \tag{21}$$

This describes the far wake region behind the line vortex.

19.4 Diffusion of a Localized Vorticity Distribution

It turns out that the nonlinear effects become negligible in the problem of diffusion of a localized vorticity distribution in what may be called the "*final period*," (i.e., $t \Rightarrow \infty$). Furthermore, the diffusion process in this limit turns out to be independent of the detailed nature of the initial condition.

The equation for the transport of vorticity ξ is

$$\frac{\partial \xi}{\partial t} + (\mathbf{v} \cdot \nabla)\xi = (\xi \cdot \nabla)\mathbf{v} + \nu \nabla^2 \xi \tag{22}$$

and for the axisymmetric case, in cylindrical coordinates, equation (22) leads to

$$\begin{aligned} &\frac{\partial}{\partial t}\left(\frac{\partial u_\theta}{\partial z}\right) + \left(u_r \frac{\partial}{\partial r} + u_z \frac{\partial}{\partial z}\right)\frac{\partial u_\theta}{\partial z} \\ &= \left[\left(\frac{\partial u_\theta}{\partial r} + \frac{u_\theta}{r}\right)\frac{\partial}{\partial z} + \frac{\partial u_\theta}{\partial z}\frac{\partial}{\partial r}\right]u_r + \\ &\quad + \nu\left(\frac{\partial^2}{\partial r^2} + \frac{1}{r}\frac{\partial}{\partial r} + \frac{\partial^2}{\partial z^2}\right)\frac{\partial u_\theta}{\partial z} \end{aligned} \tag{23}$$

$$\begin{aligned} &\frac{\partial}{\partial t}\left(\frac{\partial u_z}{\partial r} - \frac{\partial u_r}{\partial z}\right) + \left(u_r \frac{\partial}{\partial r} + u_z \frac{\partial}{\partial z}\right)\left(\frac{\partial u_z}{\partial r} - \frac{\partial u_r}{\partial z}\right) \\ &\quad = \left[\left(\frac{\partial u_\theta}{\partial r} + \frac{u_\theta}{r}\right)\frac{\partial}{\partial z} + \frac{\partial u_\theta}{\partial z}\frac{\partial}{\partial r}\right]u_\theta + \\ &\quad + \nu\left(\frac{\partial^2}{\partial r^2} + \frac{1}{r}\frac{\partial}{\partial r} + \frac{\partial^2}{\partial z^2}\right)\left(\frac{\partial u_z}{\partial r} - \frac{\partial u_r}{\partial z}\right) \end{aligned} \tag{24}$$

$$\begin{aligned} &\frac{\partial}{\partial t}\left(\frac{\partial u_\theta}{\partial r} + \frac{u_\theta}{r}\right) + \left(u_r \frac{\partial}{\partial r} + u_z \frac{\partial}{\partial z}\right)\left(\frac{\partial u_\theta}{\partial r} + \frac{u_\theta}{r}\right) \\ &\quad = \left[\left(\frac{\partial u_\theta}{\partial r} + \frac{u_\theta}{r}\right)\frac{\partial}{\partial z} + \frac{\partial u_\theta}{\partial z}\frac{\partial}{\partial r}\right]u_z + \\ &\quad + \nu\left(\frac{\partial^2}{\partial r^2} + \frac{1}{r}\frac{\partial}{\partial r} + \frac{\partial^2}{\partial z^2}\right)\left(\frac{\partial u_\theta}{\partial r} + \frac{u_\theta}{r}\right). \end{aligned} \tag{25}$$

The continuity of mass gives

$$\frac{\partial u_r}{\partial r} + \frac{u_r}{r} + \frac{\partial u_z}{\partial z} = 0. \tag{26}$$

In order to study the behavior of equations (23)-(26) in the limit $t \Rightarrow \infty$, it is natural to introduce, for a fluid with small viscosity (such as air), the following renormalized variables, (Shivamoggi, 1982)

$$\tau = \varepsilon t, \quad \tilde{r} = \sqrt{\varepsilon} r, \quad \varepsilon \equiv \frac{\nu}{LU} \ll 1. \tag{27}$$

Here, L and U denote the reference length and velocity, respectively. Let us consider a two-dimensional case with no dependence on z. Then, equation (25) becomes

$$\varepsilon\frac{\partial\zeta_z}{\partial\tau}+\left(\sqrt{\varepsilon}u_r\frac{\partial}{\partial\tilde{r}}\right)\zeta_z=\left(\sqrt{\varepsilon}\zeta_r\frac{\partial}{\partial\tilde{r}}\right)u_z+\left[\varepsilon\left(\frac{\partial^2}{\partial\tilde{r}^2}+\frac{1}{\tilde{r}}\frac{\partial}{\partial\tilde{r}}\right)\right]\zeta_z \tag{28}$$

where

$$\zeta_z\equiv\frac{\partial u_\theta}{\partial r}+\frac{u_\theta}{r}.$$

Seeking solutions of the form,

$$\begin{aligned}\zeta_z(\tilde{r},\tau;\varepsilon)&=\sum_{n=0}^{\infty}\varepsilon^n\zeta_z^{(n)}(\tilde{r},\tau)\\ u_r(\tilde{r},\tau;\varepsilon)&=\sum_{n=0}^{\infty}\varepsilon^n u_r^{(n)}(\tilde{r},\tau)\\ u_z(\tilde{r},\tau;\varepsilon)&=\sum_{n=0}^{\infty}\varepsilon^n u_z^{(n)}(\tilde{r},\tau)\end{aligned} \tag{29}$$

we obtain from equation (28)

$$\left\{\frac{\partial}{\partial\tau}-\left(\frac{\partial^2}{\partial\tilde{r}^2}+\frac{1}{\tilde{r}}\frac{\partial}{\partial\tilde{r}}\right)\right\}\zeta_z^{(0)}=0. \tag{30}$$

In order to solve equation (30), let us consider the following initial-value problem, formulated in cartesian coordinates,

$$\frac{\partial^2\xi}{\partial x^2}+\frac{\partial^2\xi}{\partial y^2}=\frac{1}{\nu}\frac{\partial\xi}{\partial t} \tag{31}$$

$$t=0:\ \xi(x,y,t)=f(x,y),x^2+y^2<a^2. \tag{32}$$

Applying the Fourier transform defined by,

$$\tilde{\xi}(k_1,k_2,t)=\frac{1}{2\pi}\int_{-\infty}^{\infty}\int_{-\infty}^{\infty}\xi(x,y,t)e^{-i(k_1x+k_2y)}dxdy \tag{33}$$

$$\xi(x,y,t)=\frac{1}{2\pi}\int_{-\infty}^{\infty}\int_{-\infty}^{\infty}\tilde{\xi}(k_1,k_2,t)e^{-i(k_1x+k_2y)}dk_1dk_2 \tag{34}$$

equation (31) gives

$$-\left(k_1^2+k_2^2\right)\tilde{\xi}=\frac{1}{\nu}\frac{\partial\tilde{\xi}}{\partial t}. \tag{35}$$

Using (32), equation (35) gives

$$\tilde{\xi}(k_1, k_2, t) = \tilde{f}(k_1, k_2)\, e^{-(k_1^2+k_2^2)\nu t} \tag{36}$$

where

$$\begin{aligned}\tilde{\xi}(k_1, k_2, 0) &= \frac{1}{2\pi}\int_{-\infty}^{\infty}\int_{-\infty}^{\infty} \xi(x, y, 0)\, e^{-i(k_1 x + k_2 y)} dx dy \\ &= \frac{1}{2\pi}\int_{-\infty}^{\infty}\int_{-\infty}^{\infty} f(x, y)\, e^{-i(k_1 x + k_2 y)} dx dy \\ &\equiv \tilde{f}(k_1, k_2).\end{aligned} \tag{37}$$

Thus

$$\begin{aligned}\xi(x, y, t) &= \int_{-\infty}^{\infty}\int_{-\infty}^{\infty} \frac{dk_1 dk_2}{2\pi} e^{-i(k_1 x + k_2 y)} \\ &\quad \int_{-\infty}^{\infty}\int_{-\infty}^{\infty} dx' dy' f(x', y') e^{-i\left(k_1 x' + k_2 y'\right) - \left(k_1^2 + k_2^2\right)\nu t} \\ &= \int_{-\infty}^{\infty}\int_{-\infty}^{\infty} dx' dy' f(x', y') \frac{\exp\left[-\frac{(x - x')^2 + (y - y')^2}{4\nu t}\right]}{4\pi\nu t}.\end{aligned} \tag{38}$$

Using (38), reverting to the original variables, and rewriting (32) as below,

$$t = 0 : \zeta_z^{(0)} = f(r), \quad r < a \tag{39}$$

equation (30) has the following solution,

$$\zeta_z(r, t) = \int_0^a 2\pi r' dr' f(r') G(r, t; r') + O(\varepsilon) \tag{40}$$

where the Green's function $G(r,t;t')$ is given by

$$G(r,t;t') = \frac{1}{2\pi}\int_{\theta}^{\theta+2\pi} \frac{\exp\left[-\frac{r^2+r'^2-2rr'\cos(\theta-\theta')}{4\nu t}\right]}{4\pi\nu t} d\theta'. \quad (41)$$

Noting that

$$\int_{\theta}^{\theta+2\pi} \exp\left[\frac{2rr'\cos(\theta-\theta')}{4\nu t}\right] d\theta' \equiv 2\int_0^{\pi} \exp\left(\frac{rr'\cos\sigma}{2\nu t}\right) d\sigma = 2\pi I_0\left(\frac{rr'}{2\nu t}\right) \quad (42)$$

(41) becomes

$$G(r,t;r') = \frac{1}{4\pi\nu t} I_0\left(\frac{rr'}{2\nu t}\right)\exp\left(-\frac{r^2+r'^2}{4\nu t}\right) \quad (43)$$

where $I_0(z)$ is a *modified Bessel function of the first kind* of zeroth order.

If the point of observation lies in the far field, for localized vorticity distribution, one may expand the quantity,

$$I_0\left(\frac{rr'}{2\nu t}\right)\exp\left(-\frac{r^2+r'^2}{4\nu t}\right)$$

as follows:

$$I_0\left(\frac{rr'}{2\nu t}\right)\exp\left(-\frac{r^2+r'^2}{4\nu t}\right) = \exp\left(-\frac{r^2}{4\nu t}\right) + \frac{r^2 r'^2}{16\nu^2 t^2}\exp\left(-\frac{r^2}{4\nu t}\right) + O\left(\frac{1}{t^3}\right). \quad (44)$$

Then, (40) can be rewritten as

$$\zeta_z(r,t) = \frac{\left[\int_0^a 2\pi r' dr' f(r')\right]}{4\pi\nu t}\exp\left(-\frac{r^2}{4\nu t}\right) + \frac{\left[\int_0^a 2\pi r'^3 dr' f(r')\right]}{64\pi\nu^3 t^3} r^2 \exp\left(-\frac{r^2}{4\nu t}\right) + O\left(\frac{1}{t^4}\right). \quad (45)$$

For $t \Rightarrow \infty$, (45) shows that the diffusion of a localized vorticity distribution corresponds dominantly to that of a line vortex placed at the origin with strength equal to the total vorticity. Furthermore, in the limit $t \Rightarrow \infty$, any lack of symmetry in the initial vorticity distribution shows up, not in the leading term but in the higher-order terms.

Note that if one assumes a single vortex filament in all at the origin, at $t = 0$, i.e.,

$$f(r) = \Gamma_0 \delta(r) \tag{46}$$

where Γ_0 denotes the circulation about the vortex filament. Then, in the limit $t \Rightarrow \infty$, we have from (45),

$$\zeta_z(r,t) = \frac{\Gamma_0}{4\pi\nu t} \exp\left(-\frac{r^2}{4\nu t}\right) \tag{47}$$

from which, the circulation is given by

$$\Gamma(r,t) = 2\pi \int_0^r \zeta_z r dr = \Gamma_0 \left[1 - \exp\left(-\frac{r^2}{4\nu t}\right)\right]. \tag{48}$$

Note that this is in agreement with (14a) for the *Lamb-Oseen vortex* (Lamb, 1916, Oseen, 1912).

The azimuthal velocity is then given by

$$u_\theta(r,t) = \frac{\Gamma}{2\pi r} = \frac{\Gamma_0}{2\pi r}\left[1 - \exp\left(-\frac{r^2}{4\nu t}\right)\right] \tag{49}$$

in agreement with (14b).

19.5 Burgers Vortex

Burgers (1948) *vortex* describes the interplay between the intensification of vorticity due to the imposed straining flow and the diffusion of vorticity due to the action of viscosity.

The Navier-Stokes equations for a steady, axisymmetric flow, with flow velocity $\mathbf{v} = (v_r, v_\theta, v_z)$, are

$$\frac{1}{r}\frac{\partial}{\partial r}(r v_r) + \frac{\partial v_z}{\partial z} = O \tag{50}$$

$$v_r \frac{\partial v_r}{\partial r} - \frac{v_\theta^2}{r} = -\frac{1}{\rho}\frac{\partial p}{\partial r} + \nu \frac{\partial}{\partial r}\left[\frac{1}{r}\frac{\partial}{\partial r}(r v_r)\right] \tag{51}$$

$$v_r \frac{\partial v_\theta}{\partial r} + \frac{v_r v_\theta}{r} = \nu \frac{\partial}{\partial r}\left[\frac{1}{r}\frac{\partial}{\partial r}(r v_\theta)\right] \tag{52}$$

$$v_r \frac{\partial v_z}{\partial r} + v_z \frac{\partial v_z}{\partial z} = -\frac{1}{\rho}\frac{\partial p}{\partial z} + \nu\left[\frac{1}{r}\frac{\partial}{\partial r}\left(r\frac{\partial v_z}{\partial r}\right) + \frac{\partial^2 v_z}{\partial z^2}\right]. \tag{53}$$

The velocity profile for the Burgers vortex, consistent with equation (50), is

$$\mathbf{v} = -\frac{a}{2}r\hat{\mathbf{i}}_r + v_\theta\hat{\mathbf{i}}_\theta + az\hat{\mathbf{i}}_z. \tag{54}$$

The vorticity associated with this velocity profile is given by

$$\omega = \omega_z\hat{\mathbf{i}}_z \tag{55a}$$

where

$$\omega_z = \frac{1}{r}\frac{\partial}{\partial r}(rv_\theta). \tag{55b}$$

We then obtain from equations (51) and (52),

$$v_r \frac{\partial \omega_z}{\partial r} + \left(\frac{\partial v_r}{\partial r} + \frac{v_r}{r}\right)\omega_z = \nu\frac{1}{r}\frac{\partial}{\partial_r}\left(r\frac{\partial \omega_z}{\partial r}\right). \tag{56}$$

Using (54), equation (56) becomes

$$-\frac{a}{\nu}\left(\frac{1}{2}r\frac{\partial \omega_z}{\partial r} + \omega_z\right) = \frac{1}{r}\frac{\partial}{\partial r}\left(r\frac{\partial \omega_z}{\partial r}\right). \tag{57a}$$

Setting,

$$\xi \equiv r^2 \tag{58}$$

equation (57a) becomes

$$\left(\xi\frac{\partial}{\partial \xi} + 1\right)\left(4\frac{\partial \omega_z}{\partial \xi} + \frac{a}{\nu}\omega_z\right) = O \tag{57b}$$

from which,

$$\omega_z = Ae^{-\frac{a}{4\nu}\xi} = Ae^{-\frac{a}{4\nu}r^2} \tag{59a}$$

where A is an arbitrary constant.

Noting that the circulation associated with the Burgers vortex is given by

$$\Gamma = \int_0^\infty \omega_z \cdot 2\pi r dr \tag{60}$$

(59a) leads to

$$A = \frac{\Gamma a}{4\pi\nu}. \tag{61}$$

Using (61), (59a) becomes

$$\omega_z = \frac{\Gamma a}{4\pi\nu} e^{-\frac{a}{4\nu}r^2}. \tag{59b}$$

Noting from (55b) that the azimuthal velocity is given by

$$v_\theta = \frac{1}{r}\int_0^r r\omega_z dr \tag{55c}$$

and using (59b), we obtain

$$v_\theta = \frac{\Gamma}{2\pi r}\left(1 - e^{-\frac{a}{4\nu}r^2}\right). \tag{62}$$

This leads to

$$v_\theta \approx \begin{cases} \left(\frac{\Gamma a}{8\pi\nu}\right) r, & r << \sqrt{\frac{4\nu}{a}} \\ \frac{\Gamma}{2\pi r}, & r >> \sqrt{\frac{4\nu}{a}} \end{cases} \tag{63}$$

which implies that, in the narrow core near the axis of the vortex, the flow resembles a rigid rotation while, away from the core, the flow is irrotational.

19.6 Flow Due to a Suddenly Accelerated Plane

In order to get further insight into the process of viscous diffusion of vorticity, consider a semi-infinite region of stationary fluid bounded by a rigid plane $y = 0$. The plane is suddenly given a velocity U parallel to the plane following which it is maintained at that velocity. The viscous stresses at the plane set the fluid into motion which is governed by

$$\frac{\partial u}{\partial t} = \nu \frac{\partial^2 u}{\partial y^2} \tag{64}$$

with the following initial and boundary conditions,

$$t \leq 0 : u = 0 \tag{65}$$

$$\left. \begin{array}{l} t > 0; \;\; y = 0 : \quad u = U \\ \qquad\quad y \to \infty : u \Rightarrow 0 \end{array} \right\} . \tag{66}$$

Here, u is the fluid velocity component in the direction of motion of the plane.
Introducing the *similarity* variables,

$$u = Uf(\eta), \;\; \eta \equiv \frac{y}{2\sqrt{\nu t}} \tag{67}$$

equation (64) leads to the ordinary differential equation,

$$f'' + 2\eta f' = 0 \tag{68}$$

while the initial and boundary conditions (65) and (66) lead to

$$\left. \begin{array}{l} \eta = 0 : \;\; f = 1 \\ \eta \Rightarrow \infty : f \Rightarrow 0. \end{array} \right\} \tag{69}$$

The boundary-value problem, given by (68) and (69), has the solution,

$$f(\eta) = 1 - \frac{2}{\sqrt{\pi}} \int_0^{\eta} e^{-\tau^2 d\tau} \tag{70}$$

and we have from (67)

$$u = U \left[1 - \frac{2}{\sqrt{\pi}} \int_0^{\eta} e^{-\tau^2 d\tau} \right]. \tag{71}$$

Note that the vorticity corresponding to this velocity field is given by

$$\zeta(y, t) = -\frac{\partial u}{\partial y} = \frac{U}{\sqrt{\pi \nu t}} \exp\left(-\frac{y^2}{4\nu t} \right). \tag{72}$$

The discontinuity in the tangential velocity arising near the boundary at $t = 0$ causes the vorticity to be concentrated initially in a thin layer of thickness δ near the plane. Then, from the conservation of vorticity in the flow, we have

$$\zeta(0,t) \cdot \delta(t) = \frac{U\delta(t)}{\sqrt{\pi\nu t}} = \int_0^\infty \zeta(y,t)\, dy. \tag{73}$$

Using (72), (73) gives

$$\delta(t) = \sqrt{\pi\nu t} \tag{74}$$

from which we have,

$$\frac{d\delta}{dt} = \frac{1}{2}\sqrt{\frac{\pi\nu}{t}}. \tag{75}$$

This gives the rate at which the vorticity, initially concentrated near the plane, diffuses into the flow. Note that the rate of diffusion decreases, as t increases, because the velocity gradient and its spatial rate of change become progressively smaller.

For the flow due to a plane set impulsively into an oscillation, we have the following initial and boundary conditions,

$$\left.\begin{array}{l} t \leq 0: \quad u = 0 \\ \quad t > 0; \quad y = 0: \quad u = U\cos\omega t \\ \qquad\qquad y \Rightarrow \infty: \quad u \Rightarrow 0. \end{array}\right\} \tag{76}$$

These conditions in conjunction with equation (64), lead to the following steady periodic state,

$$u(y,t) = U\exp\left(-\sqrt{\frac{\omega}{2\nu}}y\right)\cos\left(\omega t - \sqrt{\frac{\omega}{2\nu}}y\right). \tag{77}$$

The above expression shows that the viscous effects cause

- the usual intrinsic damping and
- a phase lag in the fluid motion.

19.7 The Round Laminar Jet: Landau-Squire Solution

In this section, we shall discuss an exact solution for an axial symmetric flow which describes a round laminar jet emerging from an orifice (Landau, 1944, Squire, 1951).

In spherical coordinates (r, θ, ϕ) with θ measured from the axis of the jet, the mass and momentum conservation equations are

$$\frac{1}{r^2}\frac{\partial}{\partial r}(r^2 u) + \frac{1}{r\sin\theta}\frac{\partial}{\partial\theta}(v\sin\theta) = 0 \tag{78}$$

$$u\frac{\partial u}{\partial r} + \frac{v}{r}\frac{\partial u}{\partial\theta} - \frac{v^2}{r} = -\frac{1}{\rho}\frac{\partial p}{\partial r} + \nu\left(\nabla^2 u - \frac{2u}{r^2} - \frac{2}{r^2}\frac{\partial v}{\partial\theta} - \frac{2v\cot\theta}{r^2}\right) \tag{79}$$

$$u\frac{\partial v}{\partial r} + \frac{v}{r}\frac{\partial v}{\partial\theta} + \frac{uv}{r} = -\frac{1}{\rho r}\frac{\partial p}{\partial\theta} + \nu\left(\nabla^2 v + \frac{2}{r^2}\frac{\partial u}{\partial\theta} - \frac{v}{r^2\sin^2\theta}\right) \tag{80}$$

where the fluid velocity $\mathbf{v}$ has components $(u, v, 0)$, and

$$\nabla^2 = \frac{1}{r^2}\frac{\partial}{\partial r}\left(r^2\frac{\partial}{\partial r}\right) + \frac{1}{r^2\sin\theta}\frac{\partial}{\partial\theta}\left(\sin\theta\frac{\partial}{\partial\theta}\right). \tag{81}$$

Introduce the stream function defined, as per equation (78), by

$$\left.\begin{aligned} u &= \frac{1}{r^2\sin\theta}\frac{\partial\Psi}{\partial\theta} \\ v &= -\frac{1}{r\sin\theta}\frac{\partial\Psi}{\partial r}. \end{aligned}\right\} \tag{82}$$

Looking for a solution of the form,

$$\Psi = \nu r f(\theta) \tag{83}$$

equation (82) leads to

$$\left.\begin{aligned} u &= \frac{\nu}{r\sin\theta}f'(\theta) \\ v &= -\frac{\nu}{r\sin\theta}f(\theta) \end{aligned}\right\}. \tag{84}$$

This implies

$$\begin{aligned} \nabla^2 u &= \frac{1}{r^2\sin\theta}\frac{\partial}{\partial\theta}\left(\sin\theta\cdot\frac{\partial u}{\partial\theta}\right) \\ \nabla^2 v &= \frac{1}{r^2\sin\theta}\frac{\partial}{\partial\theta}\left(\sin\theta\cdot\frac{\partial v}{\partial\theta}\right) \\ \frac{\partial u}{\partial r} &= -\frac{u}{r},\ \frac{\partial v}{\partial r} = -\frac{v}{r}. \end{aligned} \tag{85}$$

Using (84), equation (78) leads to

$$u + \frac{\partial v}{\partial \theta} + v \cot\theta = 0 \tag{86}$$

which is identically satisfied by (84).

Using (84)-(86), equations (79) and (80) become

$$-\frac{u^2 + v^2}{r} + \frac{v}{r}\frac{\partial u}{\partial \theta} = -\frac{1}{\rho}\frac{\partial p}{\partial r} + \frac{\nu}{r^2 \sin\theta}\frac{\partial}{\partial \theta}\left(\sin\theta \cdot \frac{\partial u}{\partial \theta}\right) \tag{87}$$

$$\frac{v}{r}\frac{\partial v}{\partial \theta} = -\frac{1}{\rho r}\frac{\partial p}{\partial \theta} + \frac{\nu}{r^2}\frac{\partial u}{\partial \theta}. \tag{88}$$

We have from equation (88),

$$\frac{p - p_0}{\rho} = -\frac{v^2}{2} + \frac{\nu u}{r} + \frac{c_1}{r^2} \tag{89}$$

where c_1 is an arbitrary constant.

Using (89), equation (87) becomes

$$-\frac{u^2}{r} + \frac{v}{r}\frac{\partial u}{\partial \theta} = \frac{\nu}{r^2}\left[2u + \frac{1}{\sin\theta}\frac{\partial}{\partial \theta}\left(\sin\theta \cdot \frac{\partial u}{\partial \theta}\right)\right] + \frac{2c_1}{r^3}. \tag{90}$$

Setting,

$$\mu \equiv \cos\theta \tag{91}$$

and using (84) in equation (90) gives

$$\left[f'(\mu)\right]^2 + f(\mu) f''(\mu) = 2f'(\mu) + \frac{d}{d\mu}\left[\left(1 - \mu^2\right) f''(\mu)\right] - \frac{2c_1}{\nu^2}. \tag{92}$$

Equation (92) can be integrated to give

$$ff' = 2f + \left(1 - \mu^2\right) f'' - \frac{2c_1 \mu}{\nu^2} - c_2 \tag{93}$$

where c_2 is an arbitrary constant. Integrating equation (93) again, we obtain

$$f^2 = 4\mu f + 2\left(1 - \mu^2\right) f' - 2\left(\frac{c_1 \mu^2}{\nu^2} + c_2 \mu + c_3\right) \tag{94}$$

where c_3 is an arbitrary constant.

Now, in order that the flow be free from singularities, one requires from (84) that[1]

$$\mu \approx \pm 1 : f \sim (1 \mp \mu). \tag{95}$$

Using (95), we have from equation (94)

$$c_1, c_2, c_3 = 0. \tag{96}$$

Using (96), we obtain the following solution to equation (94)

$$f = \frac{2(1-\mu^2)}{a+1-\mu} = \frac{2\sin^2\theta}{a+1-\cos\theta} \tag{97}$$

where a is an arbitrary constant which determines the nature of the flow described by (83) and (97).[2]

On the other hand, rewriting (97) as

$$f(s) = \frac{2s^2}{a+1-\sqrt{1-s^2}}, s \equiv \sin\theta = \sqrt{1-\mu^2} \tag{98}$$

and noting that the radius of the jet flow described by this solution is $R \sim \sin\theta = s$, we have for the throat location,

$$R = R^* \text{ or } s = s^* : \frac{\partial f}{\partial R} = 0 \text{ or } \frac{\partial f}{\partial s} = 0. \tag{99}$$

Using (98), (99) leads to

$$s_*^2 = 1 - \frac{1}{(a+1)^2}$$

or

$$\mu_* = \frac{1}{a+1}. \tag{100}$$

The flow described by (84) and (97) can be interpreted as a jet (see Figure 19.4) issuing from a nozzle coinciding with one of the streamlines up to its throat. The position of the throat is given by $\cos\theta_* = 1/(a+1)$. This jet entrains slow-moving fluid from outside. In the limit, $a \ll 1$, the throat position corresponds to $\theta_* \approx 0$ so that the jet flow is sharply peaked around $\theta = 0$. On the other hand, in the limit $a \gg 1$, the throat position corresponds to $\theta_* = \pi/2$ so that the jet

---

1 Using (91), (84) becomes

$$u = -\frac{\nu}{r} f'(\mu), v = -\frac{\nu}{r} \frac{f(\mu)}{\sqrt{1-\mu^2}}.$$

2 In particular, a helps locate the throat position for the streamlines (see discussion below (100)).

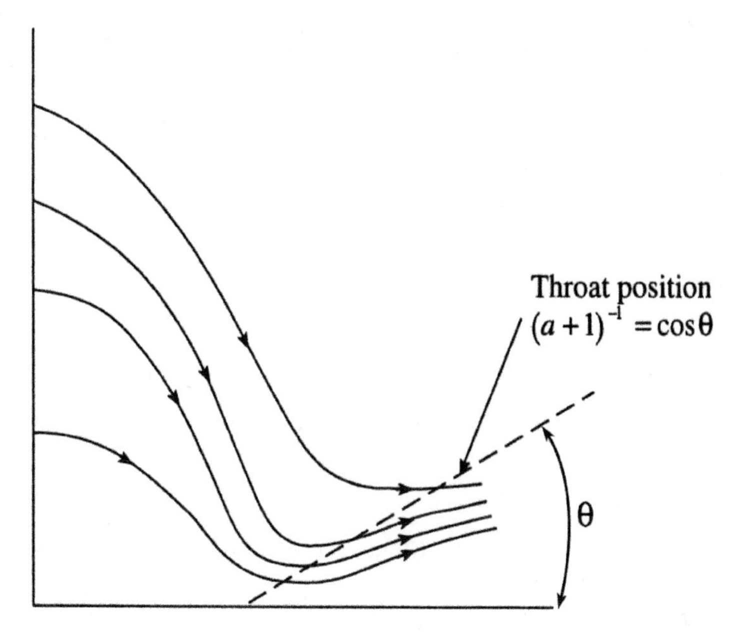

Figure 19.4 Streamline pattern for a jet issuing from a nozzle.

flow is symmetric about $\theta = \pi/2$. In order to make this jet flow possible, it must be assumed, however, that a special frictional boundary condition is satisfied on the walls of the nozzle.

19.8 Ekman Layer at a Free Surface in a Rotating Fluid

Consider a fluid bounded by a horizontal free surface at which a uniform and constant stress μS is applied (Ekman, 1905). This problem is of interest in connection with the drift of floating pieces of ice at the surface of the sea due to wind blowing over it. The drift is found to occur, thanks to the Coriolis force, somewhat to the right of the direction of the wind. We use a rectangular coordinate system rotating steadily with angular velocity $\mathbf{\Omega} = \Omega_z \hat{\mathbf{i}}_z$. The z-axis is along the vertical direction, and the x-axis is along the direction of the stress applied at the surface. The fluid velocity associated with the flow generated by the applied stress can then be taken to lie in the horizontal plane everywhere and vary only in the vertical direction. This causes the nonlinear convective derivative $(\mathbf{v} \cdot \nabla)\mathbf{v}$ to vanish identically. Then the equations of motion (see Section 9.1) give

$$-2v\Omega_z = -\frac{1}{\rho}\frac{\partial p}{\partial x} + \nu\frac{d^2u}{dz^2} \tag{101a}$$

$$2u\Omega_z = -\frac{1}{\rho}\frac{\partial p}{\partial y} + \nu\frac{d^2v}{dz^2}. \tag{101b}$$

If *geostrophic*[3] flow prevails in the limit $z \Rightarrow -\infty$, assuming the following boundary conditions,

$$z \Rightarrow -\infty : u \Rightarrow U, v \Rightarrow V \tag{102}$$

we have, from equations (101a,b)

$$-2V\Omega_z = -\frac{1}{\rho}\frac{\partial p}{\partial x} \tag{103a}$$

$$2U\Omega_z = -\frac{1}{\rho}\frac{\partial p}{\partial y}. \tag{103b}$$

Using equations (103a,b), equations (101a,b) lead to

$$-2\hat{v}\Omega_z = \nu\frac{d^2\hat{u}}{dz^2} \tag{104a}$$

$$2\hat{u}\Omega_z = \nu\frac{d^2\hat{v}}{dz^2} \tag{104b}$$

where

$$\hat{u} \equiv u - U, \hat{v} \equiv v - V. \tag{105}$$

The boundary conditions are

$$\left.\begin{array}{c} z = 0 : \dfrac{d\hat{u}}{dz} = S, \quad \dfrac{d\hat{v}}{dz} = 0 \\ z \Rightarrow -\infty : \hat{u}, \hat{v} \Rightarrow 0. \end{array}\right\} \tag{106}$$

One may combine equations (104a,b) to give

$$\nu\frac{d^2(\hat{u} + i\hat{v})}{dz^2} = 2i\Omega_z(\hat{u} + i\hat{v}) \tag{107}$$

which, on using the boundary conditions (106), leads to

$$\hat{u} + i\hat{v} = \frac{S(1-i)}{2k}e^{k(1+i)z} \tag{108}$$

---

3 *Geostrophic* approximation (see Section 9.2) refers to the balance between Coriolis force and pressure gradient transverse to the rotation axis.

where

$$k \equiv \sqrt{\frac{\Omega_z}{\nu}}.$$

The solution given by (108) yields

$$\left.\begin{aligned} \hat{u} &= \frac{S}{k\sqrt{2}} e^{kz} \cos\left(kz - \frac{\pi}{4}\right) \\ \hat{v} &= \frac{S}{k\sqrt{2}} e^{kz} \sin\left(kz - \frac{\pi}{4}\right) \end{aligned}\right\} \tag{109}$$

which are sketched in Figure 19.5.

Note that as the depth below the free surface increases, the direction of the velocity rotates due to the Coriolis force uniformly in a clockwise sense (for $\Omega_z > 0$), and its magnitude falls exponentially. This solution explains the observed discrepancy in movement between the sea surface current and floating pieces of ice.

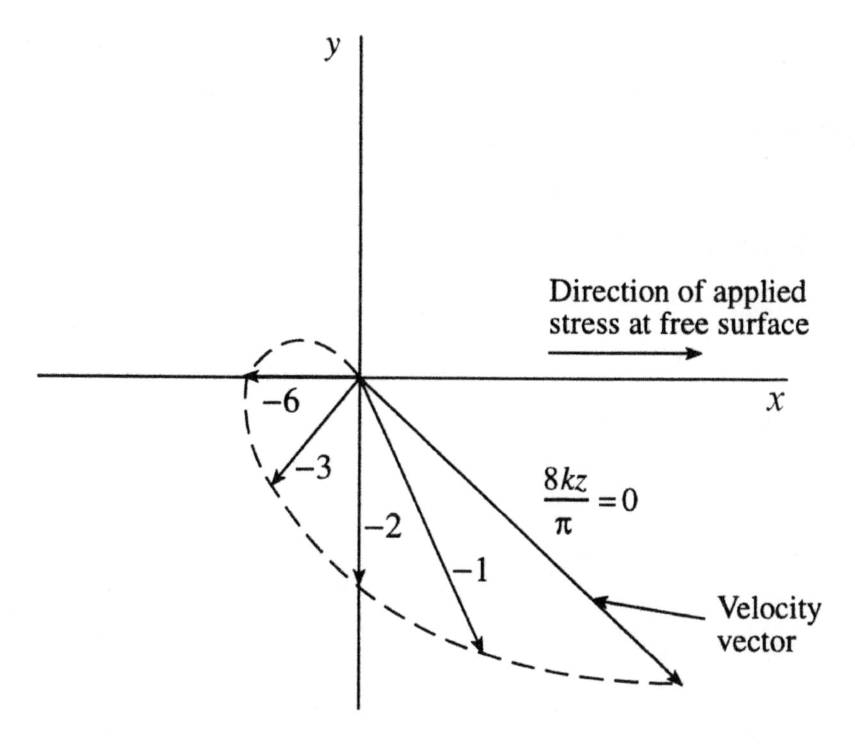

Figure 19.5 Variation of the velocity vector in the Ekman Layer (from Batchelor, 1967).

19.9 Centrifugal Flow Due to a Rotating Disk: von Kármán Solution

Consider a plane disk of large diameter rotating in its own plane with a steady angular velocity Ω in a fluid which, at infinity, is rotating rigidly (von Kármán, 1951) with a slightly smaller angular velocity Γ [4] (the latter generalization is due to Batchelor, 1951). The relative motion of the disk and the fluid leads to viscous stresses, which tend to drag the fluid around along with the disk. An exactly circular motion of fluid cannot occur near the disk. This is because the enhanced centrifugal force near the disk is too great to be balanced by the pressure gradient in the ambient fluid. Consequently, the fluid near the disk will spiral outwards. This outward radial motion near the disk leads to an axial motion toward the disk in order to ensure conservation of mass. This prevents the vorticity generated at the boundary from spreading away from it. Thus, when the disk is rotating faster ($\Omega > \Gamma$), the vorticity is confined to the vicinity of the rotating disk by convection toward the disk, induced by the centrifugal action on the fluid near the disk.

Let us look for a solution of the form (an *ansatz* suggested by von Kármán, 1951),

$$\frac{u}{r}, \frac{v}{r}, w \sim \text{ functions of } z \text{ only} \tag{110}$$

where (u, v, w) are velocity components parallel to the (r, φ, z) directions in a cylindrical coordinate system with $r = 0$ on the axis of the disk.

The equations of continuity and motion for a steady flow are

$$\frac{\partial}{\partial r}(ru) + \frac{\partial w}{\partial z} = 0 \tag{111a}$$

$$u\frac{\partial u}{\partial r} + w\frac{\partial u}{\partial z} - \frac{v^2}{r} = -\frac{1}{\rho}\frac{\partial P}{\partial r} + +\nu\left(\nabla^2 u - \frac{u}{r^2}\right) \tag{112a}$$

$$u\frac{\partial v}{\partial r} + w\frac{\partial v}{\partial z} + \frac{uv}{r} = \nu\left(\nabla^2 v - \frac{v}{r^2}\right) \tag{113a}$$

$$u\frac{\partial w}{\partial r} + w\frac{\partial w}{\partial z} = -\frac{1}{\rho}\frac{\partial P}{\partial z} + \nu\nabla^2 w \tag{114a}$$

where

$$\nabla^2 \equiv \frac{1}{r}\frac{\partial}{\partial r}\left(r\frac{\partial}{\partial r}\right) + \frac{\partial^2}{\partial z^2}, P \equiv p - \rho\Gamma^2 r^2/2.$$

Using (110), equations (111a)-(114a) lead to

$$\frac{2u}{r} + \frac{dw}{dz} = 0 \tag{111b}$$

4 This arrangement facilitates the development of an approximate analytical solution to this problem.

$$\left(\frac{u}{r}\right)^2 + w\frac{d(u/r)}{dz} - \left(\frac{v}{r}\right)^2 = \nu\frac{d^2(u/r)}{dz^2} - \Gamma^2 \tag{112b}$$

$$\frac{2uv}{r^2} + w\frac{d(v/r)}{dz} = \nu\frac{d^2(v/r)}{dz^2}. \tag{113b}$$

$$\frac{p}{\rho} = \nu\frac{dw}{dz} - \frac{1}{2}w^2 + \frac{1}{2}\Gamma^2 r^2. \tag{114b}$$

The boundary conditions are

$$\left.\begin{aligned} z = 0 : u = w = 0, \qquad & v = \Omega r \\ z \Rightarrow \infty : u \Rightarrow 0, \qquad & v = \Gamma r. \end{aligned}\right\} \tag{115}$$

Introducing the *similarity* variable ζ and the *similarity* solution according to

$$z = \left(\frac{\nu}{\Omega}\right)^{1/2}\zeta,\ \ \frac{v}{r} = \Omega g(\zeta),\ \ w = (\nu\Omega)^{1/2} h(\zeta) \tag{116}$$

equations (111b)-(113b) and (115) give

$$\frac{1}{4}h'^2 - \frac{1}{2}hh'' - g^2 = -\frac{1}{2}h''' - \left(\frac{\Gamma}{\Omega}\right)^2 \tag{117}$$

$$-gh' + g'h = g'' \tag{118}$$

$$\left.\begin{aligned} \zeta = 0 : h = h' = 0, \qquad & g = 1 \\ \zeta \Rightarrow \infty : h' \Rightarrow 0, \qquad & g \Rightarrow \frac{\Gamma}{\Omega} \end{aligned}\right\}. \tag{119}$$

Since $\Gamma/\Omega \approx 1$, it is legitimate, as per (116), to look for solutions of the form (Batchelor, 1951),

$$\Gamma = \Omega(1-\varepsilon), g = 1 + g_1, \varepsilon \ \text{ and } \ |g_1| \ll 1. \tag{120}$$

We then have from equation (118) that

$$|h| \ll 1. \tag{121}$$

Using (120) and (121), and linearizing in g_1 and h, the boundary value problem (117)-(119) gives

$$2g_1 \approx -2\varepsilon + \frac{1}{2}h''' \tag{122}$$

$$-h' = g_1'' \tag{123}$$

$$\left.\begin{aligned} \zeta = 0 : h = h' = g_1 = 0 \\ \zeta \Rightarrow \infty : h' \Rightarrow 0, g_1 \Rightarrow -\varepsilon \end{aligned}\right\} \tag{124}$$

from which we obtain[5]

$$\left.\begin{aligned} g_1(\zeta) &\approx -\varepsilon\left(1 - e^{-\zeta}\cos\zeta\right) \\ h(\zeta) &\approx -\varepsilon\left(1 - e^{-\zeta}\cos\zeta - e^{-\zeta}\sin\zeta\right) \end{aligned}\right\}. \tag{125}$$

Note from (125) that there is a net drift of fluid in the radial direction outward when $\Gamma < \Omega$, and vice versa. Such a drift as per (125) leads to an axial flow toward the disk[6] given by,

$$\zeta \Rightarrow \infty : h(\zeta) \Rightarrow -\varepsilon. \tag{126}$$

In general, equations (117) and (118) in conjunction with the boundary conditions (119) may either have no solution (Evans, 1969) or have infinitely many solutions (Zandbergen and Dijkstra, 1977).

19.10 Shock Structure: Becker's Solution

In Chapter 16, the theory of shock waves was discussed entirely in terms of an inviscid fluid. In a real fluid (as discussed in Section 15.5) however, the shock discontinuity gets spread out due to viscous and heat-conduction effects. Indeed, it turns out that the equations of conservation of mass, momentum, and energy admit a smooth, steady, one-dimensional solution for a perfect Newtonian gas of Prandtl number $Pr \equiv \mu C_p/K = 3/4$ (K being the thermal conductivity). This solution describes a shock transition in which the flow properties tend to the values corresponding to the uniform state at the limits $x \Rightarrow \pm\infty$ (Becker, 1922).

The conservation equations for a one-dimensional flow are as follows,

$$\frac{d}{dx}(\rho u) = 0 \tag{127}$$

$$\rho u \frac{du}{dx} = -\frac{dp}{dx} + \frac{4}{3}\frac{d}{dx}\left(\mu \frac{du}{dx}\right) \tag{128}$$

$$\rho u \frac{d}{dx}\left(h + \frac{1}{2}u^2\right) = \frac{d}{dx}\left(K\frac{dT}{dx}\right) + \frac{4}{3}\frac{d}{dx}\left(\mu u \frac{du}{dx}\right). \tag{129}$$

5 We use the result,

$$\left(x^4 + 4y^4\right) = \left(x^2 + 2xy + 2y^2\right)\left(x^2 - 2xy + 2y^2\right).$$

6 Related to this problem is the so-called spin-up problem where a container of fluid undergoes an impulsive change in the rotation rate. This leads to a boundary layer induced acceleration and inviscid pumping of the interior flow until a new steady state is attained.

We have, on integrating equations (127)-(129),

$$\rho u \equiv Q = const \tag{130}$$

$$\rho u^2 + p + \frac{4}{3}\mu\frac{du}{dx} \equiv P = const \tag{131}$$

$$\left(h + \frac{1}{2}u^2\right)\rho u + \frac{4}{3}\mu u\frac{du}{dx} + K\frac{dT}{dx} \equiv E = const. \tag{132}$$

Using (130) in (131) and (132), we obtain

$$\frac{4}{3}\mu\frac{du}{dx} = P - Q\left(u + \frac{\gamma - 1}{\gamma}\frac{h}{u}\right) \tag{133}$$

$$\frac{K}{C_p}\frac{dh}{dx} + \frac{4}{3}\mu u\frac{du}{dx} = E - Q\left(h + \frac{1}{2}u^2\right). \tag{134a}$$

If $\mu C_p/K = 3/4$, then (134a) becomes

$$\frac{4}{3}\mu\frac{d}{dx}\left(h + \frac{1}{2}u^2\right) = E - Q\left(h + \frac{1}{2}u^2\right) \tag{134b}$$

from which we have,

$$h + \frac{1}{2}u^2 = \frac{E}{Q} + Ae^{-\frac{3Qx}{4\mu}} \tag{135a}$$

where A is an arbitrary constant.

The boundedness of the quantity $\left(h + \frac{1}{2}u^2\right)$ implies that $A = 0$, and therefore (135a) gives

$$h + \frac{1}{2}u^2 = \frac{E}{Q} = const. \tag{135b}$$

Using (135b), (133) becomes

$$\frac{4}{3}\mu\frac{du}{dx} = P - Q\left(\frac{\gamma + 1}{2\gamma}u + \frac{\gamma - 1}{\gamma}\frac{E}{Q}\frac{1}{u}\right). \tag{136}$$

Using the boundary conditions,

$$x \Rightarrow \pm\infty : u \Rightarrow u_{1,2}, \frac{du}{dx} \Rightarrow 0 \tag{137}$$

equation (136) can be written as

$$\frac{4}{3}\mu\frac{du}{dx} = \left(\frac{\gamma + 1}{2\gamma}\right)\frac{Q}{u}(u - u_1)(u - u_2) \tag{138}$$

where

$$u_{1,2} = \left(\frac{\gamma}{\gamma+1}\right)\frac{P}{Q} \pm \left[\left(\frac{\gamma}{\gamma+1}\right)^2 \frac{P^2}{Q^2} - 2\left(\frac{\gamma-1}{\gamma+1}\right)\frac{E}{Q}\right]^{1/2}.$$

Equation (138) leads to

$$\frac{3Q}{4\mu}\left(\frac{\gamma+1}{2\gamma}\right)x = \frac{1}{(u_1-u_2)}\left[u_1 \ell n\,(u_1-u) - u_2 \ell n\,(u-u_2)\right], \quad u_2 < u < u_1 \tag{139}$$

which is sketched in Figure 19.6. This solution describes a continuous variation between two asymptotic states given by (138). The shock thickness may be given by an effective width d of this transition zone,

$$d = \frac{(u_1-u_2)}{\max\limits_{-\infty<x<\infty}\left|\frac{du}{dx}\right|}. \tag{140}$$

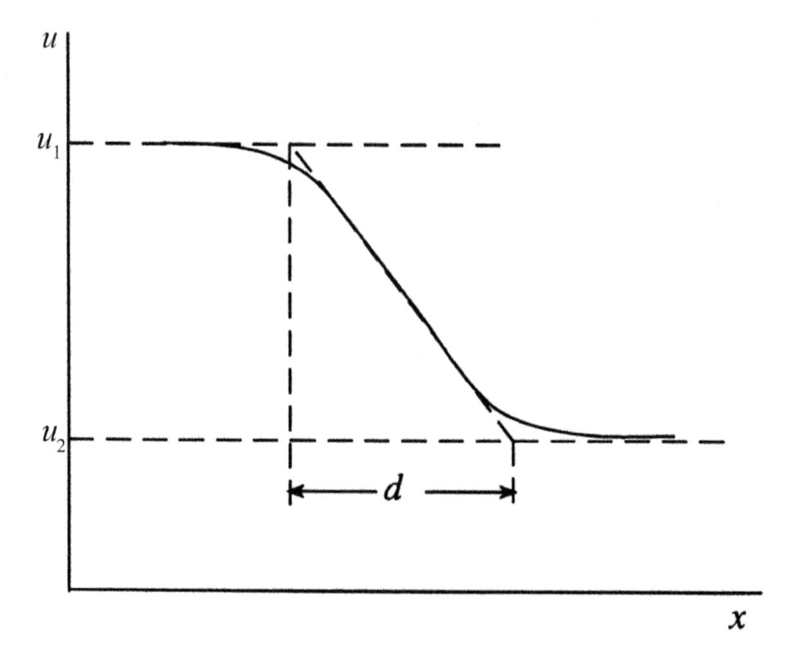

Figure 19.6 The shock structure.

19.11 Couette Flow of a Gas

Consider the plane flow of a viscous, heat-conducting gas between two infinite plates at $y = 0$ and L. The plate at $y = 0$ is at rest, and the one at $y = L$ is moving with a constant velocity U_1 in its own plane.

We nondimensionalize the various flow variables as follows;

$$\left.\begin{aligned} &y' \equiv \frac{y}{L}, \quad u' \equiv \frac{u}{U_1}, \quad v' \equiv \frac{v}{U_1} \quad p' \equiv \frac{p}{p_1} \\ &T' \equiv \frac{T}{T_1}, \quad \rho' \equiv \frac{\rho}{\rho_1} \equiv \frac{\rho}{p_1/RT_1}, \quad \mu' \equiv \frac{\mu}{\mu_1}, \\ &C_p' \equiv \frac{C_p}{C_{p_1}}, \quad K' \equiv \frac{K}{K_1} \end{aligned}\right\} \tag{141}$$

where the subscript 1 denotes the conditions at the plate at $y = L$.

The boundary conditions are given by,

$$\left.\begin{aligned} y' = 0 : \quad u', v' = 0 \\ y' = 1 : \quad u' = 1, v' = 0 \end{aligned}\right\}. \tag{142}$$

The equations governing the flow are

$$\frac{d}{dy'}(\rho' v') = 0 \tag{143}$$

$$\frac{d}{dy'}\left(\mu' \frac{du'}{dy'}\right) = 0 \tag{144}$$

$$\frac{dp'}{dy'} = 0 \tag{145}$$

$$\frac{d}{dy'}\left(\frac{\mu' C_p'}{P_r}\frac{dT'}{dy'}\right) + (\gamma_1 - 1) M_1^2 \mu' \left(\frac{du'}{dy'}\right)^2 = 0 \tag{146}$$

where

$$\gamma_1 \equiv \frac{C_{p_1}}{C_{v_1}}, P_r \equiv \frac{\mu' C_p'}{K'}, P_{r_1} \equiv \frac{\mu_1 C_{p_1}}{K_1} = 1, M_1^2 \equiv \frac{U_1^2}{\gamma_1 R T_1}.$$

We obtain from equation (143) and the boundary conditions (142)

$$v' \equiv 0 \tag{147}$$

and from equation (145)

$$p' \equiv 1. \tag{148}$$

Equation (144) yields,

$$u' = \frac{\int_0^{y'} \frac{dy'}{\mu'}}{\int_0^1 \frac{dy'}{\mu'}}. \tag{149}$$

Using (149), equation (146) gives

$$\frac{C_p'}{P_r}\frac{dT'}{dy'} + (\gamma_1 - 1)\frac{M_1^2}{2}\frac{d\left(u'^2\right)}{dy'} = \frac{B_1}{\mu'} \tag{150}$$

from which we obtain,

$$\frac{h'}{P_r} + \frac{(\gamma_1 - 1)}{2}M_1^2 u'^2 = B_1 \int_0^{y'} \frac{dy'}{\mu'} + B_2. \tag{151}$$

Using the boundary conditions,

$$\left.\begin{aligned} y' = 0: \quad h' = h_W' \\ y' = 1: \quad h' = h_1' \end{aligned}\right\} \tag{152}$$

equation (151) gives

$$h' = h_W' - \frac{(\gamma_1 - 1)P_r M_1^2}{2}u'^2 + \left[\left(h_1' - h_W'\right) + \frac{(\gamma_1 - 1)}{2}M_1^2 P_r\right]u'. \tag{153}$$

The third term on the right in (153) is due to the non-adiabatic wall condition.

Exercises

1. Calculate the steady flow between two infinitely long rotating circular cylinders of radii r_1, r_2 and angular speeds ω_1, ω_2.
2. Consider a fluid bounded by two rigid boundaries at $y = 0$ and d and initially at rest. The lower plate is suddenly brought to the steady velocity U parallel to the plate, the upper plate being held stationary. Calculate the subsequent motion of the fluid.
3. Consider a circular cylinder of radius a being rotated with steady angular velocity Ω. Calculate the motion generated from rest in the fluid contained within this cylinder.

20

Flows at Low Reynolds Numbers

Motion of a body through a fluid, such as the settling of sediment in a liquid, and the fall of mist droplets in the air belong to a class of problems called the creeping flow. These flows take place at low Reynolds numbers. A practical application of results from this regime is the use of *Stokes' formula* (Stokes, 1851) in calculating the drag of an oil drop in high school science experiments.

20.1 Dimensional Analysis

The Navier-Stokes equations for incompressible, viscous flows are

$$\nabla \cdot \mathbf{v} = 0 \tag{1}$$

$$\frac{D\mathbf{v}}{Dt} = \frac{1}{\rho}\nabla p + \nu \nabla^2 \mathbf{v}. \tag{2}$$

The boundary condition at a solid surface with which the fluid is in contact is that there is a zero slip-velocity of the fluid relative to the solid surface. Using the free-stream velocity U and a characteristic length L of the body placed in the flow, we nondimensionalize the various flow variables as follows:

$$x^* = \frac{x}{L}, \quad t^* = \frac{tU}{L}, \quad \mathbf{v}^* = \frac{\mathbf{v}}{U}, \quad p^* = \frac{(p - p_\infty)L}{\mu U}, \tag{3}$$

p_∞ being the free-stream pressure. Equations (1) and (2) then become

$$\nabla^* \cdot \mathbf{v}^* = 0 \tag{4a}$$

$$R_E \frac{D\mathbf{v}^*}{Dt^*} = -\nabla^* p^* + \nabla^{*2} \mathbf{v}^* \tag{5}$$

Introduction to Theoretical and Mathematical Fluid Dynamics, Third Edition.
Bhimsen K. Shivamoggi.

where R_E is the Reynolds number,

$$R_E \equiv \frac{\rho U L}{\mu}.$$

Note from equation (5) that one has a low-Reynolds-number flow if the flow is very slow, if the body is very small, or if the fluid is very viscous. When $R_E \ll 1$, the viscous effects will dominate the convective effects, at least in the neighbourhood of the body, which is called the *Stokes approximation*. On the mathematical side, note that the Stokes approximation also improves the solvability of the system (4) and (5), through the elimination of the nonlinear convective terms. Thus, in the limit $R_E \Rightarrow 0$, equations (4a) and (5) describing the *Stokes' flow* become

$$\nabla^* \cdot \mathbf{v}^* = 0 \tag{4a}$$

$$\nabla^* p^* = \nabla^{*2} \mathbf{v}^*. \tag{6a}$$

Reverting to dimensional variables, we obtain

$$\nabla \cdot \mathbf{v} = 0 \tag{4b}$$

$$\nabla p = \mu \nabla^2 \mathbf{v}. \tag{6b}$$

20.2 Stokes' Flow Past a Rigid Sphere: Stokes' Formula

Consider a uniform flow $U = U\hat{\mathbf{i}}_x$ past a rigid sphere of radius a centered at the origin. This problem is of interest in several physical contexts, such as the fall of mist droplets in air. Choose a spherical polar coordinate system (r, θ, ϕ) relative to which the fluid at infinity is at rest. Let the origin of this system instantaneously coincide with the center of the sphere, and the polar axis be along the x-axis. One then has the following boundary conditions:

$$\left.\begin{aligned} r = a: &\quad \mathbf{v} = U\hat{\mathbf{i}}_x \\ r \Rightarrow \infty: &\quad \mathbf{v} \Rightarrow \mathbf{0}, \quad p \Rightarrow p_\infty. \end{aligned}\right\} \tag{7}$$

Note that $\mathbf{v}$ and $(p - p_\infty)/\mu$ must be symmetrical about the x-axis, and that $\mathbf{v}$ lies in a plane through the x-axis. It follows that $(p - p_\infty)/\mu$ must be of the form $\mathbf{U} \cdot \mathbf{r} F\left(r^2/a^2\right) = UxF\left(r^2/a^2\right)$, where $r = |\mathbf{x}|$. Note, from equations (4b) and (6b) that $\left(p - p_\infty\right)/\mu$ satisfies the Laplace equation,

$$\nabla^2 \left(\frac{p - p_\infty}{\mu}\right) = 0 \tag{8}$$

and vanishes at infinity. So, it can be represented in terms of spherical solid harmonics of negative degree in r.[1]

Next, the only term in this solution set that is compatible with the form $\left(p - p_\infty\right)/\mu = UxF\left(r^2/a^2\right)$ is of degree 2 (which corresponds to a *dipole*) so that

$$\left(\frac{p - p_\infty}{\mu}\right) = C\frac{Ux}{r^3}. \tag{9}$$

Furthermore, rewriting equation (6b) in terms of the vorticity $\mathbf{\Omega} \equiv \nabla \times \mathbf{v}$, we have

$$-\nabla p = \mu \nabla \times \mathbf{\Omega} \tag{6c}$$

which yields

$$-\nabla^2 \mathbf{\Omega} = \mathbf{0}. \tag{10}$$

Noting that $\mathbf{\Omega}$ vanishes at infinity, we have from equation (10),

$$\mathbf{\Omega} = \tilde{C}\frac{U\hat{\mathbf{i}}_x \times \mathbf{r}}{r^3}. \tag{11}$$

Using (9) and (11), equation (6c) yields

$$\tilde{C} = C. \tag{12}$$

We have for the azimuthal component of vorticity,

$$\Omega_\phi = \frac{1}{r}\frac{\partial\left(r v_\theta\right)}{\partial r} - \frac{1}{r}\frac{\partial v_r}{\partial \theta} \tag{13}$$

where $\mathbf{v} = \left(v_r, v_\theta, v_\phi\right)$.

---

1 More generally, we have

$$\left(\frac{p - p_\infty}{\mu}\right) = \sum_{n=0}^{\infty} \left(a_n r^n + \frac{b_n}{r^{n+1}}\right) P_n\left(\cos\theta\right)$$

where $P_n\left(x\right)$ is the *Legendre polynomial* of order n.

Introducing the stream function Ψ for the axisymmetric flow under consideration we have,

$$v_r = \frac{1}{r^2 \sin\theta}\frac{\partial\Psi}{\partial\theta}, \quad v_\theta = -\frac{1}{r\sin\theta}\frac{\partial\Psi}{\partial r}. \tag{14}$$

Using (11) and (14) in (13), we obtain

$$\frac{\partial^2\Psi}{\partial r^2} + \frac{\sin\theta}{r^2}\frac{\partial}{\partial\theta}\left(\frac{1}{\sin\theta}\frac{\partial\Psi}{\partial\theta}\right) = -C\frac{U\sin^2\theta}{r}. \tag{15}$$

Separating the variables in equation (15) as follows,

$$\Psi(r,\theta) = f(r)\cdot U\sin^2\theta \tag{16a}$$

which implies

$$\mathbf{v} = \mathbf{U}\left(\frac{1}{r}\frac{df}{dr}\right) + \mathbf{x}\left(\frac{\mathbf{x}\cdot\mathbf{U}}{r^2}\right)\left(\frac{2f}{r^2} - \frac{1}{r}\frac{df}{dr}\right). \tag{16b}$$

Equation (15) then gives,

$$\frac{d^2 f}{dr^2} - \frac{2f}{r^2} = -\frac{C}{r} \tag{17}$$

the solution of which is

$$f(r) = \frac{1}{2}Cr + \frac{A}{r} + Br^2. \tag{18}$$

Using (16a) and (18), (14) gives

$$v_r = 2U\cos\theta\left(\frac{C}{2r} + \frac{A}{r^3} + B\right), v_\theta = -U\sin\theta\left(\frac{C}{2r} - \frac{A}{r^3} + 2B\right). \tag{19}$$

The boundary conditions (7) give

$$\left.\begin{aligned} r = a:\quad & v_r = U\cos\theta \\ & v_\theta = -U\sin\theta \\ r \Rightarrow \infty:\ & v_r, v_\theta \Rightarrow 0. \end{aligned}\right\} \tag{20}$$

Using (19), (20) leads to

$$A = -\frac{a^3}{4}, B = 0, C = \frac{3a}{2}. \tag{21}$$

Using (18) and (21), (16a) becomes

$$\Psi = Ur^2\sin^2\theta\cdot\left(\frac{3a}{4r} - \frac{a^3}{4r^3}\right) \tag{22}$$

and (19) becomes

$$\left.\begin{aligned} v_r &= -2\left(\frac{Ua^3}{4r^3} - \frac{3Ua}{4r}\right)\cos\theta \\ v_\theta &= -\left(\frac{Ua^3}{4r^3} + \frac{3Ua}{4r}\right)\sin\theta. \end{aligned}\right\} \tag{23}$$

The streamline-pattern given by (22) is shown in Figure 20.1. These streamlines are symmetric about the equatorial plane normal to the x-axis.

The resultant force acting on the surface of the sphere is, by symmetry, in the x-direction, and is given by

$$F = 2\pi a^2 \int_0^\pi (\tau_{rx}) \Big|_{r=a} \sin\theta \cdot d\theta \tag{24}$$

where

$$\tau_{rx} = \tau_{rr}\cos\theta - \tau_{r\theta}\sin\theta, \tag{25}$$

and

$$\tau_{rr} = \left(-p + 2\mu\frac{\partial v_r}{\partial r}\right), \tau_{r\theta} = \mu\left[r\frac{\partial}{\partial r}\left(\frac{v_\theta}{r}\right) + \frac{1}{r}\frac{\partial v_r}{\partial \theta}\right]. \tag{26}$$

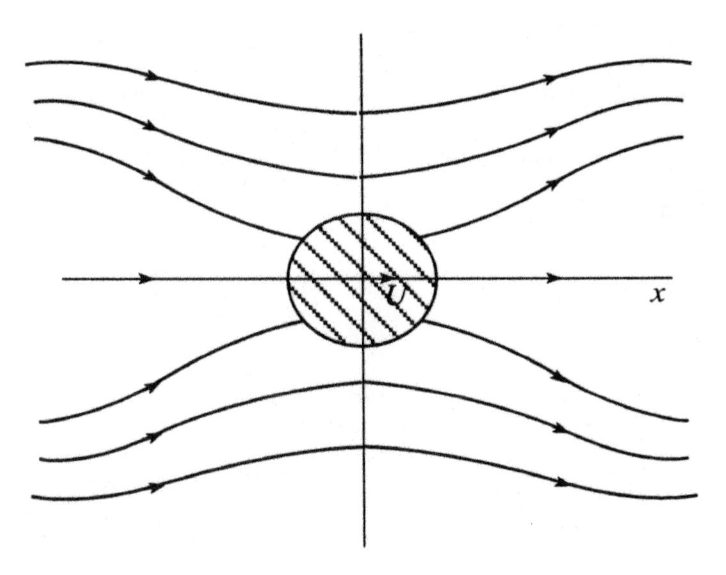

Figure 20.1 Stokes' flow past a sphere.

Using (9), (21), and (23), in (26) leads to

$$\left.\begin{aligned}\tau_{rr}\Big|_{r=a} &= -\frac{3\mu U\cos\theta}{2a}\\ \tau_{r\theta}\Big|_{r=a} &= -\frac{3\mu U\sin\theta}{2a}.\end{aligned}\right\} \tag{27}$$

Using (25) and (27), in (24) gives the *Stokes' formula*,

$$\mathbf{F} = 6\pi\mu a U\hat{\mathbf{i}}_x. \tag{28}$$

Figure 20.2 shows a comparison of (28) with experimental observations. This shows that departures from Stokes' formula (28) occur when the Reynolds number becomes high.

An obvious application of Stokes' formula (28) is in determining the coefficient of viscosity by measuring the drag of small spheres dropped in the fluid. Millikan's famous oil drop experiment to determine the charge on an electron is a case in point.

20.3 Stokes' Flow Past a Spherical Drop

Consider a spherical drop of radius a translating with velocity $\mathbf{U}$ in a fluid. We suppose that the two fluids are immiscible. Let the surface tension at the interface be strong enough to withstand the deforming tendencies of viscous forces, keeping the drop nearly spherical. Let the motion both outside and inside the drop occur at small Reynolds numbers. One determines the solution to the flow outside the drop as in the previous section except that the boundary conditions are now somewhat different. One has inside the drop (denoted by hat) the following governing equations,

$$\nabla\cdot\hat{\mathbf{v}} = 0 \tag{29}$$

$$\nabla\hat{p} = \mu\nabla\times\hat{\mathbf{\Omega}}. \tag{30}$$

Equation (30) leads to

$$\nabla^2\hat{p} = 0, \quad \nabla^2\hat{\mathbf{\Omega}} = \mathbf{0}. \tag{31a,b}$$

The boundary conditions for this flow are

$$\hat{\mathbf{v}} \text{ and } (\hat{p} - p_\infty) \text{ finite inside the drop} \tag{32}$$

$$\left.\begin{aligned} r = a:\ \hat{\mathbf{n}}\cdot\mathbf{v} = \hat{\mathbf{n}}\cdot\hat{\mathbf{v}} &= \hat{\mathbf{n}}\cdot\mathbf{U}\\ \hat{\mathbf{n}}\times\mathbf{v} &= \hat{\mathbf{n}}\times\hat{\mathbf{v}}\\ \hat{\mathbf{n}}\times(\hat{\mathbf{n}}\cdot\boldsymbol{\tau}) &= \hat{\mathbf{n}}\times(\hat{\mathbf{n}}\cdot\hat{\boldsymbol{\tau}})\end{aligned}\right\} \tag{33a,b,c}$$

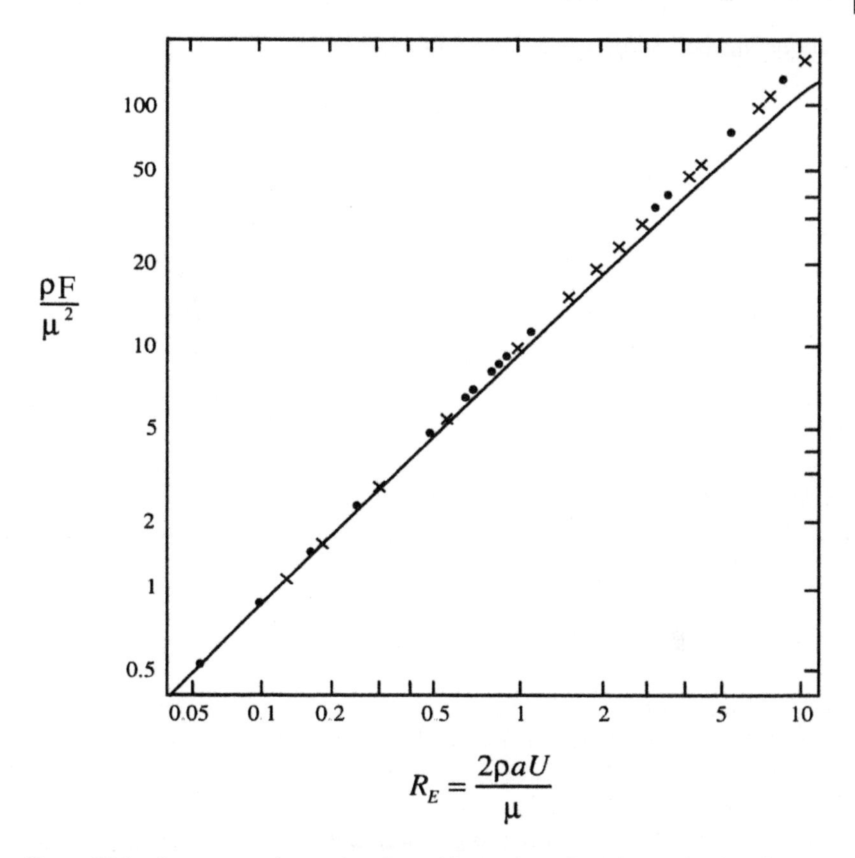

Figure 20.2 Drag on a sphere a low Reynolds numbers. Experimental points from Liebster (1927) (x) and Schmiedel (1928) (•), both using the falling sphere method. The line represents equation (28), plotted by Tritton (1988).

where $\hat{\mathbf{n}}$ denotes the outward drawn unit normal to the drop. Conditions (33a) and (33b) imply that there can be no relative motion of the two fluids at the interface. Condition (33c) sets forth the fact that the tangential stresses at the interface on the two sides are equal and opposite.

From equations (30) and (31), we obtain

$$\frac{\hat{p} - p_\infty}{\mu} = \hat{C}\mathbf{U} \cdot \mathbf{x}, \quad \hat{\boldsymbol{\Omega}} = -\frac{1}{2}\hat{C}\mathbf{U} \times \mathbf{x}. \tag{34}$$

Introducing the stream function Ψ, as in (16),

$$\Psi(r, \theta) = \hat{f}(r) \cdot U \sin^2 \theta. \tag{35}$$

Proceeding as in Section 20.2, we obtain

$$\hat{f}'' - \frac{2\hat{f}}{r^2} = \frac{1}{2}\hat{C}r^2 \tag{36}$$

the solution to which is given by,

$$\hat{f} = \frac{1}{20}\hat{C}r^4 + \frac{\hat{A}}{r} + \hat{B}r^2. \tag{37}$$

Removal of the singularity at $r = 0$ requires $\hat{A} = 0$. Using (35) and (37), we then obtain from the kinematic condition (33a),

$$\hat{B} = \frac{1}{2} - \frac{1}{20}\hat{C}a^2. \tag{38}$$

Thus using (35) and (37) in (16b), we obtain

$$\mathbf{v} = \mathbf{U} - \frac{1}{10}\hat{C}\left[\mathbf{U}\left(a^2 - 2r^2\right) + \mathbf{x}(\mathbf{U}\cdot\mathbf{x})\right]. \tag{39}$$

Using (23) and (39), the kinematic conditions (33a,b) give

$$C - \frac{1}{2}a = \frac{1}{10}\hat{C}a^3 + a. \tag{40}$$

Next, the ith component of the force per unit area exerted on the drop at the position $\mathbf{x} = a\hat{\mathbf{n}}$ is given by

$$\begin{aligned}
\left(\hat{n}_j\tau_{ij}\right)_{r=a} &= \hat{n}_j\left[-p\delta_{ij} + \mu\left(\frac{\partial v_i}{\partial x_j} + \frac{\partial v_j}{\partial x_i}\right)\right]_{r=a} \\
&= \left[-p\hat{n}_i + \mu\hat{n}_i\mathbf{U}\cdot\hat{\mathbf{n}}\left(-\frac{f''}{r} + \frac{6f'}{r^2} - \frac{10f}{r^3}\right)\right. \\
&\quad \left. + \mu U_i\left(\frac{f''}{r} - \frac{2f'}{r^2} + \frac{2f}{r^3}\right)\right]_{r=a} \\
&= n_i\left[-p_0 + 3\mu\frac{\mathbf{U}\cdot\hat{\mathbf{n}}}{a}\left(\frac{2C}{a} - 3\right)\right] + 3\mu\frac{U_i}{a}\left(1 - \frac{C}{a}\right).
\end{aligned} \tag{41}$$

Using (41) in the dynamic condition (33c) gives

$$\frac{3\mu}{2}(a - C) = \frac{3}{10}\hat{\mu}a\hat{C}. \tag{42}$$

Equations (40) and (42) lead to

$$C = \frac{1}{2} a \frac{2\mu + 3\hat{\mu}}{\mu + \hat{\mu}}, \quad \hat{C} = -\frac{5}{a^2} \frac{\mu}{\mu + \hat{\mu}}. \tag{43}$$

The drag force exerted on the drop by the external flow is then given by

$$\begin{aligned} F_i &= \int_0^{2\pi} \hat{n}_j \left(\tau_{ij}\right)_{r=a} a d\theta \\ &= -4\pi\mu U_i C = -6\pi a \mu U_i \left[\frac{1 + \frac{2}{3}(\mu/\hat{\mu})}{1 + (\mu/\hat{\mu})} \right]. \end{aligned} \tag{44}$$

The dependence of $|F_i|$ on $(\mu/\hat{\mu})$ is sketched in Figure 20.3. Note that the case of the flow past a rigid sphere is recovered in the limit $\mu/\hat{\mu} \Rightarrow 0$. The case of a gas bubble moving through the liquid corresponds to $\mu/\hat{\mu} \Rightarrow \infty$.

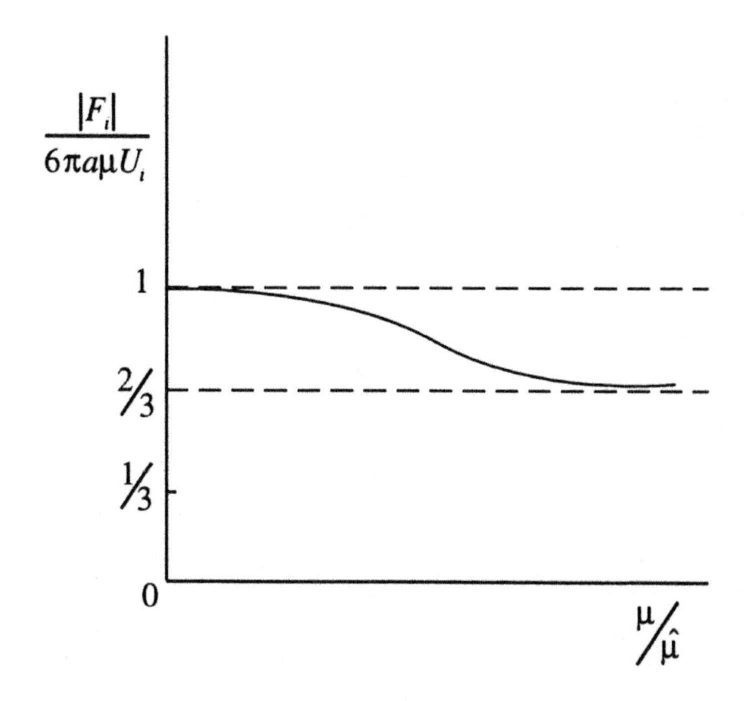

Figure 20.3 Variation of drag force with viscosity of the fluid in the drop.

20.4 Stokes' Flow Past a Rigid Circular Cylinder: Stokes' Paradox

Consider a circular cylinder of radius a moving with a velocity $\mathbf{U}$ in the x-direction normal to its axis in a fluid. The flow is now two-dimensional. Using the plane-polar coordinates (r, θ), one has, as in Section 20.3,

$$\frac{p - p_\infty}{\mu} = C\frac{Ux}{r^2}, \quad \boldsymbol{\Omega} = C\frac{U\hat{\mathbf{i}}_x \times \mathbf{r}}{r^2}. \tag{45}$$

We have

$$\Omega_z = \left[\frac{1}{r}\frac{\partial}{\partial r}(rv_\theta) - \frac{\partial v_r}{r\partial\theta}\right] \tag{46}$$

which, on using,

$$v_r = \frac{\partial\Psi}{r\partial\theta}, \quad v_\theta = -\frac{\partial\Psi}{\partial r} \tag{47}$$

and separating the variables as follows

$$\Psi(r,\theta) = f(r) \cdot U\sin\theta \tag{48}$$

leads to the following equation for $f(r)$,

$$\frac{d^2 f}{dr^2} + \frac{1}{r}\frac{df}{dr} - \frac{f}{r^2} = -\frac{C}{r}. \tag{49}$$

Equation (49) has the solution,

$$f(r) = -\frac{1}{2}Cr\,\ell n\, r + Ar + \frac{B}{r}. \tag{50}$$

The boundary conditions (7), on using (47) and (48), lead to

$$r = a: \frac{f}{r} = 1, \quad \frac{df}{dr} = 1 \tag{51}$$

$$r \Rightarrow \infty: \frac{f}{r} \Rightarrow 0. \tag{52}$$

Using (50) in (51), we obtain

$$A = 1 + \frac{C}{4} + \frac{C}{2}\ell n a, \quad B = -\frac{a^2 C}{4}. \tag{53}$$

However, using (50) and (53) in (52), we see that no choice of C satisfies the upstream infinity condition. Thus, the above solution is not valid at infinity. For two-dimensional flow of an unbounded fluid past a body, the solutions of the Stokes equations satisfying the proper ambient conditions do not exist. This is called *Stokes' paradox*. This is simply due to the fact that the Stokes approximation (i.e., neglect of the convective terms in equation (5)) is not uniformly valid in space. There is thus a region in which inertia forces become significant, and this region gets closer to the cylinder as the Reynolds number increases. The remedy for this is to go back and include the convective terms in equation (5) in at least some approximate form. One then obtains the so-called *Oseen flow*[2] (Oseen, 1910).

20.5 Oseen's Flow Past a Rigid Sphere

Oseen's equations are obtained by linearizing the Navier-Stokes equations (1) and (2) about the free-stream velocity, say, $\mathbf{v}_\infty = -U\hat{\mathbf{i}}_x$. By constrast, Stokes' equations can be viewed as a linearization about zero velocity. Therefore, Oseen's equations are more accurate than Stokes' equations in the region away from the sphere where the flow velocity is close to the free-stream velocity. Stokes' equations on the other hand, are accurate in the region near the sphere where the flow velocity is close to zero. Once again, fixing the coordinate system with respect to the fluid which is at rest at infinity, we have for this flow,

$$\nabla \cdot \mathbf{v} = 0. \tag{54}$$

$$\frac{\partial \mathbf{v}}{\partial t} + (\mathbf{v} \cdot \nabla)\mathbf{v} = -\frac{1}{\rho}\nabla p + \nu \nabla^2 \mathbf{v}. \tag{55}$$

If the body is moving with steady velocity $\mathbf{U}$ (so the flow relative to the body is steady), then the inertial term $\partial \mathbf{v}/\partial t$ at a fixed point is simply due to the motion of the sphere relative to the fixed point. Thus we have,

$$\frac{\partial \mathbf{v}}{\partial t} = -\mathbf{U} \cdot \nabla \mathbf{v}. \tag{56}$$

This inertial term dominates the other inertial term in equation (55) at large distances from the body. Oseen (1910) therefore proposed to keep only this inertial term and drop the other one in equation (55), which then becomes

$$-U\frac{\partial \mathbf{v}}{\partial x} = -\frac{1}{\rho}\nabla p + \nu \nabla^2 \mathbf{v}. \tag{57}$$

---

2 The *Stokes limit* corresponds to infinite viscosity while the *Oseen limit* corresponds to infinitesimal body size.

The boundary conditions are

$$\left.\begin{aligned} &r = a : \mathbf{v} = U\hat{\mathbf{i}}_x \\ &r \Rightarrow \infty : \mathbf{v} \Rightarrow \mathbf{0}, \quad p \Rightarrow p_\infty. \end{aligned}\right\} \tag{58}$$

Let us introduce the *Helmoholtz decomposition*,

$$\mathbf{v} = -\nabla\phi + \mathbf{v}'. \tag{59}$$

Equations (54) and (57) then lead to

$$\nabla^2\phi = 0 \tag{60}$$

$$p = -\rho U \frac{\partial \phi}{\partial x} \tag{61}$$

$$\left(\nabla^2 + 2k\frac{\partial}{\partial x}\right)\mathbf{v}' = \mathbf{0} \tag{62}$$

where

$$k \equiv \frac{U}{2\nu}.$$

Equation (62) implies[3]

$$\mathbf{v}' = \nabla\chi + 2k\chi\hat{\mathbf{i}}_x \tag{63}$$

where χ satisfies

$$\nabla^2\chi + 2k\frac{\partial\chi}{\partial x} = 0. \tag{64a}$$

Equation (64a) may be rewritten via a *Liouville transformation* as

$$(\nabla^2 - k^2)(e^{kx}\chi) = 0 \tag{64b}$$

one solution to which is

$$\chi = const = B \tag{65a}$$

while the other solution is

$$e^{kx}\chi = C\frac{1}{r}e^{-kr}. \tag{65b}$$

So, the general solution is

$$\chi = C\frac{1}{r}e^{-k(r+x)} + B. \tag{65c}$$

3 Note that (63) also reflects the axisymmetric nature of the flow.

In order to deal with the boundary conditions (58), note that (65c), in the low Reynolds number limit $kr \ll 1$, becomes

$$\chi \approx \frac{C}{r}\left[1 - k\left(r + x\right)\right] + B. \tag{66a}$$

One takes

$$B = Ck \tag{65d}$$

to obtain a series in descending powers of r, in the low Reynolds number limit $kr \ll 1$ given by,

$$\chi = C\left(\frac{1}{r} - \frac{kx}{r} + \cdots\right), kr \ll 1. \tag{66b}$$

Outside the sphere we obtain, for the potential flow component ϕ,

$$\phi = \frac{A_0}{r} + \frac{A_1}{r^2}P_1\left(\cos\theta\right) + \frac{A_2}{r^3}P_2\left(\cos\theta\right) + \cdots \tag{67}$$

where the polar axis is again taken to be along the direction of motion of the sphere, and

$$\cos\theta = \frac{x}{r}$$

and $P_n\left(\cos\theta\right)$ is the *Legendre polynomial* of order n.

Using (59), (63), (66b), and (67) in the boundary conditions (58), we obtain, for $ka = Ua/2\nu \ll 1$,

$$C = \frac{3aU}{2}, A_0 = \frac{3a\nu}{2}, A_1 = -\frac{1}{4}Ua^3. \tag{68}$$

Note from (59) and (63), that we have

$$v_r = -\frac{\partial\phi}{\partial r} + \frac{\partial\chi}{\partial r} + 2k\chi\cos\theta. \tag{69}$$

The stream function is given by

$$\Psi = r^2\int_0^\theta v_r\sin\theta \cdot d\theta. \tag{70a}$$

Using (65), (67)-(69), in (70a) we obtain

$$\Psi \approx -\frac{Ua^3}{4r}\sin^2\theta + \frac{3a\nu}{2}\left(1 - \cos\theta\right)\left[1 - e^{-kr(1+\cos\theta)}\right]. \tag{70b}$$

Near the sphere, where $kr \ll 1$, (70b) gives

$$\Psi \approx \frac{3Ua}{4}\left(-\frac{a^2}{3r} + r\right)\sin^2\theta, \tag{71}$$

which agrees with the Stokes solution (22), as expected.

The streamline pattern given by (71) is sketched in Figure 20.4. Note that the streamlines are no longer symmetrical about the plane $\theta = \pi/2$. The flow tends to become radial far away from the sphere as though it were created by a source of fluid at the sphere, except within a wake directly behind it. Note that in the far field limit $kr \gg 1$, the flow has different forms depending on whether $(1 + \cos\theta)$ is small compared to unity or is of the order of unity. For the latter case, we obtain from (70)

$$\theta \neq \pi : \Psi \approx \frac{3Ua}{4k}(1 - \cos\theta). \tag{72}$$

This describes the radial flow from a source at the origin of strength $3\pi aU/k$. For the former case (i.e., within the wake), we obtain from (70)

$$\theta \approx \pi : \Psi \approx \frac{3Ua}{2k}\left\{1 - \exp\left[-\frac{kr}{2}(\pi - \theta)^2\right]\right\}. \tag{73}$$

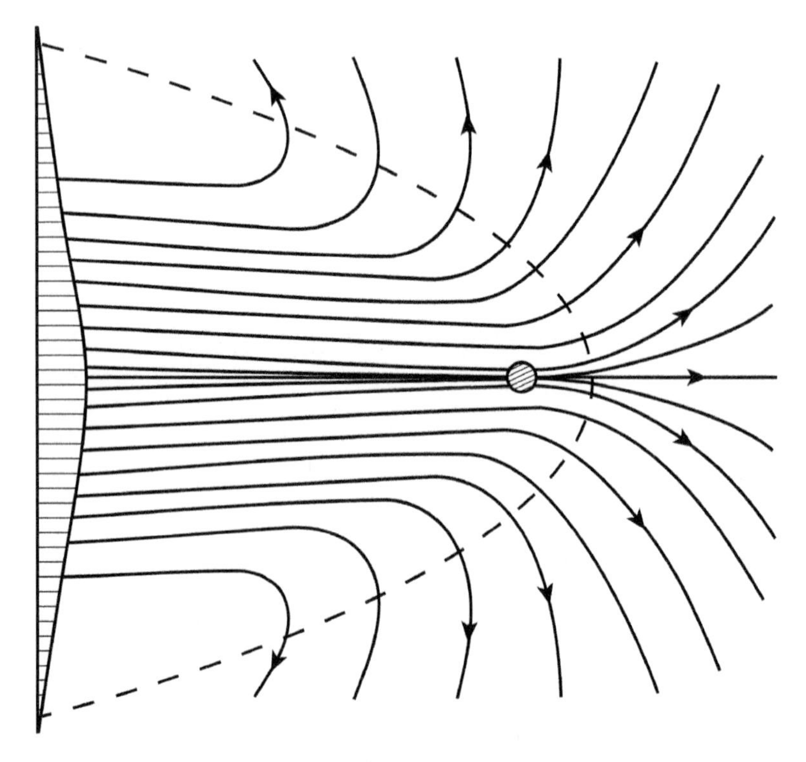

Figure 20.4 Oseen's flow past a sphere.

Therefore, (72) and (73) indicate that in the Oseen approximation, far away from the sphere, the vorticity is zero in the source-flow region and is confined to the wake. On the other hand, in the Stokes approximation, the vorticity diffuses out in all directions from the sphere, which is assumed to be effectively at rest.

20.6 Oseen's Approximation for Periodically Oscillating Wakes

The wake behind a body which is moving slowly through a fluid can be represented by the von Kármán's vortex street model. Periodic oscillating flows are well described by this model (see Section 8.4). In this theory, the motion of discrete vortices in a perfect fluid is considered. At considerable distances downstream, Oseen's approximation may be applied because viscous forces must have had a significant influence on the vortices.

Consider a small periodically oscillating disturbance superposed on a uniform stream with velocity U. The linearized vorticity equation describing this small disturbance is

$$\frac{\partial \zeta}{\partial t} + U\frac{\partial \zeta}{\partial x} = \nu\left(\frac{\partial^2}{\partial x^2} + \frac{\partial^2}{\partial y^2}\right)\zeta, \tag{74}$$

where ζ is the z-component of vorticity,

$$\zeta \equiv (\partial v/\partial x) - (\partial u/\partial y).$$

Seeking solutions with a harmonic time-dependence of the form,

$$\zeta(x, y, t) = f(x, y)\, e^{i\omega t}, \tag{75}$$

we obtain from equation (74)

$$\left(\frac{\partial^2}{\partial x^2} + \frac{\partial^2}{\partial y^2}\right) f = \frac{1}{\nu}\left(U\frac{\partial f}{\partial x} + i\omega f\right). \tag{76}$$

Setting

$$f(x, y) = e^{kx} g(x, y) \tag{77}$$

where

$$k \equiv \frac{U}{2\nu}$$

equation (76) gives

$$\left(\frac{\partial^2}{\partial x^2} + \frac{\partial^2}{\partial y^2} - \alpha^2\right) g = 0 \tag{78}$$

where

$$\alpha^2 \equiv k^2 (1 + 8\pi i\beta), \quad \beta \equiv \frac{\nu\omega}{2\pi U^2}.$$

The solution of equation (78) is

$$g(r,\theta) = \sum_{n=0}^{\infty} (A_n \cos n\theta + B_n \sin n\theta) K_n(\alpha r) \tag{79}$$

where

$$x = r\cos\theta, \quad y = r\sin\theta$$

and $K_n(\alpha r)$ is the modified Bessel function, with the asymptotic property,

$$r \Rightarrow \infty : K_n(\alpha r) \sim \sqrt{\frac{\pi}{2\alpha r}} e^{-\alpha r}.$$

Using (77) and (79) in (75) gives

$$\zeta(x,y,t) = e^{i\omega t} \sum_{n=0}^{\infty} K_n(\alpha r) [A_n \cos n\theta + B_n \sin n\theta] e^{kx} \tag{80}$$

from which we obtain,

$$\alpha r \Rightarrow \infty : \zeta(x,y,t) \sim \sqrt{\frac{\pi}{2\alpha r}} e^{i\omega t} \sum_{n=0}^{\infty} [A_n \cos n\theta + B_n \sin n\theta] e^{(kx-\alpha r)}. \tag{81}$$

This vorticity distribution vanishes exponentially with increasing distance, except for when the real part of $(kx - \alpha r)$ is small. Now, assuming that $|\beta| \ll 1$, we have

$$kx - \alpha r \approx k(x-r) - i\frac{\omega}{U}r = kr(\cos\theta - 1) - i\frac{\omega}{U}r.$$

So, ζ is finite at large distances only for a parabolic region where θ is of the order of $\left(1/\sqrt{r}\right)$ - *wake*. In this region, one may make the approximation,

$$r \approx x, \quad \theta \approx \frac{y}{x}, \quad kr\left(1-\cos\theta\right) \approx \frac{\eta^2}{2} \tag{82}$$

where

$$\eta \equiv y\sqrt{U/2\nu x}.$$

Then, (81) becomes

$$\zeta\left(x,y,t\right) \sim \frac{1}{\sqrt{x}}\exp\left(-\frac{\eta^2}{2}\right)\exp\left[i\omega\left(t-\frac{x}{U}\right)\right]\left\{A+\frac{B\eta}{\sqrt{x}}\right\}. \tag{83}$$

Note that (83) has a symmetrical part,

$$\zeta_s \sim \frac{A}{\sqrt{x}}\exp\left(-\frac{\eta^2}{2}\right)\exp\left[i\omega\left(t-\frac{x}{U}\right)\right]$$

and an antisymmetrical part,

$$\zeta_a \sim \frac{B}{x}\eta\exp\left(-\frac{\eta^2}{2}\right)\exp\left[i\omega\left(t-\frac{x}{U}\right)\right].$$

We shall now discuss various flows represented by the superposition of such solutions.

(i) For $\omega = 0$, the symmetrical vorticity part represents a steady shear layer flow with velocity difference (Figure 20.5),

$$\Delta u \sim -\int_{-\infty}^{\infty}\zeta_s dy = -\sqrt{\frac{4\pi\nu}{U}}A.$$

(ii) For $\omega = 0$, the antisymmetrical vorticity part represents the wake solution with velocity distribution (Figure 20.5),

$$u \sim \frac{B'}{\sqrt{x}}\exp\left(-\frac{\eta^2}{2}\right), B' \equiv B\sqrt{\frac{2\nu}{U}}.$$

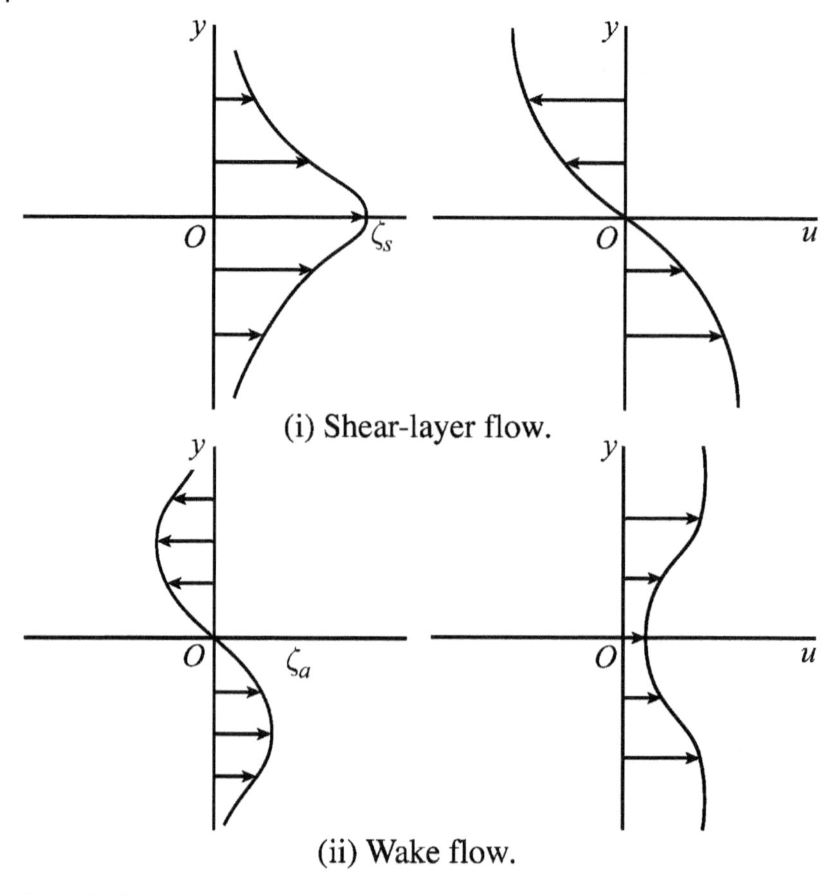

Figure 20.5 Free shear layer and wake flows.

(iii) A double row of vortices arranged like the von Kármán vortex street may be represented by

$$\zeta = \frac{A}{\sqrt{x}} \exp\left[-\frac{1}{2}(\eta - \eta_0)^2\right]\left[1 + \lambda \sin \omega\left(t - \frac{x}{U}\right)\right] + \\ -\frac{A}{\sqrt{x}} \exp\left[-\frac{1}{2}(\eta + \eta_0)^2\right]\left[1 - \lambda \sin \omega\left(t - \frac{x}{U}\right)\right]$$

where

$$\eta_0 \equiv y_0 \sqrt{\frac{U}{2\nu x}}.$$

For small values of x, the vortices are nearly discrete and are located on the lines $y = \pm y_0$. The initial spacing between the two rows is

$$h_0 \equiv 2y_0 = 2\eta_0\sqrt{\frac{2\nu x}{U}}.$$

and that downstream is given by,

$$h \equiv 2y = 2\eta\sqrt{\frac{2\nu x}{U}}.$$

Exercises

1. Show that, in a slow, steady, two-dimensional flow of an incompressible viscous fluid, the stream function Ψ satisfies the *biharmonic* equation,

$$\nabla^4\Psi = 0.$$

   Furthermore, deduce that $x\phi, y\phi$ and $(x^2 + y^2)\phi$ are solutions to this equation provided $\phi(x, y)$ is a harmonic function.
2. Calculate the terminal velocity which a sphere would have when falling freely under gravity through *a* fluid.
3. Consider the low Reynolds number flow of a fluid between two parallel flat plates which are fixed at a small distance h apart (*Hele-Shaw*, 1898). Show that this flow is governed, to a leading approximation, by the following equations:

$$0 = -\frac{\partial p}{\partial x} + \mu\frac{\partial^2 u}{\partial z^2}$$

$$0 = -\frac{\partial p}{\partial y} + \mu\frac{\partial^2 v}{\partial z^2}$$

$$0 = \frac{\partial p}{\partial z}$$

$$\frac{\partial u}{\partial x} + \frac{\partial v}{\partial y} + \frac{\partial w}{\partial z} = 0,$$

   with the boundary conditions,

$$z = 0 \text{ and } h : u, v, w = 0.$$

   Hence, show that

$$u = -\frac{1}{2\mu}\frac{\partial p}{\partial x}z(h - z), \quad v = -\frac{1}{2\mu}\frac{\partial p}{\partial y}z(h - z),$$

   where

$$\frac{\partial^2 p}{\partial x^2} + \frac{\partial^2 p}{\partial y^2} = 0.$$

This leads to the mean velocity in the plane of the Hele-Shaw (1898) cell given by

$$\overline{u} \equiv \frac{1}{h}\int_0^h u dz = -\frac{h^2}{12\mu}\frac{\partial p}{\partial x}$$

$$\overline{v} \equiv \frac{1}{h}\int_0^h v dz = -\frac{h^2}{12\mu}\frac{\partial p}{\partial y}$$

which predicts that the mean velocity field for a Hele-Shaw flow corresponds to a potential flow in two dimensions with the "mean" velocity potential given by $-\frac{h^2}{12\mu}p$ satisfying the Laplace equation!

21

Flows at High Reynolds Numbers

Large Reynolds number flows correspond to fluids having small viscosity like those discussed in Chapters 5–10. The effects of fluid viscosity are crucial to the calculation of quantities like skin friction. For flows past streamlined bodies at large Reynolds numbers, Prandtl (1904) proposed that it suffices to recognize the effects of viscosity only in a thin *boundary layer* adjacent to the body and that the rest of the flow may be considered inviscid. As a first approximation, the inviscid-flow equations are solved with appropriate boundary conditions, ignoring the presence of the boundary layer. However, in general, the inviscid flow will not satisfy the condition of no-slip of the fluid at the body. Therefore, it is necessary to introduce a boundary layer between the inviscid flow and the body to adjust the inviscid solution so that the no-slip condition at the body is satisfied. The vorticity that is generated along the surface of the body is diffused across and convected along the boundary layer. Consequently, the flow is not irrotational within the boundary layer. Besides, the presence of the boundary layer helps explain the common phenomenon of separation of flow behind many bodies placed in the flow.

21.1 Prandtl's Boundary-Layer Concept

Consider the vorticity generated at the surface of a body placed in a flow with velocity U. Certain qualitative features of boundary layers may be explained by considering the relative importance of convection and diffusion of this vorticity. The mechanism for the continual generation of a sheet of concentrated vorticity at the surface of a body is provided by the no-slip condition at the body. This vorticity is convected downstream with a velocity of $O(U)$. The problem of flow due to a suddenly accelerated plane discussed in Section 19.6 showed that this vorticity will diffuse outward with an effective velocity $\sqrt{\nu/t}$. One may then

Introduction to Theoretical and Mathematical Fluid Dynamics, Third Edition.
Bhimsen K. Shivamoggi.

expect the existence of an effective region, called the *boundary layer* (Prandtl, 1904) where vorticity is confined. This is similar to the Mach cone in inviscid supersonic flows (see Section 18.2). The width of this region is approximately given by (see equation (74) in Chapter 19),

$$\delta(x) \sim \sqrt{\nu t} \sim \sqrt{\frac{\nu x}{U}}$$

or

$$\frac{\delta(x)}{L} \sim \sqrt{\frac{x}{L}}\sqrt{\frac{1}{R_E}}$$

where R_E is the Reynolds number,

$$R_E \equiv \frac{UL}{\nu}$$

and L is a characteristic length of body. Thus, for $R_E \gg 1$, the convective effects on the vorticity outweigh the diffusive effects. Consequently, the vorticity layer is very thin relative to a typical dimension of the body, hence the name - *boundary layer*.

Downstream of the body, the vorticity which has been carried by convection is essentially confined to a region called the wake. At a distance d far away downstream from the body, the wake has a width of $O\left(\sqrt{\nu d/U}\right)$. Outside the wake and the vorticity layer near the body, the flow is essentially irrotational. Note that the boundary layer thickness δ increases as the square root of the distance from the leading edge of the plate. This is caused by progressive retardation of fluid, as flow occurs past the plate. Furthermore, there is no flow velocity component convecting vorticity toward the plate to counter diffusion of vorticity away from it.

Prandtl's boundary-layer theory can be embedded in a systematic scheme of successive approximations via the *method of matched asymptotic expansions* (Kaplun, 1954, Lagerstrom and Cole, 1955). The solution in question is represented by an infinite series in powers of $R_E^{-1/2}$, which turns out to be asymptotic (Shivamoggi, 2003).

21.2 The Method of Matched Asymptotic Expansions

In cases where a small parameter multiplies the highest derivative in a differential equation, there occurs a sharp change in the dependent variable in a certain region of the domain. This region is called the *boundary layer* of the independent variable. Such sharp changes in the dependent variable inside the boundary layer maybe characterized by a modified scale for the independent variable. This is different from the scale characterizing the behavior of the dependent variable outside the boundary layer. In other words, one represents the uniformly valid solution by two different asymptotic expansions. These are described using the independent variable x (for the *"outer"* region) and x/ε (for the boundary layer or the *"inner"* region). This is called the *method of matched asymptotic expansions*. Since they are different asymptotic representations of the same function, they should be related to each other in a rational manner. This leads to the *asymptotic matching principle* which makes the two representations completely determinate. A unified (or *"composite"*) expansion is then constructed that is everywhere asymptotic to the *exact* solution.

Example 1: Consider the boundary value problem,

$$\left.\begin{aligned} &\varepsilon y'' + y' + y = 0 \quad 0 \le x \le 1, \quad \varepsilon \ll 1 \\ &x = 0 : y = a \\ &x = 1 : y = b \end{aligned}\right\} \tag{1}$$

the exact solution of which is

$$y = \frac{(ae^{s_2} - b)\,e^{s_1 x} + (b - ae^{s_1})\,e^{s_2 x}}{(e^{s_2} - e^{s_1})} \tag{2}$$

where

$$s_{1,2} = \frac{-1 \pm \sqrt{1 - 4\varepsilon}}{2\varepsilon} \approx -1, -\frac{1}{\varepsilon} + 1.$$

The exact solution (2) may therefore be approximated by a uniformly valid expression given by,

$$y = be^{1-x} + (a - be)\,e^{x-(x/\varepsilon)} + O(\varepsilon). \tag{3}$$

Note that this expansion cannot be obtained keeping either x or x/ε fixed.[1] In the former case, we obtain

$$y^{(0)} = be^{1-x} + O(\varepsilon), x \neq 0 \tag{4a}$$

1 The singular perturbation aspect is apparent by noting that

$$e^{-\frac{x}{\varepsilon}} \sim \begin{cases} o(\varepsilon^n), \text{ as } \varepsilon \Rightarrow 0, \forall n, \text{ if } x = O(1) \\ O(1), \text{ as } \varepsilon \Rightarrow 0, \text{ if } x = O(\varepsilon). \end{cases}$$

which is not valid in the boundary layer near $x = 0$, since it misses the boundary condition there,[2]

$$y^{(0)}(0) = be \neq a. \tag{5a}$$

In the latter case, we obtain

$$y^{(i)} = be + (a - be)\,e^{-x/\varepsilon} + O(\varepsilon) \tag{4b}$$

which is not valid as $x \Rightarrow 1$, since it misses the boundary condition there,

$$y^{(i)}(1) = be \neq b. \tag{5b}$$

$y, y^{(0)}$, and $y^{(i)}$ are sketched in Figure 21.1.

This suggests that we represent the solution by two different asymptotic expansions using the variables x and x/ε. However, these asymptotic expansions have an overlapping domain of validity because

$$\lim_{x \Rightarrow 0} y^{(0)} = \lim_{x/\varepsilon \Rightarrow \infty} y^{(i)} \tag{5c}$$

as revealed by (5a) and (5b). The occurence of the term $e^{-x/\varepsilon}$ underlies the singular nature of this perturbation problem because $e^{-x/\varepsilon}$ cannot be expanded in a power series near $\varepsilon = 0$ (see footnote 1).

Thus, we seek an outer expansion (valid away from $x = 0$) given by,

$$y^{(0)}(x;\varepsilon) \sim \sum_{n=0}^{N-1} \varepsilon^n y_n^{(0)}(x) + O\left(\varepsilon^N\right) \tag{6}$$

where, in accordance with the outer limit process, we have

$$y_m^{(0)}(x) = \lim_{\substack{\varepsilon \Rightarrow 0 \\ x \text{ fixed}}} \frac{y^{(0)} - \sum_{n=0}^{m-1} \varepsilon^n y_n^{(0)}(x)}{\varepsilon^m}.$$

Substituting (6) in equation (1), and equating the coefficients of equal powers of ε, we obtain

$$\left.\begin{aligned} &O(1): \quad y_0^{(0)'} + y_0^{(0)} = 0 \\ &O(\varepsilon): \quad y_1^{(0)'} + y_1^{(0)} = -y_0^{(0)''} \\ &etc. \end{aligned}\right\} \tag{7}$$

2 From (4a) and (4b), we see that

$$\lim_{\varepsilon \to 0}\left(\lim_{x \to 0} y\right) = a \neq \lim_{x \to 0}\left(\lim_{\varepsilon \to 0}\right) = be.$$

So, the two limits are noncommutable.

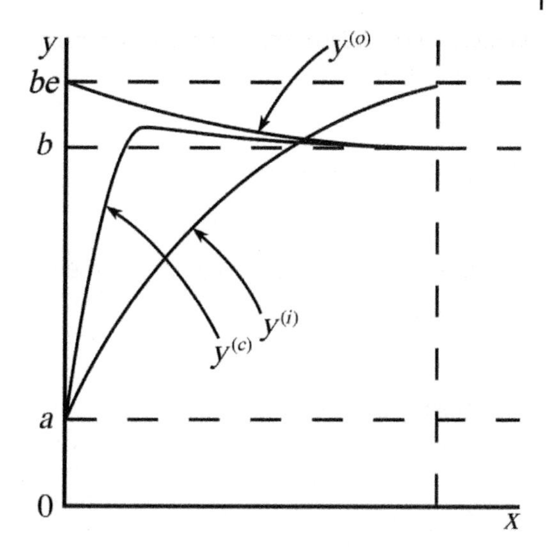

Figure 21.1 The inner, outer, and composite solutions.

Note that a small parameter ε multiplies the highest derivative in equation (1). A regular perturbation scheme, such as (6), misses that derivative in the first approximation so that the order of the differential equation (1) is reduced. Therefore, the system (7) cannot, in general, take on both of the boundary conditions, and one of these boundary conditions, namely, $y(0) = a$ should be dropped. This means that the outer solution (6) is valid everywhere, except in the boundary layer $x = O(\varepsilon)$. So, we have the boundary conditions,

$$\left.\begin{aligned} &y_0^{(0)}(1) = b \\ &y_1^{(0)}(1) = 0 \\ &etc. \end{aligned}\right\} \tag{8}$$

We obtain the following solutions for equation (7),

$$\left.\begin{aligned} &y_0^{(0)} = be^{1-x} \\ &y_1^{(0)} = b(1-x)e^{1-x} \\ &etc. \end{aligned}\right\}. \tag{9}$$

So, the outer expansion is given by

$$y^{(0)} \sim b\left[1 + \varepsilon(1-x)\right]e^{1-x} + O\left(\varepsilon^2\right). \tag{10}$$

For small ε, $y^{(0)}$ is close to the exact solution (3) everywhere, except in a small interval at $x = 0$. In this interval, the exact solution (3) changes rapidly (see

Figure 21.1) in order to retrieve the boundary condition at $x = 0$ which is about to be lost.

In order to determine an expansion valid in the boundary layer in $x = O(\varepsilon)$, one magnifies the independent variable as follows

$$\xi \equiv \frac{x}{\varepsilon} \tag{11}$$

so that the width of the boundary-layer region becomes independent of ε, in the limit $\varepsilon \Rightarrow 0$. This artifice leads to the retention of the highest derivative in equation (1), as $\varepsilon \Rightarrow 0$. This is essential to represent the rapidly varying behavior of the solution in the boundary layer. Using (11), equation (1) becomes

$$\frac{d^2y}{d\xi^2} + \frac{dy}{d\xi} + \varepsilon y = 0. \tag{12}$$

Now, let us seek an inner expansion (valid near $x = 0$),

$$y^{(i)}(\xi;\varepsilon) \sim \sum_{n=0}^{N-1} \varepsilon^n y_n^{(i)}(\xi) + O\left(\varepsilon^N\right) \tag{13}$$

where, in accordance with the inner limit process, we have

$$y_m^{(i)}(\xi) = \lim_{\substack{\varepsilon \Rightarrow 0 \\ \xi \text{ fixed}}} \frac{y^{(i)} - \sum_{n=0}^{m-1} \varepsilon^n y_n^{(i)}(\xi)}{\varepsilon^m}. \tag{14}$$

Substituting (13) in equation (12), and equating the coefficients of equal powers of ε, we obtain

$$\left.\begin{aligned} O(1):& \quad \frac{d^2y_0^{(i)}}{d\xi^2} + \frac{dy_0^{(i)}}{d\xi} = 0 \\ O(\varepsilon):& \quad \frac{d^2y_1^{(i)}}{d\xi^2} + \frac{dy_1^{(i)}}{d\xi} = -y_0^{(i)} \\ \text{etc.}& \end{aligned}\right\} \tag{15}$$

Noting that $y^{(i)}$ is valid only in the region $x = O(\varepsilon)$, we have the following boundary conditions on $y^{(i)}$,

$$\left.\begin{aligned} & y_0^i(0) = a \\ & y_1^{(i)}(0) = 0, \\ & \text{etc.} \end{aligned}\right\}. \tag{16}$$

We then obtain the following solutions for equation (15),

$$\left.\begin{aligned} &O(1): \ y_0^{(i)} = a - A_0\left(1-e^{-\xi}\right) \\ &O(\varepsilon): \ y_1^{(i)} = A_1\left(1-e^{-\xi}\right) - \left[a - A_0\left(1+e^{-\xi}\right)\right]\xi \\ &etc. \end{aligned}\right\} \tag{17}$$

where $A_0, A_1, \ldots$ are arbitrary constants.

So, the inner solution is given by

$$y^{(i)} \sim a - A_0\left(1-e^{-\xi}\right) + \varepsilon\left[A_1\left(1-e^{-\xi}\right) - \left\{a - A_0\left(1+e^{-\xi}\right)\right\}\xi\right] + O\left(\varepsilon^2\right). \tag{18}$$

$y^{(0)}$ and $y^{(i)}$ are different asymptotic representations of y. So, they should be related to each other in an overlapping region where both expansions are valid. One version of the *asymptotic matching principle* (Shivamoggi, 1978) states

$$\left\{\begin{array}{l}\text{“The } n\text{-term } \textit{formal} \text{ Laurent} \\ \text{series expansion of the outer} \\ \text{expansion about the inner} \\ \text{boundary written in terms of} \\ \text{the inner variable”}\end{array}\right\} = \left\{\begin{array}{l}\text{The } n\text{-term } \textit{formal} \text{ outer} \\ \text{limit of the inner expansion}\end{array}\right\}$$

Thus, in the neighbourhood of $x = 0$, we write

$$y^{(0)}(x) = y^{(0)}(0) + xy^{(0)'}(0) + O\left(x^2\right)$$

or

$$y^{(0)}(x) = y_0^{(0)}(0) + \varepsilon\left[\frac{x}{\varepsilon}y_0^{(0)'}(0) + y_1^{(0)}(0)\right] + O\left(\varepsilon^2\right).$$

When written in terms of the inner variable ξ we have,

$$y^{(0)}(x) = y_0^{(0)}(0) + \varepsilon\left[\xi y_0^{(0)'}(0) + y_1^{(0)}(0)\right] + O\left(\varepsilon^2\right). \tag{19}$$

Using (10) and (18), we obtain according to the above asymptotic matching principle,

$$be + \varepsilon\left[be - be\xi\right] + O\left(\varepsilon^2\right) = \left(a - A_0\right) + \varepsilon\left[A_1 - \left\{a - A_0\right\}\xi\right] + O\left(\varepsilon^2\right) \tag{20}$$

from which, we have

$$\left.\begin{aligned} &A_0 = a - be, \quad A_1 = be \\ &etc. \end{aligned}\right\} \tag{21}$$

Using (21) in (18), we obtain

$$\left.\begin{aligned} y^{(i)} =& be + (a - be)\,e^{-\xi} + \\ &+ \varepsilon\left\{be\left(1 - e^{-\xi}\right) - \left[be - (a - be)\,e^{-\xi}\right]\xi\right\} + O\left(\varepsilon^2\right). \end{aligned}\right\} \quad (22)$$

$y^{(0)}$ is valid everywhere except in a boundary layer of width $O(\varepsilon)$ near the origin while $y^{(i)}$ is valid only in this boundary layer. Although $y^{(0)}$ and $y^{(i)}$ have overlapping domains, one needs to switch from one expansion to the other to cover the entire interval. However, the switching location is not precisely known. This difficulty can be circumvented by combining both expansions into a single composite expansion $y^{(c)}$,

$$y^{(c)} = y^{(0)} + y^{(i)} - y^{(0)_i}\left(\text{or } y^{(i)_0}\right) \quad (23)$$

where $y^{(0)_i}$ represents the inner limit of the outer expansion, and $y^{(i)_0}$ represents the outer limit of the inner expansion.

Note

$$\left.\begin{aligned} y^{(c)_0} &= y^{(0)_0} + y^{(i)_0} - y^{(i)_{00}} = y^{(0)} + y^{(i)_0} - y^{(i)_0} = y^{(0)} \\ y^{(c)_i} &= y^{(0)_i} + y^{(i)_i} - y^{(0)_{ii}} = y^{(0)_i} + y^{(i)} - y^{(0)_i} = y^{(i)} \end{aligned}\right\} \quad (24a,b)$$

so that $y^{(c)}$ reproduces $y^{(0)}$ in the outer domain while it reproduces $y^{(i)}$ in the inner domain, and is therefore, valid everywhere.

For the present example, using (10) and (18), we have for the composite expansion,

$$\begin{aligned} y^{(c)} \sim\, & b\left[1 + \varepsilon(1 - x)\right]e^{1-x} + O\left(\varepsilon^2\right) + be + (a - be)\,e^{-\xi} \\ & + \varepsilon\left\{be\left(1 - e^{-\xi}\right) - \left[be - (a - be)\,e^{-\xi}\right]\xi\right\} \\ & - \left[be + \varepsilon\left(be - be\xi\right)\right] + O\left(\varepsilon^2\right) \end{aligned}$$

or

$$y^{(c)} \sim b\left[1 + \varepsilon(1 - x)\right]e^{1-x} + \left[(a - be)(1 + x) - \varepsilon be\right]e^{-x/\varepsilon} + O\left(\varepsilon^2\right) \quad (25a)$$

which may be rewritten as,

$$y^{(c)} \sim be^{1-x} + (a - be)\,e^{x-(x/\varepsilon)} + O(\varepsilon) \quad (25b)$$

in agreement with (3).

21.3 Location and Nature of the Boundary Layers

Consider an elliptic equation of the form

$$\varepsilon\left[\alpha_{11}\frac{\partial^2 u}{\partial x^2}+2\alpha_{12}\frac{\partial^2 u}{\partial x\partial y}+\alpha_{22}\frac{\partial^2 u}{\partial y^2}\right]=a\frac{\partial u}{\partial x}+b\frac{\partial u}{\partial y},\varepsilon\ll 1 \tag{26}$$

where the α's, a, and b are constants. Let the α's satisfy the condition,

$$\alpha_{12}^2-\alpha_{11}\alpha_{22}<0 \; and \; \alpha_{11}>0.$$

In order to determine a unique solution $u\,(x,y;\varepsilon)$ to equation (26), it is sufficient to prescribe one boundary condition on u or its normal derivative, or a combination of these two on a closed boundary - *Dirichlet* or *Neumann* problem.

Consider an interior boundary-value problem with $u = u_B\,(\boldsymbol{x}_B)$ prescribed on a closed boundary curve (see Figure 21.2).

The curves

$$\xi\equiv bx-ay=const \tag{27}$$

are the characteristics of equation (26). In the limit $\varepsilon\Rightarrow 0$ these are called the *subcharacteristics* of equation (1) (Section 14.5). Introducing another canonical independent variable,

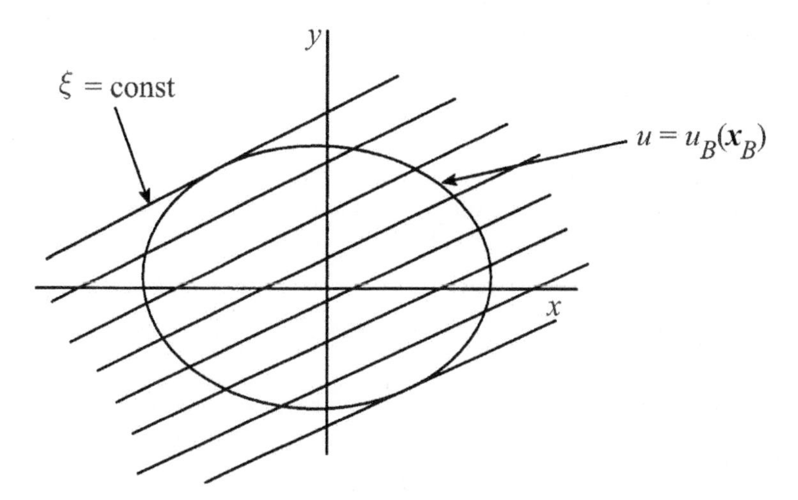

Figure 21.2 The subcharacteristics.

$$\eta \equiv ax + by \tag{28}$$

and transforming the independent variables x, y to ξ, η, equation (26) becomes

$$\varepsilon \left[A_{11} u_{\xi\xi} + 2A_{12} u_{\xi\eta} + A_{22} u_{\eta\eta} \right] = u_{\eta}. \tag{29}$$

Here,

$$A_{11} \equiv \frac{\alpha_{11} b^2 - 2\alpha_{12} ab + \alpha_{22} a^2}{a^2 + b^2}$$

$$= \frac{\alpha_{11} \left(b - \frac{\alpha_{12}}{\alpha_{11}} a \right)^2 + a^2 \left(\alpha_{11} \alpha_{22} - \alpha_{12}^2 \right)}{\alpha_{11} \left(a^2 + b^2 \right)} > 0$$

$$A_{12} \equiv \frac{\alpha_{11} ab + \alpha_{12} \left(b^2 - a^2 \right) - \alpha_{22} ab}{a^2 + b^2}$$

$$A_{22} \equiv \frac{\alpha_{11} a^2 + 2\alpha_{12} ab + \alpha_{22} b^2}{a^2 + b^2}.$$

It may be verified that $\left(A_{11} A_{22} - A_{12}^2 \right) > 0$, so that equation (29) remains elliptic.

Now, consider an outer limit process,

$$\lim_{\substack{\varepsilon \Rightarrow 0 \\ \xi, \eta \text{ fixed}}} u(\xi, \eta; \varepsilon) \Rightarrow u^{(0)}(\xi) \tag{30}$$

where the boundary condition on one side of the domain (see Figure 21.3) is sufficient to determine $u^{(0)}(\xi)$ uniquely in the entire domain. However, in general, $u^{(0)}(\xi)$ does not satisfy the boundary condition on the other side of the domain, say $\eta = \eta_B(\xi)$. Therefore, one has to introduce a boundary layer there. In order to deal with this boundary layer region, we introduce a new independent variable,

$$\eta^* \equiv \frac{\eta - \eta_B(\xi)}{\delta(\varepsilon)} \tag{31}$$

with η^* and ξ held fixed in the associated limit process $\varepsilon \Rightarrow 0$. The retention of the highest-order derivatives in equation (29) then requires $\delta(\varepsilon) = \varepsilon$. Seeking an inner solution of the form,

$$u^{(i)}(\xi, \eta^*; \varepsilon) \sim u_0^{(i)}(\xi, \eta^*) + O(\varepsilon) \tag{32}$$

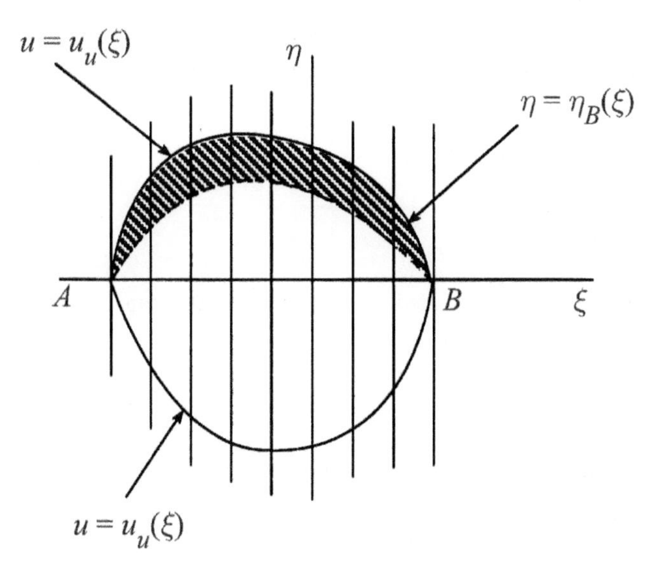

Figure 21.3 Production of a boundary layer on $u = u_u(\xi)$ (from Kevorkian and Cole, 1980).

equation (29) gives, in the limit $\varepsilon \Rightarrow 0$,

$$\kappa(\xi)\, u^{(i)}_{0\eta^*\eta^*} = u^{(i)}_{o\eta^*}. \tag{33}$$

Here, thanks to the elliptic nature of equation (29), we have

$$\begin{aligned}\kappa(\xi) &\equiv A_{11}\eta_{B\xi}^2 - 2A_{12}\eta_{B\xi} + A_{22} \\ &= \frac{1}{A_{11}}\left[A_{11}^2\left(\eta_{B_\xi} - \frac{A_{12}}{A_{11}}\right)^2 + \left(A_{11}A_{22} - A_{12}^2\right)\right] > 0.\end{aligned}$$

We obtain from equation (33) the solution,

$$u_0^{(i)}(\xi, \eta^*) = A(\xi) + B(\xi)\exp\left(\frac{\eta^*}{\kappa(\xi)}\right). \tag{34}$$

Matching (34) asymptotically to the outer solution $u_0^{(0)}(\xi)$, we obtain

$$u_0^{(i)}(\xi, \eta^*) = u_0^{(0)}(\xi) + \left[u_u(\xi) - u_0^{(0)}(\xi)\right]\exp\left(\frac{\eta^*}{\kappa(\xi)}\right) \tag{36}$$

where the subscript u refers to values on $\eta = \eta_B(\xi)$.

This solution becomes invalid for the case when the boundary is a subcharacteristic, say $\xi = \xi_s$. We then introduce a new independent variable,

$$\xi^* \equiv \frac{\xi - \xi_s}{\sqrt{\varepsilon}} \tag{37}$$

and assume an inner expansion,

$$u^{(i)}\left(\xi^*, \eta; \varepsilon\right) \sim u_0^{(i)}\left(\xi^*, \eta\right) + O\left(\varepsilon\right) \tag{38}$$

with ξ^* and η held fixed, in the associated limit process $\varepsilon \Rightarrow 0$. Then, equation (29) gives, in the limit $\varepsilon \Rightarrow 0$,

$$A_{11}\frac{\partial^2 u_0^{(i)}}{\partial \xi^{*2}} = \frac{\partial u_0^{(i)}}{\partial \eta}. \tag{39}$$

Since $A_{11} > 0, \eta$ is a time-like coordinate, which means that one requires the boundary condition,

$$\xi^* = 0 : u = u_s\left(\eta\right). \tag{40}$$

Furthermore, the requirement of matching of $u_0^{(i)}$ with the outer solution $u_0^{(0)}$ gives

$$\xi^* \Rightarrow \infty : u_0^{(i)}\left(\xi^*, \eta\right) \Rightarrow u_0^{(0)}\left(\xi\right). \tag{41}$$

Thus, the boundary layers arising on the subcharacteristics are characterized by a *diffusion-like* behavior, as we saw in Section 14.5.

For the problem of a viscous flow past a body, the boundary layer is along the body surface which is a streamline of the inviscid flow and hence a subcharacteristic of the full problem. The characteristic surfaces in general are the loci of possible discontinuities, and the streamlines of an inviscid flow can support a discontinuity in vorticity. In the inviscid limit, in which the ambient flow is irrotational, such a discontinuity occurs at the surface of the body. Here, the velocity component tangential to the body jumps so as to meet the no-slip condition at the surface of the body.

21.4 Incompressible Flow Past a Flat Plate

We consider here, the problem of steady two-dimensional viscous incompressible flow past a flat plate. It turns out that the external inviscid flow is associated with an outer limit process, and the boundary layer, with an inner-limit process. In the outer limit, the order of the differential equation is lowered and the boundary condition of no-slip of the flow at the plate is lost. The problem under consideration is therefore, one of *singular-perturbation* type.

Let x measure the distance along the plate from the leading edge and y, the distance normal to the plate. In terms of the stream function $\Psi(x, y)$, the boundary-value problem in nondimensionalized flow variables is given by

$$\left(\Psi_y \frac{\partial}{\partial x} - \Psi_x \frac{\partial}{\partial y} - \frac{1}{R_E}\nabla^2\right)\nabla^2\Psi = 0 \tag{42}$$

$$\left.\begin{aligned} & y = 0 : \Psi = 0, \Psi_y = 0, 0 < x < 1 \text{ or } \infty \\ & \text{upstream}: \quad \Psi \sim y \end{aligned}\right\} \tag{43}$$

where R_E is the Reynolds number,

$$R_E \equiv \frac{\rho\, UL}{\mu}$$

Here, U is the velocity of the fluid in the free stream, and L is some reference length (the plate is semi-infinite, so there is no natural length in the problem).

(i) The Outer Expansion

Let us seek an outer asymptotic expansion, as $R_E \Rightarrow \infty$, of the form

$$\Psi^{(0)}(x, y; R_E) \sim \Psi_1^{(0)}(x, y) + \delta_2^{(0)}(R_E)\Psi_2^{(0)}(x, y) + \cdots \tag{44}$$

with x and y fixed in the associated outer limit process $R_E \Rightarrow \infty$. Then, equation (42) gives

$$\left(\Psi_{1y}^{(0)} \frac{\partial}{\partial x} - \Psi_{1x}^{(0)} \frac{\partial}{\partial y}\right)\nabla^2\Psi_1^{(0)} = 0 \tag{45}$$

from which we obtain,

$$\nabla^2\Psi_1^{(0)} = -\zeta_1^{(0)}\left(\Psi_1^{(0)}\right) \tag{46}$$

ζ_1 being the vorticity. If the oncoming stream is irrotational, equation (46) gives

$$\nabla^2\Psi_1^{(0)} = 0. \tag{47}$$

The boundary conditions (43) give

$$\left.\begin{aligned} & y = 0 : \Psi_1^{(0)} = 0 \\ & \text{upstream}: \Psi_1^{(0)} \sim y. \end{aligned}\right\} \tag{48a}$$

Note that here the no-slip condition,

$$y = 0 : \Psi_{1y}^{(0)} = 0 \tag{48b}$$

has been discarded since, in the outer limit, the order of the differential equation (42) drops.

From (47) and (48a), we obtain

$$\Psi_1^{(0)}(x, y) = y \tag{49}$$

which confirms the fact that, in the limit $R_E \Rightarrow \infty$, a flat plate causes no disturbance.

(ii) The Inner Expansion

Due to the loss of the no-slip condition, the basic inviscid solution is not valid close to the flat plate. Therefore, we assume an inner expansion valid within the boundary layer given by,

$$\Psi^{(i)}(x, y; R_E) \sim \delta_1^{(i)}(R_E)\Psi_1^{(i)}(x, Y) + \delta_2^{(i)}(R_E)\Psi_2^{(i)}(x, Y) + \cdots \tag{50a}$$

with x and Y fixed in the associated outer limit process $R_E \Rightarrow \infty$ where,

$$Y \equiv \frac{y}{\delta_1^{(i)}(R_E)}. \tag{50b}$$

$\delta_1^{(i)}(R_E)$ is a function to be determined as follows. Using (50a, b), equation (42) gives

$$\left(\Psi_{1Y}^{(i)}\frac{\partial}{\partial x} - \Psi_{1x}^{(i)}\frac{\partial}{\partial Y}\right)\Psi_{1YY}^{(i)} = \lim_{R_E \Rightarrow \infty}\left[\frac{1}{R_E\delta_1^{(i)2}(R_E)}\right]\Psi_{1YYYY}^{(i)}. \tag{51}$$

So, the retention of the highest derivative (the term on the right-hand side in equation (51)) requires

$$R_E\delta_1^{(i)2}(R_E) \sim 1 \text{ or } \delta_1^{(i)}(R_E) \sim \frac{1}{\sqrt{R_E}}. \tag{52}$$

Using (52), (50b) becomes

$$Y = \sqrt{R_E}y \tag{53}$$

and equation (51) leads to

$$\left(\frac{\partial^2}{\partial Y^2} - \Psi_{1Y}^{(i)}\frac{\partial}{\partial x} + \Psi_{1x}^{(i)}\frac{\partial}{\partial Y}\right)\Psi_{1YY}^{(i)} = 0 \tag{54}$$

which may be rewritten as

$$\frac{\partial}{\partial Y}\left(\Psi_{1YYY}^{(i)} + \Psi_{1x}^{(i)}\Psi_{1YY}^{(i)} - \Psi_{1Y}^{(i)}\Psi_{1xY}^{(i)}\right) = 0. \tag{55}$$

So, we have

$$\Psi_{1YYY}^{(i)} + \Psi_{1x}^{(i)}\Psi_{1YY}^{(i)} - \Psi_{1Y}^{(i)}\Psi_{1xY}^{(i)} = f(x) \tag{56}$$

where $f(x)$ is proportional to the pressure gradient impressed on the plate by the inviscid flow. Equations (55) and (56) imply that the pressure is almost constant across the boundary layer.

The asymptotic matching between the outer and the inner solutions, using (44) and (50a), gives

$$Y \Rightarrow \infty : \delta_1^{(i)}\Psi_1^{(i)}(x, Y) + \cdots \approx \Psi_1^{(0)}(x, 0) + \delta_1^{(i)} Y \Psi_{1y}^{(0)}(x, 0) + \cdots \tag{57a}$$

which, using (49), leads to

$$\Psi_{1Y}^{(i)}(x, \infty) = \Psi_{1y}^{(0)}(x, 0) = 1. \tag{57b}$$

This simply implies that the tangential velocity of the boundary-layer flow approaches the inviscid flow-velocity $\Psi_{1y}^{(0)}(x, 0)$ at the outer edge of the boundary layer, $y \Rightarrow \infty$.

Using the outer inviscid solution (44), equation (56) becomes

$$\Psi_{1YYY}^{(i)} + \Psi_{1x}^{(i)}\Psi_{1YY}^{(i)} - \Psi_{1Y}^{(i)}\Psi_{1xY}^{(i)} = -\Psi_{1y}^{(0)}(x, 0)\,\Psi_{1xy}^{(0)}(x, 0). \tag{58}$$

Notice that the boundary-layer equation is parabolic with x acting as a time-like variable although the original Navier-Stokes equations are elliptic. This is in agreement with the result in the previous subsection that the boundary layers arising on the subcharacteristics are characterized by a diffusion-like behaviour. This means that the upstream influence is lost so that the first-order boundary-layer solution on a flat plate is not affected by the trailing edge (if the plate is finite) and the wake beyond.

For a flat plate, equation (58) becomes

$$\Psi_{1YYY}^{(i)} + \Psi_{1x}^{(i)}\Psi_{1YY}^{(i)} - \Psi_{1Y}^{(i)}\Psi_{1xY}^{(i)} = 0. \tag{59}$$

The boundary conditions (43) lead to

$$\left.\begin{aligned} &Y = 0 : \Psi_1^{(i)} = 0, \Psi_{1Y}^{(i)} = 0 \text{ for } 0 < x < 1 \text{ or } \infty \\ &Y \Rightarrow \infty : \Psi_Y^{(i)} = 1. \end{aligned}\right\} \tag{60a}$$

For a finite flat plate, the boundary layers at the top and bottom surfaces merge at the trailing edge and leave the plate as a wake without separating. As a result, the boundary-layer approximation continues to be valid in the wake.

One then needs only to replace the no-slip of the flow at $y = 0$ by a symmetry requirement of the form,

$$Y = 0 : \Psi_{1YY}^{(i)} = 0, \quad x > 1. \tag{60b}$$

Now, the problem given by (59) and (60a) is invariant under the *scaling* transformation,

$$\Psi_1^{(i)} \Rightarrow c\Psi_1^{(i)}, \quad x \Rightarrow c^2 x, \quad Y \Rightarrow cY. \tag{61}$$

So, it becomes possible to look for *self-similar* solutions to $\Psi_1^{(i)}$, of the form,

$$\Psi_1^{(i)}(x, Y) = \sqrt{2x} f_1(\eta), \quad \eta \equiv \frac{Y}{\sqrt{2x}}. \tag{62a}$$

This solution turns equation (59) into an ordinary differential equation (called the *Blasius* (1908) *equation*),

$$f_1''' + f_1 f_1'' = 0 \tag{63}$$

while the boundary conditions (60a) lead to

$$\left.\begin{aligned} &\eta = 0 : f_1 = 0, \quad f_1' = 0 \\ &\eta \Rightarrow \infty : f_1' \Rightarrow 1 \end{aligned}\right\} \tag{64}$$

where the primes denote differentiation with respect to η. Note that the flow velocity components are given by

$$u = f_1'(\eta), \quad v = (\eta f_1' - f_1) \tag{62b}$$

which implies that the velocity profile in the boundary layer is the same for all x, except for the change of scale associated with Y.

In order to find the asymptotic behavior of the solution to the boundary value problem given by (63) and (64), note first that equation (63) admits a solution of the form,

$$f_1 \sim \frac{1}{\eta}. \tag{65}$$

This implies that equation (63) has the following *scaling group* (Shivamoggi and Rollins, 1999),

$$\overline{f}_1 = \alpha^{-1} f_1, \quad \overline{\eta} = \alpha\eta. \tag{66}$$

We may therefore introduce the following *canonical* coordinates,

$$s = f_1\eta, \quad t = \eta^2 \frac{df_1}{d\eta}. \tag{67}$$

The transformation from (f, η) to (t, s) is given differentially by

$$\frac{ds}{d\eta} = \frac{1}{\eta}(t+s). \tag{68}$$

The transformation rules of the various derivatives are as follows,

$$\left.\begin{aligned}\frac{d^2 f_1}{d\eta^2} &= -\frac{2}{\eta^3}t + \frac{1}{\eta^3}\frac{dt}{ds}(t+s)\\ \frac{d^3 f_1}{d\eta^3} &= \frac{6}{\eta^4}t - \frac{5}{\eta^4}(t+s)\frac{dt}{ds} + \frac{1}{\eta^4}\frac{d^2t}{ds^2}(t+s)^2 + \frac{1}{\eta^4}(t+s)\left(\frac{dt}{ds}\right)\left(\frac{dt}{ds}+1\right)\end{aligned}\right\} \tag{69}$$

In terms of the new coordinates (t, s), the boundary-value problem given by (63) and (64) becomes

$$\begin{aligned}&(t+s)^2\frac{d^2t}{ds^2} + (t+s)\left(\frac{dt}{ds}+1\right)\frac{dt}{ds} - 5(t+s)\frac{dt}{ds}\\ &\quad + 6t + s\left[-2t + (t+s)\frac{dt}{ds}\right] = 0\end{aligned} \tag{70}$$

$$\left.\begin{aligned}&s = 0 : t = 0\\ &s \Rightarrow \infty : t \Rightarrow \infty\end{aligned}\right\}. \tag{71}$$

Near $s = 0$, equation (70) shows that

$$t \approx \lambda s \tag{72}$$

where the parameter λ is given by,

$$\lambda(\lambda+1)^2 - 5\lambda(\lambda+1) + 6\lambda \approx 0 \tag{73a}$$

from which we have,

$$\lambda = 1, 2. \tag{73b}$$

The root $\lambda = 1$ turns out to be a spurious one because, from (68), it leads to

$$s \sim \eta^2. \tag{74a}$$

Hence, we have from (67),

$$\eta \Rightarrow 0 : f_1 \sim \eta \tag{74b}$$

which does not satisfy the boundary condition (64). On the other hand, the root $\lambda = 2$ from (68) leads to

$$s \sim \eta^3 \tag{75}$$

and hence, we have from (67)

$$\eta \Rightarrow 0 : f_1 \approx \eta^2. \tag{76}$$

Next, near $s \Rightarrow \infty$, equation (70) again shows that

$$t \approx \lambda s \tag{77}$$

where λ is given by,

$$-2\lambda + \lambda(\lambda + 1) \approx 0 \tag{78a}$$

from which we have,

$$\lambda = 1. \tag{78b}$$

Equation using (68), this leads to

$$s \approx \eta^2. \tag{79}$$

Hence, we have from (67),

$$\eta \Rightarrow \infty : f_1 \approx \eta - \beta_1 \tag{80}$$

as expected! Here, β_1 is an arbitrary constant.

Figure 21.4 shows the numerical solution of equation (63), due to Schlichting (1972), compared with the experimental data, for several values of Reynolds numbers. The agreement between the two shows the validity of the various approximations and assumptions made in the boundary layer theory. Moreover, the velocity profile is seen to preserve its shape as one moves downstream despite the fact that the boundary layer thickness is changing.

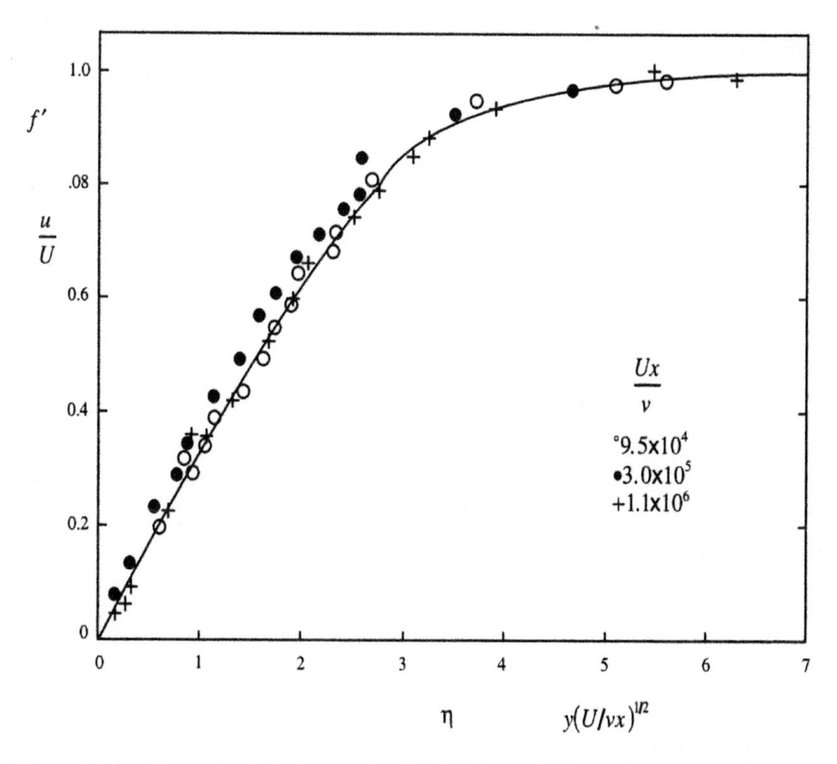

Figure 21.4 Theoretical Blasius profile (Schlichting, 1972) and experimental confirmation (Dhawan, 1952), plotted by Tritton (1988).

(iii) Flow Due to Displacement Thickness

Combining the outer solution ((44) and (49)) with the inner solution ((50a), (52), (59), (62a), (76) and (80)), the *asymptotic matching* gives

$$
\begin{aligned}
&\Psi_1^{(0)}(x,0) + \delta_2^{(0)}(R_E)\Psi_2^{(0)}(x,0) + \cdots + \frac{Y}{\sqrt{R_E}}\Psi_{1y}^{(0)}(x,0) + \cdots \\
&= \frac{1}{\sqrt{R_E}}\Psi_1^{(i)}(x,\infty) + \cdots \\
&\approx \frac{1}{\sqrt{R_E}}\left[\sqrt{2x}f_1\left(\frac{Y}{\sqrt{2x}}\right)\right] + \cdots, \quad \text{as } Y \Rightarrow \infty \\
&\approx \frac{Y}{\sqrt{R_E}} - \frac{1}{\sqrt{R_E}}\beta_1\sqrt{2x} + \cdots, \quad \text{as } Y \Rightarrow \infty.
\end{aligned}
\tag{81}
$$

Thus we obtain from (81),

$$\delta_2^{(0)}(R_E) = \frac{1}{\sqrt{R_E}} \tag{82}$$

$$y = 0 : \Psi_2^{(0)} = -\beta_1\sqrt{2x}. \tag{83}$$

Using (49), (82), and (83) in (44) gives

$$y \approx 0 : \Psi^{(0)}(x, y) \sim y - \frac{\beta_1}{\sqrt{R_E}}\sqrt{2x}. \tag{84}$$

This, in turn, leads to

$$y = \frac{\beta_1}{\sqrt{R_E}}\sqrt{2x} : \Psi^{(0)} \approx 0. \tag{85}$$

This implies that the presence of a boundary layer endows a certain thickness to the plate. This then displaces the outer inviscid flow like a solid parabola of nose radius $\beta_1^2/2R_E$.

For a semi-infinite flat plate, if the oncoming stream is irrotational, we then have

$$\nabla^2\Psi_2^{(0)} = 0 \tag{86}$$

$$\left.\begin{aligned} y = 0 : \Psi_2^{(0)} &= 0, \quad x < 0 \\ \Psi_2^{(0)} &= -\beta_1\sqrt{2x}, \quad x > 0 \\ \text{upstream } \Psi_2^{(0)} &= o(y) \end{aligned}\right\} \tag{87}$$

which corresponds to the linearized flow for a thin body given by $y = \beta_1\left(\sqrt{2x/R_E}\right)$, and we have

$$\Psi_2^{(0)}(x, y) = -\beta_1 Re\left[\sqrt{2(x + iy)}\right]. \tag{88}$$

Thus, even though the flow outside the wake and the boundary layer is essentially irrotational, it is not accurately described by the solution for potential flow past the given body. One has to then take into account the apparent change of shape of the body caused by the *displacement-thickness* effect of the boundary layer.

Note that the foregoing theory is not valid within a distance of $O(\nu/U)$ from the leading edge of the plate where the thickness of the boundary layer is comparable with the distance from the leading edge.

21.5 Separation of Flow in a Boundary Layer: Landau's Theory

Inviscid theory provides agreement with experimental results for streamlined bodies that is useful in applications. Indeed, the irrotational flow theory for a rigid body moving in an unbounded fluid is directly applicable to flows at large Reynolds numbers. This is contingent upon the separation of flow not occurring in the boundary layer. For such flows, the boundary layers play a passive role by simply effecting a smooth transition between a given ambient flow and the no-slip condition at the wall. However, sometimes the boundary layers can exert a controlling influence on the flow as a whole. One such case occurs with bodies of other shapes especially those of blunt form such as cylinders, wherein an adverse pressure gradient arises (i.e., the pressure increases in the flow-direction along the body). The retardation of the fluid particles under these circumstances leads to an abrupt thickening of the boundary layer which may then separate from the body. Mathematically, the only way in which this can happen is via a breakdown of the solution of the boundary-layer equations.[3]

Downstream of this point (or line) of breakaway, the original boundary-layer fluid passes over a region of recirculating flow. Thus, separation involves the existence of a region in which the vorticity has opposite sign from that associated with the flow as a whole. The point at which the thin boundary layer breaks away from the surface is known as the separation point. This point divides the region of downstream-directed flow from the region of recirculating flow. For a two-dimensional steady flow over a fixed wall, the point of *vanishing shear* coincides with the separation point. However, this is not true for the cases of two-dimensional steady flow over a moving wall, two dimensional unsteady flow, and three-dimensional steady flow. Two different types of post separation behavior are known to exist. In some cases, the original boundary layer passes over the region of recirculating fluid and reattaches to the body at some point downstream, trapping a bubble of recirculating fluid beneath it. In other cases, the original boundary-layer never reattaches to the body but passes downstream, mixing with recirculating fluid, to form a *wake*. It is obvious that the

---

3 This may be seen mathematically by generalizing equation (63) to take account of a varying ambient velocity. If $\partial \Psi_1^{(i)} / \partial y \Rightarrow U(x)$ as $y \Rightarrow \infty$ and $U(x) \sim x^m$, then a self-similar solution can still exist, with equation (63) replaced by

$$f_1''' + (m+1) f_1 f_1'' = 2m\left(f_1^2 - 1\right).$$

Numerical work (Schlichting, 1972) shows that this equation has a unique solution for $m > 0$, and two solutions (one of which represents reverse flow) for $-0.0904 < m < 0$ and no solution for $m < -0.0904$.

recirculating flow alters the effective body shape and hence the inviscid ambient flow about the body so that separation is the controlling feature of many fluid flows.

Consider a boundary layer on a plane wall. One has for the flow in the boundary layer the following governing equations,

$$\frac{\partial u}{\partial t} + u\frac{\partial u}{\partial x} + v\frac{\partial u}{\partial y} = -\frac{1}{\rho}\frac{dp}{dx} + \nu\frac{\partial^2 u}{\partial y^2} \tag{89}$$

$$\frac{\partial u}{\partial x} + \frac{\partial v}{\partial y} = 0. \tag{90}$$

The flow near the separation point is controlled by the pressure gradient. In fact, equations (89) and (90) show that, in the presence of an adverse pressure gradient $dp/dx > 0$, we have

$$\left.\frac{\partial^2 u}{\partial y^2}\right|_{y=0} > 0. \tag{91}$$

Consider now, the flow upstream of the point where $\left.\partial u/\partial y\right|_{y=0} = 0$. The adverse pressure gradient condition (91) then implies that the boundary layer velocity profile must have an inflection point.

As one approaches the separation point (or line), the flow moves away from the boundary and toward the interior of the fluid. Then the velocity component normal to the wall ceases to be small compared to the velocity component tangential to the wall. It increases to be of an order that is at least of the same magnitude as the velocity component tangential to the wall. This implies an essentially unlimited magnification of the normal velocity component as one approaches the interior of the fluid from the separation point. This implies that $\partial v/\partial y \Rightarrow \infty$ at the separation point. Using this in equation (90), we infer that $\partial u/\partial x \Rightarrow \infty$ at the separation point. Thus, one may write in the neighborhood of the separation point[4] (Landau and Lifshitz, 1987),

$$(x_0 - x) = (u_0 - u)^2 f(y,t) + O(u_0 - u)^3 \tag{92}$$

where the subscript 0 denotes the conditions at the separation point.

---

4 In fact, differentiating equations (89) and (90), one obtains, when $y = 0$,

$$\frac{\partial}{\partial x}\left[\frac{1}{2}\left(\frac{\partial u}{\partial y}\right)^2\right] = \nu\frac{\partial^4 u}{\partial y^4}.$$

If $\partial^4 u/\partial y^4$ is nonzero at the separation point, the above equation implies

Equivalently, one may write

$$u(x,y,t) = u_0(\Psi,t) + \frac{\partial \beta(\Psi,t)}{\partial \Psi}\sqrt{\chi} + \frac{\partial \beta_1(\Psi,t)}{\partial \Psi}\chi + \cdots \tag{93}$$

where

$$\chi \equiv x_0(t) - x, \quad \text{and} \quad \Psi = y - y_0(t).$$

One then finds from equation (90) that,

$$v(x,y,t) = \frac{\beta(\Psi,t)}{2\sqrt{\chi}} + \beta_1(\Psi,t) + \cdots. \tag{94}$$

Using (93) and (94) in equation (89) gives

$$\frac{\partial \beta}{\partial \Psi}\frac{dx_0}{dt} - u_0\frac{\partial \beta}{\partial \Psi} + \beta\frac{\partial u_0}{\partial \Psi} = 0. \tag{95}$$

From this we have,

$$\beta(\Psi,t) = A(t)\left[u_0(\Psi,t) - U_s(t)\right] \tag{96}$$

where

$$U_s(t) \equiv \frac{dx_0}{dt}.$$

Using (96), (93) and (94) become

$$u(x,y,t) = u_0(\Psi,t) + A(t)\frac{\partial u_0(\Psi,t)}{\partial \Psi}\sqrt{\chi} + \frac{\partial \beta_1(\Psi,t)}{\partial \Psi}\chi + \cdots \tag{97}$$

$$v(x,y,t) = A(t)\frac{\left[u_0(\Psi,t) - U_s(t)\right]}{2\sqrt{\chi}} + \beta_1(\Psi,t) + \cdots. \tag{98}$$

Thus, we have, at the point of separation,

$$y = 0: \; v_0 = 0, \;\; u_0 = U_s, \;\; \frac{\partial u_0}{\partial y} = 0. \tag{99}$$

We observe from (99) that a point on the surface of the body where $\partial u/\partial y = 0$ corresponds to the separation point.

The pressure gradient in the separated boundary layer is determined by the ambient flow which is in turn strongly influenced by the pressure gradient. Therefore, for a satisfactory formulation of separated flows, an extension

---

$$\left.\frac{\partial u}{\partial y}\right|_{y=0} \sim (x_0 - x)^{1/2}$$

in agreement with (92)!

of classical boundary layer theory incorporating this interaction between the ambient pressure gradient and the separating boundary layer is necessary. So, the pressure gradient in the separated boundary layer should be determined as part of the solution, once the back reaction of the separated boundary layer on the ambient flow has been accounted for.

21.6 Boundary Layers in Compressible Flows

The main differences between boundary layers in incompressible and compressible flows arise from thermal effects. In a compressible boundary layer flow, the transport coefficients of the fluid (like viscosity and thermal conductivity) are functions of temperature T of the fluid. This is caused by the viscous dissipation in the fluid and heat transfer at the wall which vary considerably across the boundary layer. Consequently, the equations of motion and energy reduce to a pair of coupled equations whose solution is usually difficult to find.

The rise in temperature of the fluid causes its density ρ to diminish and its viscosity μ to increase. These effects lead to a thickening of the boundary layer and a decrease in the velocity gradient in the boundary layer. Consequently, the skin friction given by

$$\tau_s \equiv \mu \frac{\partial u}{\partial y}$$

changes slowly. In fact, if $\mu \sim T$, and the free-stream speed $U = const$, then the skin friction is independent of the free-stream Mach number M_∞ and the plate temperature.

Furthermore, there is no qualitative difference between supersonic and subsonic boundary layers. This is because the pressure is not the dominant force controlling the flow in the boundary layer. On the other hand, the characteristics of inviscid flows change dramatically as the speed of sound is surpassed. Nevertheless, differences are likely to occur in how the two boundary layers interact with the exterior inviscid flow. This is because the free-stream Mach number M_∞ enters the problem through the boundary conditions as well. For example, regions of influence and domains of dependence materialize again via the displacement effect.

Consider a viscous, compressible, perfect gas flowing steadily past a body. The equations for the flow are

$$\frac{\partial}{\partial x}(\rho u) + \frac{\partial}{\partial y}(\rho v) = 0 \tag{100}$$

$$\rho\left(u\frac{\partial u}{\partial x}+v\frac{\partial u}{\partial y}\right)=-\frac{\partial p}{\partial x}+\frac{\mu}{3}\frac{\partial}{\partial x}\left(\frac{\partial u}{\partial x}+\frac{\partial v}{\partial y}\right)-\frac{2}{3}\frac{\partial\mu}{\partial x}\left(\frac{\partial u}{\partial x}+\frac{\partial v}{\partial y}\right)+$$
$$+2\frac{\partial\mu}{\partial x}\frac{\partial u}{\partial x}+\frac{\partial\mu}{\partial y}\left(\frac{\partial u}{\partial y}+\frac{\partial v}{\partial x}\right)+\mu\left(\frac{\partial^2 u}{\partial x^2}+\frac{\partial^2 u}{\partial y^2}\right) \tag{101}$$

$$\rho\left(u\frac{\partial v}{\partial x}+v\frac{\partial v}{\partial y}\right)=-\frac{\partial p}{\partial y}+\frac{\mu}{3}\frac{\partial}{\partial y}\left(\frac{\partial u}{\partial x}+\frac{\partial v}{\partial y}\right)-\frac{2}{3}\frac{\partial\mu}{\partial y}\left(\frac{\partial u}{\partial x}+\frac{\partial v}{\partial y}\right)+$$
$$+2\frac{\partial\mu}{\partial y}\frac{\partial v}{\partial y}+\frac{\partial\mu}{\partial x}\left(\frac{\partial u}{\partial y}+\frac{\partial v}{\partial x}\right)+\mu\left(\frac{\partial^2 v}{\partial x^2}+\frac{\partial^2 v}{\partial y^2}\right) \tag{102}$$

$$\rho C_p\left(u\frac{\partial T}{\partial x}+v\frac{\partial T}{\partial y}\right)=u\frac{\partial p}{\partial x}+v\frac{\partial p}{\partial y}+\frac{\partial}{\partial x}\left(K\frac{\partial T}{\partial x}\right)+\frac{\partial}{\partial y}\left(K\frac{\partial T}{\partial y}\right)+\Phi \tag{103}$$

$$p=\rho RT \tag{104}$$

where Φ is the viscous dissipation term given by,

$$\Phi\equiv 2\mu\left(\frac{\partial u}{\partial x}\right)^2+2\mu\left(\frac{\partial v}{\partial y}\right)^2+\mu\left(\frac{\partial v}{\partial x}+\frac{\partial u}{\partial y}\right)^2-\frac{2}{3}\mu\left(\frac{\partial u}{\partial x}+\frac{\partial v}{\partial y}\right)^2,$$

R is the universal-gas constant and K is the thermal conductivity of the fluid.

Following a procedure like the one we saw in Section 21.4, using equations (101)-(103), we arrive at the following equations for the flow in the compressible boundary layer (Stewartson, 1949),

$$\rho\left(u\frac{\partial u}{\partial x}+v\frac{\partial u}{\partial y}\right)=-\frac{\partial p}{\partial x}+\frac{\partial}{\partial y}\left(\mu\frac{\partial u}{\partial y}\right) \tag{105}$$

$$0=-\frac{\partial p}{\partial y} \tag{106}$$

$$\rho C_p\left(u\frac{\partial T}{\partial x}+v\frac{\partial T}{\partial y}\right)=u\frac{\partial p}{\partial x}+\frac{\partial}{\partial y}\left(K\frac{\partial T}{\partial y}\right)+\mu\left(\frac{\partial u}{\partial y}\right)^2 \tag{107}$$

while equations (100) and (104) remain unchanged.

The boundary conditions at the body are given by

$$y = 0 : u, v = 0, \frac{\partial T}{\partial y} = 0 \text{ or } T = T_w \tag{108}$$

depending on whether the wall is thermally insulated or isothermal. Next, as $y \Rightarrow \infty$, the boundary-layer solution must be matched to the inviscid-flow solution valid outside the boundary layer. This leads to

$$y \Rightarrow \infty : u \Rightarrow U(x), \quad T \Rightarrow T_\infty(x), \quad p \Rightarrow p_\infty(x). \tag{109}$$

Using (106), one obtains for the inviscid flow outside the boundary layer,

$$-\frac{dp_\infty}{dx} = \rho_\infty U \frac{dU}{dx}. \tag{110}$$

If this inviscid flow is *isoenergetic*, then we have

$$C_p T_\infty + \frac{1}{2} U^2 = const. \tag{111}$$

(i) Crocco's Integral

If

$$T = T(u) \tag{112}$$

then equation (107) becomes

$$\rho C_p T_u \left(u \frac{\partial u}{\partial x} + v \frac{\partial u}{\partial y} \right) = u \frac{dp}{dx} + \frac{\partial}{\partial y} \left(K T_u \frac{\partial u}{\partial y} \right) + \mu \left(\frac{\partial u}{\partial y} \right)^2. \tag{113}$$

Using equation (105), equation (113) becomes

$$C_p T_u \left[-\frac{dp}{dx} + \frac{\partial}{\partial y} \left(\mu \frac{\partial u}{\partial y} \right) \right] = u \frac{dp}{dx} + T_u \frac{\partial}{\partial y} \left(K \frac{\partial u}{\partial y} \right) + (T_{uu} K + \mu) \left(\frac{\partial u}{\partial y} \right)^2$$

from which we have,

$$\begin{aligned} -\frac{dp}{dx} (C_p T_u + u) + T_u \left[C_p \frac{\partial}{\partial y} \left(\mu \frac{\partial u}{\partial y} \right) - \frac{\partial}{\partial y} \left(K \frac{\partial u}{\partial y} \right) \right] \\ = (T_{uu} K + \mu) \left(\frac{\partial u}{\partial y} \right)^2. \end{aligned} \tag{114}$$

For the case of flow over a flat plate,

$$\frac{dp}{dx} = 0,$$

and the *Prandtl number*,

$$P_r \equiv \frac{\mu C_p}{K} = 1 \tag{115}$$

equation (114) gives

$$T_{uu} = -\frac{\mu}{K} = -\frac{1}{C_p}. \tag{116}$$

Equation (116) leads to Crocco's (1939) integral,

$$T = -\frac{u^2}{2C_p} + Au + B. \tag{117}$$

Considering an insulated wall, for which the boundary conditions are

$$\left.\begin{array}{r} y = 0 : u = 0, \dfrac{\partial T}{\partial y} = 0 \ \text{or} \ \dfrac{\partial T}{\partial u} = 0 \\ y \Rightarrow \infty : u \Rightarrow U, \ \ T \Rightarrow T_\infty \end{array}\right\}, \tag{118}$$

(117) becomes

$$T = T_\infty + \frac{U^2 - u^2}{2C_p}. \tag{119}$$

On the other hand, considering an isothermal wall, for which the boundary conditions are

$$\left.\begin{array}{r} y = 0 : u = 0, T = T_w \\ y \Rightarrow \infty : u \Rightarrow U, \ \ T \Rightarrow T_\infty \end{array}\right\}, \tag{120}$$

(117) becomes

$$T = -\left(T_w - T_\infty - \frac{U^2}{2C_p}\right)\frac{u}{U} + T_w - \frac{u^2}{2C_p}. \tag{121}$$

Note, from (121) that, depending on

$$T_w - T_\infty \gtrless \frac{U^2}{2C_p}$$

there is a heat transfer from the wall to the fluid or vice versa.

(ii) Flow Past a Flat Plate: Howarth-Dorodnitsyn Transformation

For the flow in a boundary layer on a flat plate, equation (100), (105), and (107) become

$$\frac{\partial}{\partial x}(\rho u) + \frac{\partial}{\partial y}(\rho v) = 0 \tag{122}$$

$$\rho\left(u\frac{\partial u}{\partial x} + v\frac{\partial u}{\partial y}\right) = \frac{\partial}{\partial y}\left(\mu\frac{\partial u}{\partial y}\right) \tag{123}$$

$$\rho\left(u\frac{\partial T}{\partial x} + v\frac{\partial T}{\partial y}\right) = \frac{1}{P_r}\frac{\partial}{\partial y}\left(\mu\frac{\partial T}{\partial y}\right) + \frac{\mu}{C_p}\left(\frac{\partial u}{\partial y}\right)^2. \tag{124}$$

Introducing (Howarth, 1948, Dorodnitsyn, 1942)

$$y_1 \equiv \int_0^y \frac{\rho}{\rho_\infty}dy, \quad x_1 = x \tag{125}$$

the stream-function relations, derived from equation (122),

$$\rho u = \rho_\infty\frac{\partial \Psi}{\partial y}, \quad \rho v = -\rho_\infty\frac{\partial \Psi}{\partial x} \tag{126}$$

give

$$u = \frac{\partial \Psi}{\partial y_1}, \quad \rho v = -\rho_\infty\left[\frac{\partial \Psi}{\partial x} + u\left(\frac{\partial y_1}{\partial x}\right)_y\right]. \tag{127}$$

Using (125) and (127), equations (123) and (124) can be rewritten as

$$\frac{\partial \Psi}{\partial y_1}\frac{\partial^2 \Psi}{\partial x_1 \partial y_1} - \frac{\partial \Psi}{\partial x_1}\frac{\partial^2 \Psi}{\partial y_1^2} = \frac{\partial}{\partial y_1}\left(\frac{\mu\rho}{\rho_\infty^2}\frac{\partial^2 \Psi}{\partial y_1^2}\right) \tag{128}$$

$$\frac{\partial \Psi}{\partial y_1}\frac{\partial T}{\partial x_1} - \frac{\partial \Psi}{\partial x_1}\frac{\partial T}{\partial y_1} = \frac{1}{P_r}\frac{\partial}{\partial y_1}\left(\frac{\mu\rho}{\rho_\infty^2}\frac{\partial T}{\partial y_1}\right) + \frac{\mu\rho}{C_p\rho_\infty^2}\left(\frac{\partial^2 \Psi}{\partial y_1^2}\right)^2. \tag{129}$$

Setting

$$\Psi = \sqrt{2U\nu_\infty x_1}\, f(\eta), \quad T = T(\eta)$$
$$\eta = y_1\sqrt{\frac{U}{2\nu_\infty x_1}} \tag{130}$$

equations (128) and (129) give

$$ff'' + \left[\left(\frac{\mu\rho}{\mu_\infty\rho_\infty}\right)f''\right]' = 0 \tag{131}$$

$$\left[\left(\frac{\mu\rho}{\mu_\infty\rho_\infty}\right)T'\right]' + P_r fT' = -P_r\left(\frac{\mu\rho}{\mu_\infty\rho_\infty}\right)\frac{U^2}{C_p}\left(f''\right)^2. \tag{132}$$

The boundary conditions (108) and (109) give

$$\begin{aligned} &\eta = 0 : f, f' = 0, T = T_w \\ &\eta \Rightarrow \infty : f' \Rightarrow 1, T \Rightarrow T_\infty. \end{aligned} \tag{133}$$

If the fluid behaves like a perfect gas, and

$$\mu \sim T,\ \ P_r = 1,\ \ T_w = const \tag{134}$$

then equations (131) and (132) become

$$ff'' + f''' = 0 \tag{135}$$

$$T'' + fT' = -(\gamma - 1)M_\infty^2 T_\infty \left(f''\right)^2 \tag{136}$$

where

$$M_\infty^2 \equiv \frac{U^2}{\gamma R T_\infty},\quad \gamma \equiv \frac{C_p}{C_v}.$$

Note that equation (135) is the same as the Blasius equation (63) for incompressible flows. This establishes a correlation between the two boundary layers.

Applying the boundary conditions (133), equation (136) may be integrated to give,

$$\frac{T}{T_\infty} = 1 + \frac{\gamma - 1}{2}M_\infty^2\left(1 - f'^2\right) + \left(\frac{T_w}{T_\infty} - 1 - \frac{\gamma - 1}{2}M_\infty^2\right)\left(1 - f'\right). \tag{137}$$

21.7 Flow in a Mixing Layer between Two Parallel Streams

Consider the flow in a mixing layer between two parallel streams (see Figure 21.5).

We adopt the boundary-layer model for this flow, and hence make the following assumptions:

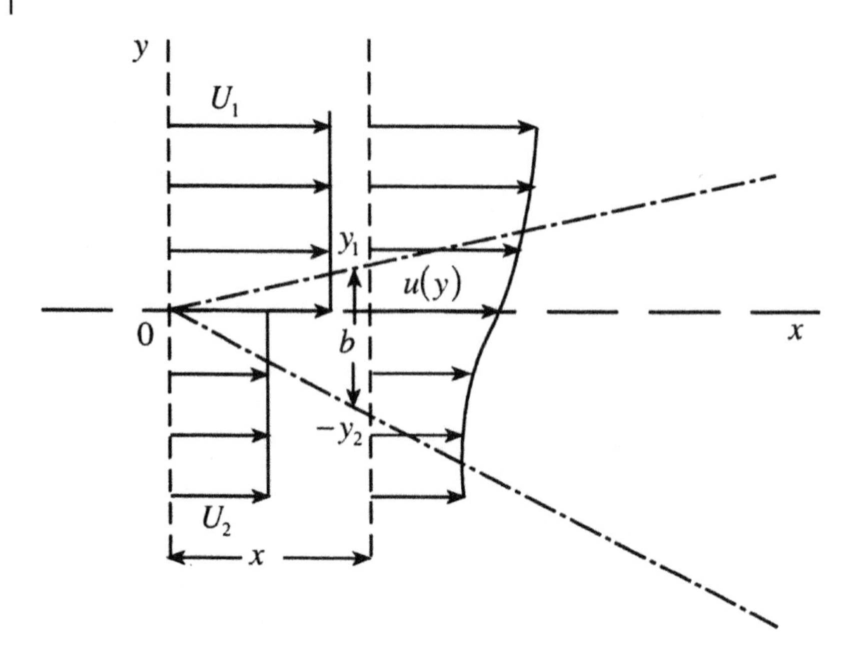

Figure 21.5 The mixing layer between two streams.

(i) the change in velocity from that of one stream to the other takes place in a mixing region of thickness that is small compared to the length of mixing;
(ii) the normal component of the velocity is small compared to the velocity component parallel to the main flow.

Note that there are only two apparent boundary conditions namely,

$$\left.\begin{array}{l} y \Rightarrow \infty : u \Rightarrow U_1 \\ y \Rightarrow -\infty : u \Rightarrow U_2 \end{array}\right\}. \tag{138}$$

The absence of a third boundary condition leads to the admission of an infinite number of solutions to the mixing layer. This is because the dividing streamline remains indeterminate to $O\left(1/\sqrt{R_E}\right)$, as $R_E \Rightarrow \infty$. However, the difference between any two solutions is simply equivalent to a shift in the velocity profile as a whole in the y-direction.

Consider the mixing flow for the case, when $U_1 \approx U_2$. One has for this flow,

$$\frac{\partial \Psi}{\partial y}\frac{\partial^2 \Psi}{\partial x \partial y} - \frac{\partial \Psi}{\partial x}\frac{\partial^2 \Psi}{\partial y^2} = \nu \frac{\partial^3 \Psi}{\partial y^3}. \tag{139}$$

Setting

$$\Psi = \sqrt{2U\nu x} f(\eta), \quad \eta \approx y\sqrt{\frac{U}{2\nu x}} \tag{140}$$

where

$$U \equiv \frac{U_1 + U_2}{2}$$

and using (140), equation (139) gives again the Blasius equation

$$ff'' + f''' = 0. \tag{141}$$

The boundary conditions (138) give

$$\eta \Rightarrow \pm\infty : f' = 1 \pm \lambda \tag{142}$$

where

$$\lambda \equiv \frac{U_1 - U_2}{U_1 + U_2} \ll 1.$$

Placing the dividing streamline arbitrarily at $y = 0$ (which is indeed plausible, if $\lambda \ll 1$), one has

$$\eta = 0 : f = 0. \tag{143}$$

Seeking solutions of the form

$$f \sim \eta + \lambda f_1 + O(\lambda^2) \tag{144}$$

the boundary value problem (141)-(143) gives

$$f_1''' + \eta f_1'' = 0 \tag{145}$$

$$\left.\begin{array}{r} \eta = 0 : f_1 = 0 \\ \eta \Rightarrow \pm\infty : f_1' = \pm 1 \end{array}\right\} \tag{146}$$

from which, we have

$$f_1' = \frac{2}{\sqrt{\pi}} \int_0^{\eta} e^{-\xi^2} d\xi. \tag{147}$$

Using (144) and (147) in (140) gives

$$u = \frac{U_1 + U_2}{2} \left[1 + \frac{U_1 - U_2}{U_1 + U_2} \frac{2}{\sqrt{\pi}} \int_0^{\eta} e^{-\xi^2} d\xi \right]. \tag{148}$$

(i) Geometrical Characteristics of the Mixing Flow

The conservation of mass in the mixing layer (see Figure 21.5) gives

$$\int_{-y_2}^{y_1} \rho u dy = \rho\left(U_1 y_1 - V_1 x\right) + \rho\left(-U_2 y_2 + V_2 x\right) \tag{149}$$

where (U, V) are the velocity components in the (x, y)-directions, respectively.
The momentum conservation gives

$$\int_{-y_2}^{y_1} \rho u^2 dy = \rho U_1\left(U_1 y_1 - V_1 x\right) + \rho U_2\left(-U_2 y_2 + V_2 x\right). \tag{150}$$

We multiply equation (149) by U_1 and subtract it from equation (150), so that

$$\int_{-y_2}^{y_1} \rho u\left(U_1 - u\right) dy = \rho\left(U_1 - U_2\right)\left(-U_2 y_2 + V_2 x\right). \tag{151}$$

Similarly, multiplying equation (149) by U_2 and subtracting it from equation (150), we obtain

$$\int_{-y_2}^{y_1} \rho u\left(u - U_2\right) dy = \rho\left(U_1 - U_2\right)\left(U_1 y_1 - V_1 x\right). \tag{152}$$

Now, noting that the momentum lost by one stream is equal to the momentum gained by the other, we have

$$\int_{-y_2}^{y_1} \rho u\left(U_1 - u\right) dy = \int_{-y_2}^{y_1} \rho u\left(u - U_2\right) dy. \tag{153}$$

Using (151) and (152) in (153) gives

$$-U_2 y_2 + V_2 x = U_1 y_1 - V_1 x. \tag{154}$$

Using equation (154) in equation (149) gives

$$\rho\left(-U_2 y_2 + V_2 x\right) = \rho\left(U_1 y_1 - V_1 x\right) = \frac{1}{2}\int_{-y_2}^{y_1} \rho u dy \equiv \frac{M}{2} \tag{155}$$

from which, we have

$$V_1 = \frac{1}{\rho x}\left(\rho U_1 y_1 - \frac{M}{2}\right) \tag{156a}$$

and

$$V_2 = \frac{1}{\rho x}\left(\rho m U_1 y_2 + \frac{M}{2}\right) \tag{156b}$$

where

$$m \equiv \frac{U_2}{U_1}.$$

The transverse equilibrium of the mixing flow, on using equation (149), gives

$$\int_{-y_2}^{y_1} \rho u v dy = \rho V_1 (U_1 y_1 - V_1 x) + \rho V_2 (-U_2 y_2 + V_2 x) = 0 \tag{157}$$

from which, using equation (154), we obtain

$$V_1 = -V_2. \tag{158}$$

Using equations (156a, b) in equation (158) gives

$$y_1 = -m y_2 \tag{159}$$

or

$$|y_1| = m|y_2|$$

or

$$|y_1| < |y_2|, \text{ if } m < 1. \tag{160}$$

This inequality shows that the mixing layer penetrates more into the lower-speed stream than it does into the higher-speed stream. If one assumes $y = 0$ to be the dividing streamline, equation (159) implies that the two streams contribute equally to the total amount of mass flux through the mixing layer.

21.8 Narrow Jet: Bickley's Solution

We now consider another example of a boundary-layer type flow that occurs in a narrow jet. Here, the steep gradients of velocity originate at an orifice with the total change of velocity across the layer being zero. Consider a steady, two-dimensional jet discharged from an orifice in the form of a long slit. The pressure in the surrounding nearly stationary fluid is uniform, and the equation of motion becomes

$$u\frac{\partial u}{\partial x} + v\frac{\partial u}{\partial y} = \nu\frac{\partial^2 u}{\partial y^2}. \tag{161}$$

the x-axis being taken along the axis of the jet. The force per unit area acting on the fluid at the origin shows up as momentum flux across a surface surrounding the origin. At a section $x = const$, it is given by

$$F \approx \rho \int_{-\infty}^{\infty} u^2 dy = const \tag{162}$$

which is independent of x because of the momentum conservation of the jet as a whole. This follows from the fact that, on using equation (161), we have

$$\begin{aligned} \frac{dF}{dx} &= 2\rho \int_{-\infty}^{\infty} u\frac{\partial u}{\partial x} dy = 2\rho \int_{-\infty}^{\infty} \left[-v\frac{\partial u}{\partial y} + \nu\frac{\partial^2 u}{\partial y^2}\right] dy \\ &= -2\rho \int_{-\infty}^{\infty} u\frac{\partial u}{\partial x} dy + 2\rho\left[-uv + \nu\frac{\partial u}{\partial y}\right]\Bigg|_{-\infty}^{\infty} = 0. \end{aligned} \tag{163}$$

Note however, that the rate of efflux from the orifice is positive,

$$\frac{d}{dx}\int_{-\infty}^{\infty} \rho u dy > 0 \tag{164}$$

because the amount of the fluid transported downstream by the jet increases with the distance downstream. This is caused by the entrainment of the ambient fluid by the jet.

Let us now look for a self-similar solution in terms of the stream function (Schlichting, 1933),

$$\Psi(x, y) = 6\nu x^p f(\eta), \eta \equiv \frac{y}{x^q}. \tag{165}$$

Equations (161) and (162) then give

$$p + q = 1, \quad 2p - q = 0 \tag{166a}$$

or

$$p = \frac{1}{3}, \quad q = \frac{2}{3}. \tag{166b}$$

Using equation (166b), (165) leads to

$$u = \frac{\partial \psi}{\partial y} = 6\nu x^{-1/3} f', v = -\frac{\partial \psi}{\partial x} = 2\nu x^{-2/3} (2\eta f' - f) \tag{167}$$

and equation (161) gives

$$f''' + 2ff'' + 2f'^2 = 0. \tag{168}$$

The boundary conditions are

$$\eta \Rightarrow \pm\infty : f' \Rightarrow 0 \tag{169a}$$

and the symmetry conditions are

$$f'(\eta) = f'(-\eta). \tag{169b}$$

The solution to the boundary value problem (167)-(169a) is (Bickley, 1937)

$$f = \alpha \tanh \alpha\eta \tag{170}$$

α being an arbitrary constant, which turns out to be related to F, as seen in the following. Using (170), (167) leads to

$$u = 6\nu x^{-1/3} \alpha^2 \text{sech}^2 \alpha\eta. \tag{171}$$

Using (171), (162) becomes

$$F = 36\rho\nu^2\alpha^4 \int_{-\infty}^{\infty} \text{sech}^4 \alpha\eta d\eta = 48\rho\nu^2\alpha^3. \tag{172}$$

Using (172) in (171) we obtain

$$u = \left(\frac{3F^2}{32\rho^2 x\nu} \right)^{1/3} \text{sech}^2 \left(\frac{F}{48\rho\nu^2} \right)^{1/3} \eta \tag{173}$$

which describes the spreading and smoothing of the velocity profile as one proceeds downstream. Laboratory experiments (Andrade, 1939) suggested that a jet with an arbitrary initial velocity profile approaches the similarity form (171) asymptotically, as $x \Rightarrow \infty$. However, this form may not be observable because of the instability of the jet flow.

21.9 Wakes

Wake refers to the region of vorticity on the downstream side of a body placed in an otherwise uniform flow. Although the velocity distribution in the wake is complicated near the body, the streamlines for a steady flow far downstream become nearly straight and parallel again. In this case a boundary-layer-type flow prevails asymptotically. If we assume that the departures from the free-stream velocity U (taken in the x-direction) are small in this region, then we have, for the asymptotic wake flow (i.e., as $x \Rightarrow \infty$),

$$U\frac{\partial u}{\partial x} = \nu\frac{\partial^2 u}{\partial y^2} \tag{174}$$

with the boundary conditions,

$$y \Rightarrow \pm\infty : u \Rightarrow U. \tag{175}$$

We have, from the boundary value problem (174) and (175),

$$x \Rightarrow \infty : U - u \Rightarrow \frac{Q\sqrt{U}}{\sqrt{4\pi\nu x}}\exp\left(-\frac{Uy^2}{4\nu x}\right) \tag{176}$$

where Q is the *velocity deficit* given by

$$Q \equiv \int_{-\infty}^{\infty}(U - u)\,dy = const.$$

Note that, the total drag on the body giving rise to the wake is related to this velocity defect. For a nonlifting body moving steadily through the fluid, it is given by

$$D = \rho U\int_{-\infty}^{\infty}(U - u)\,dy = \rho UQ. \tag{177}$$

21.10 Periodic Boundary Layer Flows

In problems of unsteady motion of bodies through a fluid, viscous effects are generally limited to a thin boundary layer region only for small values of the time since the onset of motion. In general, a low level of vorticity persists beyond the boundary layer.

We have for a two-dimensional boundary-layer flow in the x-direction past the plane $y = 0$,

$$\frac{\partial u}{\partial x} + \frac{\partial v}{\partial y} = 0 \tag{178}$$

$$\frac{\partial u}{\partial t} + u\frac{\partial u}{\partial x} + v\frac{\partial u}{\partial y} = \frac{\partial U}{\partial t} + U\frac{\partial U}{\partial x} + \nu\frac{\partial^2 u}{\partial y^2} \tag{179}$$

where $U(x,t)$ represents the ambient flow. If L denotes a representative length in the x-direction, U_∞ the reference velocity at infinity, and T a representative time interval, the validity of equations (178) and (179) requires

$$\frac{TU_\infty}{L} \sim O(1)$$

or

$$\frac{TU_\infty}{L(U_\infty L/\nu)} = \frac{\nu T}{L^2} \ll 1. \tag{180}$$

This means that vorticity will be confined to a thin boundary layer only within a comparatively small time interval from the onset of motion. It does not imply the complete breakdown of the boundary layer approximation at large times. However, it indicates that vorticity has diffused or has been convected beyond the boundary layer region, where it is no longer negligible.

Consider a periodic flow produced by the oscillations of an infinite plate in a direction parallel to itself in a fluid which is at rest at infinity. If the amplitude of the motion is assumed to be small, we may assume an expansion of the form (Shivamoggi, 1979),

$$\left.\begin{aligned} u &= u_0 + \varepsilon u_1 + \varepsilon^2 u_2 + \cdots \\ v &= \varepsilon v_0 + \varepsilon^2 v_1 + \cdots \end{aligned}\right\} \varepsilon \ll 1. \tag{181}$$

ε characterizes deviations from boundary-layer approximation in this system.

In the frame of reference moving with the plate, we have

$$U(x,t) = U_0(x)\,e^{i\sigma t}. \tag{182}$$

Let us assume that $\partial u/\partial x \sim O(\varepsilon)$. Then, equation (179) gives

$$\frac{\partial u_0}{\partial t} - \nu\frac{\partial^2 u_0}{\partial y^2} = \frac{\partial U}{\partial t} \tag{183}$$

$$\frac{\partial u_1}{\partial t} - \nu\frac{\partial^2 u_1}{\partial y^2} = U\frac{\partial U}{\partial x} - u_0\frac{\partial u_0}{\partial x} - \nu_0\frac{\partial u_0}{\partial y} \tag{184}$$

etc.

The boundary conditions are

$$
\begin{aligned}
&y = 0 : u_0, u_1, \cdots = 0 \\
&y = \infty : u_0 = U(x,t); u_1, u_2, \cdots = 0.
\end{aligned} \tag{185}
$$

We may satisfy equation (178) identically by introducing the stream function $\Psi_o(x,y,t)$, as per

$$
\left.
\begin{aligned}
&u_0 = \frac{\partial \psi_0}{\partial y}, \quad v_0 = -\frac{\partial \psi_0}{\partial x} \\
&\psi_0 = \sqrt{\frac{2\nu}{\sigma}} U_0(x) f_0(\eta) e^{i\sigma t}, \quad \eta = y\sqrt{\frac{\sigma}{2\nu}}.
\end{aligned}
\right\} \tag{186}
$$

Equation (183) then gives

$$i f_0' - \frac{1}{2} f_0''' = i \tag{187}$$

from which we obtain

$$f_0' = 1 - e^{-(1+i)\eta}. \tag{188}$$

Using (188) in (186), we obtain

$$u_0 = U_0(x)\left[\cos\sigma t - e^{-\eta}\cos(\sigma t - \eta)\right] \tag{189}$$

which shows that at high frequencies the fluid motion is in phase with that of the plate.

Next, let us introduce for the $O(\varepsilon)$ problem, a stream function given by,

$$\psi_1 = \sqrt{\frac{2\nu}{\sigma}} U_0 U_0' \frac{1}{\sigma}\left[f_1(\eta) e^{2i\sigma t} + g_1(\eta)\right] \tag{190}$$

so that equation (184) gives

$$
\begin{aligned}
2i f_1' e^{2i\sigma t} - \frac{1}{2}\left(f_1''' e^{2i\sigma t} + g_1'''\right) = &\, Re\left(U_0 e^{i\sigma t}\right) Re\left(U_0' e^{i\sigma t}\right) - \left[Re\left(f_0' e^{i\sigma t}\right)\right]^2 + \\
&+ Re\left(f_0 e^{i\sigma t}\right) Re\left(f_0'' e^{i\sigma t}\right)
\end{aligned} \tag{191}
$$

from which we have,

$$
\begin{aligned}
2i f_1' - \frac{1}{2} f_1''' &= \frac{1}{2}\left(1 - f_0'^2 + f_0 f_0''\right) \\
\frac{1}{2} g_1''' &= \frac{1}{2}\left(1 - f_0' f_0'^*\right) + \frac{1}{4}\left(f_0'' f_0^* + f_0 f_0^{*''}\right)
\end{aligned} . \tag{192}
$$

The boundary conditions are

$$\left.\begin{array}{l}\eta = 0 : f_1, f_1' = 0 \\ \qquad g_1, g_1' = 0 \\ \eta \Rightarrow \infty : f_1' \Rightarrow 0.\end{array}\right\} \tag{193}$$

Therefore, the solution to the boundary value problem (192) and (193) is given by,

$$f_1' = -\frac{1}{2} i e^{-(1+i)\sqrt{2}\eta} + \frac{1}{2} i e^{-(1+i)\eta} - \frac{1}{2}(i-1)\,\eta e^{-(1+i)\eta} \tag{194}$$

$$g_1' = -\frac{3}{4} + \frac{1}{4} e^{-2\eta} + 2e^{-\eta}\sin\eta + \frac{1}{2} e^{-\eta}\cos\eta - \frac{1}{2}\eta e^{-\eta}(\cos\eta - \sin\eta). \tag{195}$$

Note the existence of a steady component of velocity $U_0 U_0' g_1'(\eta)/\sigma$ induced by the oscillatory potential flow. Such a steady motion can lead to extensive migration of fluid elements in an apparently purely oscillatory system (Lighthill, 1954).

Exercises

1. Using x and ψ as the independent variables instead of x and y, show that the boundary-layer equation can be reduced to a diffusion equation.
2. Consider a columnar jet formed by forcing a fluid through a small circular hole in a wall. Take the x-axis along the axis of the jet, and use cylindrical coordinates (r, θ, x). Make arguments concerning the relative sizes of velocity gradients in the jet, similar to those used for a two-dimensional jet. Show that the equations of motion and continuity for the columnar-jet flow may be approximated by

$$\left.\begin{array}{l}u\dfrac{\partial u}{\partial x} + v\dfrac{\partial u}{\partial r} = \dfrac{\nu}{r}\dfrac{\partial}{\partial r}\left(r\dfrac{\partial u}{\partial r}\right) \\ \dfrac{\partial}{\partial x}(ru) + \dfrac{\partial}{\partial r}(rv) = 0\end{array}\right\}$$

u, v being the velocity components along the x, r-directions, respectively. The boundary conditions for this flow are

$$\left.\begin{array}{ll}r = 0: & v, \dfrac{\partial u}{\partial r} = 0 \\ r \Rightarrow \infty: & u \Rightarrow 0.\end{array}\right\}$$

Show that the flux of momentum of the fluid forced through the hole, namely,

$$M = 2\pi\rho \int_{0}^{\infty} u^2 r dr$$

is constant. Introducing a Stokes stream function defined by

$$u = \frac{1}{r}\frac{\partial \psi}{\partial r}, \quad v = -\frac{1}{r}\frac{\partial \psi}{\partial x}$$

determine a *similarity* solution of the form $\psi = x^p f(rx^q)$.

3. Consider the compressible flow in a boundary layer on a flat plate. Using Howarth-Dorodnitsyn transformation, and assuming a constant-plate temperature, find the integral for the energy equation (i.e., the counterpart of (137)) for the case $P_r \neq 1$.
4. Consider a semi-infinite region of a stationary perfect gas which is bounded by a rigid plane. The plane is suddenly given a velocity U in its own plane, and thereafter is maintained at that velocity. Using Howarth-Dorodnitsyn transformation calculate the ensuing flow.

22

Jeffrey-Hamel Flow

The Jeffrey-Hamel flow refers to a two-dimensional flow in the region between two intersecting plane walls. The steady flow between the stationary walls is caused by the presence of a source or sink of fluid at the point of intersection of the walls. This example affords an impressive illustration of the combined effects of convection and diffusion of vorticity generated at a rigid boundary. In addition, it plays a role in the development of viscous flow theory similar to that played by the two-dimensional *Ising* (1925) *model* in developing the theory of *critical phenomena* (Stanley, 1971).

22.1 The Exact Solution

Let us use the cylindrical polar coordinates (r, θ) with the plane walls located at $\theta = \pm\alpha$ (Figure 22.1). The equations governing this flow are

$$\rho\left(u_r\frac{\partial u_r}{\partial r} + u_\theta\frac{\partial u_r}{r\partial\theta} - \frac{u_\theta^2}{r}\right) = -\frac{\partial p}{\partial r} + \mu\left(\nabla^2 u_r - \frac{u_r}{r^2} - 2\frac{\partial u_\theta}{r^2\partial\theta}\right) \tag{1}$$

$$\rho\left(u_r\frac{\partial u_\theta}{\partial r} + u_\theta\frac{\partial u_\theta}{r\partial\theta} + \frac{u_r u_\theta}{r}\right) = -\frac{\partial p}{r\partial\theta} + \mu\left(\nabla^2 u_\theta - \frac{u_\theta}{r^2} + 2\frac{\partial u_r}{r^2\partial\theta}\right) \tag{2}$$

$$\frac{\partial u_r}{\partial r} + \frac{u_r}{r} + \frac{1}{r}\frac{\partial u_\theta}{\partial\theta} = 0 \tag{3}$$

where

$$\nabla^2 = \frac{\partial^2}{\partial r^2} + \frac{1}{r}\frac{\partial}{\partial r} + \frac{\partial^2}{r^2\partial\theta^2}.$$

Let us consider a purely radial flow, and look for solutions of the form,

$$u_r = \frac{f(\theta)}{r}, \quad u_\theta \equiv 0 \tag{4}$$

Introduction to Theoretical and Mathematical Fluid Dynamics, Third Edition.
Bhimsen K. Shivamoggi.

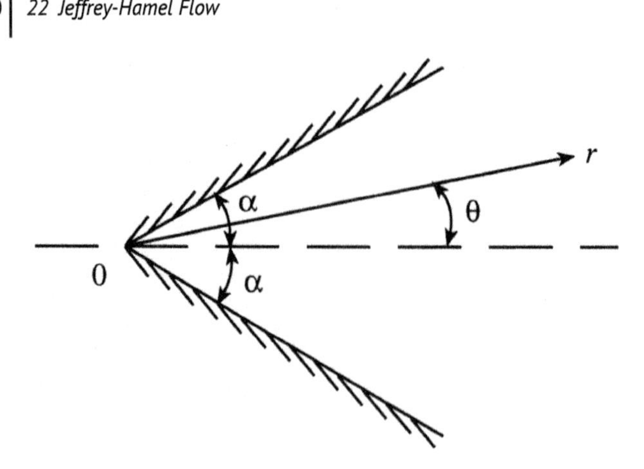

Figure 22.1 The Jeffrey-Hamel flow.

so that equation (3) is identically satisfied, and equations (1) and (2) give

$$-\frac{f^2}{r^3} = \frac{-1}{\rho}\frac{\partial p}{\partial r} + \frac{\nu}{r^3}f'' \tag{5}$$

$$0 = -\frac{1}{\rho r}\frac{\partial p}{\partial \theta} + \frac{2\nu}{r^3}f' \tag{6}$$

where the primes denote differentiation with respect to θ. The boundary conditions on f are

$$\theta = \pm\alpha : f = 0. \tag{7}$$

We obtain from equation (6)

$$\frac{p}{\rho} = \frac{2\nu}{r^2}f + F(r). \tag{8}$$

Using (8), equation (5) gives

$$\nu f'' + f^2 + 4\nu f = r^3\frac{dF}{dr} = const = A \tag{9}$$

from which we have,

$$F(r) = -\frac{A}{r^2} + B \tag{10}$$

where A and B are arbitrary constants.

We multiply equation (9) by f' and integrate between $-\alpha$ and θ. Applying the boundary conditions given by (7), we obtain

$$\frac{1}{3}f^3 + 2\nu f^2 + \frac{\nu}{2}f'^2 = Af + C_1 \tag{11}$$

where C_1 is an arbitrary constant. Rewriting (11), we have

$$f'^2 = -\frac{2}{3\nu}f^3 - 4f^2 + \frac{2Af}{\nu} + \frac{2}{\nu}C_1 \equiv -\frac{2}{3\nu}G(f) \tag{12}$$

from which,

$$\theta = \pm \int_{f(0)}^{f} \frac{df}{\sqrt{-\frac{2}{3\nu}G(f)}}. \tag{13}$$

Let us write

$$G(f) = (f - e_1)(f - e_2)(f - e_3). \tag{14}$$

We then have from (12)

$$e_1 + e_2 + e_3 = -6\nu. \tag{15}$$

Equation (15) shows that at least one of the e_k's must be negative. We now have two cases:

(i) Only e_1 Is Real and Positive

$G(f)$ then looks as shown in Figure 22.2. We then have pure outflow, and (13) becomes

$$\theta = \sqrt{\frac{3\nu}{2}} \int_{f}^{e_1} \frac{df}{\sqrt{-G(f)}} \tag{16}$$

which may be rewritten as

$$\theta = \sqrt{\frac{3\nu}{2}} \int_{f}^{e_1} \frac{df}{\sqrt{(e_1 - f)\left[(f-\rho)^2 + \sigma^2\right]}} \tag{17}$$

where

$$(f-\rho)^2 + \sigma^2 \equiv (f - e_2)(f - e_3)$$

$$\rho \equiv \frac{e_2 + e_3}{2}, \sigma^2 \equiv e_2 e_3 - \left(\frac{e_2 + e_3}{2}\right)^2.$$

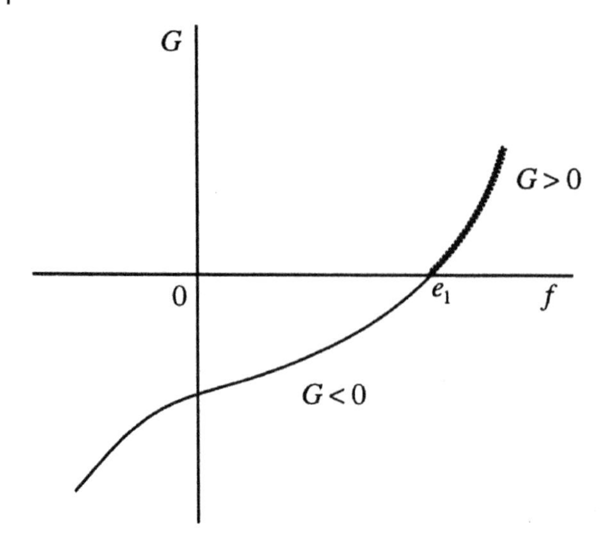

Figure 22.2 Variation of the function G with f for the case wherein only e_1 is real and positive.

Thus,

$$\theta = \frac{1}{\sqrt{M}} cn^{-1} \left[\frac{M^2 - (e_1 - f)}{M^2 + (e_1 - f)}, \chi \right] \tag{18}$$

where

$$\left. \begin{aligned} M^2 &\equiv \frac{2}{3\nu} (e_1 - e_2)(e_1 - e_3) = \frac{2}{3\nu} \left[(\rho - e_1)^2 + \sigma^2 \right] \\ \chi^2 &\equiv \frac{1}{2} + \frac{1}{2} \frac{e_1 - \rho}{(3\nu/2) M^2} = \frac{1}{2} + \frac{e_1 + 2\nu}{2\nu M^2}. \end{aligned} \right\}$$

Equation (18) leads to

$$f = e_1 - \frac{3M^2\nu}{2} \frac{1 - cn(M\theta, \chi)}{1 + cn(M\theta, \chi)}. \tag{19}$$

(ii) e_1, e_2, and e_3 Are Real and Distinct

$G(f)$ then looks as shown in Figure 22.3. We then have, for inflow,

$$\theta = \sqrt{\frac{3\nu}{2}} \int_{e_2}^{f} \frac{df}{\sqrt{-G(f)}} \tag{20}$$

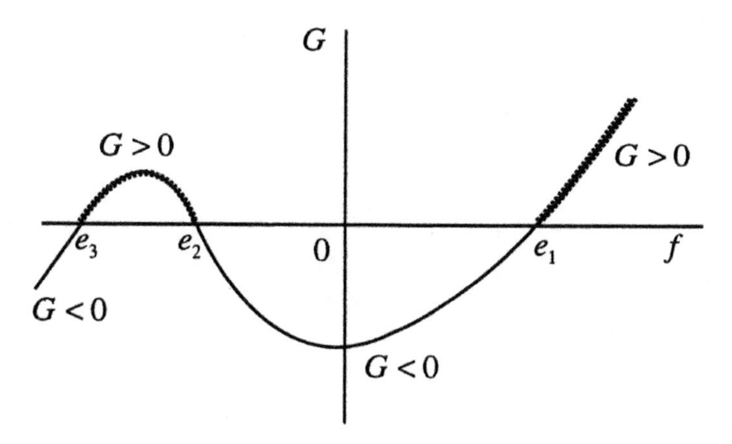

Figure 22.3 Variation of the function G with f for the case wherein e_1, e_2, and e_3 is real and distinct.

or

$$\theta = \sqrt{\frac{3\nu}{2}} \int_{e_2}^{e_1} \frac{df}{\sqrt{-G(f)}} - \sqrt{\frac{3\nu}{2}} \int_{f}^{e_1} \frac{df}{\sqrt{-G(f)}} \tag{21}$$

so,

$$\int_{f}^{e_1} \frac{df}{\sqrt{-G(f)}} = K(k^2) - m\theta \tag{22}$$

where $K(k^2)$ is the *complete elliptic integral of the first kind,*

$$K(k^2) \equiv \int_0^1 \frac{dt}{\sqrt{t(1-t)(1-k^2 t)}}, t \equiv \frac{f - e_2}{e_1 - e_2}$$

$$k^2 \equiv \frac{e_1 - e_2}{e_1 - e_3}, m^2 \equiv \frac{e_1 - e_3}{6\nu}.$$

Noting that

$$\int_{f}^{e_1} \frac{df}{\sqrt{-G(f)}} = sn^{-1}\left(\sqrt{\frac{e_1 - f}{e_1 - e_2}}, k\right) \tag{23}$$

equation (22) leads to

$$f = e_1 - (e_1 - e_2)\, sn^2(K - m\theta, k^2) \tag{24}$$

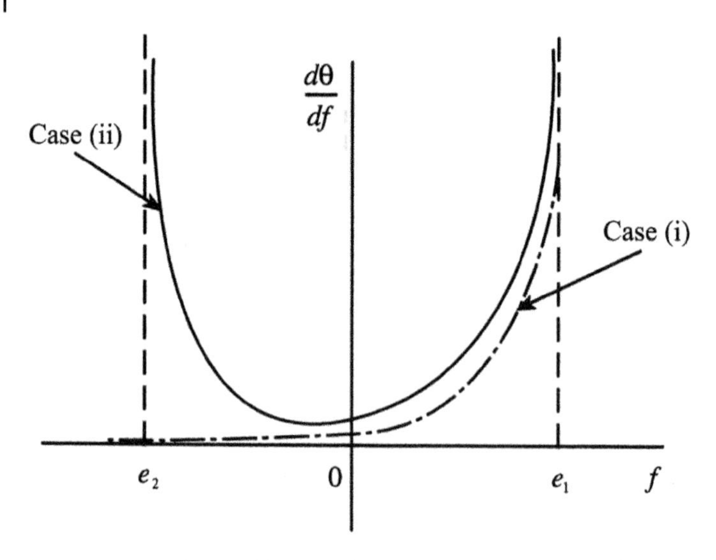

Figure 22.4 Variation of f' with f for the two cases illustrated in Figures 22.2 and 22.3.

or

$$f = e_1 - 6\nu k^2 m^2 sn^2 (K - m\theta, k). \tag{25}$$

For outflow,

$$\theta = \sqrt{\frac{3\nu}{2}} \int_f^{e_1} \frac{df}{\sqrt{-G(f)}} \tag{26}$$

which similarly leads to

$$f = e_1 - 6\nu k^2 m^2 sn^2 (m\theta, k) \tag{27}$$

f, f' for cases (i) and (ii) are sketched in Figures 22.4 and 22.5. Note that $f = f(\theta)$ is not periodic for case (i), while it is periodic with period $2(\beta + \alpha)$ for case (ii), where

$$\beta \equiv \int_0^{e_1} \frac{df}{\sqrt{-\frac{2}{3\nu} G(f)}}, \quad \alpha \equiv \int_{e_2}^{0} \frac{df}{\sqrt{-\frac{2}{3\nu} G(f)}}.$$

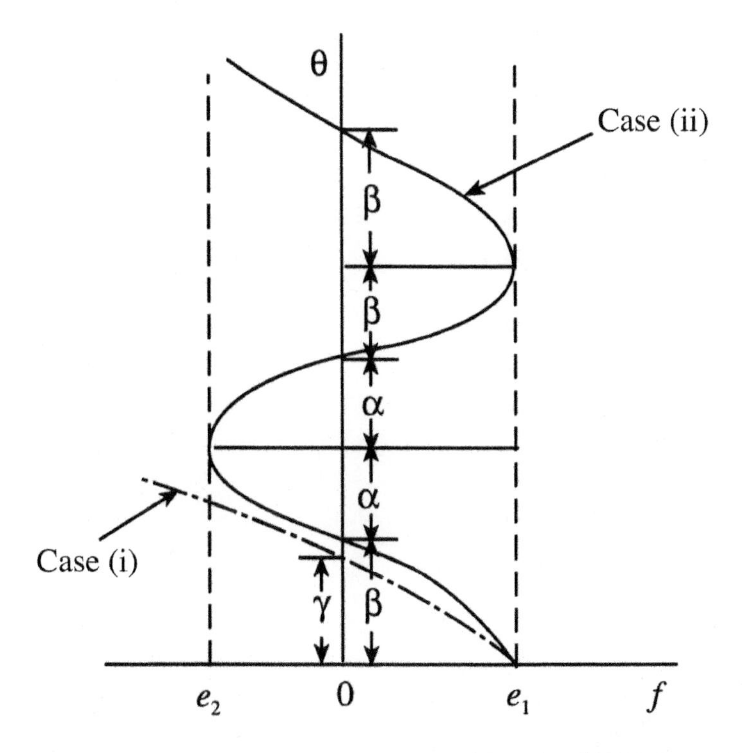

Figure 22.5 Variation of f with θ for the two cases illustrated in Figures 22.2 and 22.3.

The flows corresponding to cases (i) and (ii) are shown in Figure 22.6. Note that

$$f(0) = \begin{cases} e_1, \\ e_2, \end{cases} \quad \text{if } u_r \gtrless 0. \tag{28}$$

As the parameter $\alpha f(0)/\nu$ increases, it is seen from Figures 22.5 and 22.6 that the possibility of finding compound flows with zero values of f at $\theta = \pm\alpha$ grows.

22.2 Flows at Low Reynolds Numbers

Nondimensionalizing f as follows

$$\bar{f} \equiv \frac{f}{f(0)} \tag{29}$$

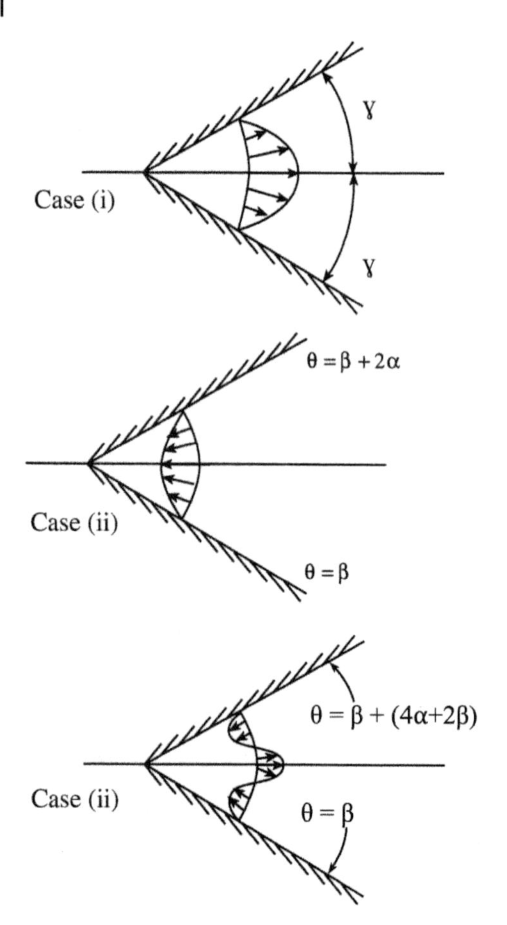

Figure 22.6 Diverging and converging flows at low Reynolds numbers.

equation (9) and boundary condition (7) give

$$\frac{d^2\overline{f}}{d\theta^2} + 4\overline{f} + R_E\overline{f}^2 = \overline{A} \tag{30}$$

$$\left.\begin{array}{l} \theta = \pm\alpha : \overline{f} = 0 \\ \theta = 0 : \overline{f} = 1 \end{array}\right\} \tag{31}$$

where $R_E \equiv \dfrac{f(0)}{\nu} = \dfrac{e_{1,2}}{\nu}$, is the Reynolds number based on the local maximum velocity and

$$\overline{A} \equiv \frac{A}{\nu f(0)} = \frac{A}{\nu e_{1,2}}.$$

For $|R_E| \ll 1$, let us seek solutions of the form,

$$\overline{f}(\theta) \sim \overline{f}_0(\theta) + R_E \overline{f}_1(\theta) + O\left(R_E^2\right) \tag{32}$$

$$\overline{A} \sim \overline{A}_0 + R_E \overline{A}_1 + O\left(R_E^2\right) \tag{33}$$

so that we have from (30) and (31),

$$\left.\begin{aligned} O(1):\ & \frac{d^2\overline{f}_0}{d\theta^2} + 4\overline{f}_0 = \overline{A}_0 \\ & \theta = \pm\alpha : \overline{f}_0 = 0 \\ & \theta = 0 : \overline{f}_0 = 1 \end{aligned}\right\} \tag{34}$$

$$\left.\begin{aligned} O(R_E):\ & \frac{d^2\overline{f}_1}{d\theta^2} + 4\overline{f}_1 = -\overline{f}_0^{\,2} + A_1 \\ & \theta = \pm\alpha : \overline{f}_1 = 0 \\ & \theta = 0 : \overline{f}_1 = 0 \\ & etc. \end{aligned}\right\} \tag{35}$$

We have, from (34),[1]

$$\overline{f}_0 = a\left(b - \cos 2\theta\right) \tag{36}$$

where

$$a \equiv \frac{\overline{A}_0/4}{\cos 2\alpha}, \quad b \equiv \cos 2\alpha.$$

---

1 For very small α, when the two walls become nearly parallel, (36) leads to the parabolic velocity profile, as expected,

$$\overline{f}_0 \sim \theta^2 - \alpha^2.$$

Using (36), (35) becomes

$$\left.\begin{aligned} &\frac{d^2\overline{f}_1}{d\theta^2} + 4\overline{f}_1 = -a^2\left(b - \cos 2\theta\right)^2 + \overline{A}_1 \\ &\theta = \pm\alpha : \overline{f}_1 = 0 \\ &\theta = 0 : \overline{f}_1 = 0. \end{aligned}\right\} \tag{37}$$

We have, from (37),

$$\overline{f}_1(\theta) = C_2(\cos 2\theta - 1) + \frac{a^2 b}{2}\theta \sin 2\theta + \frac{a^2}{24}(\cos 4\theta - 1) \tag{38}$$

where

$$C_2 \equiv -\left[\frac{a^2 b}{2}\alpha \sin 2\alpha + \frac{a^2}{24}(\cos 4\alpha - 1)\right]\frac{1}{\cos 2\alpha - 1}.$$

The solutions (36), (38), and (32) are sketched in Figure 22.7. For purely convergent flow, an increase in $|R_E|$ produces a flatter velocity profile at the center of the channel with steep velocity gradients near the walls. However, the effect of an increase in $|R_E|$ in purely divergent flow is to concentrate the fluid flux at the center of the channel with smaller velocity gradients at the walls. This suggests that one may expect the boundary-layer flows to arise near the walls in purely convergent flows. This development is verified in detail below.

Figure 22.7 shows the occurrence of backflow in a diverging channel as R_E increases. In order to investigate this backflow, let us note from (7) and (12) that,

$$C_1 = \frac{\nu}{2}\left[\left(\frac{df}{d\theta}\right)^2_{\theta=\pm\alpha}\right] \tag{39}$$

so that the onset of backflow is characterized by

$$C_1 = 0. \tag{40}$$

Note that, from (12) and (14), for case (i)

$$\theta = 0 : f = e_1, \quad f' = 0. \tag{41}$$

So, we have, from (12),

$$e_1^3 + 6\nu e_1^2 - 3Ae_1 - 3C_1 = 0. \tag{42}$$

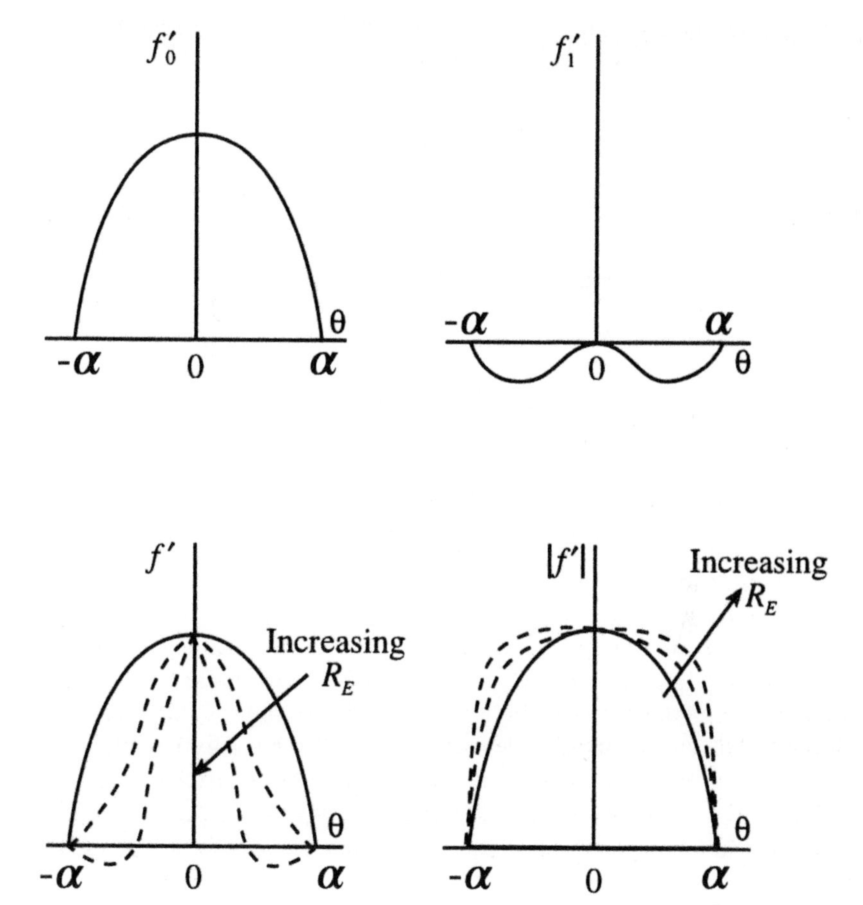

Figure 22.7 Diverging and converging flows at low Reynolds numbers.

Using (42), (12) may be rewritten as

$$f'^2 = \frac{2}{3\nu}(e_1 - f)\left[f^2 + f(6\nu + e_1) + \frac{3C_1}{e_1}\right] \tag{43}$$

from which, corresponding to the onset of backflow, i.e., $C_1 = 0$, we obtain for the critical value of the angle between the walls,[2]

$$\begin{aligned}
\theta_{crit} &= \int_0^{e_1} \frac{df}{\sqrt{\frac{2}{3\nu}(e_1 - f)\left[f^2 + f(6\nu + e_1)\right]}}. \\
&= \sqrt{\frac{3\nu}{2e_1}} \int_0^1 \frac{dt}{\sqrt{t(1-t)\left[t + \left(1 + \frac{6\nu}{e_1}\right)\right]}}, t \equiv f/e_1 \\
&= \sqrt{\frac{3\nu}{3\nu + e_1}} K\left[\frac{1}{2}\left(\frac{e_1}{3\nu + e_1}\right)\right] \\
&= 2\sqrt{\frac{3\nu}{2e_1}} \int_0^{\pi/2} \frac{d\phi}{\sqrt{\cos^2\phi + 1 + 6\nu/e_1}}, t \equiv \cos^2\phi \\
&= 2\sqrt{\frac{3\nu}{2e_1}} k \int_0^{\pi/2} \frac{d\phi}{\sqrt{1 - k^2 \sin^2\phi}}, k^2 \equiv \frac{e_1/2}{e_1 + 3\nu}.
\end{aligned} \tag{44}$$

2 Alternatively, for the onset of backflow, we have from (12), $e_2 = 0$, and (27) gives

$$0 = e_2 - e_1 sn^2(m\theta_{crit}, k)$$

which leads to

$$e_1 cn^2(m\theta_{crit}, k) = 0$$

so,

$$m\theta_{crit} = K(k)$$

where

$$\begin{aligned}
m^2 &\equiv \frac{e_1 - e_3}{6\nu} = \frac{e_1 - (-6\nu - e_1)}{6\nu} = \frac{e_1 + 3\nu}{3\nu} \\
k^2 &\equiv \frac{e_1}{e_1 - e_3} = \frac{e_1}{e_1 - (-6\nu - e_1)} = \frac{e_{1/2}}{e_1 + 3\nu}.
\end{aligned}$$

Therefore,

$$\theta_{crit} = \sqrt{\frac{3\nu}{3\nu + e_1}} K\left(\frac{e_{1/2}}{e_1 + 3\nu}\right).$$

In the limit of large ν, noting that $K(0) = \pi/2$, (44) leads to $\theta_{crit} \approx \pi/2$. Therefore, a purely divergent flow becomes impossible when the angle between the channel walls exceeds π.

22.3 Flows at High Reynolds Numbers

We mentioned above that as the Reynolds number R_E becomes large, the flow tends to become uniform except in boundary layers near the walls. This may be seen by referring to (25). In this limit m becomes large, and it follows, from the finiteness of α, that $K(k^2)$ must be large or $k \sim 1$. The latter implies that $e_2 \approx e_3$, and $snt \sim \tanh t, e_1 = -6\nu - 2e_2 \approx -2e_2$, and (12) becomes

$$\begin{aligned} f'^2 &\approx -\frac{2}{3\nu}(f - e_2)(f^2 + e_2 f - 2e_2^2) \\ &= -\frac{2}{3\nu}(f - e_2)^2 (f + 2e_2). \end{aligned} \tag{45}$$

Setting

$$\hat{f} \equiv -f, \hat{e}_2 \equiv -e_2$$

and dropping the hats, (45) leads to

$$\sqrt{\frac{2}{3\nu}} \int_{\alpha}^{\theta} d\theta \approx \int_{0}^{f} \frac{df}{(f - e_2)\sqrt{f + 2e_2}}. \tag{46}$$

Setting

$$\tilde{f} \equiv \frac{f}{e_2} \tag{47}$$

we obtain from (46)

$$\sqrt{\frac{2e_2}{3\nu}}(\alpha - \theta) \approx \int_{0}^{f} \frac{d\tilde{f}}{(1 - \tilde{f})\sqrt{\tilde{f} + 2}}. \tag{48}$$

This yields

$$\sqrt{\frac{e_2}{2\nu}}(\alpha - \theta) \approx \tanh^{-1}\sqrt{\frac{\tilde{f} + 2}{3}} - \tanh^{-1}\sqrt{\frac{2}{3}} \tag{49}$$

from which we have,

$$\tilde{f} \approx 3 \tanh^2 \left[\sqrt{\frac{e_2}{2\nu}} (\alpha - \theta) + \tanh^{-1} \sqrt{\frac{2}{3}} \right] - 2. \tag{50}$$

Thus,

$$f = e_2 \left[3 \tanh^2 \left\{ \sqrt{-\frac{e_2}{2\nu}} (\alpha - \theta) + \delta \right\} - 2 \right] \tag{51}$$

where

$$\delta \equiv \tanh^{-1} \sqrt{\frac{2}{3}}.$$

Therefore, f is approximately equal to $-e_2$, except in the boundary layer of thickness proportional to $1/\sqrt{-e_2/\nu}$.

In order to verify these features in detail, write equation (30) as

$$\frac{d^3 \overline{f}}{d\theta^3} + 4 \frac{d\overline{f}}{d\theta} + 2R_E \overline{f} \frac{d\overline{f}}{d\theta} = 0. \tag{52}$$

In the limit $R_E \Rightarrow \infty$, corresponding to the outer core flow, equation (52) gives

$$\overline{f}^{(0)} \frac{d\overline{f}^{(0)}}{d\theta} = 0. \tag{53}$$

The boundary condition (31) gives

$$\theta = 0 : \overline{f}^{(0)} = 1. \tag{54}$$

We have, from (53) and (54),

$$\overline{f}^{(0)} \equiv 1. \tag{55}$$

For $R_E \gg 1$, (55) implies that, the radial velocity is nearly uniform over $\theta \in (-\alpha, \alpha)$, except in boundary layers close to the wall. Near the wall it varies rapidly in order to recover the no-slip condition at $\theta = \underline{+}\alpha$, which is about to be lost in the inviscid limit.

In order to consider the flow near the wall, let us set

$$\Theta \equiv \sqrt{|R_E|} (\alpha - \theta) \tag{56}$$

so that equation (52) becomes

$$\frac{d^3 \overline{f}}{d\Theta^3} + \frac{4}{|R_E|} \frac{d\overline{f}}{d\Theta} + 2 \frac{R_E}{|R_E|} \overline{f} \frac{d\overline{f}}{d\Theta} = 0. \tag{57}$$

In the limit $R_E \Rightarrow \infty$, equation (57) gives, for the flow in a convergent channel,

$$\frac{d^2\overline{f}^{(i)}}{d\Theta^2} - \left(\overline{f}^{(i)}\right)^2 = -1 \tag{58}$$

and for the flow in a divergent channel,

$$\frac{d^2\overline{f}^{(i)}}{d\Theta^2} + \left(\overline{f}^{(i)}\right)^2 = -1. \tag{59}$$

Thus, setting,

$$\overline{f}^{(i)} = 1 + \overline{h}^{(i)}(\Theta), |\overline{h}^{(i)}(\Theta)| \ll | \tag{60}$$

and noting that the boundary-layer nature of the flow near the wall requires $\overline{f}^{(i)}$ to match asymptotically with the outer-core flow given by (55), we have

$$\Theta \Rightarrow \infty : \overline{h}^{(i)} \Rightarrow 0. \tag{61}$$

From equations (58) and (59), for a convergent channel we obtain,

$$\frac{d^2\overline{h}^{(i)}}{d\Theta^2} - 2\overline{h}^{(i)} = 0$$

or

$$\overline{h}^{(i)} = B_1 e^{-\sqrt{2}\Theta} \tag{62}$$

which satisfies (61), as required for the existence of a boundary layer near the wall. Physically, the latter is produced by the effect of convection which, for a convergent channel, tends to oppose the diffusion of vorticity away from the wall.

For a divergent channel, we have, on the other hand, from equations (58) and (59),

$$\frac{d^2\overline{h}^{(i)}}{d\Theta^2} + 2\overline{h}^{(i)} = 0$$

or

$$\overline{h}^{(i)} = D_1 \sin\sqrt{2}\Theta + D_2 \cos\sqrt{2}\Theta \tag{63}$$

which does not satisfy (61) unlike the convergent channel solution (62). It is, therefore, obvious that there can be no boundary-layer flow in a divergent channel.

References

Part 1

Holm, D. D., Marsden, J. E., Ratiu, I. and Weinstein, A.: *Phys. Rep.* **123**, 1, (1985).
Kelvin, Lord: *Trans. Roy. Soc. Edinburgh* **25**, 217, (1868).
Kuznetsov, E. A. and Mikhailov, A. M.: *Phys. Lett. A* **77**, 37, (1980).
Larichev, V. D. and Rezink, G. M.: *Dokl. Akad. Nauk.* **231**, 1077,(1976).
Moffatt, H. K.: *J. Fluid Mech.* **35**, 117, (1969).
Olver, P. J. : *J. Math. Anal. Applics.* **89**, 233, (1982).
Salmon, R.: *Ann. Rev. Fluid Mech.* **20**, 225, (1988).
Shepherd, T. G.: *Adv. Geophys.* **32**, 287, (1990).

Part 2

Bateman, H.: Proc. *Roy. Soc. (Lond.) A***125**, 598, (1929).
Benjamin, T. B.: *Proc. Roy. Soc. (Lond.)* **A299**, 59, (1967).
Benjamin, T. B. and Feir, J. E.: *J. Fluid Mech.* **27**, 417, (1967).
Benjamin, T. B.: *Proc. Roy. Soc.* (Lond.), **A328**, 153, (1972).
Benjamin, T. B.: *J. Fluid Mech.* **14**, 593, (1962).
Biot, J. B. and Savart, F.: *J. Phys.* (Paris) **91**, 151, (1820).
Blasius, H.: Z. *Math. Phys.* **58**, 90, (1910).
Da Rios, L. S. : *Rend. Circ. Mat.* Palerno **22**, 117, (1906).
Dodd, R. K., Eilbeck, J. C., Gibbon, J. D. and Morris, H. C.: *Solitons and Nonlinear Wave Equations*, Academic Press, (1982).
Fermi, E., Pasta, J. and Ulam, S.: *Los Alamos Sci. Lab.* Report LA-1940, (1955).
Glauert, M. B.: Proc. *Roy. Soc.* (Lond.) **A242**, 108, (1957).
Harvey, J. K.: *J. Fluid Mech.* **14**, 585, (1962).
Helmholtz, H. von: *V Crelle J.* **55**, (1858).
Hill, M. J. M.: *Phil. Trans. Roy. Soc.* **A185**, (1894).

Introduction to Theoretical and Mathematical Fluid Dynamics, Third Edition.
Bhimsen K. Shivamoggi.

Hirota, R.: *Phys. Rev. Lett.* **27**, 1192, (1971).
Janssen, P. A. E. M.: *Phys. Fluids* **24**, 23, (1981).
Jeffrey, A. and Kakutani, T.: *SIAM Rev.* **14**, 582, (1972).
Joukowski, N. E.: *Trans. Phys. Soc.*, Imp. Soc. Friends Natl. Sci. Moscow **23**, (1906).
Kadomtsev, B. B. and Petviashvili, V. I.: *Sov. Phys. Dokl.* **15**, 539, (1970).
Kelvin, Lord: Math. *Phys. Papers* **1**, 107, (1849).
Kirchoff, G.: *J. Reine Angew. Math.* **70**, 289, (1869).
Kutta, W. M.: *Ill. Aero. Mitt.* **6**, 133, (1902).
Lake, B. M., Yuen, H. C., Rungaldier, H., and Fergusson, W. E.: *J. Fluid Mech.* **83**, 49, (1977).
Lighthill, M. J.: *Phil. Trans. Roy. Soc. A* **252**, 397, (1960).
Lighthill, M. J.: *J. Fluid Mech.* **26**, 411, (1966).
Longuet-Higgins, M. S.: *J. Fluid Mech.* **12**, 321, (1962).
Luke, J. C.: *J. Fluid Mech.* **27**, 395, (1967).
Magnus, G.: *Poggendorf's Ann. der Phys.* U. Chemie **88**, 1, (1853).
Manley, J. M. and Rowe, H. E.: *Proc. IRE* **47**, 2115, (1959).
Miles, J.: *J. Fluid Mech.* **79**, 157, (1977).
Milne-Thomson, L. M.: *Proc. Cambridge Phil. Soc.*, **36**, 246, (1940).
Oikawa, J., Satsuma, J. and Yojima, N.: *J. Phys. Soc.* Japan **35**, 511, (1974).
Phillips, O. M.: *J. Fluid Mech.* **9**, 193, (1960).
Prandtl, L.: *J. Roy. Aero. Soc.* **31**, 730, (1927).
Proudman, J.: *Proc. Roy. Soc.* (Lond.) **A92**, 408, (1916).
Rao, A. R.: *J. Math. Phys. Sci.* **1**, 223, (1967).
Rossby, C. G.: *J. Mar. Res.* **2**, 38, (1939).
Segur, H. and Finkel, A.: *Stud. Appl. Math.* **73**, 183, (1985).
Shivamoggi, B. K.: *Acta Mech.* **62**, 29, (1986).
Shivamoggi, B. K.: *J. Phys.* **A 23**, 4289, (1990).
Shivamoggi, B. K. and Uberoi, M. S.: *Acta Mech.* **41**, 211, (1981).
Stokes, G.: *Trans. Cambridge Phil. Soc.* **8**, 441, (1847).
Taylor, G. I.: *Proc. Roy. Soc.* (Lond.) **A100**, 114, (1921); **A102**, 180, (1922); **A104**, 213, (1923).
Theodorsen, T.: *NACA Rep.* 411, (1932). NACA Rep. 496, (1935).
Zabusky, N. J. and Kruskal, M. D.: *Phys. Rev. Lett.* **15**, 240, (1965).

Part 3

Ackeret, J.: *Helv. Phys. Acta* **1**, 301, (1928).
Burgers, J. M.: *Adv. Appl. Mech.* **1**, 171, (1948).
Chaplygin, S. A.: *On Gas Jets*, Imperial Univ. Moscow, (1904).
Cole, J. D.: *Q. Appl. Math.* **9**, 225, (1951).
Crocco, L.: *Z. Angew. Math. Mech.* **17**, 1, (1937).

Fox, P. A.: *J. Math and Phys.* **34**, 133, (1955).
Fubini-Ghiron: *Alta Frequenza* **4**, 530, (1935).
Hopf, E.: *Comm. Pure Appl. Math.* **3**, 201, (1950).
Hugoniot, H.: *J. Ecole Polytech.* (1) **57**, 1, (1887); Also, (1) **58**, 1, (1889).
Karman, T. von and Tsien, H. S.: *J. Aero. Sci.* **6**, 399, (1939).
Lighthill, M. J.: *Phil. Mag.* **40**, 1179, (1949).
Lin, C. C.: *J. Math. Phys.* **33**, 117, (1954).
Manley, J. M. and Rowe, H. E.: *Proc. IRE* **47**, 2115, (1959).
Molenbrock, P.: *Arch. Math. Phys.* **9**, 157, (1890).
Murmon, E. M. and Cole, J. D.: *AIAA J.* **9**, 114, (1971).
Parker, E. N.: *Astrophys. J.* **128**, 664, (1958).
Parker, E. N.: in *The Century of Space Science*, Eds. J. A. Bleeker, J. Geiss & M. Huber, Springer-Verlag, (2001).
Possio, C.: *Aerotecnica* **18**, 441, (1938).
Prandtl, L.: *NACA TM* 805, (1936).
Prandtl, L. and Busemann, A.: *Stodola Festschrift* **85**, 499, (1929).
Pritulo, M. F. : *J. Appl. Math. Mech.* **26**, 661, (1972).
Rankine, W. J. M.: *Phil. Trans. Roy. Soc.* **160**, 277, (1870).
Riemann, B.: *Math. Phys.* Klasse **8**, 43, (1860).
Ringleb, F.: *Z. Angew. Math. Mech.* **20**, 185, (1940).
Sedov, L.I.: *Dokl, Akad. Nauk.* SSSR **52**, 17, (1946).
Shivamoggi, B. K.: *J. Sound Vib.* **55**, 594, (1978).
Shivamoggi, B. K.: *J. Sound Vib.* **176**, 271, (1994).
Shivamoggi B. K.: *Phys. Plasmas* **27**, 012902, (2020).
Taylor, G.I.: *Proc. Roy. Soc.* (Lond.) **A201**, 159, (1950).
Tidjeman, H. and Seebass, R.: *Ann. Rev. Fluid Mech.* **12**, 181, (1980).
Stewartson, K.: *Q. J. Mech. Appl. Math.* **3**, 182, (1950).
Whitham, G. B.: *Comm. Pure Appl. Math.* **5**, 301, (1952).

Part 4

Andrade, E. N. da C: *Proc. Phys. Soc.* **51**, 784, (1939).
Batchelor G. K.: *Q. J. Mech. Appl. Math.* **4**, 29, (1951).
Becker, R.: *Z. Phys.* **8**, 321, (1922).
Bickley, W. G.: *Phil. Mag.* **23 (7)**, 727, (1939).
Blasius, H.: *Z. Math. Phys.* **56**, 1, (1908).
Burgers, J. M.: *Adv. Appl. Mech.* **1**, 171, (1948).
Crocco. L.: *Aero. Res. Council Rep.* **4**, 582, (1939).
Dhawan, S.: *NACA Tech.* Note 2567, (1952).
Dorodnitsyn, A. A.: *Frikl. Mat. Mekh.* **6 (6)**, 449, (1942).
Ekman, V. W.: *Ark. Mat. Astr. Fys.* **2**, 1, (1905).

Evans, D. J.: *Q. J. Mech. Appl. Math.* **22**, 467, (1969).

Hamel, G.: *Jahresbericht der Deutschen Mathematicker Vereinigung* **25**, 34, (1927).

Hele-Shaw, H. S.: *Nature* **58**, 34, (1898).

Howarth, L.: *Proc. Roy. Soc.* (Lond.) **A194**. 16, (1948).

Ising, E.: *Z. Phys.* **31**, 253, (1925).

Jeffrey, G. B.: *Phil. Mag.* **29 (172)**, 455, (1915).

Kaplun, S.: *Z. Angew. Math. Phys.* **5**, 111, (1954).

Kaplun, S. and Lagerstrom, P. A.: *J. Math. Mech.* **6**, 585, (1957).

Karman, T. von: *Z. Angew. Math. Mech.* **4**, 29, (1951).

Lamb, H.: *Hydrodynamics*, Cambridge Univ. Press, (1916).

Landau, L. D.: *Dokl. Acad. Sci.* URSS **43**, 286, (1944).

Landau, L. D. and Lifshitz, E. M.: *Fluid Mechanics*, Second Ed., Elsevier, (1987).

Lighthill, M. J.: *Proc. Roy. Soc.* (Lond.) **A224**, 1, (1954).

Oseen, C. W.: *Flussigkeit. Ark. Mat. Astro. Fys.* **7**, 14, (1912).

Oseen, C. W.: *Ark. Mat. Abstr. Fys.* **6**, 29, (1910).

Prandtl, L.: *Proc. III Inter. Math. Congr.*, Heidelberg, (1904); Also, NACA TM-452, (1928).

Proudman, I. and Pearson, J. R. A.: *J. Fluid Mech.* **2**, 237, (1957).

Schlichting, H.: *Z. Angew. Math. Mech.* **13**, 260, (1933).

Shivamoggi, B. K.: *Z. Angew. Math. Mech.* **58**, 354, (1978).

Shivamoggi, B. K.: *Acta Mech.* **33**, 303, (1979).

Shivamoggi, B. K.: *Mec. Appliquee* **27**, 101, (1982).

Shivamoggi, D. K. and Rollins, D. K.: *J. Math. Phys.* **40**, 3372, (1999).

Squire, H. B.: *Q. J. Mech. Appl. Math.* **4**, 321, (1951).

Stanley, H. E.: *Introduction to Phase Transitions and Critical Phenomena*, Oxford Univ. Press, (1971).

Stewartson. K.: *Proc. Roy. Soc.* (Lond.) **A200**, 84, (1949).

Stokes, G. G.: *Trans. Camb. Phil. Soc.* **9**, 8, (1851).

Zandbergen, P. J. and Dijkstra, D.: *J. Eng. Math.* **11**, 167, (1977).

Bibliography

My indebtedness to information and benefit received from many sources in connection with this book material cannot be adequately documented. In particular, I have extensively consulted the following books:

Batchelor G. K.: *Introduction to Fluid Dynamics*, Cambridge Univ. Press, (1967).

Bisplinghoff, R. L., Ashley, H., and Halfman, R.L.: *Introduction to Aeroelasticity*, Addison-Wesley Publishing Co., (1955).

Chorin, A. J. and Marsden, J. E.: *A Mathematical Introduction to Fluid Mechanics*, Springer-Verlag, (1990).

Drazin, P. G. and Johnson, R. S.: *Solitons*, Cambridge Univ. Press, (1989).

Faber, T. E.: *Fluid Dynamics for Physicists*, Cambridge Univ. Press, (1995).

Falkovich, G.: *Fluid Mechanics*, Second Edition, Cambridge Univ. Press, (2018).

Fung, Y. C.: *Introduction to the Theory of Aeroelasticity*, John Wiley and Sons, (1956).

Greenspan, H. P. : *Theory of Rotating Fluids*, Cambridge Univ. Press, (1968).

Karamcheti, K.: *Principles of Ideal Fluid Aerodynamics*, John Wiley and Sons, (1966).

Kevorkian, J. and Cole, J. D.: *Perturbation Methods in Applied Mathematics*, Springer-Varlag, (1980).

Lagerstrom, P. A.: Laminar Flow Theory, in *Theory of Laminar Flows*, Ed. F. K, Moore, Princeton Univ. Press, (1964).

Lamb, H.: *Hydrodynamics*, Cambridge Univ. Press, (1945).

Landau, L. D. and Lifshitz, E. M.: *Fluid Mechanics,* Second Edition, Elsevier, (1987).

Liepmann, H. W. and Pucket, A. E.: *Aerodynamics of a Compressible Fluid*, John Wiley and Sons, (1949).

Liepmann, H. W. and Roshko, A.: *Elements of Gas Dynamics*, John Wiley and Sons, (1957).

Introduction to Theoretical and Mathematical Fluid Dynamics, Third Edition.
Bhimsen K. Shivamoggi.

Lighthill, M. J.: *Waves in Fluids*, Cambridge Univ. Press, (1978).
Milne-Thomson, L. M. : *Theoretical Hydrodynamics*, Fifth Edition, Macmillan, (1967).
O'Neil, M. E. and Chorlton, F.: *Ideal and Incompressible Fluid Dynamics*, Ellis Horwood, (1986).
O'Neil, M. E. and Chorlton, F.: *Viscous and Compressible Fluid Dynamics*, Ellis Horwood, (1987).
Rutherford, D. E.: *Fluid Dynamics*, Oliver & Boyd, (1959).
Schlichting, H.: *Boundary Layer Theory*, McGraw Hill Publishing Co., (1972).
Serrin, J.: Mathematical Principles of Classical Fluid Mechanics, in *Handbuch der Physik* **8**, pg.125, Spring-Verlag, (1959).
Shivamoggi, B. K.: *Perturbation Methods for Differential Equations*, Birkhauser Verlag, (2003).
Shivamoggi, B. K.: *Nonlinear Dynamics and Chaotic Phenomena*, Second Edition, Springer Verlag, (2014).
Tritton, D. J.: *Physical Fluid Dynamics*, Clarendon Press, (1988).
Tsien, H. S.: The Equations of Gas Dynamics in *Fundamentals of Gas Dynamics*, Ed. H. W. Emmons, Princeton Univ. Press, (1958).
Van Dyke, M. D.: *Perturbation Methods in Fluid Mechanics*, Parabolic Press, Stanford (1975).
Whitham, G. B.: *Linear and Nonlinear Waves*, Wiley-Interscience, New York, (1974).

Index

Introduction to Theoretical and Mathematical Fluid Dynamics, Third Edition.
Bhimsen K. Shivamoggi.

m

n

o

p

r

z